16th Workshop on Biomedical Natural Language Processing (BioNLP 2017)

Held at ACL 2017

Vancouver, Canada
4 August 2017

ISBN: 978-1-5108-4573-2

BioNLP 2017

SIGBioMed Workshop on Biomedical Natural Language Processing

Proceedings of the 16th BioNLP Workshop

August 4, 2017
Vancouver, Canada

Biomedical natural language processing in 2017:
The view from computational linguistics

Kevin Bretonnel Cohen, Dina Demner-Fushman,
Sophia Ananiadou, and Jun-ichi Tsujii

According to the Association for Computational Linguistics guidelines on special interest groups (SIGs), *The function of a SIG is to encourage interest and activity in specific areas within the ACL's field*[1]. Is the SIGBioMed special interest group "within the ACL's field"? The titles of this year's papers suggest that it is, in that the current interest in deep learning in its many and varied manifestations is mirrored in those titles. Do those papers cover a specific area? They do, and in doing so, they demonstrate one of the great satisfactions of working in biomedical natural language processing.

One of the joys of involvement in the biomedical natural language processing community is seeing the development of research with clinical applications. As examples of such work being presented at BioNLP 2017, we would like to point out the two papers that discuss the application of natural language processing to the diagnosis of neurological disorders. Bhatia et al.[2] describe an approach to using speech processing in the assessment of patients with amyotrophic lateral sclerosis (also known as Lou Gehrig's disease), one of the more horrific motor neuron diseases. Good assessment of amyotrophic lateral sclerosis patients is important for a number of reasons, including the fact that accurate tracking of the inevitable deterioration that is a hallmark of this disease gives patients and their families the possibility of purposeful planning for the attendant disability and death. However, current methodologies for evaluating the status of amyotrophic lateral sclerosis patients necessarily involve expensive equipment and highly trained personnel; when further developed, this methodology could make such evaluation much more, and more frequently, available to ALS patients. The fact that the work reported here involves a speech modality is especially exciting, as speech-related indicators of future ALS can be present long before diagnosis. The paper uses measurements of phonological features of speech and their divergence from a baseline, and demonstrates correlation with physiological measures.

Adams et al.[3] describe work on detecting and categorizing word production errors associated with anomia, a particular kind of inability to find words. Screening for anomia is important because anomia is a symptom of stroke, but it is difficult and time-consuming to do, and therefore is not done as often as it should be. Automatic detection of anomia could be a nice enabler of improved care for stroke victims, but it is made difficult due to the subtlety of the phonological and semantic judgments that have to be made when assessing the phenomenon. The paper uses a combination of language modeling and phonologically-based edit distance calculation to approach the task, applying these techniques to data from the AphasiaBank collection of transcribed aphasic and healthy speech.

Although we have summarized only these two examples that address neurological disorders, there are several other papers on the use of natural language processing in clinical applications: patient-produced content in dementia [4], and health records ([5] on sepsis, [6] on e-cig use, [7] on pain and confusion); in the aggregate, these papers illustrate very nicely the potential for natural language processing to contribute to human well-being. Additionally, the current interest in the potential of natural language processing for social media is reflected in papers on studying medication intake via Twitter [8] and on monitoring dementia via blog posts [9]. Linguistics and language resources are represented in this year's papers, as well, including work on comparative structures [10] and a corpus construction effort [11].

The work in biomedical NLP was dominated by applications of deep learning to: punctuation restoration [12], text classification [13], relation extraction [14], [15], [16], information retrieval [17], and similarity judgments [18], among other exciting progress in biomedical language processing.

These are just a few examples of the high-quality research presented in BioNLP 2017.

In addition to the excellent submissions to the BioNLP workshop, this year features equally strong submissions to BioASQ challenge on large-scale biomedical semantic indexing and question answering, a shared task affiliated with BioNLP 2017. This year, the BioASQ challenge, which started in 2013, had three tasks:

- Large-Scale Online Biomedical Semantic Indexing
- Biomedical Semantic Question Answering
- Funding Information Extraction From Biomedical Literature

An overview of the tasks and the results of the challenge [19] are presented in an invited talk. The invited speaker, George Paliouras, is a senior researcher and head of the Intelligent Information Systems division of the Institute of Informatics and Telecommunications at NCSR "Demokritos", Greece. He holds a PhD in Machine Learning and has performed basic and applied research in Artificial Intelligence for the last 20 years. He is interested in the development of novel methods for addressing challenging big and small data analysis problems, such as learning complex models from structured relational data, learning from noisy and sparse data, learning from multiple heterogeneous data streams, and discovering patterns in hypergraphs. His research is motivated by the real-world problems. George has contributed to solving a variety of such problems, ranging from spam filtering and Web personalization to biomedical information retrieval. He has co-founded the spin-off company em i-sieve Technologies, which provides online reputation monitoring services.

Among various contributions to the research community, George Paliouras has served as board member in national and international scientific societies; he is serving on the editorial boards of international journals, and has chaired international conferences. He is involved in several research projects, in the role of scientific coordinator/principal investigator in some of them. In particular, he has coordinated and provided the infrastructure for the BioASQ project that was funded by the European Commission. He is currently coordinating iASiS, another project funded by the European Commission to develop big data integration and analysis methods that will provide insight to public health policy-making for personalized medicine.

Acknowledging the community

As always, the organizers thank the authors who submitted their work to BioNLP 2017 —without them, there would be no meeting, no opportunity to share the progress and the pain of the past year with the community. We have listed above only a few of the exceptional submissions that were accepted for oral (20) and poster (28) presentations.

The distribution of scores this year suggests that a large amount of excellent work was submitted for review and resulted in 77% acceptance ratio. At the same time, the distribution suggests that the reviewers were careful and thorough, and the organizers thank them for that, and for thoroughly reviewing up to five papers on a very tight schedule.

We greatly appreciate the BioNLP core authors and program committee members who have been building up the community and the workshop for the past sixteen years. We are also happy to see the excellent new submissions and the new reviewers, and hope they will continue contributing to BioNLP.

References

[1] SIG Creation Guidelines *https://goo.gl/yIQCHo* 2017.

[2] Archna Bhatia, Bonnie Dorr, Kristy Hollingshead, Samuel L. Phillips and Barbara McKenzie *Characterization of Divergence in Impaired Speech of ALS Patients* 2017.

[3] Joel Adams, Steven Bedrick, Gerasimos Fergadiotis, Kyle Gorman and Jan van Santen *Target word prediction and paraphasia classification in spoken discourse* 2017.

[4] Vaden Masrani, Gabriel Murray, Thalia Field and Giuseppe Carenini *Detecting Dementia through Retrospective Analysis of Routine Blog Posts by Bloggers with Dementia* 2017.

[5] Emilia Apostolova and Tom Velez *Toward Automated Early Sepsis Alerting: Identifying Infection Patients from Nursing Notes* 2017.

[6] Danielle Mowery, Brett South, Olga Patterson, Shu-Hong Zhu and Mike Conway *Investigating the Documentation of Electronic Cigarette Use in the Veteran Affairs Electronic Health Record: A Pilot Study* 2017.

[7] Hans Moen, Kai Hakala, Farrokh Mehryary, Laura-Maria Peltonen, Tapio Salakoski, Filip Ginter and Sanna Salanterä *Detecting mentions of pain and acute confusion in Finnish clinical text* 2017.

[8] Ari Klein, Abeed Sarker, Masoud Rouhizadeh, Karen O'Connor and Graciela Gonzalez *Detecting Personal Medication Intake in Twitter: An Annotated Corpus and Baseline Classification System* 2017.

[9] Vaden Masrani, Gabriel Murray, Thalia Field and Giuseppe Carenini *Detecting Dementia through Retrospective Analysis of Routine Blog Posts by Bloggers with Dementia* 2017.

[10] Samir Gupta, A.S.M. Ashique Mahmood, Karen Ross, Cathy Wu and K. Vijay-Shanker *Identifying Comparative Structures in Biomedical Text* 2017.

[11] Rezarta Islamaj Dogan, Andrew Chatr-aryamontri, Sun Kim, Chih-Hsuan Wei, Yifan Peng, Donald Comeau and Zhiyong Lu *BioCreative VI Precision Medicine Track: creating a training corpus for mining protein-protein interactions affected by mutations* 2017.

[12] Wael Salloum, Greg Finley, Erik Edwards, Mark Miller and David Suendermann-Oeft *Deep Learning for Punctuation Restoration in Medical Reports* 2017.

[13] Simon Baker and Anna Korhonen *Initializing neural networks for hierarchical multi-label text classification* 2017.

[14] Chen Lin, Timothy Miller, Dmitriy Dligach, Steven Bethard and Guergana Savova *Representations of Time Expressions for Temporal Relation Extraction with Convolutional Neural Networks* 2017.

[15] Masaki Asada, Makoto Miwa and Yutaka Sasaki *Extracting Drug-Drug Interactions with Attention CNNs* 2017.

[16] Yifan Peng and Zhiyong Lu *Deep learning for extracting protein-protein interactions from biomedical literature* 2017.

[17] Sunil Mohan, Nicolas Fiorini, Sun Kim and Zhiyong Lu *Deep Learning for Biomedical Information Retrieval: Learning Textual Relevance from Click Logs* 2017.

[18] Bridget McInnes and Ted Pedersen *Improving Correlation with Human Judgments by Integrating Semantic Similarity with Second–Order Vectors* 2017.

[19] Anastasios Nentidis, Konstantinos Bougiatiotis, Anastasia Krithara, Georgios Paliouras and Ioannis Kakadiaris *Results of the fifth edition of the BioASQ Challenge* 2017.

Organizers:

Kevin Bretonnel Cohen, University of Colorado School of Medicine, USA
Dina Demner-Fushman, US National Library of Medicine
Sophia Ananiadou, National Centre for Text Mining and University of Manchester, UK
Jun-ichi Tsujii, National Institute of Advanced Industrial Science and Technology, Japan

Program Committee:

Sophia Ananiadou, National Centre for Text Mining and University of Manchester, UK
Ion Androutsopoulos, Athens University of Economics and Business, Greece
Emilia Apostolova, Language.ai, USA
Eiji Aramaki, University of Tokyo, Japan
Alan Aronson, US National Library of Medicine
Asma Ben Abacha, US National Library of Medicine
Olivier Bodenreider, US National Library of Medicine
Kevin Bretonnel Cohen, University of Colorado School of Medicine, USA
Leonardo Campillos Llanos, LIMSI - CNRS, France
Kevin Bretonnel Cohen, University of Colorado School of Medicine, USA
Nigel Collier, University of Cambridge, UK
Dina Demner-Fushman, US National Library of Medicine
Filip Ginter, University of Turku, Finland
Graciela Gonzalez, University of Pennsylvania, USA
Cyril Grouin, LIMSI - CNRS, France
Antonio Jimeno Yepes, IBM, Melbourne Area, Australia
Halil Kilicoglu, US National Library of Medicine
Aris Kosmopoulos, NCSR Demokritos, Greece
Robert Leaman, US National Library of Medicine
Chris Lu, US National Library of Medicine
Zhiyong Lu, US National Library of Medicine
Juan Miguel Cejuela, Technische Universität München, Germany
Timothy Miller, Children's Hospital Boston, USA
Makoto Miwa, Toyota Technological Institute, Japan
Danielle L Mowery, VA Salt Lake City Health Care System, USA
Diego Molla, Macquarie University, Australia
Jim Mork, National Library of Medicine, USA
Yassine Mrabet, US National Library of Medicine
Henning Müller, University of Applied Sciences, Switzerland
Claire Nédellec, INRA, France
Anastasios Nentidis, NCSR Demokritos, Athens, Greece
Aurélie Névéol, LIMSI - CNRS, France
Mariana Neves, Hasso Plattner Institute and University of Potsdam, Germany
Nhung Nguyen, The University of Manchester, Manchester
Naoaki Okazaki, Tohoku University, Japan
Georgios Paliouras, NCSR Demokritos, Athens, Greece
Ioannis Partalas, Viseo group, France
John Prager, Thomas J. Watson Research Center, IBM, USA
Sampo Pyysalo, University of Cambridge, UK

Francisco J. Ribadas-Pena, University of Vigo, Spain
Fabio Rinaldi, University of Zurich, Switzerland
Angus Roberts, The University of Sheffield, UK
Kirk Roberts, The University of Texas Health Science Center at Houston, USA
Hagit Shatkay, University of Delaware, USA
Pontus Stenetorp, University College London, UK
Karin Verspoor, The University of Melbourne, Australia
Ellen Voorhees, National Institute of Standards and Technology, USA
Byron C. Wallace, University of Texas at Austin, USA
W John Wilbur, US National Library of Medicine
Hai Zhao, Shanghai Jiao Tong University, Shanghai
Pierre Zweigenbaum, LIMSI - CNRS, France

Additional Reviewers:

Moumita Bhattacharya, University of Delaware, USA
Louise Deléger, INRA - MaIAGE, France
Lenz Furrer, Institute of Computational Linguistics, UZH, Zurich, Switzerland
Genevieve Gorrell, Sheffield University, UK
Ari Klein, University of Pennsylvania School of Medicine
Yifan Peng, US National Library of Medicine
Vassiliki Rentoumi, National Centre for Scientific Research Demokritos, Athens, Greece
Masoud Rouhizadeh, University of Pennsylvania, USA
Abeed Sarker, University of Pennsylvania, USA
Xingyi Song, University of Sheffield, UK
Tasnia Tahsin, Arizona State University, USA
Hegler Tissot, Federal University of Parana, Brazil
Ken Yano, Nara Institute of Science and Technology, Japan

Invited Speaker:

George Paliouras, National Centre for Scientific Research Demokritos, Athens, Greece

Conference Program

Friday August 4, 2017

8:30–8:45	**Opening remarks**

8:45–10:30 Session 1: Prediction and relation extraction

8:45–9:00   *Target word prediction and paraphasia classification in spoken discourse*
Joel Adams, Steven Bedrick, Gerasimos Fergadiotis, Kyle Gorman and Jan van Santen

9:00–9:15   *Extracting Drug-Drug Interactions with Attention CNNs*
Masaki Asada, Makoto Miwa and Yutaka Sasaki

9:15–9:30   *Insights into Analogy Completion from the Biomedical Domain*
Denis Newman-Griffis, Albert Lai and Eric Fosler-Lussier

9:30–9:45   *Deep learning for extracting protein-protein interactions from biomedical literature*
Yifan Peng and Zhiyong Lu

9:45–10:00   *Stacking With Auxiliary Features for Entity Linking in the Medical Domain*
Nazneen Fatema Rajani, Mihaela Bornea and Ken Barker

10:00–10:30 Invited Talk: "Results of the 5th edition of BioASQ Challenge" – Georgios Paliouras

Results of the fifth edition of the BioASQ Challenge
Anastasios Nentidis, Konstantinos Bougiatiotis, Anastasia Krithara, Georgios Paliouras and Ioannis Kakadiaris

10:30–11:00   *Coffee Break*

Friday August 4, 2017 (continued)

14:00–15:30   Session 3: From bio to clinical NLP

14:00–14:15   *Improving Correlation with Human Judgments by Integrating Semantic Similarity with Second–Order Vectors*
Bridget McInnes and Ted Pedersen

14:15–14:30   *Proactive Learning for Named Entity Recognition*
Maolin Li, Nhung Nguyen and Sophia Ananiadou

14:30–14:45   *Biomedical Event Extraction using Abstract Meaning Representation*
Sudha Rao, Daniel Marcu, Kevin Knight and Hal Daumé III

14:45–15:00   *Detecting Personal Medication Intake in Twitter: An Annotated Corpus and Baseline Classification System*
Ari Klein, Abeed Sarker, Masoud Rouhizadeh, Karen O'Connor and Graciela Gonzalez

15:00–15:15   *Unsupervised Context-Sensitive Spelling Correction of Clinical Free-Text with Word and Character N-Gram Embeddings*
Pieter Fivez, Simon Suster and Walter Daelemans

15:15–15:30   *Characterization of Divergence in Impaired Speech of ALS Patients*
Archna Bhatia, Bonnie Dorr, Kristy Hollingshead, Samuel L. Phillips and Barbara McKenzie

15:30–16:00   *Coffee Break*

16:00–16:30   Session 4 More clinical NLP

16:00–16:15   *Deep Learning for Punctuation Restoration in Medical Reports*
Wael Salloum, Greg Finley, Erik Edwards, Mark Miller and David Suendermann-Oeft

16:15–16:30   *Unsupervised Domain Adaptation for Clinical Negation Detection*
Timothy Miller, Steven Bethard, Hadi Amiri and Guergana Savova

16:30–18:00   Poster Session

BioCreative VI Precision Medicine Track: creating a training corpus for mining protein-protein interactions affected by mutations
Rezarta Islamaj Dogan, Andrew Chatr-aryamontri, Sun Kim, Chih-Hsuan Wei, Yi-fan Peng, Donald Comeau and Zhiyong Lu

Painless Relation Extraction with Kindred
Jake Lever and Steven Jones

Noise Reduction Methods for Distantly Supervised Biomedical Relation Extraction
Gang Li, Cathy Wu and K. Vijay-Shanker

Role-Preserving Redaction of Medical Records to Enable Ontology-Driven Processing
Seth Polsley, Atif Tahir, Muppala Raju, Akintayo Akinleye and Duane Steward

Annotation of pain and anesthesia events for surgery-related processes and outcomes extraction
Wen-wai Yim, Dario Tedesco, Catherine Curtin and Tina Hernandez-Boussard

Identifying Comparative Structures in Biomedical Text
Samir Gupta, A.S.M. Ashique Mahmood, Karen Ross, Cathy Wu and K. Vijay-Shanker

Tagging Funding Agencies and Grants in Scientific Articles using Sequential Learning Models
Subhradeep Kayal, Zubair Afzal, George Tsatsaronis, Sophia Katrenko, Pascal Coupet, Marius Doornenbal and Michelle Gregory

Deep Learning for Biomedical Information Retrieval: Learning Textual Relevance from Click Logs
Sunil Mohan, Nicolas Fiorini, Sun Kim and Zhiyong Lu

Target word prediction and paraphasia classification in spoken discourse

Joel Adams[1], **Steven Bedrick**[1], **Gerasimos Fergadiotis**[2], **Kyle Gorman**[3] and **Jan van Santen**[1]

[1]Center for Spoken Language Understanding, Oregon Health & Science University, Portland, OR
[2]Speech & Hearing Sciences Department, Portland State University, Portland, OR
[3]Google, Inc., New York, NY

Abstract

We present a system for automatically detecting and classifying phonologically anomalous productions in the speech of individuals with aphasia. Working from transcribed discourse samples, our system identifies neologisms, and uses a combination of string alignment and language models to produce a lattice of plausible words that the speaker may have intended to produce. We then score this lattice according to various features, and attempt to determine whether the anomalous production represented a phonemic error or a genuine neologism. This approach has the potential to be expanded to consider other types of paraphasic errors, and could be applied to a wide variety of screening and therapeutic applications.

1 Introduction

Aphasia is an acquired neurogenic language disorder in which an individual's ability to produce or comprehend language is compromised. It can be caused by a number of different underlying pathologies, but can generally be traced back to physical damage to the individual's brain: tissue damage following ischemic or hemorrhagic stroke, lesions caused by a traumatic brain injury or infection, etc. It can also be associated with various neurodegenerative diseases, as in the case of Primary Progressive Aphasia. According to the National Institute of Neurological Disorders and Stroke, approximately 1,000,000 people in the United States suffer from aphasia, and aphasia is a common consequence of strokes (prevalence estimates for aphasia among stroke patients vary, but are typically in the neighborhood of 30% (Engelter et al., 2006)).

Anomia is a the inability to access and retrieve words during language production, and is a common manifestation of aphasia (Goodglass and Wingfield, 1997). An anomic individual will experience difficulty producing words and naming items, which can cause substantial difficulties in day-to-day communication.

The process of screening for, diagnosing, and assessing anomia is typically manual in nature, and requires substantial time, labor, and expertise. Compared to other neuropsychological assessment instruments, aphasia-related assessments are particularly difficult to computerize, as they typically depend on subtle and complex linguistic judgments about the phonological and semantic similarity of words, and also require the examiner to interpret phonologically disordered speech. Furthermore, the most commonly used assessments focus for practical reasons on relatively constrained tasks such as picture naming, which may lack ecological validity (Mayer and Murray, 2003).

In this work, we describe an approach to automatically detecting and analyzing certain categories of word production errors characteristic of anomia in connected speech. Our approach is a first step towards an automated anomia assessment tool that could be used cost effectively in both clinical and research settings,[1] and could also be applied to other disorders of speech production. The method we propose uses statistical language models to identify possible errors, and employs a phonologically-informed edit distance model to determine phonological similarity between the subject's utterance and a set of plausible "intended words." We then apply machine learning techniques to determine which of several categories a given erroneous production may fall into. We

---

[1]As in the computer-administered (but manually-scored) assessments developed by Fergadiotis and colleagues (Fergadiotis et al., 2015; Hula et al., 2015).

1

Proceedings of the BioNLP 2017 workshop, pages 1–8,
Vancouver, Canada, August 4, 2017. ©2017 Association for Computational Linguistics

show results on intrinsic evaluations comparable to state-of-the-art sentence completion, as well as an extrinsic measure of classification well above a "most-frequent" baseline strategy.

1.1 Anomia and Paraphasias

Anomia can take several different forms, but in this work we are concerned with *paraphasias*, which are unintended errors in word production.[2]

There are several categories of paraphasic error. *Semantic errors* arise when an individual unintentionally produces a semantically-related word to their original, intended word (their "target word"). A classic semantic error would be saying "cat" when one intended to say "dog."

Phonemic (sometimes called "formal") errors occur when the speaker produces an unrelated word that is *phonemically related* to their target: "mat" for "cat", for example. It is also possible for an erroneous production to be *mixed*, that is both semantically and phonemically related to the target word: "rat" for "cat." Individuals with anomia also produce *unrelated* errors, which are words that are neither semantically or phonemically related to their intended target word: for example, producing "skis" instead of "zipper."

Each of these categories shares the commonality that the word produced by the individual is a "real" word. There is another family of anomic errors, *neologisms*, in which the individual produces *non-word* productions. A neologistic production may be phonemically related to the target, but containing phonological errors: "[dɑɪnoʊsɔɹ]" for "dinosaur." These are often referred to as *phonological* paraphasias. Alternatively, the individual may produce *abstruse neologisms*, in which the produced phonemes bear no discernable similarity to any "real" lexical item ("[æpməl]" for "comb"[3]).

The present work focuses exclusively on neologisms, both of the phonological variety as well as the abstruse variety. However, our fundamental approach can be extended to include other forms, as described in section 6.

Typical methods of diagnosing, staging, and otherwise characterizing anomia involve determining the number and kinds of paraphasias produced by an individual while undergoing some structured language elicitation process, for example a confrontation naming test (see (Kendall et al., 2013) and (Brookshire et al., 2014) for examples of such a study). As alluded to previously, producing these counts and classifications is a complex and laborious process. Furthermore, it is also often an inherently subjective process: are "carrot" and "banana" semantically related? What about "hose" and "rope"?

Reliability estimates of expert human performance at paraphasia classification in confrontation naming scenarios reflect the difficulty in this task. One recent study reported a kappa-equivalent score of 0.76 — a score that that is certainly acceptable, but that leaves much room for disagreement on the status of specific erroneous productions (Minkina et al., 2015). Other reported scores fall in a similar range (Kristensson et al., 2015), including when the productions are from neurotypical individuals (Nicholas et al., 1989). Automating this aspect of the task would not only improve efficiency, but would also decrease scoring variability. Having a reliable, automated method to analyze paraphasic errors would also expand the scope of what is currently possible in terms of assessment methodologies.

Notably, the approach we outline in this paper is explicitly designed to work on samples of natural, connected speech. It builds upon previous work by Fergadiotis et al. (2016) on automated analysis of errors produced in confrontation naming tests, and extends it into the realm of naturalistic discourse. It is our hope that, by enabling automated calculation of error frequencies and types on narrative speech, we might make using such material far easier in practice than it is today.

2 Data

For this work, we use the data set provided by the AphasiaBank project (MacWhinney et al., 2011), which has assembled a large database of transcribed interactions between examiners and people with aphasia, nearly all of whom have suffered a stroke. Notably, AphasiaBank also includes transcribed sessions with neurotypical controls. Each interaction follows a common protocol and script,

---

[2]Note that individuals *without* any sort of language disorder do occasionally produce errors in their speech; this fact has led to a truly shocking amount of study by linguists. Frisch & Wright (2002) provide a reasonable overview of the background and phonology of the phenomenon.

[3]This example was taken from a corpus of responses to a confrontation naming test (Mirman et al., 2010), in which the subject is shown a picture and required to name its contents. As such, in the case of this specific error, we have *a priori* knowledge of what the target word "should" have been. Obviously, in a more naturalistic task or setting, we would not have this advantage.

and is transcribed in great detail using a standardized set of annotation guidelines. The transcripts include word-level error codes, according to a detailed taxonomy of errors and associated annotations. In the case of semantic, formal, and phonemic errors, the word-level annotations include a "best guess" on the part of the transcriber as to what the speaker's intended production may have been.

Each transcribed session consists of a prescribed sequence of language elicitation activities, including a set of personal narratives (e.g.,"Do you remember when you had your stroke? Please tell me about it."), standardized picture description tasks, a story retelling task (involving the story of *Cinderella*), and a procedural discourse task.

We noted that the distribution of errors within sentences seems to obey the power law , with the majority of error-containing sentences containging a single error, followed somewhat distantly by sentences containing two errors, with a relatively steep dropoff thereafter. For the present study, we restricted our analysis to sentences that contained a single error. Our reasoning for this restriction was that we do not presently have a theoretically-informed model of what, if any, relationship there may be between multiple errors within a sentence. However, it seems quite likely that the errors occurring in a sentence containing (for instance) five paraphasic errors might be somehow related to one another. We anticipate exploring this phenomenon in the future (see section 6).

We chose to restrict our data to the story retelling task due to the constrained and focused vocabulary of the Cinderella story. This resulted in $\approx$ 1000 sentences from 385 individuals. We then identified sentences containing instances of our errors of interest: phonological paraphasia (AphasiaBank codes "p:n", "p:m") or abstruse neologism ("n:uk" and "n:k").

3   Methods

We first tokenized the AphasiaBank data using a modified version of the Penn Treebank tokenizer which left contractions as a single word and simply removed the connecting apostrophe, as these occasionally appear as target words and thus we needed to treat them as a single token. We left stopwords intact, and case-folded all sentences to upper-case. Cardinal numbers were collapsed into a category token, as were ordinal numbers and dates (each category was given its own token). The Aphasia-

Bank transcripts include detailed IPA-encoded representations of neologistic productions, but any "real-world" usage scenario for our algorithm is unlikely to benefit from such high-quality transcription. We therefore translated the non-lexical productions into a simulated "best-guess" orthographic representation of the subject's non-lexical productions.

We next turned to the question of identifying neologisms in our sentences. Simply using a standard dictionary to determine lexicality could result in numerous "false positives," driven largely by proper names of people, brands, etc. To avoid this, we used the SUBTLEX-US corpus (Brysbaert and New, 2009) to identify neologisms. SUBTLEX-US was build using subtitles from English-language television shows and movies, and Brysbaert and New have demonstrated that it correlates with a number of psycholinguistic behavior measures (most notably, naming latencies) better than better-known frequency norms such as those derived from the Brown corpus or CELEX-2.

Upon identifying a possible non-word production, recall that our next goal is to determine whether it represents a *phonemic* error (substituting "[dɑɪnoʊsɔɪ]" for "dinosaur") or an *abstruse neologism* (a completely novel sequence of phonemes that does not correspond to an actual word). To help accomplish this, we train a language model to identify plausible words that *could* fit in the slot occupied by the erroneous production, and produce a lattice of these candidate target words (i.e., words that the subject may have been intending to produce, given what we know about the context in which they were speaking).

Our language models for this study were built using the New York Times section of the Gigaword newswire corpus (Parker et al., 2011). After success in preliminary experiments, we filtered this corpus by first training a Latent Dirichlet Allocation (LDA) topic model on the corpus using Gensim (Řehůřek and Sojka, 2010) over 20 topics. We then projected the text of each of the Cinderella narrative samples into the resulting topic space, and calculated the centroids for the narrative task. This yielded a subset of the larger corpus of New York Times articles that was "most similar" to the Cinderella retellings, and we used these to build our language models.

We investigated two different language modeling approaches: a traditional FST-encoded ngram language model, and a neural-net based log-bilinear (LBL) language model. For the FST representation, we used the the OpenGrm-NGram language modeling toolkit (Roark et al., 2012) and used an n-gram order of 4, with Kneser-Ney smoothing (Kneser and Ney, 1995). For the LBL approach, we used a Python implementation[4] of the language model described by Mnih and Teh (Mnih and Teh, 2012). We used word embeddings of dimension 100, and a 5-gram context window. In both cases we trained two language models: one trained on the "task-specific" subset of Gigaword, and another trained on the AphasiaBank control data. We combined these with a simple mixing coefficient, λ as shown in Equation 1 where $P_{GW}(w)$ is the language model probability of word w computed against the Gigaword corpus and the $P_{AB}(w)$ is the language model probability trained on the AphasiaBank controls.

$$P(w) = \lambda \cdot P_{AB}(w) + (1 - \lambda) \cdot P_{GW}(w) \quad (1)$$

We evaluate non-lexical productions as follows. First, we use the Phonetisaurus grapheme-to-phoneme toolkit (Novak et al., 2012) to translate our orthographic representation into an estimated phoneme sequence. We then calculate a phonologically-aware edit distance between each non-word production and every word in our lexicon up to some maximum edit distance (in our case 4.0). Phonemes from a related class (e.g. vowels) are considered lower cost replacements than those from another class (e.g. unvoiced fricatives). This gives us a set of candidates which are phonologically similar to the production.

We next used our language models to produce lattices representing a set of possible sentences that the subject could plausibly have been intending to produce. At the point in the produced sentence where our error detection system indicated that a non-word production occurred, we represent the anomaly by the union of all possible words in our edit-distance constrained lexicon (see figure 3 for an example sentence lattice). Finally, we use the language models to score the resulting sentence lattice so as to be able to rank the candidate words, and use the estimated sentence-level probability for each candidate word (i.e., the hypothesized intended utterance featuring that word). Put simply,

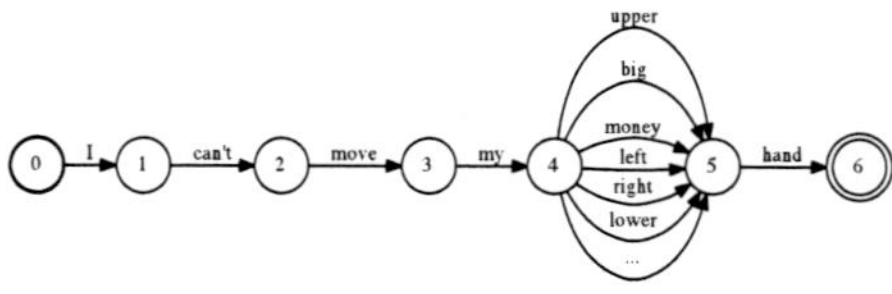

Figure 1: An example candidate word lattice for the production "I can't move my [vɑɪ] hand."

for each candidate intended word, we produce a version of the subject's utterance with that hypothesized word in place of the anomalous utterance, and score this hypothesized utterance with the language model.

At this point in the process, we have the following information about each erroneous production: a best-guess orthographic transcription of what the individual actually produced, and a ranked list of plausible words that they could potentially have been attempting to produce, together with probability estimates for each hypothesized production.

To determine the category of our error productions— again, between productions representing phonological errors such as "[dɑɪnoʊsɔɹ]" for "dinosaur", and productions representing abstruse neologisms— we trained a binary classifier using features representing the characteristics of the candidate word space surrounding the erroneous production. Our intuition is that phonemic errors were much more likely than abstruse neologisms to have highly-ranked candidate target words that were *also* phonologically similar to the subject's actual production.

To capture this, we performed the following procedure for each error-containing utterance. We first divide our list of candidate intended words into buckets by edit distance (0.5, 1.0, 1.5, etc.[5]). Each bucket can now be thought of as a ranked list of probabilities, each representing a possible hypothesized intended utterance featuring a word within that bucket's edit distance of the actual (anomalous) utterance.

We next represent each bucket with a feature vector consisting of the count of words in that

---

[4] https://github.com/ddahlmeier/neural_lm

[5] Recall that our phonological edit distance metric allows for partial costs for related phonological substitutions.

bucket, as well as descriptive statistics regarding the distribution of language model probabilities in that bucket (min, max, etc.). We then concatenate each bucket's features together into a master feature vector for the utterance. Our expectation is that these features will be relatively evenly distributed across buckets in the case of utterances containing abstruse neologisms, whereas utterances featuring phonological paraphasias will vary according to phonological edit distance.

Once we have computed feature vectors for each utterance, we used the Scikit-learn Python machine learning library (Pedregosa et al., 2011) to train a Support Vector Machine classifier to distinguish between utterances phonological and abstruse neologisms. We evaluate its performance using leave-one-out cross-validation.

4 Results

We perform two evaluations of our model: an intrinsic evaluation of how often our system includes the target word in the top-n ranked candidates, and an extrinsic evaluation where we attempt to classify a paraphasia between phonological errors and abstruse neologisms.

Our motivation for evaluating our system's performance on target word prediction is tied to our classification assumptions. In an ideal case for a phonological error, the target word should fall within one of the buckets representing a low edit distance. If our language model successfully rates the target as likely, we would see an high maximum probability within that bucket, which is a feature in our classifier.

The performance of our language models on the top-n ranked evaluation can be seen in Table 1. The log-bilinear model outperformed the FST in all cases. This finding is similar to state of the art results for automatic sentence completion systems–particularly for phonemic errors–as we'll discuss in greater detail in Section 5. Both systems did a better job of predicting the target word for phonemic errors than they did for abstruse neologisms. It's not immediately clear what the reason for this is. However, anecdotally, sentences including abstruse neologisms are also often agrammatical.

For the evaluation of our classification, we created a simple majority class baseline classifier that always chooses the largest class of errors (phonemic errors in this case). This baseline classifier has

Error	n	FST	LBL
Phonemic	1	.43	.52
Phonemic	5	.54	.66
Phonemic	10	.59	.69
Phonemic	20	.67	.77
Phonemic	50	.72	.81
Abstruse Neo.	1	.29	.35
Abstruse Neo.	5	.41	.49
Abstruse Neo.	10	.44	.51
Abstruse Neo.	20	.51	.59
Abstruse Neo.	50	.54	.60

Table 1: Accuracy of language model predicting the correct target word within the first n results.

Features	FST	LBL
count, mean	.612	.661
count, mean, max	.621	**.680**
count, mean, max, min	.610	.659
count, mean, max, dist.	.610	.659

Table 2: Classification accuracy by model. Baseline accuracy of choosing the most common error type is .510.

a classification accuracy of .51. The results of classification can be seen in Table 2. Both of our systems handily outperformed baseline: the FST by a relative 20% improvement, and the LBL nearly 33%. As we expected from the top-n results, classification based on the LBL outperformed that based on the FST.

The "dist" feature listed in table 2 is the edit distance of a given bucket normalized by the number of phonemes in the actual error production. It was not found to be productive as a feature, nor was the minimum language model probability of words in a given bucket ("min" in the table). The best results for both systems were a combination of count of candidate terms per bucket ("count") concatenated with the maximum and mean language model probabilities within a bucket ("max" and "min", respectively).

We varied the mixing-coefficient (λ) from Equation 1 in both the FST and LBL approaches. As long as the resulting model includes a non-trivial weighting of the Cinderella corpus (typically anything better than $\lambda = 3$), the actual value of the mixing coefficient was irrelevant to either of our evaluations. In this, it appears to work as designed, with the Gigaword corpus providing background probabilities, and the AphasiaBank Cinderella control retellings increasing the weight of topically important words that are otherwise rare (like "Cinderella" and "carriage").

5 Related Work & Discussion

As far back as Shannon's word-guessing game (Shannon, 1951), researchers have sought to leverage the statistical regularities in natural language to predict missing or subsequent words. In practice, however, this proves to be a surprisingly challenging problem. Language occurs at levels beyond simply choosing lexical items, and local statistical characteristics of language often fail to capture syntactic and semantic patterns. Zweig & Burges (2012) provide an enlightening discussion on the limitations of relying on n-gram guessing for syntactically complex tasks such as "identify the missing word in the sentence," and also describe a very challenging language model evaluation task built around this problem. They tested a variety of language modeling approaches using their task, and report that well-trained generative n-gram models achieve correct predictions $\approx$ 30% of the time. State-of-the-art performance on the their word prediction task using recurrent neural network language models,[6] report highest scores are in the mid-50% range (Mirowski and Vlachos, 2015; Mnih and Kavukcuoglu, 2013).

In our case, the nature of our data renders this task even more challenging. Our sentences are often short and agrammatical (often missing or misusing determiners, for example), and are produced by individuals with impaired language ability.

As such, our results are actually quite similar to those reported in recent literature. Our average accuracy of our FST n-gram (over both classes of errors) selecting the appropriate word is $\approx$ 35% while our LBL model's performance of $\approx$ 43% is comparable to the 5-gram LBL performance of 49.3 reported by Mnih and Teh on the MSR Sentence Completion Challenge dataset (Mnih and Teh, 2012).

6 Conclusion & Future Work

While the system's performance is quite good on both intrinsic and extrinsic evaluation, there remains much interesting work left to do on the problem.

We currently only evaluate sentences with a single error, and more generally have not investigated whether sentences with multiple errors are different from mono-error sentences in terms of error distribution or structure. However, our approach *should* be able to generalize to sentences with additional errors, and we will be investigating this in future work.

Additionally, the AphasiaBank transcripts include phrasal dependency and part-of-speech tags which we are currently not using. In future work we will investigate including these as features in language modelling, as there is some evidence that this improves the conceptually related task of contextual spellcheck(Fossati and Di Eugenio, 2008).

There is quite a bit of work that can be done on the language models as well. A more nuanced approach to topic adaptation is worth investigating, and we plan to experiment with using non-newswire corpora. Despite our attempts to focus the corpus via LDA, there is a major difference between the written language of the New York Times, and the spoken dialogue between the aphasic subjects and their clinicians.

The most exciting area for further research is the inclusion of semantic information in our classification. While our topic-specific language model provides our model with some implicit semantic information, a more principled approach to semantic relevance could potentially improve the classification of phonemic errors versus abstruse neologisms by determining whether a given candidate word is semantically relevant in context. More intriguingly, it would give us a way to start investigating semantic errors, and those errors that include "real" words (for example, the previously discussed error of replacing "dog" with "cat").

One major limitation of our current system is its reliance on human-produced transcriptions of speech samples. In practice, transcription is rarely feasible in clinical settings, and even in research settings is often challenging, which may seem to limit the applicability of our approach. Notably, however, our system does not require detailed *phonetic* transcription, and merely requires "best-guess" orthographic transcription of neologisms. As such, one could in principle use automatic speech recognition (ASR) to analyze a recording of a patient or research subject, and produce a transcript on which our methods could be

---

[6]See De Mulder et al. (2015) for a recent review on this subject.

run.[7] Fraser et al. (2015) have had some success at applying ASR to samples of aphasic speech and performing downstream analysis on the resulting transcripts, and we anticipate experimenting with similar techniques in the future.

Acknowledgments

We thank the BioNLP reviewers for their helpful comments and advice. This material is based upon work supported in part by the National Institute on Deafness and Other Communication Disorders of the National Institutes of Health under awards R01DC012033 and R03DC014556. The content is solely the responsibility of the authors and does not necessarily represent the official views of the granting agencies or any other individual.

References

C. E. Brookshire, T. Conway, R. H. Pompon, M. Oelke, and D. L. Kendall. 2014. Effects of intensive phonomotor treatment on reading in eight individuals with aphasia and phonological alexia. *American Journal of Speech-Language Pathology* 23(2):S300–S311.

M. Brysbaert and B. New. 2009. Moving beyond Kučera and Francis: A critical evaluation of current word frequency norms and the introduction of a new and improved word frequency measure for American English. *Behavior Research Methods* 41(4):977–990.

Wim De Mulder, Steven Bethard, and Marie-Francine Moens. 2015. A survey on the application of recurrent neural networks to statistical language modeling. *Computer Speech & Language* 30(1):61–98.

S. T. Engelter, M. Gostynski, S. Papa, M. Frei, C. Born, V. Ajdacic-Gross, F. Gutzwiller, and P. A. Lyrer. 2006. Epidemiology of aphasia attributable to first ischemic stroke: Incidence, severity, fluency, etiology, and thrombolysis. *Stroke* 37(6):1379–1384.

G. Fergadiotis, S. Kellough, and W. D. Hula. 2015. Item Response Theory modeling of the Philadelphia Naming Test. *Journal of Speech, Language, and Hearing Research* 58(3):865–877.

Gerasimos Fergadiotis, Kyle Gorman, and Steven Bedrick. 2016. Algorithmic Classification of Five Characteristic Types of Paraphasias. *American Journal of Speech-Language Pathology* 25(4S):S776–12.

Davide Fossati and Barbara Di Eugenio. 2008. I saw tree trees in the park: How to correct real-word spelling mistakes. In *LREC*.

K. C. Fraser, N. Ben-David, G. Hirst, N. Graham, and E. Rochon. 2015. Sentence segmentation of aphasic speech. In *ACL*. pages 862–871.

S. A. Frisch and R. Wright. 2002. The phonetics of phonological speech errors: An acoustic analysis of slips of the tongue. *Journal of Phonetics* 30(2):139–162.

H. Goodglass and A. Wingfield. 1997. *Anomia: Neuroanatomical and cognitive correlates*. Academic Press, New York.

W. D. Hula, S. Kellough, and G. Fergadiotis. 2015. Development and simulation testing of a computerized adaptive version of the Philadelphia Naming Test. *Journal of Speech, Language, and Hearing Research* 58(3):878–890.

D. L. Kendall, R. H. Pompon, C. E. Brookshire, I. Minkina, and L. Bislick. 2013. An analysis of aphasic naming errors as an indicator of improved linguistic processing following phonomotor treatment. *American Journal of Speech-Language Pathology* 22(2):S240–S249.

R Kneser and H Ney. 1995. Improved backing-off for M-gram language modeling. In *1995 International Conference on Acoustics, Speech, and Signal Processing*. IEEE, pages 181–184.

J. Kristensson, I. Behrns, and C. Saldert. 2015. Effects on communication from intensive treatment with semantic feature analysis in aphasia. *Aphasiology* 29(4):466–487.

B. MacWhinney, D. Fromm, M. Forbes, and A. Holland. 2011. AphasiaBank: Methods for studying discourse. *Aphasiology* 25(11):1286–1307.

J. Mayer and L. Murray. 2003. Functional measures of naming in aphasia: Word retrieval in confrontation naming versus connected speech. *Aphasiology* 17(5):481–497.

I. Minkina, M. Oelke, L. P. Bislick, C. E. Brookshire, R. Hunting Pompon, J. P. Silkes, and D. L. Kendall. 2015. An investigation of aphasic naming error evolution following phonomotor treatment. *Aphasiology* epub ahead of print.

D. Mirman, T. J. Strauss, A. Brecher, G. M. Walker, P. Sobel, G. S. Dell, and M. F. Schwartz. 2010. A large, searchable, web-based database of aphasic performance on picture naming and other tests of cognitive function. *Cognitive Neuropsychology* 27(6):495–504.

Piotr Mirowski and Andreas Vlachos. 2015. Dependency Recurrent Neural Language Models for Sentence Completion. In *Proceedings of the 53rd Annual Meeting of the Association for Computational Linguistics and the 7th International Joint Conference on Natural Language Processing (Volume 2: Short Papers)*. Association for Computational Linguistics, Beijing, China, pages 511–517.

---

[7]Depending on the specifics of the ASR system, it may in fact be possible to retain phonological information, which, while not necessary, certainly could be helpful to our system.

Andriy Mnih and Koray Kavukcuoglu. 2013. Learning word embeddings efficiently with noise-contrastive estimation. In C J C Burges, L Bottou, M Welling, Z Ghahramani, and K Q Weinberger, editors, *Advances in Neural Information Processing Systems 26*, Curran Associates, Inc., pages 2265–2273.

Andriy Mnih and Yee Whye Teh. 2012. A fast and simple algorithm for training neural probabilistic language models. *arXiv preprint arXiv:1206.6426* .

L. E. Nicholas, R. H.and Maclennan D. L. Brookshire, J. G. Schumacher, and S. A. Porrazzo. 1989. Revised administration and scoring procedures for the Boston Naming Test and norms for non-brain-damaged adults. *Aphasiology* 3(6):569–580.

J. R. Novak, N. Minematsu, and K. Hirose. 2012. WFST-based grapheme-to-phoneme conversion: Open source tools for alignment, model-building and decoding. In *International Workshop on Finite State Methods and Natural Language Processing*. pages 45–49.

R. Parker, D. Graff, J. Kong, K. Chen, and K. Maeda. 2011. English Gigaword 5th Edition. Linguistic Data Consortium: LDC2011T07.

F. Pedregosa, G. Varoquaux, A. Gramfort, V. Michel, B. Thirion, O. Grisel, M. Blondel, P. Prettenhofer, R. Weiss, V. Dubourg, J. Vanderplas, A. Passos, D. Cournapeau, M. Brucher, M. Perrot, and E. Duchesnay. 2011. Scikit-learn: Machine learning in Python. *Journal of Machine Learning Research* 12:2825–2830.

B. Roark, R. Sproat, C. Allauzen, M. Riley, J. Sorensen, and T. Tai. 2012. The OpenGrm open-source finite-state grammar software libraries. In *ACL*. pages 61–66.

C. Shannon. 1951. Prediction and entropy of printed English. *Bell System Technical Journal* 50:50–64.

Geoffrey Zweig and Chris J C Burges. 2012. A Challenge Set for Advancing Language Modeling. In *Proceedings of the NAACL-HLT 2012 Workshop: Will We Ever Really Replace the N-gram Model? On the Future of Language Modeling for HLT*. Association for Computational Linguistics, Montréal, Canada, pages 29–36.

R. Řehůřek and P. Sojka. 2010. Software framework for topic modelling with large corpora. In *LREC*. pages 45–50.

Extracting Drug-Drug Interactions with Attention CNNs

Masaki Asada, **Makoto Miwa** and **Yutaka Sasaki**
Computational Intelligence Laboratory
Toyota Technological Institute
2-12-1 Hisakata, Tempaku-ku, Nagoya, Aichi, 468-8511, Japan
`{sd17402, makoto-miwa, yutaka.sasaki}@toyota-ti.ac.jp`

Abstract

We propose a novel attention mechanism for a Convolutional Neural Network (CNN)-based Drug-Drug Interaction (DDI) extraction model. CNNs have been shown to have a great potential on DDI extraction tasks; however, attention mechanisms, which emphasize important words in the sentence of a target-entity pair, have not been investigated with the CNNs despite the fact that attention mechanisms are shown to be effective for a general domain relation classification task. We evaluated our model on the Task 9.2 of the DDIExtraction-2013 shared task. As a result, our attention mechanism improved the performance of our base CNN-based DDI model, and the model achieved an F-score of 69.12%, which is competitive with the state-of-the-art models.

1 Introduction

When drugs are concomitantly administered to patients, the effects of the drugs may be enhanced or weakened, which may cause side effects. These kinds of interactions are called Drug-Drug Interactions (DDIs). Several drug databases, such as DrugBank (Law et al., 2014), Therapeutic Target Database (Yang et al., 2016), and PharmGKB (Thorn et al., 2013), have been provided to summarize drug and DDI information for researchers and professionals; however, many newly discovered or rarely reported interactions are not covered in the databases, and they are still buried in biomedical texts. Therefore, studies on automatic DDI extraction that extract DDIs from texts are expected to support maintenance of databases with high coverage and quick update to help medical experts deepen their understanding of DDIs.

For the DDI extraction, deep neural network-based methods have recently drawn a considerable

attention (Liu et al., 2016; Zhao et al., 2016; Sahu and Anand, 2017). Deep neural networks have been widely used in the NLP field. They show high performance on several NLP tasks without requiring manual feature engineering. Convolutional Neural Networks (CNNs) and Recurrent Neural Networks (RNNs) are often employed for the network architectures. Among these, CNNs have an advantage that they can be easily parallelized and the calculation is thus fast with recent Graphical Processing Units (GPUs).

Liu et al. (2016) showed that CNN-based model can achieve a high accuracy on the task of DDI extraction. Sahu and Anand (2017) proposed an RNN-based model with attention mechanism to tackle the DDI extraction task and show the state-of-the-art performance. The integration of an attention mechanism into a CNN-based relation extraction is proposed by Wang et al. (2016). This is applied to a general domain relation extraction task SemEval 2010 Task 8 (Hendrickx et al., 2009). Their model showed the state-of-the-art performance on the task. CNNs with attention mechanisms, however, are not evaluated on the task of DDI extraction.

In this study, we propose a novel attention mechanism that is integrated into a CNN-based DDI extraction model. The attention mechanism extends attention mechanism by Wang et al. (2016) in that it deals with anonymized entities separately from other words and incorporates a smoothing parameter. We implement a CNN-based relation extraction model and integrate the novel mechanism into the model. We evaluate our model on the Task 9.2 of the DDIExtraction-2013 shared task (Segura Bedmar et al., 2013).

The contribution of this paper is as follows. First, this paper proposes a novel attention mechanism that can boost the performance on CNN-based DDI extraction. Second, the DDI extraction model with the attention mechanism achieves

9

Proceedings of the BioNLP 2017 workshop, pages 9–18,
Vancouver, Canada, August 4, 2017. ©2017 Association for Computational Linguistics

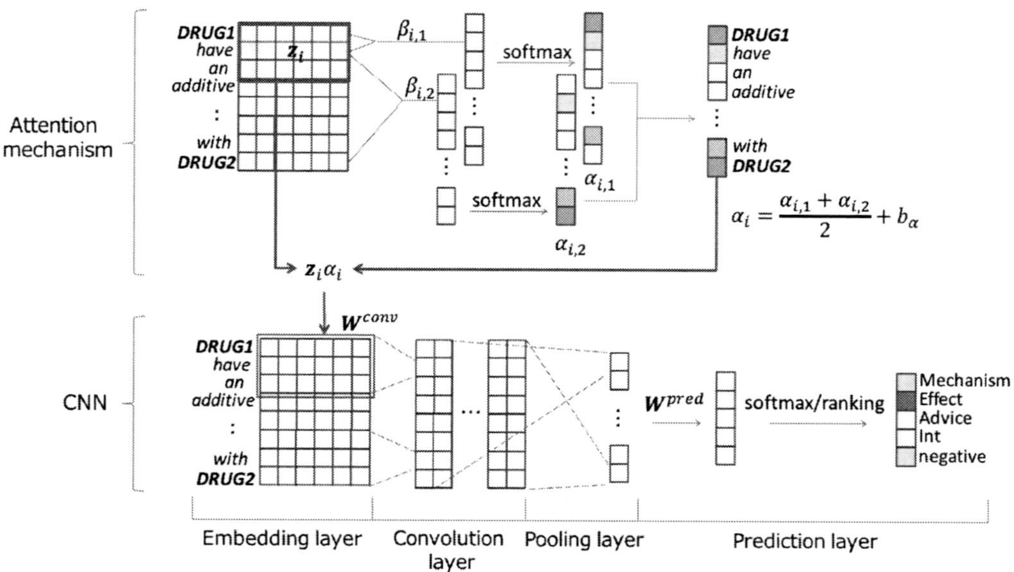

Figure 1: Overview of our model

the performance with an F-score of 69.12% that is competitive with other state-of-the-art DDI extraction models when we compare the performance without negative instance filtering (Chowdhury and Lavelli, 2013).

2   Methods

We propose a novel attention mechanism for a CNN-based DDI extraction model. We illustrate the overview of the proposed DDI extraction model in Figure 1. The model extracts interactions from sentences with drugs are given. In this section, we first present preprocessing of input sentences. We then introduce the base CNN model and explain the attention mechanism. Finally, we explain the training method.

2.1   Preprocessing

Before processing a drug pair in a sentence, we replace the mentions of the target drugs in the pair with "*DRUG1*" and "*DRUG2*" according to their order of appearance. We also replace other mentions of drugs with "*DRUGOTHER*".

Table 1 shows an example of preprocessing when an input sentence *Exposure to oral S-ketamine is unaffected by itraconazole but greatly increased by ticlopidine* is given with a target entity pair. By performing preprocessing, it is possible to prevent the DDI extraction model to be

specialized for the surface forms of the drugs in a training data set and to perform DDI extraction using the information of the whole context.

2.2   Base CNN model

The base CNN model for extracting DDIs is one by Zeng et al. (2014). In addition to their original objective function, we employ an ranking-based objective function by dos Santos et al. (2015). The model consists of four layers: embedding, convolution, pooling, and prediction layers. We show the CNN model at the bottom half of Figure 1.

2.2.1   Embedding layer

In the embedding layer, each word in the input sentence is mapped to a real-valued vector representation using an embedding matrix that is initialized with pre-trained embeddings. Given an input sentence $S = (w_1, \cdots, w_n)$ with drug entities e_1 and e_2, we first convert each word w_i into a real-valued vector $\boldsymbol{w}_i^w$ by an embedding matrix $\boldsymbol{W}^{emb} \in \mathbb{R}^{d_w \times |V|}$ as follows:

$$\boldsymbol{w}_i^w = \boldsymbol{W}^{emb} \boldsymbol{v}_i^w, \tag{1}$$

where d_w is the number of dimensions of the word embeddings, V is the vocabulary in the training data set and the pre-trained word embeddings, and $\boldsymbol{v}_i^w$ is a one hot vector that represents the index of word embedding in $\boldsymbol{W}^{emb}$. $\boldsymbol{v}_i^w$ thus extracts the corresponding word embedding from $\boldsymbol{W}^{emb}$.

Entity1	Entity2	Preprocessed input sentence
S-ketamine	*itraconazole*	*Exposure to oral **DRUG1** is unaffected by **DRUG2** but greatly increased by DRUGOTHER.*
S-ketamine	*ticlopidine*	*Exposure to oral **DRUG1** is unaffected by DRUGOTHER but greatly increased by **DRUG2**.*
itraconazole	*ticlopidine*	*Exposure to oral DRUGOTHER is unaffected by **DRUG1** but greatly increased by **DRUG2**.*

Table 1: An example of preprocessing on the sentence "*Exposure to oral S-ketamine is unaffected by itraconazole but greatly increased by ticlopidine*" for each target pair.

The word embedding matrix $\boldsymbol{W}^{emb}$ is fine-tuned during training.

We also prepare d_{wp}-dimensional word position embeddings $\boldsymbol{w}^{p}_{i,1}$ and $\boldsymbol{w}^{p}_{i,2}$ that correspond to the relative positions from first and second target entities, respectively. We concatenate the word embedding $\boldsymbol{w}^{w}_i$ and these word position embeddings $\boldsymbol{w}^{p}_{i,1}$ and $\boldsymbol{w}^{p}_{i,2}$ as in the following Equation (2), and we use the resulting vector as the input to the subsequent convolution layer:

$$\boldsymbol{w}_i = [\boldsymbol{w}^{w}_i; \boldsymbol{w}^{p}_{i,1}; \boldsymbol{w}^{p}_{i,2}]. \tag{2}$$

2.2.2 Convolution layer

We define a weight tensor for convolution as $\boldsymbol{W}^{conv}_k \in \mathbb{R}^{d_c \times (d_w + 2d_{wp}) \times k}$ and we represent the j-th column of $\boldsymbol{W}^{conv}_k$ as $\boldsymbol{W}^{conv}_{k,j} \in \mathbb{R}^{(d_w + 2d_{wp}) \times k}$. Here, d_c denotes the number of filters for each window size, k is a window size, and K is a set of the window sizes of the filters. We also introduce $\boldsymbol{z}_{i,k}$ that is concatenated k word embeddings:

$$\boldsymbol{z}_{i,k} = [\boldsymbol{w}^{\mathrm{T}}_{\lfloor i-(k-1)/2 \rfloor}; \dots; \boldsymbol{w}^{\mathrm{T}}_{\lfloor i-(k+1)/2 \rfloor}]^{\mathrm{T}}. \tag{3}$$

We apply the convolution to the embedding matrix as follows:

$$m_{i,j,k} = f(\boldsymbol{W}^{conv}_{k,j} \odot \boldsymbol{z}_{i,k} + b), \tag{4}$$

where $\odot$ is an element-wise product, b is the bias term, and f is the ReLU function defined as:

$$f(x) = \begin{cases} x, & \text{if } x > 0 \\ 0, & \text{otherwise.} \end{cases} \tag{5}$$

2.2.3 Pooling layer

We employ the max pooling (Boureau et al., 2010) to convert the output of each filter in the convolution layer into a fixed-size vector as follows:

$$\boldsymbol{c}_k = [c_{1,k}, \dots, c_{d_c,k}], \; c_{j,k} = \max_i m_{i,j,k}. \tag{6}$$

We then obtain the d_p-dimensional output of this pooling layer, where d_p equals to $d_c \times |K|$, by concatenating the obtained outputs $\boldsymbol{c}_k$ for all the window sizes $k_1, \dots, k_K (\in K)$:

$$\boldsymbol{c} = [\boldsymbol{c}_{k_1}; \dots; \boldsymbol{c}_{k_i}; \dots; \boldsymbol{c}_{k_K}]. \tag{7}$$

2.2.4 Prediction layer

We predict the relation types using the output of the pooling layer. We first convert $\boldsymbol{c}$ into scores using a weight matrix $\boldsymbol{W}^{pred} \in \mathbb{R}^{o \times d_p}$:

$$\boldsymbol{s} = \boldsymbol{W}^{pred}\boldsymbol{c}, \tag{8}$$

where o is the total number of relationships to be classified and $\boldsymbol{s} = [s_1, \dots, s_o]$. We then employ the following two different objective functions for prediction.

Softmax We convert $\boldsymbol{s}$ into the probability of possible relations $\boldsymbol{p}$ by a softmax function:

$$\boldsymbol{p} = [p_1, \dots, p_o], \; p_j = \frac{\exp(s_j)}{\sum_{l=1}^{o} \exp(s_l)}. \tag{9}$$

The loss function $L_{softmax}$ is defined as in the Equation (10) when the gold type distribution $\boldsymbol{y}$ is given. $\boldsymbol{y}$ is a one-hot vector where the probability of the gold label is 1 and the others are 0.

$$L_{softmax} = -\sum \boldsymbol{y} \log \boldsymbol{p} \tag{10}$$

Ranking We employ the ranking-based objective function following dos Santos et al. (2015). Using the scores $\boldsymbol{s}$ in the Equation (8), the loss is calculated as follows:

$$L_{ranking} = \log(1 + \exp(\gamma(m^+ - s_y))) \\ + \log(1 + \exp(\gamma(m^- + s_c))), \tag{11}$$

where m^+ and m^- are margins, γ is a scaling factor, y is a gold label, and c ($\neq y$) is a negative label with the highest score in $\boldsymbol{s}$. We set γ to 2, m^+ to 2.5 and m^- to 0.5 following dos Santos et al. (2015).

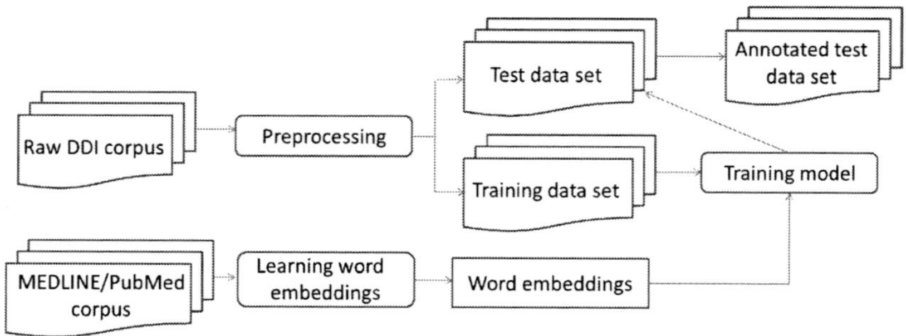

Figure 2: Workflow of DDI extraction

2.3 Attention mechanism

Our attention mechanism is based on the input attention by Wang et al. (2016)[1]. The proposed attention mechanism is different from the base one in that we prepare separate attentions for entities and we incorporate a bias term to adjust the smoothness of attentions. We illustrate the attention mechanism at the upper half of Figure 1.

We define the word index of the first and second target drug entities in the sentence as e_1 and e_2, respectively. We also denote by $E = \{e_1, e_2\}$ the set of indices and by $j \in \{1, 2\}$ the index of the entities. We calculate our attentions using these:

$$\beta_{i,j} = \boldsymbol{w}_{e_j} \cdot \boldsymbol{w}_i \tag{12}$$

$$\alpha_{i,j} = \begin{cases} \frac{\exp(\beta_{i,j})}{\sum_{1 \le l \le n, l \notin E} \exp(\beta_{l,j})}, & \text{if } i \notin E \\ a_{drug}, & \text{otherwise} \end{cases} \tag{13}$$

$$\alpha_i = \frac{\alpha_{i,1} + \alpha_{i,2}}{2} + b_\alpha. \tag{14}$$

Here, a_{drug} is an attention parameter for entities and b_α is the bias term. a_{drug} and b_α are tuned during training. If we set E to empty and b_α to zero, the attention will be the same as one by Wang et al. (2016). We incorporate the attentions α_i into the CNN model by replacing the Equation (4) with the following equation:

$$m_{i,j,k} = f(\boldsymbol{W}_j^{conv} \odot \boldsymbol{z}_{i,k} \alpha_i + b). \tag{15}$$

2.4 Training method

We use L2 regularization to avoid over-fitting. We use the following objective functions L'_* ($L'_{softmax}$ or $L'_{ranking}$) by incorporating the L2 regularization on weights to the Equation (10).

$$L'_* = L_* + \lambda(||\boldsymbol{W}^{emb}||_F^2 + ||\boldsymbol{W}^{conv}||_F^2 \tag{16}$$
$$+ ||\boldsymbol{W}^{pred}||_F^2)$$

Here, λ is a regularization parameter and $|| \cdot ||_F$ denotes the Frobenius norm. We update all the parameters including the weights $\boldsymbol{W}^{emb}$, $\boldsymbol{W}^{conv}$, and $\boldsymbol{W}^{pred}$, biases b and b_α, and the attention parameter a_{drug} to minimize L'_*. We use the adaptive moment estimation (Adam) (Kingma and Ba, 2015) for the optimizer. We randomly shuffle training data set and divide them into mini-batch samples in each epoch.

3 Experimental settings

We illustrate the workflow of the DDI extraction in Figure 2. As preprocessing, we performed word segmentation of the input sentences using the GENIA tagger (Tsuruoka et al., 2005). In this section, we explain the settings for the data sets, tasks, initial embeddings, and hyper-parameter tuning.

3.1 Data set

We used the data set from the DDIExtraction-2013 shared task (SemEval-2013 Task 9) (Segura Bedmar et al., 2013; Herrero-Zazo et al., 2013) for the evaluation. This data set is composed of documents annotated with drug mentions and their relationships. The data set consists of two parts: MEDLINE and DrugBank. MEDLINE consists of abstracts in PubMed articles, and DrugBank consists of the descriptions of drug interactions in the DrugBank database. This data set annotates the following four types of interactions.

- *Mechanism*: A sentence describes pharmacokinetic mechanisms of a DDI, e.g., "***Grepafloxacine*** *may inhibit the metabolism of* ***theobromine***."

- *Effect*: A sentence represents the effect of a DDI, e.g., "***Methionine*** *may protect against the ototoxic effects of* ***gentamicin***."

---

[1] We do not incorporate the attention-based pooling in Wang et al. (2016). We leave this for future work.

	Train		Test	
	DrugBank	MEDLINE	DrugBank	MEDLINE
No. of documents	572	142	158	33
No. of sentences	5,675	1,301	973	326
No. of pairs	26,005	1,787	5,265	451
No. of positive DDIs	3,789	232	884	95
No. of negative DDIs	22,216	1,555	4,381	356
No. of *Mechanism* pairs	1,257	62	278	24
No. of *Effect* pairs	1,535	152	298	62
No. of *Advice* pairs	818	8	214	7
No. of *Int* pairs	179	10	94	2

Table 2: Statistics for the DDIExtraction-2013 shared task data set

Parameter	Value
Word embedding size	200
Word position embeddings size	20
Convolutional window size	[3, 4, 5]
Convolutional filter size	100
Initial learning rate	0.001
Mini-batch size	100
L2 regularization parameter	0.0001

Table 3: Hyperparamters

	Counts
Sentences	1,404
Pairs	4,998
Mechanism pairs	232
Effect pairs	339
Advice pairs	132
Int pairs	48

Table 4: Statistics of the development data set

- *Advice*: A sentence represents a recommendation or advice on the concomitant use of two drugs, e.g., "**Alpha-blockers** *should not be combined with* **uroxatral**."

- *Int*: A sentence simply represents the occurrence of a DDI without any information about the DDI, e.g., "*The interaction of* **omeprazole** *and* **ketoconazole** *has established.*"

The statistics of the data set is shown in Table 2. As shown in this table, the number of pairs that have no interaction (negative pairs) is larger than that of pairs that have interactions (positive pairs).

3.2 Task settings

We followed the task setting of Task 9.2 in the DDIExtraction-2013 shared task (SemEval task 9). The task is to classify a given pair of drugs into the four interaction types or no interaction. We evaluated the performance with precision (P), recall (R), and F-score (F) on each interaction type as well as micro-averaged precision, recall, and F-score on all the interaction types. We used the official evaluation script provided by the task organizers and report the averages of 10 runs. Please note that we took averages of precision, recall and F-scores individually, so F-scores cannot be calculated from precision and recall.

3.3 Initializing embeddings

Skip-gram (Mikolov et al., 2013) was employed for the pre-training of word embeddings. We used 2014 MEDLINE/PubMed baseline distribution, and the size of vocabulary was 1,630,978. The embedding of the drugs, i.e., "*DRUG1*", "*DRUG2*" and "*DRUGOTHER*" are initialized with the pre-trained embedding of the word "*drug*". The embeddings of training words that did not appear in the pre-trained embeddings, as well as the word position embeddings, are initialized with the random values drawn from a uniform distribution and normalized to unit vectors. Words whose frequencies are one in the training data were replaced with an "*UNK*" word during training, and the embedding of words in the test data set that did not appear in both training and pre-trained embeddings were set to the embedding of the "*UNK*" word.

3.4 Hyperparameter tuning

We split the official training data set into two parts: training and development data sets. We tuned the hyper-parameters on the development data set on the softmax model without attentions. Table 3 shows the best hyperparameters on the softmax model without attentions. We applied the same

13

Type	P (%)	R (%)	F (%)
Softmax without attention			
Mechanism	76.24 (±4.48)	57.58 (±4.41)	65.31 (±1.76)
Effect	67.84 (±3.56)	63.61 (±4.95)	65.39 (±1.38)
Advice	82.26 (±7.04)	66.65 (±9.07)	72.75 (±2.72)
Int	**78.99** (±6.87)	33.55 (±2.62)	**47.05** (±1.71)
All (micro)	73.69 (±3.00)	59.92 (±3.73)	65.93 (±1.21)
Softmax with attention			
Mechanism	76.34 (±4.20)	**64.43** (±5.72)	67.86 (±4.10)
Effect	66.84 (±3.12)	65.98 (±2.63)	65.58 (±2.09)
Advice	80.98 (±6.14)	70.83 (±2.72)	76.28 (±1.40)
Int	73.21 (±6.30)	**38.44** (±9.82)	46.11 (±3.96)
All (micro)	73.74 (±1.88)	63.05 (±1.39)	67.94 (±0.70)
Ranking without attention			
Mechanism	78.41 (±3.99)	58.17 (±5.10)	66.51 (±2.61)
Effect	68.16 (±3.30)	65.75 (±3.22)	66.80 (±1.46)
Advice	**84.49** (±3.55)	67.14 (±4.68)	74.61 (±1.82)
Int	73.95 (±7.09)	33.43 (±1.18)	45.91 (±1.23)
All (micro)	74.79 (±2.41)	60.99 (±2.65)	67.10 (±1.09)
Ranking with attention			
Mechanism	**80.75** (±2.76)	61.09 (±3.03)	**69.45** (±1.45)
Effect	**69.73** (±2.64)	**66.63** (±2.93)	**68.05** (±1.29)
Advice	83.86 (±2.29)	**71.81** (±2.61)	**77.30** (±1.13)
Int	74.20 (±8.95)	33.02 (±1.40)	45.50 (±1.51)
All (micro)	**76.30** (±2.18)	**63.25** (±1.71)	**69.12** (±0.71)

Table 5: Performance of softmax/ranking CNN models with and without our attention mechanism. The highest scores are shown in bold.

hyperparameters to the other models. The statistics of our development data set is shown in Table 4. We set the sizes of the convolution windows to [3, 4, 5] that are the same as in Kim (2014). We chose the word position embedding size from {10, 20, 30, 40, 50}, the convolutional filter size from {10, 50, 100, 200}, the learning rate of Adam from {0.01, 0.001, 0.0001}, the mini-batch size from {10, 20, 50, 100, 200}, and the L2 regularization parameter λ from {0.01, 0.001, 0.0001, 0.00001}.

4 Results

In this section, we first summarize the performance of the proposed models and compare the performance with existing models. We then compare attention mechanisms and finally illustrate some results for the analysis of the attentions.

4.1 Performance analysis

The performance of the base CNN models with two objective functions, as well as with or without the proposed attention mechanism, is summa-rized in Table 5. The incorporation of the attention mechanism improved the F-scores by about 2 percent points (pp) on models with both objective functions. Both improvements were statistically significant (p < 0.01) with t-test. This shows that the attention mechanism is effective for both models. The improvement of F-scores from the least performing model (softmax objective function without our attention mechanism) to the best performing model (ranking objective function with our attention mechanism) is 3.19 pp (69.12% versus 65.93%), and this shows both objective function and attention mechanism are key to improve the performance. When looking into the individual types, ranking function with our attention mechanism archived the best F-scores on *Mechanism*, *Effect*, *Advice*, while the base CNN model achieved the best F-score on *Int*.

4.2 Comparison with existing models

We show comparison with the existing state-of-the-art models in Table 6. We mainly compare

Methods	P (%)	R (%)	F (%)
No negative instance filtering			
CNN (Liu et al., 2016)	75.29	60.37	67.01
MCCNN (Quan et al., 2016)	-	-	67.80
SCNN (Zhao et al., 2016)	68.5	61.0	64.5
Joint AB-LSTM (Sahu and Anand, 2017)	71.82	66.90	69.27
Proposed model	76.30	63.25	69.12
With negative instance filtering			
FBK-irst (Chowdhury and Lavelli, 2013)	64.6	65.6	65.1
Kim et al. (2015)	-	-	67.0
CNN (Liu et al., 2016)	75.72	64.66	69.75
MCCNN (Quan et al., 2016)	75.99	65.25	70.21
SCNN (Zhao et al., 2016)	72.5	65.1	68.6
Joint AB-LSTM (Sahu and Anand, 2017)	73.41	69.66	71.48

Table 6: Comparison with existing models

	P (%)	R (%)	F (%)
No attention	74.79 ($\pm$2.41)	60.99 ($\pm$2.65)	67.10 ($\pm$1.09)
Input attention by Wang et al. (2016)	73.48 ($\pm$1.96)	59.58 ($\pm$1.51)	65.77 ($\pm$0.80)
Our attention	76.30 ($\pm$2.66)	63.25 ($\pm$2.59)	69.12 ($\pm$0.71)
Our attention without separate attentions a_{drug}	74.03 ($\pm$2.11)	63.30 ($\pm$2.41)	68.17 ($\pm$0.71)
Our attention without the bias term b_α	71.56 ($\pm$2.18)	64.19 ($\pm$2.21)	67.62 ($\pm$0.96)

Table 7: Comparison of attention mechanisms on CNN models with ranking objective function

the performance without negative instance filtering, which omits some apparent negative instance pairs with rules (Chowdhury and Lavelli, 2013), since we did not incorporate it. We also show the performance of the existing models with negative instance filtering for reference.

In the comparison without negative instance filtering, our model outperformed the existing CNN models (Liu et al., 2016; Quan et al., 2016; Zhao et al., 2016). The model was competitive with Joint AB-LSTM model (Sahu and Anand, 2017) that was composed of multiple RNN models.

When considering negative instance filtering, our model showed lower performance than the state-of-the-art. However we believe we can get similar performance with theirs if we incorporate negative instance filtering. Still, the model outperformed several models such as Kim et al. (2015), Chowdhury and Lavelli (2013) and SCNN model even if we consider negative instance filtering.

4.3 Comparison of attention mechanisms

We compare the proposed attention mechanism with the input attention of Wang et al. (2016) to show the effectiveness of our attention mechanism. Table 7 shows the comparison of the attention mechanisms. We also show the base CNN-based model with ranking loss for reference, and the results of ablation tests. As is shown in the table, the attention mechanism by Wang et al. (2016) did not work in DDI extraction. However, our attention improved the performance. This result shows that the proposed extensions are crucial for modeling attentions in DDI extraction. The ablation test results show that both extensions to our attention mechanism, i.e., separate attentions for entities and incorporation of the bias term, are effective for the task.

4.4 Visual analysis

Figure 3 shows visualization of attentions on some sentences with DDI pairs using our attention mechanism. In the first sentence, *"DRUG1"* and *"DRUG2"* have a *Mechanism* interaction. The attention mechanism successfully highlights the keyword *"concentration"*. In the second sentence, which have an *Effect* interaction, the attention mechanism put high weights on *"increase"* and *"effects"*. The word *"necessary"* has a high weight on the third sentence with an *Advice* interaction. For an *Int* interaction in the last sentence, the word *"interaction"* is most highlighted.

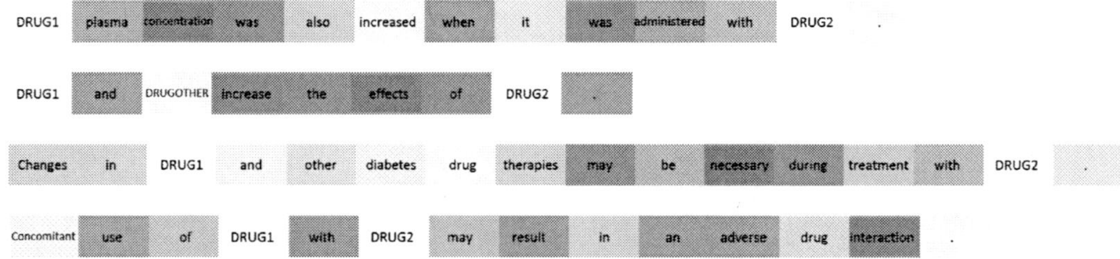

Figure 3: Visualization of attention

5 Related work

Various feature-based methods have been proposed during and after the DDIExtraction-2013 shared task (Segura Bedmar et al., 2013). Björne et al. (2013) tackled with DDI extraction using Turku Event Extraction System (TEES), which is an event extraction system based on the Support Vector Machines (SVMs). Thomas et al. (2013) and Chowdhury and Lavelli (2013) proposed two-phase processing models that first detected DDIs and then classified the extracted DDIs into one of the four proposed types. Thomas et al. (2013) used the ensembles of several kernel methods, while Chowdhury and Lavelli (2013) proposed hybrid kernel-based approach with negative instance filtering. The negative instance filtering is employed by all the subsequent models except for ours. Kim et al. (2015) proposed a two-phase SVM-based approach that employed a linear SVM with rich features including word features, word pairs, dependency relations, parse tree structures, and noun phrase-based constraint features. Our model does not use features and instead employs CNNs.

Deep learning-based models recently dominated the DDI extraction task. Among these, CNN-based models have been often employed and RNNs has received less attention. Liu et al. (2016) built a CNN-based model on word embedding and word position embeddings. Zhao et al. (2016) proposed Syntax CNN (SCNN) that employs syntax word embeddings with the syntactic information of a sentence as well as features of POS tags and dependency trees. Liu et al. (2016) tackled DDI extraction using Multi-Channel CNN (MCCNN) that enables the fusion of multiple word embeddings. Our work is different from theirs in that we employed an attention mechanism.

As for RNN-based approach, Sahu and Anand (2017) proposed an RNN-based model named Joint AB-LSTM (Long Short-Term Memory). Joint AB-LSTM is composed of the concatenation of two RNN-based models: bidirectional LSTM (Bi-LSTM) and attentive pooling Bi-LSTM. The model showed the state-of-the-art performance on the DDIExtraction-2013 shared task data set. Our model is a single model with a CNN and attention mechanism, and it performed comparable to theirs as shown in Table 6.

Wang et al. (2016) proposed muli-level attention CNNs and applied it to a general domain relation classification task SemEval 2010 Task 8 (Hendrickx et al., 2009). Their attention mechanism improved the macro F1 score by 1.9pp (from 86.1% to 88.0%), and their model achieved the state-of-the-art performance on the task.

6 Conclusions

In this paper, we proposed a novel attention mechanism for the extraction of DDIs. We built base CNN-based DDI extraction models with two different objective functions, softmax and ranking, and we incorporated the attention mechanism into the models. We evaluated the performance on the Task 9.2 of the DDIExtraction-2013 shared task, and we showed that both attention mechanism and ranking-based objective function are effective for the extraction of DDIs. Our final model achieved an F-score of 69.12% that is competitive with the state-of-the-art model when we compared the performance without negative instance filtering.

As future work, we would like to incorporate an attention mechanism in the pooling layer (Wang et al., 2016) and adopt negative instance filtering (Chowdhury and Lavelli, 2013) for the further performance improvement and fair comparison with the state-of-the-art methods.

Acknowledgments

This work was supported by JSPS KAKENHI Grant Number 17K12741.

References

Jari Björne, Suwisa Kaewphan, and Tapio Salakoski. 2013. UTurku: drug named entity recognition and drug-drug interaction extraction using SVM classification and domain knowledge. In *Second Joint Conference on Lexical and Computational Semantics (* SEM)*. volume 2, pages 651–659.

Y-Lan Boureau, Jean Ponce, and Yann LeCun. 2010. A theoretical analysis of feature pooling in visual recognition. In *Proceedings of the 27th international conference on machine learning (ICML-10)*. pages 111–118.

Md Faisal Mahbub Chowdhury and Alberto Lavelli. 2013. FBK-irst: A multi-phase kernel based approach for drug-drug interaction detection and classification that exploits linguistic information. *Atlanta, Georgia, USA* 351:53.

Cicero Nogueira dos Santos, Bing Xiang, and Bowen Zhou. 2015. Classifying relations by ranking with convolutional neural networks. In *Proceedings of the 53rd Annual Meeting of the Association for Computational Linguistics. Association for Computational Linguistics*. volume 1, pages 626–634.

Iris Hendrickx, Su Nam Kim, Zornitsa Kozareva, Preslav Nakov, Diarmuid Ó Séaghdha, Sebastian Padó, Marco Pennacchiotti, Lorenza Romano, and Stan Szpakowicz. 2009. SemEval-2010 Task 8: Multi-way classification of semantic relations between pairs of nominals. In *Proceedings of the Workshop on Semantic Evaluations: Recent Achievements and Future Directions*. Association for Computational Linguistics, pages 94–99.

María Herrero-Zazo, Isabel Segura-Bedmar, Paloma Martínez, and Thierry Declerck. 2013. The DDI corpus: An annotated corpus with pharmacological substances and drug–drug interactions. *Journal of biomedical informatics* 46(5):914–920.

Sun Kim, Haibin Liu, Lana Yeganova, and W John Wilbur. 2015. Extracting drug–drug interactions from literature using a rich feature-based linear kernel approach. *Journal of biomedical informatics* 55:23–30.

Yoon Kim. 2014. Convolutional neural networks for sentence classification. In *Proceedings of the 2014 Conference on Empirical Methods in Natural Language Processing, EMNLP 2014, October 25-29, 2014, Doha, Qatar, A meeting of SIGDAT, a Special Interest Group of the ACL*. pages 1746–1751.

Diederik Kingma and Jimmy Ba. 2015. Adam: A method for stochastic optimization. In *The International Conference on Learning Representations (ICLR), San Diego, 2015*.

Vivian Law, Craig Knox, Yannick Djoumbou, Tim Jewison, An Chi Guo, Yifeng Liu, Adam Maciejewski, David Arndt, Michael Wilson, Vanessa Neveu, et al. 2014. DrugBank 4.0: shedding new light on drug metabolism. *Nucleic acids research* 42(D1):D1091–D1097.

Shengyu Liu, Buzhou Tang, Qingcai Chen, and Xiaolong Wang. 2016. Drug-drug interaction extraction via convolutional neural networks. *Computational and mathematical methods in medicine* 2016.

Tomas Mikolov, Ilya Sutskever, Kai Chen, Greg S Corrado, and Jeff Dean. 2013. Distributed representations of words and phrases and their compositionality. In *Advances in neural information processing systems*. pages 3111–3119.

Chanqin Quan, Lei Hua, Xiao Sun, and Wenjun Bai. 2016. Multichannel convolutional neural network for biological relation extraction. *BioMed Research International* 2016.

Sunil Kumar Sahu and Ashish Anand. 2017. Drug-drug interaction extraction from biomedical text using long short term memory network. *arXiv preprint arXiv:1701.08303* .

Isabel Segura Bedmar, Paloma Martínez, and María Herrero Zazo. 2013. SemEval-2013 Task 9: Extraction of drug-drug interactions from biomedical texts (DDIExtraction 2013). In *Proceedings of the 7th International Workshop on Semantic Evaluations*. Association for Computational Linguistics, pages 341–350.

Philippe Thomas, Mariana Neves, Tim Rocktäschel, and Ulf Leser. 2013. WBI-DDI: drug-drug interaction extraction using majority voting. In *Second Joint Conference on Lexical and Computational Semantics (* SEM)*. volume 2, pages 628–635.

Caroline F Thorn, Teri E Klein, and Russ B Altman. 2013. PharmGKB: the pharmacogenomics knowledge base. *Pharmacogenomics: Methods and Protocols* pages 311–320.

Yoshimasa Tsuruoka, Yuka Tateishi, Jin-Dong Kim, Tomoko Ohta, John McNaught, Sophia Ananiadou, and Junichi Tsujii. 2005. Developing a robust part-of-speech tagger for biomedical text. In *Panhellenic Conference on Informatics*. Springer, pages 382–392.

Linlin Wang, Zhu Cao, Gerard de Melo, and Zhiyuan Liu. 2016. Relation classification via multi-level attention CNNs. In *Proceedings of the 54th Annual Meeting of the Association for Computational Linguistics. Association for Computational Linguistics*. pages 1298–1307.

Hong Yang, Chu Qin, Ying Hong Li, Lin Tao, Jin Zhou, Chun Yan Yu, Feng Xu, Zhe Chen, Feng Zhu, and Yu Zong Chen. 2016. Therapeutic target database update 2016: enriched resource for bench to clinical drug target and targeted pathway information. *Nucleic acids research* 44(D1):D1069–D1074.

Daojian Zeng, Kang Liu, Siwei Lai, Guangyou Zhou, Jun Zhao, et al. 2014. Relation classification via convolutional deep neural network. In *COLING*. pages 2335–2344.

Zhehuan Zhao, Zhihao Yang, Ling Luo, Hongfei Lin, and Jian Wang. 2016. Drug drug interaction extraction from biomedical literature using syntax convolutional neural network. *Bioinformatics* 32(22):3444–3453.

Insights into Analogy Completion from the Biomedical Domain

Denis Newman-Griffis[†,‡] and **Albert M. Lai**[♣] and **Eric Fosler-Lussier**[†]

[†]The Ohio State University, Columbus, OH

[‡]National Institutes of Health, Clinical Center, Bethesda, MD

[♣]Washington University in St. Louis, St Louis, MO

`newman-griffis.1@osu.edu, amlai@wustl.edu,`
`fosler@cse.ohio-state.edu`

Abstract

Analogy completion has been a popular task in recent years for evaluating the semantic properties of word embeddings, but the standard methodology makes a number of assumptions about analogies that do not always hold, either in recent benchmark datasets or when expanding into other domains. Through an analysis of analogies in the biomedical domain, we identify three assumptions: that of a *Single Answer* for any given analogy, that the pairs involved describe the *Same Relationship*, and that each pair is *Informative* with respect to the other. We propose modifying the standard methodology to relax these assumptions by allowing for multiple correct answers, reporting MAP and MRR in addition to accuracy, and using multiple example pairs. We further present BMASS, a novel dataset for evaluating linguistic regularities in biomedical embeddings, and demonstrate that the relationships described in the dataset pose significant semantic challenges to current word embedding methods.

1 Introduction

Analogical reasoning has long been a staple of computational semantics research, as it allows for evaluating how well implicit semantic relations between pairs of terms are represented in a semantic model. In particular, the recent boom of research on learning vector space models (VSMs) for text (Turney and Pantel, 2010) has leveraged analogy completion as a standalone method for evaluating VSMs without using a full NLP system. This is due largely to the observations of "linguistic regularities" as linear off-

sets in context-based semantic models (Mikolov et al., 2013c; Levy and Goldberg, 2014; Pennington et al., 2014).

In the analogy completion task, a system is presented with an example term pair and a query, e.g., *London:England::Paris:_____*, and the task is to correctly fill in the blank. Recent methods consider the vector difference between related terms as representative of the relationship between them, and use this to find the closest vocabulary term for a target analogy, e.g., *England - London + Paris ≈ France*. However, recent analyses reveal weaknesses of such offset-based methods, including that the use of cosine similarity often reduces to just reflecting nearest neighbor structure (Linzen, 2016), and that there is significant variance in performance between different kinds of relations (Köper et al., 2015; Gladkova et al., 2016; Drozd et al., 2016).

We identify three key assumptions encoded in the standard offset-based methodology for analogy completion: that a given analogy has only one correct answer, that all relationships between the example pair and the query-target pair are the same, and that the example pair is sufficiently informative with respect to the query-target pair. We demonstrate that these assumptions are violated in real-world data, including in existing analogy datasets. We then propose several modifications to the standard methodology to relax these assumptions, including allowing for multiple correct answers, making use of multiple examples when available, and reporting mean average precision (MAP) and mean reciprocal rank (MRR) to give a more complete picture of the implicit ranking used in finding the best candidate for completing a given analogy.

Furthermore, we present the BioMedical Analogic Similarity Set (BMASS), a novel dataset for

Proceedings of the BioNLP 2017 workshop, pages 19–28,
Vancouver, Canada, August 4, 2017. ©2017 Association for Computational Linguistics

analogical reasoning in the biomedical domain. This new resource presents real-world examples of semantic relations of interest for biomedical natural language processing research, and we hope it will support further research into biomedical VSMs (Chiu et al., 2016; Choi et al., 2016).[1]

2 Related work

Analogical reasoning has been studied both on its own and as a component of downstream tasks, using a range of systems. Early work used rule-based systems for world knowledge (Reitman, 1965) and syntactic (Federici and Pirelli, 1997) relationships. Supervised models were used for SAT (Scholastic Aptitude Test) analogies (Veale, 2004), and later for synonymy, antonymy, and some world knowledge (Turney, 2008; Herdağdelen and Baroni, 2009). Analogical reasoning has also been used in support of downstream tasks, including word sense disambiguation (Federici et al., 1997) and morphological analysis (Lepage and Goh, 2009; Lavallée and Langlais, 2010; Soricut and Och, 2015).

Recent work on analogies has largely focused on their use as an intrinsic evaluation of the properties of a VSM. The analogy dataset of Mikolov et al. (2013a), often referred to as the Google dataset, has become a standard evaluation for general-domain word embedding models (Pennington et al., 2014; Levy and Goldberg, 2014; Schnabel et al., 2015; Faruqui et al., 2015), and includes both world knowledge and morphosyntactic relations. Other datasets include the MSR analogies (Mikolov et al., 2013c), which describe morphological relations only; and BATS (Gladkova et al., 2016), which includes both morphological and semantic relations. The semantic relations from SemEval-2012 Task 2 (Jurgens et al., 2012) have also been used to derive analogies; however, as with the lexical Sem-Para dataset of Köper et al. (2015), the semantic relationships tend to be significantly more challenging for embedding-based methods (Drozd et al., 2016). Additionally, Levy et al. (2015b) demonstrate that even for some lexical relations where embeddings appear to perform well, they are actually learning prototypicality as opposed to relatedness.

3 Analogy completion task

3.1 Standard methodology

Given an analogy $a{:}b{::}c{:}d$, the evaluation task is to guess d out of the vocabulary, given a, b, c as evidence. Recent methods for this involve using the vector difference between embedded representations of the related pairs to rank all terms in the vocabulary by how well they complete the analogy, and choosing the best fit. The vector difference is most commonly used in one of three ways, where cos is cosine similarity:

$$argmax_{d \in V} \left(cos(d, b - a + c) \right) \quad (1)$$

$$argmax_{d \in V} \left(cos(d - c, b - a) \right) \quad (2)$$

$$argmax_{d \in V} \frac{cos(d, b)cos(d, c)}{cos(d, a) + \epsilon} \quad (3)$$

Following the terminology of Levy and Goldberg (2014), we refer to Equation 1 as 3COSADD, Equation 2 as PAIRWISEDISTANCE, and Equation 3 (which is equivalent to 3COSADD with log cosine similarities) as 3COSMUL.

In order to generate analogy data for this task, recent datasets have followed a similar process (Mikolov et al., 2013a,c; Köper et al., 2015; Gladkova et al., 2016). First, relations of interest were manually selected for the target domains: syntactic/morphological, lexical (e.g., hypernymy, synonymy), or semantic (e.g., *CapitalOf*). Then, for each relation, example word pairs were manually selected or automatically generated from existing resources (e.g., WordNet). The final analogies were then generated by exhaustively combining the sets of word pairs within each relation.

3.2 Assumptions

Several key assumptions are inherent in this standard methodology that are not reflected in recent benchmark analogy datasets. The first we refer to as the *Single-Target* assumption: namely, that there is a single correct answer for any given analogy. Since the target d is chosen via argmax, if we consider the following two analogies:

flu:nausea::fever:?cough
flu:nausea::fever:?light-headedness

we must necessarily get at least one answer wrong. Gladkova et al. (2016) convert these analogies into a single case:

flu:nausea::fever:?[cough, light-headedness]

Pair	Relations
brother:sister	**FemaleCounterpart** *SiblingOf*
husband:wife	**FemaleCounterpart** *MarriedTo*

Table 1: Binary semantic relations in "brother is to sister as husband is to wife." The target common relation is shown in bold.

where either *cough* or *lightheadedness* is a correct guess. However, this still misses our desire to get both correct answers, if possible. Relations with multiple correct targets are present in all of Google, BATS, and Sem-Para.

The second key assumption is that all the information relating a to b also relates c to d. While the pairs are chosen based on a single common relationship, each pair may actually pertain to multiple relationships. An example from the Google dataset is *brother:sister::husband:wife*; Table 1 shows the semantic relations involved in this analogy. While the target relation *FemaleCounterpart* is present in both pairs, by comparing the offsets $sister - brother$ and $wife - husband$, we assume that either all ways in which each pair is related are present in both, or that *FemaleCounterpart* dominates the offset. We refer to this as the *Same-Relationship* assumption.

Finally, it is not sufficient for two pairs to share a common relationship label; that relationship must be both representative and informative for analogies to make sense (the *Informativity* assumption). Relation labels may be sufficiently broad as to be meaningless, as we encountered when drawing unfiltered binary relations from the Unified Medical Language System (UMLS) Metathesaurus. One sample analogy from the *RO:Null* relation (indicating "related in some way") was *socks:stockings::Finns:Finnish language*. While both pairs are of related terms, they are in no way related to one another.

Furthermore, even when two pairs are examples of the same kind of clearly-defined relation, they may still be relatively uninformative. For example, in the Sem-Para *Meronym* analogy *apricot:stone::trumpet:mouthpiece* the meronymic relationship between *apricot* and *stone* could plausibly identify a number of parts of a trumpet: *mouthpiece, valves, slide*, etc.[2] The extremely

high-level nature of several of the Sem-Para relations (hypernymy, antonymy, and synonymy) suggests that some of the difficulty observed by Köper et al. (2015) is due to violations of Informativity.

4 BMASS

We present BMASS (the BioMedical Analogic Similarity Set), a dataset of biomedical analogies, generated using the expert-curated knowledge in the Unified Medical Language System (UMLS)[3] (Bodenreider, 2004) in order to identify medical term pairs sharing the same relationships. We followed the standard process for dataset generation outlined in Section 3.1, with some adjustments for the assumptions in Section 3.2.

The UMLS Metathesaurus is centered around normalized *concepts*, represented by Concept Unique Identifiers (CUIs). Each concept can be represented in textual form by one or more *terms* (e.g., C0009443 → "Common cold", "acute rhinitis"). These terms may be multi-word expressions (MWEs); in fact, many concepts in the UMLS have no unigram terms.

The Metathesaurus also contains ⟨*subject, relation, object*⟩ triples describing binary relationships between concepts. These relationships are specified at two levels: relationship types (RELs), such as *broader-than* and *qualified-by*, and specific relationships (RELAs) within each type, e.g., *tradename-of* and *has-finding-site*. For this work, we used the 721 unique REL/RELA pairings as our source relationships, and treated the ⟨*subject, object*⟩ pairs linked within each of these relationships as candidates for generating analogies.

To enable a word embedding–based evaluation, we first identified terms that appeared at least 25 times in the 2016 PubMed baseline collection of biomedical abstracts,[4] and removed all ⟨*subject, object*⟩ pairs involving concepts that did not correspond to these frequent terms. Most relationships in the Metathesaurus are many-to-many (i.e., each subject can be paired with multiple objects and

---

[2]While this is similar to the Single-Target assumption, it bears separate consideration in that Single-Target refers to multiple valid objects of a specific relationship, while this is an issue of multiple valid relationships being described.

[3]We use the 2016AA release of the UMLS.

[4]We chose 25 as our minimum frequency to ensure that each term appeared often enough to learn reasonable embeddings for its component words. To determine term frequency, we first lowercased and stripped punctuation from both the PubMed corpus and the term list extracted from UMLS, then searched the corpus for exact term matches.

vice versa), and thus may challenge Single-Target and Informativity assumptions; we therefore next identified relations that had at least 50 1:1 instances, i.e., a subject and object that are only paired with one another within a specific relationship. Since 1:1 instances are not sufficient to guarantee Informativity, we then manually reviewed the remaining relations to identify those those that we deemed to satisfy Informativity constraints. For example, the *is-a* relationship between *tongue muscles* and *head muscle* is not specific enough to suggest that *carbon monoxide* should elicit *gasotransmitters* as its corresponding answer. However, for *associated-with*, sampled pairs such as *leg injuries : leg* and *histamine release : histamine* were sufficiently consistent that we deemed it Informative. This gave us a final set of 25 binary relations, listed in Table 2.[5]

We follow Gladkova et al. (2016) in generating a balanced dataset, to enable a more robust comparative analysis between relations. We randomly sampled 50 ⟨subject, object⟩ pairs from each relation, again restricting to concepts with strings appearing frequently in PubMed. For each subject concept that we sampled, we collected all valid object concepts and bundled them as a single ⟨subject, objects⟩ pair. We then exhaustively combined each concept pair with the others in its relation to create 2,450 analogies, giving us a total dataset size of 61,250 analogies. Finally, for each concept, we chose a single frequent term to represent it, giving us both CUI and string representations of each analogy.

5 Evaluation

We assess how well biomedical word embeddings can perform on our dataset, and explore modifications to the standard evaluation methodology to relax the assumptions described in Section 3.2. We use the skip-gram embeddings trained by Chiu et al. (2016) on the PubMed citation database, one set using a window size of 2 (PM-2) and another set with window size 30 (PM-30). All other word2vec hyperparameters were tuned by Chiu et al. on a combination of similarity and relatedness and named entity recognition tasks.

Additionally, we use the hyperparameters they identified (minimum frequency=5, vector dimension=200, negative samples=10, sample=1e-4,

---

[5]Examples of each relation, along with their mappings to UMLS REL/RELA values, are available online.

ID	Name	Amb
Lab/Rx		
L1	form-of	1.0
L2	has-lab-number	1.1
L3	has-tradename	1.5
L4	tradename-of	1.3
L5	associated-substance	1.6
L6	has-free-acid-or-base-form	1.0
L7	has-salt-form	1.1
L8	measured-component-of	1.3
Hierarchical		
H1	refers-to	1.0
H2	same-type	10.4
Morphological		
M1	adjectival-form-of	1.1
M2	noun-form-of	1.0
Clinical		
C1	associated-with-malfunction-of-gene-product	2.6
C2	gene-product-malfunction-associated-with-disease	1.5
C3	causative-agent-of	4.6
C4	has-causative-agent	2.0
C5	has-finding-site	1.9
C6	associated-with	1.2
Anatomy		
A1	anatomic-structure-is-part-of	1.6
A2	anatomic-structure-has-part	5.4
A3	is-located-in	1.4
Biology		
B1	regulated-by	1.0
B2	regulates	1.0
B3	gene-encodes-product	1.1
B4	gene-product-encoded-by	2.4

Table 2: List of the relations kept after manual filtering; Amb is the average ambiguity, i.e., the average number of correct answers per analogy.

α=0.05, window size=2) to train our own embeddings on a subset of the 2016 PubMed Baseline (14.7 million documents, 2.7 billion tokens). We train word2vec (Mikolov et al., 2013a) samples with the continuous bag-of-words (CBOW) and skip-gram (SGNS) models, trained for 10 iterations, and GloVe (Pennington et al., 2014) samples, trained for 50 iterations.

We performed our evaluation with each of 3CosAdd, PairwiseDistance, and 3CosMul as the scoring function over the vocabulary. In contrast to the prior findings of Levy and Goldberg (2014) on the Google dataset, performance on BMASS is roughly equivalent among the three methods, often differing by only one or two correct answers. We therefore only report results with 3COSADD, since it is the most familiar method.

5.1 Modifications to the standard method

We consider 3COSADD under three settings of the analogies in our dataset. For a given analogy $a{:}b{:}{:}c{:}?d$, we refer to $\langle a, b \rangle$ as the exemplar pair and $\langle c, d \rangle$ as the query pair; $?d$ signifies the target answer.

Single-Answer puts analogies in $a{:}b{:}{:}c{:}d$ format, with a single example object b and a single correct object d, by taking the first object listed for each term pair. This enforces the Single-Answer assumption.

Multi-Answer takes the first object listed for the exemplar term pair, but keeps all valid answers, i.e. $a{:}b{:}{:}c{:}[d_1,d_2,\dots]$; this is similar to the approach of Gladkova et al. (2016). There are approximately 16k analogies in our dataset with multiple valid answers.

All-Info keeps all valid objects for both the exemplar and query pairs. The exemplar offset is then calculated over $B = [b_1, b_2, \dots]$ as

$$a - B = \frac{1}{|B|} \sum_i a - b_i$$

Though this is superficially similar to 3COSAVG (Drozd et al., 2016), we average over objects for a specific subject, as opposed to averaging over all subject-object pairs.

We report a relaxed accuracy (denoted Acc_R), in which the guess is correct if it is in the set of correct answers. (In the Single-Answer case, this reduces to standard accuracy.) Acc_R, as with standard accuracy, necessitates ignoring a, b, or c if they are the top results (Linzen, 2016).

In order to capture information about all correct answers, we also report Mean Average Precision (MAP) and Mean Reciprocal Rank (MRR) over the set of correct answers in the vocabulary, as ranked by Equation 1. Since MAP and MRR do not have a cutoff in terms of searching for the correct answer in the ranked vocabulary, they can be used without the adjustment of ignoring a, b, and c; thus, they can give a more accurate picture of how close the correct terms are to the calculated guesses.

5.2 MWEs and candidate answers

As noted in Section 4, the terms in our analogy dataset may be multi-word expressions (MWEs). We follow the common baseline approach of representing an MWE as the average of its component words (Mikolov et al., 2013b; Chen et al., 2013; Wieting et al., 2016). For phrasal terms containing one or more words that are out of our embedding vocabulary, we only consider the in-vocabulary words: thus, if "parathyroid" is not in the vocabulary, then the embedding of *parathyroid hypertensive factor* will be

$$\frac{hypertensive + factor}{2}$$

For any individual analogy $a{:}b{:}{:}c{:}?d$, the vocabulary of candidate phrases to complete the analogy is derived by calculating averaged word embeddings for each UMLS term appearing in PubMed abstracts at least 25 times. Terms for which none of the component words are in vocabulary are discarded. This yields a candidate set of 229,898 phrases for the PM-2 and PM-30, and 263,316 for our CBOW, SGNS, and GloVe samples.

Since prior work on analogies has primarily been concerned with unigram data, we also identified a subset of our data for which we could find single-word string realizations for all concepts in an analogy, using the full vocabulary of our trained embeddings. Even in the All-Info setting, we could only identify 606 such analogies; Table 3 shows MAP results for PM-2 and CBOW embeddings on the three relations with at least 100 unigram analogies. The unigram analogies are slightly better captured than the full MWE data for *has-lab-number* (L2) and *has-tradename* (L3); however, lower performance on the unigram subset in *tradename-of* (L4) shows that unigram analogies are not always easier. We see a small effect from the much larger set of candidate answers in the unigram case (>1m unigrams), as shown by the slightly higher MAP numbers in the Uni_M case. In general, it is clear that the difficulty of some of the relations in our dataset is not due solely to using MWEs in the analogies.

5.3 Metric comparison

Figure 1 shows Acc_R, MAP, and MRR results for each relation in BMASS, using PM-2 embeddings in the Multi-Answer setting. Overall, performance varies widely between relations, with all

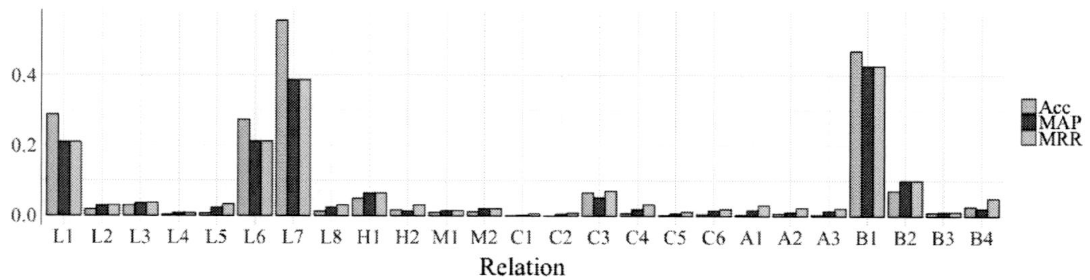

Figure 1: Acc_R, MAP, and MRR for each relation, using PM-2 embeddings under the Multi-Answer setting. Note that MAP is calculated using the position of all correct answers in the ranked list, while MRR reflects only the position of the first correct answer found in the ranked list for each individual query.

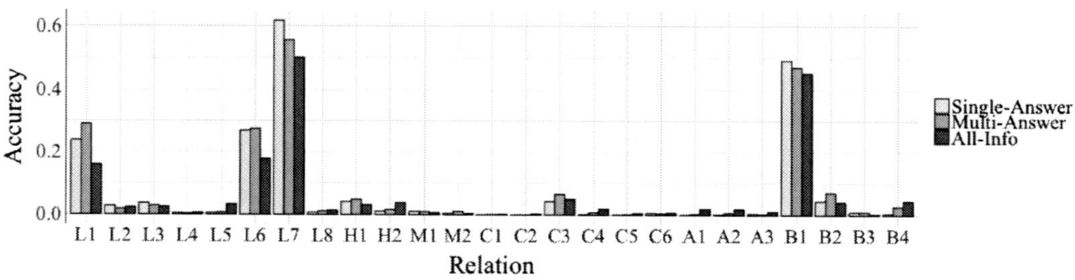

Figure 2: Acc_R per relation for PM-2 on BMASS, under Single-Answer, Multi-Answer, and All-Info settings.

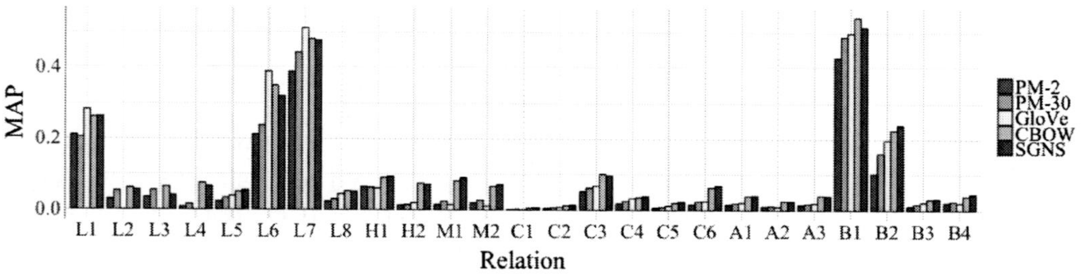

Figure 3: Per-relation MAP for all embeddings under the Multi-Answer setting.

three metrics staying under 0.1 in the majority of cases; this mirrors previous findings on other analogy datasets (Levy and Goldberg, 2014; Gladkova et al., 2016; Drozd et al., 2016).

MAP further fleshes out these differences by reporting performance over all correct answers for a given analogy. This lets us distinguish between relations like *has-salt-form* (L7), where noticeably lower MAP numbers reflect a wider distribution of the multiple correct answers, and relations like *regulates* (B2) or *associated-with* (C6), where a low Acc_R reflects many incorrect answers, but a higher MAP indicates that the correct answers are relatively near the guess.

MRR, on the other hand, more optimistically reports how close we got to finding any correct answer. Thus, for the *has-causative-agent* (C4) relation, low Acc_R is belied by a noticeably higher MRR, suggesting that even when we guess wrong, the correct answer is close. This contrasts with relations like *refers-to* (H1) or *causative-agent-of* (C3), where MRR is more consistent with Acc_R, indicating that wrong guesses tend to be farther from the truth. Since most of our analogies (45,178 samples, or about 74%) have only a single correct answer, MAP and MRR tend to be highly similar. However, in high-ambiguity relations like *same-type* (H2), higher MRR numbers give a better sense of our best case performance.

5.4 Analogy settings

To compare across the Single-Answer, Multi-Answer, and All-Info settings, we first look at Acc_R for each relation in BMASS, shown for PM-2 embeddings in Figure 2 (the observed patterns are similar with the other embeddings). Unsurprisingly, allowing for multiple answers in Multi-Answer and All-Info slightly raises Acc_R in most cases. What is surprising, however, is that including more sample exemplar objects in the All-Info setting had widely varying results. In some cases, such as *same-type* (H2), *associated-substance* (L5), and *gene-product-encoded-by* (B4), the additional exemplars gave a noticeable improvement in accuracy. In others, accuracy actually went down: *form-of* (L1) and *has-free-acid-or-base-form* (L6) are the most striking examples, with absolute decreases of 4% and 8% respectively from the Multi-Answer case for PM-2 (the decreases are similar with other embeddings). Thus, it seems that multiple examples may help with Informativity in some cases, but confuse it in others. Taken together with the improvements seen in Drozd et al. (2016) from using 3CosAvg, this is another indication that any single subject-object pair may not be sufficiently representative of the target relationship.

5.5 Embedding methods

Averaging over all relations, the five embedding settings we tested behaved roughly the same, with our trained embeddings slightly outperforming the pretrained embeddings of Chiu et al. (2016); summary Acc_R, MAP, and MRR performances are given in Table 4. At the level of individual relations, Figure 3 shows MAP performance in the Multi-Answer setting. The four word2vec sam-

Rel	PM-2			CBOW		
	Uni	Uni$_M$	MWE	Uni	Uni$_M$	MWE
L2	0.05	0.04	0.03	0.11	0.13	0.06
L3	0.10	0.12	0.04	0.12	0.17	0.06
L4	0.00	0.00	0.01	0.04	0.05	0.07

Table 3: MAP performance on the three BMASS relations with ≥ 100 unigram analogies. Uni is using unigram embeddings on unigram data, Uni$_M$ is using MWE embeddings on unigram data, and MWE is performance with MWE embeddings over the full MWE data.

ples tend to behave similarly, with some inconsistent variations. Interestingly, CBOW outperforms the other embeddings by a large margin in several relations, including *regulated-by* (B1) and *has-tradename* (L3).

GloVe varies much more widely across the relations, as reflected in the higher standard deviations in Table 4. While GloVe consistently outperforms word2vec embeddings on *has-free-acid-or-base-form* (L6) and *has-salt-form* (L7), it significantly underperforms on the morphological and hierarchical relations, among others. Most notably, while the word2vec embeddings show minor differences in performance between the Multi-Answer and All-Info settings, GloVe Acc_R performance falls drastically on *form-of* (L1) and *has-free-acid-or-base-form* (L6), as shown in Table 5. However, its MAP and MRR numbers stay similar, suggesting that there is only a reshuffling of results closest to the guess.

5.6 Error analysis

Several interesting patterns emerge in reviewing individual *a:b::c:?d* predictions. A number of errors follow directly from our word averaging approach to MWEs: words that appear in b or c often appear in the predictions, as in *gosorelin:ici 118630::letrozole:*ici 164384*. Prefix substitutions also occurred, as with *mianserin hydrochloride:mianserin::scopolamine hydrobromide:*scopolamine methylbromide*.

Often, the b term(s) would outweigh c, leading to many of the top guesses being variants on b. In one analogy, *sodium acetylsalicyclate:aspirin::intravenous immunoglobulins:?immunoglobulin g*, the top guesses were: **aspirin prophylaxis*, **aspirin*, **aspirin antiplatelet*, and **low-dose aspirin*.

In other cases, related to the nearest neighborhood over-reporting observed by Linzen (2016), we saw guesses very similar to c, regardless of a or b, as with *acute inflammations:acutely inflamed::endoderm:*embryonic endoderm*; other near guesses included **endoderm cell* and *epiblast*.

Finally, we found several analogies where the incorrect guesses made were highly related to the correct answer, despite not matching. One such analogy was *oropharyngeal suctioning:substances::thallium scan:?radioisotopes*; the top guess was **radioactive substances*, and

Setting	Single-Answer			Multi-Answer			All-Info		
	Acc_R	MAP	MRR	Acc_R	MAP	MRR	Acc_R	MAP	MRR
PM-2	.08 (.16)	.07 (.13)	.07 (.13)	.08 (.15)	.07 (.12)	.08 (.11)	.07 (.13)	.07 (.11)	.07 (.11)
PM-30	.08 (.16)	.08 (.13)	.08 (.13)	.09 (.16)	.08 (.13)	.09 (.13)	.08 (.13)	.08 (.12)	.09 (.12)
GloVe	.11 (.22)	.09 (.16)	.09 (.16)	.11 (.22)	.09 (.16)	.10 (.15)	.10 (.18)	.09 (.16)	.10 (.15)
CBOW	.11 (.18)	.11 (.14)	.11 (.14)	.12 (.18)	.12 (.14)	.12 (.14)	.11 (.17)	.12 (.14)	.13 (.14)
SGNS	.11 (.18)	.11 (.14)	.11 (.14)	.11 (.18)	.11 (.14)	.12 (.13)	.11 (.17)	.12 (.14)	.12 (.14)

Table 4: Average performance over all relations in the dataset, for each set of embeddings. Results are reported as "Mean (Standard deviation)" for each metric.

Metric	L1			L6		
	SA	MA	AI	SA	MA	AI
Acc_R	0.49	0.49	0.24	0.62	0.62	0.39
MAP	0.28	0.28	0.28	0.39	0.39	0.39
MRR	0.28	0.28	0.28	0.39	0.39	0.39

Table 5: Acc_R, MAP, and MRR performance variation between Single-Answer (SA), Multi-Answer (MA), and All-Info (AI) settings for GloVe embeddings on *form-of* (L1) and *has-free-acid-or-base-form* (L6)

gallium compounds was two guesses farther down. Showing some mixed effect from the neighborhood of b, *performance-enhancing substances* was the next-ranked candidate.

6 Discussion

Relaxing the Single-Answer, Same-Relationship, and Informativity assumptions by including multiple correct answers and multiple exemplar pairs and by reporting MAP and MRR in addition to accuracy paints a more complete picture of how well word embeddings are performing on analogy completion, but leaves a number of questions unanswered. While we can more clearly see the relations where we correctly complete analogies (or come close), and contrast with relations where a vector arithmetic approach completely misses the mark, what distinguishes these cases remains unclear. Some more straightforward relationships, such as *gene-encodes-product* (B3) and its inverse *gene-product-encoded-by* (B4), show surprisingly poor results, while the very broad synonymy of *refers-to* (H1) is captured comparatively well. Additionally, in contrast to prior work with morphological relations, *adjectival-form-of* (M1) and *noun-form-of* (M2) are much more challenging in the biomedical domain, as we see non-morphological related pairs such as *predisposed:disease susceptibility* and *venous lumen:endovenous*, in addition to more normal pairs like *sweating:sweaty* and *muscular:muscle*. Fur-

ther analysis may provide some insight into specific challenges posed by the relations in our dataset, as well as why performance with PAIRWISEDISTANCE and 3COSMUL did not noticeably differ from 3COSADD.

In terms of specific model errors, we did not evaluate the effects of any embedding hyperparameters on performance in BMASS, opting to use hyperparameter settings tuned for general-purpose use in the biomedical domain. Levy et al. (2015a) and Chiu et al. (2016), among others, show significant impact of embedding hyperparameters on downstream performance. Exploring different settings may be one way to get a better sense of exactly what incorrect answers are being highly-ranked, and why those are emerging from the affine organization of the embedding space. Additionally, the higher variance in per-relation performance we observed with GloVe embeddings suggests that there is more to unpack as to what the GloVe model is capturing or failing to capture compared to word2vec approaches.

Finally, while we considered Informativity during the generation of BMASS, and relaxed the Single-Answer assumption in our evaluation, we have not really addressed the Same-Relationship assumption. Using multiple exemplar pairs is one attempt to reduce the impact of confusing extraneous relationships, but in practice this helps some relations and harms others. Drozd et al. (2016) tackle this problem with the LRCos method; however, their findings of mis-applied features and errors due to very slight mis-rankings show that there is still room for improvement. One question is whether this problem can be addressed at all with non-parametric models like the vector offset approaches, to retain the advantages of evaluating directly from the word embedding space, or if a learned model (like LRCos) is necessary to separate out the different aspects of a related term pair.

7 Conclusions

We identified three key assumptions in the standard methodology for analogy-based evaluations of word embeddings: Single-Answer (that there is a single correct answer for an analogy), Same-Relationship (that the exemplar and query pairs are related in the same way), and Informativity (that the exemplar pair is informative with respect to the query pair). We showed that these assumptions do not hold in recent benchmark datasets or in biomedical data. Therefore, to relax these assumptions, we modified analogy evaluation to allow for multiple correct answers and multiple exemplar pairs, and reported Mean Average Precision and Mean Reciprocal Recall over the ranked vocabulary, in addition to accuracy of the highest-ranked choice.

We also presented the BioMedical Analogic Similarity Set (BMASS), a novel analogy completion dataset for the biomedical domain. In contrast to existing datasets, BMASS was automatically generated from a large-scale database of ⟨subject, relation, object⟩ triples in the UMLS Metathesaurus, and represents a number of challenging real-world relationships. Similar to prior results, we find wide variation in word embedding performance on this dataset, with accuracies above 50% on some relationships such as *has-salt-form* and *regulated-by*, and numbers below 5% on others, e.g., *anatomic-structure-is-part-of* and *measured-component-of*.

Finally, we are able to address the Single-Answer assumption by modifying the analogy evaluation to accommodate multiple correct answers, and we consider Informativity in generating our dataset and using multiple example pairs. However, the Same-Relationship assumption remains a challenge, as does a more automated approach to either evaluating or relaxing Informativity. These offer promising directions for future work in analogy-based evaluations.

Acknowledgments

We would like to thank the CLLT group at Ohio State and the anonymous reviewers for their helpful comments. Denis is a pre-doctoral fellow at the National Institutes of Health Clinical Center, Bethesda, MD.

References

Olivier Bodenreider. 2004. The Unified Medical Language System (UMLS): integrating biomedical terminology. *Nucleic Acids Research* 32(90001):D267–D270.

Danqi Chen, Richard Socher, Christopher D. Manning, and Andrew Y. Ng. 2013. Learning New Facts From Knowledge Bases With Neural Tensor Networks and Semantic Word Vectors. *arXiv preprint arXiv:1301.3618* pages 1–4.

Billy Chiu, Gamal Crichton, Anna Korhonen, and Sampo Pyysalo. 2016. How to Train Good Word Embeddings for Biomedical NLP. *Proceedings of the 15th Workshop on Biomedical Natural Language Processing* pages 166–174.

Edward Choi, Mohammad Taha Bahadori, Elizabeth Searles, Catherine Coffey, Michael Thompson, James Bost, Javier Tejedor-Sojo, and Jimeng Sun. 2016. Multi-layer Representation Learning for Medical Concepts. In *Proceedings of the 22nd ACM SIGKDD International Conference on Knowledge Discovery and Data Mining*. ACM, San Francisco, California, USA, KDD '16, pages 1495–1504.

Aleksandr Drozd, Anna Gladkova, and Satoshi Matsuoka. 2016. Word Embeddings, Analogies, and Machine Learning: Beyond king - man + woman = queen. In *Proceedings of COLING 2016, the 26th International Conference on Computational Linguistics: Technical Papers*. The COLING 2016 Organizing Committee, Osaka, Japan, pages 3519–3530.

Manaal Faruqui, Jesse Dodge, Sujay Kumar Jauhar, Chris Dyer, Eduard Hovy, and Noah A Smith. 2015. Retrofitting Word Vectors to Semantic Lexicons. In *Proceedings of the 2015 Conference of the North American Chapter of the Association for Computational Linguistics: Human Language Technologies*. Association for Computational Linguistics, Denver, Colorado, pages 1606–1615.

Stefano Federici, Simonetta Montemagni, and Vito Pirelli. 1997. Inferring Semantic Similarity from Distributional Evidence: An Analogy-Based Approach to Word Sense Disambiguation. *Automatic Information Extraction and Building of Lexical Semantic Resources for NLP Applications* pages 90–97.

Stefano Federici and Vito Pirelli. 1997. Analogy, computation, and linguistic theory. In *New Methods in Language Processing*, UCL Press, pages 16–34.

Anna Gladkova, Aleksandr Drozd, and Satoshi Matsuoka. 2016. Analogy-based Detection of Morphological and Semantic Relations With Word Embeddings: What Works and What Doesn't. *Proceedings of the NAACL Student Research Workshop* pages 8–15.

Amaç Herdağdelen and Marco Baroni. 2009. Bag-Pack: A General Framework to Represent Semantic Relations. In *Proceedings of the Workshop on Geometrical Models of Natural Language Semantics*. Association for Computational Linguistics, Athens, Greece, pages 33–40.

David Jurgens, Saif Mohammad, Peter Turney, and Keith Holyoak. 2012. SemEval-2012 Task 2: Measuring Degrees of Relational Similarity. In **SEM 2012, Volume 2: Proceedings of the Sixth International Workshop on Semantic Evaluation (SemEval 2012)*. Association for Computational Linguistics, Montréal, Canada, pages 356–364.

Maximilian Köper, Christian Scheible, and Sabine im Walde. 2015. Multilingual Reliability and "Semantic" Structure of Continuous Word Spaces. In *Proceedings of the 11th International Conference on Computational Semantics*. Association for Computational Linguistics, London, UK, pages 40–45.

Jean-François Lavallée and Philippe Langlais. 2010. *Unsupervised Morphological Analysis by Formal Analogy*, Springer Berlin Heidelberg, Berlin, Heidelberg, pages 617–624.

Yves Lepage and Chooi Ling Goh. 2009. Towards automatic acquisition of linguistic features. In *Proceedings of the 17th Nordic Conference of Computational Linguistics (NODALIDA 2009)*. Northern European Association for Language Technology (NEALT), Odense, Denmark, pages 118–125.

Omer Levy and Yoav Goldberg. 2014. Linguistic Regularities in Sparse and Explicit Word Representations. In *Proceedings of the Eighteenth Conference on Computational Natural Language Learning*. Association for Computational Linguistics, Ann Arbor, Michigan, pages 171–180.

Omer Levy, Yoav Goldberg, and Ido Dagan. 2015a. Improving Distributional Similarity with Lessons Learned from Word Embeddings. *Transactions of the Association for Computational Linguistics* 3:211–225.

Omer Levy, Steffen Remus, Chris Biemann, and Ido Dagan. 2015b. Do Supervised Distributional Methods Really Learn Lexical Inference Relations? In *Proceedings of the 2015 Conference of the North American Chapter of the Association for Computational Linguistics: Human Language Technologies*. Association for Computational Linguistics, Denver, Colorado, pages 970–976.

Tal Linzen. 2016. Issues in evaluating semantic spaces using word analogies. In *Proceedings of the 1st Workshop on Evaluating Vector-Space Representations for NLP*. pages 13–18.

Tomas Mikolov, Kai Chen, Greg Corrado, and Jeffrey Dean. 2013a. Efficient Estimation of Word Representations in Vector Space. *arXiv preprint arXiv:1301.3781* pages 1–12.

Tomas Mikolov, Ilya Sutskever, Kai Chen, Greg S Corrado, and Jeff Dean. 2013b. Distributed Representations of Words and Phrases and Their Compositionality. In *Advances in Neural Information Processing Systems 26*. Curran Associates, Inc., NIPS '13, pages 3111–3119.

Tomas Mikolov, Wen-tau Yih, and Geoffrey Zweig. 2013c. Linguistic Regularities in Continuous Space Word Representations. In *Proceedings of the 2013 Conference of the North American Chapter of the Association for Computational Linguistics: Human Language Technologies*. Association for Computational Linguistics, Atlanta, Georgia, pages 746–751.

Jeffrey Pennington, Richard Socher, and Christopher D. Manning. 2014. Glove: Global Vectors for Word Representation. In *Proceedings of the 2014 Conference on Empirical Methods in Natural Language Processing*. Association for Computational Linguistics, Doha, Qatar, pages 1532–1543.

Walter R Reitman. 1965. *Cognition and Thought: An Information Processing Approach*. John Wiley and Sons, New York, NY.

Tobias Schnabel, Igor Labutov, David Mimno, and Thorsten Joachims. 2015. Evaluation methods for unsupervised word embeddings. In *Proceedings of the 2015 Conference on Empirical Methods in Natural Language Processing*. Association for Computational Linguistics, Lisbon, Portugal, pages 298–307.

Radu Soricut and Franz Och. 2015. Unsupervised Morphology Induction Using Word Embeddings. In *Proceedings of the 2015 Conference of the North American Chapter of the Association for Computational Linguistics: Human Language Technologies*. Association for Computational Linguistics, Denver, Colorado, pages 1627–1637.

Peter D Turney. 2008. A Uniform Approach to Analogies, Synonyms, Antonyms, and Associations. In *Proceedings of the 22nd International Conference on Computational Linguistics - Volume 1*. Association for Computational Linguistics, Stroudsburg, PA, USA, COLING '08, pages 905–912.

Peter D Turney and Patrick Pantel. 2010. From Frequency to Meaning: Vector Space Models of Semantics. *Journal of Artificial Intelligence Research* 37:141–188.

Tony Veale. 2004. WordNet Sits the S.A.T. A Knowledge-based Approach to Lexical Analogy. In *Proceedings of the 16th European Conference on Artificial Intelligence*. IOS Press, Amsterdam, The Netherlands, The Netherlands, ECAI'04, pages 606–610.

John Wieting, Mohit Bansal, Kevin Gimpel, and Karen Livescu. 2016. Towards Universal Paraphrastic Sentence Embeddings. In *Proceedings of the 4th International Conference on Learning Representations*.

Deep learning for extracting protein-protein interactions from biomedical literature

Yifan Peng　　　**Zhiyong Lu**
National Center for Biotechnology Information
National Library of Medicine
National Institutes of Health
Bethesda, MD 20894
{yifan.peng, zhiyong.lu}@nih.gov

Abstract

State-of-the-art methods for protein-protein interaction (PPI) extraction are primarily feature-based or kernel-based by leveraging lexical and syntactic information. But how to incorporate such knowledge in the recent deep learning methods remains an open question. In this paper, we propose a multichannel dependency-based convolutional neural network model (McDepCNN). It applies one channel to the embedding vector of each word in the sentence, and another channel to the embedding vector of the head of the corresponding word. Therefore, the model can use richer information obtained from different channels. Experiments on two public benchmarking datasets, AIMed and BioInfer, demonstrate that McDepCNN compares favorably to the state-of-the-art rich-feature and single-kernel based methods. In addition, McDepCNN achieves 24.4% relative improvement in F1-score over the state-of-the-art methods on cross-corpus evaluation and 12% improvement in F1-score over kernel-based methods on "difficult" instances. These results suggest that McDepCNN generalizes more easily over different corpora, and is capable of capturing long distance features in the sentences.

1 Introduction

With the growing amount of biomedical information available in the textual form, there has been considerable interest in applying natural language processing (NLP) techniques and machine learning (ML) methods to the biomedical litera-ture (Huang and Lu, 2015; Leaman and Lu, 2016; Singhal et al., 2016; Peng et al., 2016). One of the most important tasks is to extract protein-protein interaction relations (Krallinger et al., 2008).

Protein-protein interaction (PPI) extraction is a task to identify interaction relations between protein entities mentioned within a document. While PPI relations can span over sentences and even cross documents, current works mostly focus on PPI in individual sentences (Pyysalo et al., 2008; Tikk et al., 2010). For example, "ARFTS" and "XIAP-BIR3" are in a PPI relation in the sentence "ARFTS$_{\text{PROT1}}$ specifically binds to a distinct domain in XIAP-BIR3$_{\text{PROT2}}$".

Recently, deep learning methods have achieved notable results in various NLP tasks (Manning, 2015). For PPI extraction, convolutional neural networks (CNN) have been adopted and applied effectively (Zeng et al., 2014; Quan et al., 2016; Hua and Quan, 2016). Compared with traditional supervised ML methods, the CNN model is more generalizable and does not require tedious feature engineering efforts. However, how to incorporate linguistic and semantic information into the CNN model remains an open question. Thus previous CNN-based methods have not achieved state-of-the-art performance in the PPI task (Zhao et al., 2016a).

In this paper, we propose a multichannel dependency-based convolutional neural network, McDepCNN, to provide a new way to model the syntactic sentence structure in CNN models. Compared with the widely-used one-hot CNN model (e.g., the shortest-path information is firstly transformed into a binary vector which is zero in all positions except at this shortest-path's index, and then applied to CNN), McDepCNN utilizes a separate channel to capture the dependencies of the sentence syntactic structure.

To assess McDepCNN, we evaluated our

Proceedings of the BioNLP 2017 workshop, pages 29–38,
Vancouver, Canada, August 4, 2017. ©2017 Association for Computational Linguistics

model on two benchmarking PPI corpora, AIMed (Bunescu et al., 2005) and BioInfer (Pyysalo et al., 2007). Our results show that McDepCNN performs better than the state-of-the-art feature- and kernel-based methods.

We further examined McDepCNN in two experimental settings: a cross-corpus evaluation and an evaluation on a subset of "difficult" PPI instances previously reported (Tikk et al., 2013). Our results suggest that McDepCNN is more generalizable and capable of capturing long distance information than kernel methods.

The rest of the manuscript is organized as follows. We first present related work. Then, we describe our model in Section 3, followed by an extensive evaluation and discussion in Section 4. We conclude in the last section.

2   Related work

From the ML perspective, we formulate the PPI task as a binary classification problem where discriminative classifiers are trained with a set of positive and negative relation instances. In the last decade, ML-based methods for the PPI tasks have been dominated by two main types: the feature-based vs. kernel based method. The common characteristic of these methods is to transform relation instances into a set of features or rich structural representations like trees or graphs, by leveraging linguistic analysis and knowledge resources. Then a discriminative classifier is used, such as support vector machines (Vapnik, 1995) or conditional random fields (Lafferty et al., 2001).

While these methods allow the relation extraction systems to inherit the knowledge discovered by the NLP community for the pre-processing tasks, they are highly dependent on feature engineering (Fundel et al., 2007; Van Landeghem et al., 2008; Miwa et al., 2009b; Bui et al., 2011). The difficulty with feature-based methods is that data cannot always be easily represented by explicit feature vectors.

Since natural language processing applications involve structured representations of the input data, deriving good features is difficult, time-consuming, and requires expert knowledge. Kernel-based methods attempt to solve this problem by implicitly calculating dot products for every pair of examples (Erkan et al., 2007; Airola et al., 2008; Miwa et al., 2009a; Kim et al., 2010; Chowdhury et al., 2011). Instead of extracting fea-

ture vectors from examples, they apply a similarity function between examples and use a discriminative method to label new examples (Tikk et al., 2010). However, this method also requires manual effort to design a similarity function which can not only encode linguistic and semantic information in the complex structures but also successfully discriminate between examples. Kernel-based methods are also criticized for having higher computational complexity (Collins and Duffy, 2002).

Convolutional neural networks (CNN) have recently achieved promising results in the PPI task (Zeng et al., 2014; Hua and Quan, 2016). CNNs are a type of feed-forward artificial neural network whose layers are formed by a convolution operation followed by a pooling operation (LeCun et al., 1998). Unlike feature- and kernel-based methods which have been well studied for decades, few studies investigated how to incorporate syntactic and semantic information into the CNN model. To this end, we propose a neural network model that makes use of automatically learned features (by different CNN layers) together with manually crafted ones (via domain knowledge), such as words, part-of-speech tags, chunks, named entities, and dependency graph of sentences. Such a combination in feature engineering has been shown to be effective in other NLP tasks also (e.g. (Shimaoka et al., 2017)).

Furthermore, we propose a multichannel CNN, a model that was suggested to capture different "views" of input data. In the image processing, (Krizhevsky et al., 2012) applied different RGB (red, green, blue) channels to color images. In NLP research, such models often use separate channels for different word embeddings (Yin and Schütze, 2015; Shi et al., 2016). For example, one could have separate channels for different word embeddings (Quan et al., 2016), or have one channel that is kept static throughout training and the other that is fine-tuned via backpropagation (Kim, 2014). Unlike these studies, we utilize the head of the word in a sentence as a separate channel.

3   CNN Model

3.1   Model Architecture Overview

Figure 1 illustrates the overview of our model, which takes a complete sentence with mentioned entities as input and outputs a probability vector (two elements) corresponding to whether there is a relation between the two entities. Our model mainly consists of three layers: a multichannel embedding layer, a convolution layer, and a fully-connected layer.

3.2 Embedding Layer

In our model, as shown in Figure 1, each word in a sentence is represented by concatenating its word embedding, part-of-speech, chunk, named entity, dependency, and position features.

3.2.1 Word embedding

Word embedding is a language modeling techniques where words from the vocabulary are mapped to vectors of real numbers. It has been shown to boost the performance in NLP tasks. In this paper, we used pre-trained word embedding vectors (Pyysalo et al., 2013) learned on PubMed articles using the word2vec tool (Mikolov et al., 2013). The dimensionality of word vectors is 200.

3.2.2 Part-of-speech

We used the part-of-speech (POS) feature to extend the word embedding. Similar to (Zhao et al., 2016b), we divided POS into eight groups. Then each group is mapped to an eight-bit binary vector. In this way, the dimensionality of the POS feature is 8.

3.2.3 Chunk

We used the chunk tags obtained from Genia Tagger for each word (Tsuruoka and Tsujii, 2005). We encoded the chunk features using a one-hot scheme. The dimensionality of chunk tags is 18.

3.2.4 Named entity

To generalize the model, we used four types of named entity encodings for each word. The named entities were provided as input by the task data. In one PPI instance, the types of two proteins of interest are PROT1 and PROT2 respectively. The type of other proteins is PROT, and the type of other words is O. If a protein mention spans multiple words, we marked each word with the same type (we did not use a scheme such as IOB). The dimensionality of named entity is thus 4.

3.2.5 Dependency

To add the dependency information of each word, we used the label of "incoming" edge of that word in the dependency graph. Take the sentence from Figure 2 as an example, the dependency of "ARFTS" is "nsubj" and the dependency of "binds" is "ROOT". We encoded the dependency features using a one-hot scheme, and their dimensionality is 101.

3.2.6 Position feature

In this work, we consider the relationship of two protein mentions in a sentence. Thus, we used the position feature proposed in (Sahu et al., 2016), which consists of two relative distances, $d1$ and $d2$, for representing the distances of the current word to PROT1 and PROT2 respectively. For example in Figure 2, the relative distances of the word "binds" to PROT1 ("ARFTs") and PROT2 ("XIAP-BIR3") are 2 and -6, respectively. Same as in Table S4 of (Zhao et al., 2016b), both $d1$ and $d2$ are non-linearly mapped to a ten-bit binary vector, where the first bit stands for the sign and the remaining bits for the distance.

3.3 Multichannel Embedding Layer

A novel aspect of McDepCNN is to add the "head" word representation of each word as the second channel of the embedding layer. For example, the second channel of the sentence in Figure 2 is "binds binds ROOT binds domain domain binds domain" as shown in Figure 1. There are several advantages of using the "head" of a word as a separate channel.

First, it intuitively incorporates the dependency graph structure into the CNN model. Compared with (Hua and Quan, 2016) which used the shortest path between two entities as the sole input for CNN, our model does not discard information outside the scope of two entities. Such information was reported to be useful (Zhou et al., 2007). Compared with (Zhao et al., 2016b) which used the shortest path as a bag-of-word sparse 0-1 vector, our model intuitively reflects the syntactic structure of the dependencies of the input sentence.

Second, together with convolution, our model can better capture longer distance dependencies than the sliding window size. As shown in Figure 2, the second channel of McDepCNN breaks the dependency graph structure into structural <head word, child word> pairs where each word is a modifier of its previous word. In this way, it reflects the skeleton of a constituent where the second channel shadows the detailed information of all sub-constituents in the first channel. From the perspective of the sentence string, the second channel is similar to a gapped n-gram or a skipped

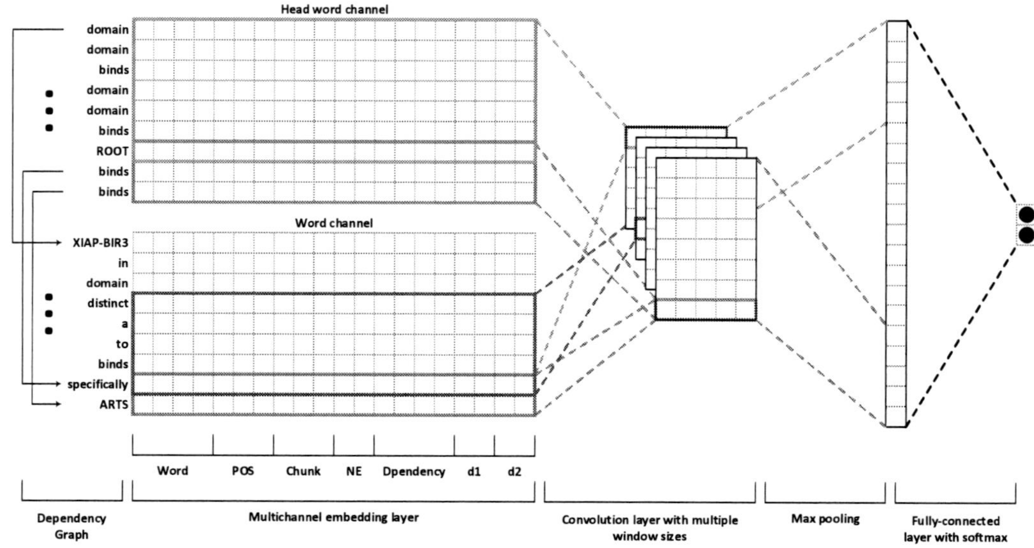

Figure 1: Overview of the CNN model.

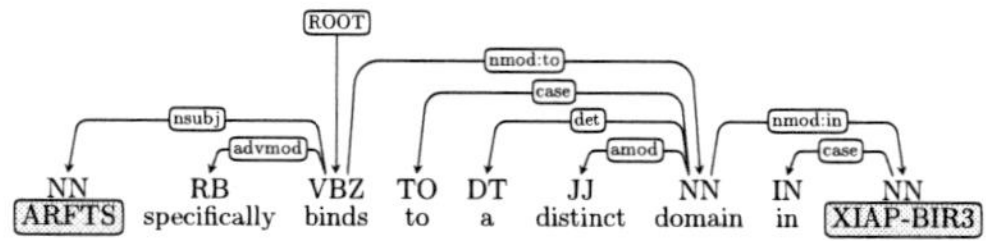

Figure 2: Dependency graph.

n-gram where the skipped words are based on the structure of the sentence.

3.4 Convolution

We applied convolution to input sentences to combine two channels and get local features (Collobert et al., 2011). Consider $x_1, \ldots, x_n$ to be the sequence of word representations in a sentence where

$$x_i = E_{word} \oplus \cdots \oplus E_{poistion}, i = 1, \ldots, n \quad (1)$$

Here $\oplus$ is concatenation operation so $x^i \in \mathbb{R}^d$ is the embedding vector for the ith word with the dimensionality d. Let $x_{i:i+k-1}^c$ represent a window of size k in the sentence for channel c. Then the output sequence of the convolution layer is

$$con_i = f(\sum_c w_k^c x_{i:i+k-1}^c + b_k) \quad (2)$$

where f is a rectify linear unit (ReLU) function and b_k is the biased term. Both w_k^c and b_k are the learning parameters.

1-max pooling was then performed over each map, i.e., the largest number from each feature map was recorded. In this way, we obtained fixed length global features for the whole sentence. The underlying intuition is to consider only the most useful feature from the entire sentence.

$$m_k = \max_{1 \leq i \leq n-k+1} (con_i) \quad (3)$$

3.5 Fully Connected Layer with Softmax

To make a classifier over extracted global features, we first applied a fully connected layer to the feature vectors of multichannel obtained above.

$$O = w_o(m_3 \oplus m_5 \oplus m_7) + b_o \quad (4)$$

The final softmax then receives this vector O as input and uses it to classify the PPI; here we assume binary classification for the PPI task and hence depict two possible output states.

$$p(ppi|x, \theta) = \frac{e^{O_{ppi}}}{e^{O_{ppi}} + e^{O_{other}}} \quad (5)$$

Here, θ is a vector of the hyper-parameters of the model, such as w_k^c, b_k, w_o, and b_o. Further, we used dropout technique in the output of the max pooling layer for regularization (Srivastava et al., 2014). This prevented our method from overfitting by randomly "dropping" with probability $(1 - p)$ neurons during each forward/backward pass while training.

3.6 Training

To train the parameters, we used the log-likelihood of parameters on a mini-batch training with a batch size of m. We use the Adam algorithm to optimize the loss function (Kingma and Ba, 2015).

$$J(\theta) = \sum_m p(ppi^{(m)}|x^{(m)}, \theta) \qquad (6)$$

3.7 Experimental setup

For our experiments, we used the Genia Tagger to obtain the part-of-speech, chunk tags, and named entities of each word (Tsuruoka and Tsujii, 2005). We parsed each sentence using the Bllip parser with the biomedical model (Charniak, 2000; Mc-Closky, 2009). The universal dependencies were then obtained by applying the Stanford dependencies converter on the parse tree with the *CCProcessed* and *Universal* options (De Marneffe et al., 2014).

We implemented the model using Tensor-Flow (Abadi et al., 2016). All trainable variables were initialized using the Xavier algorithm (Glorot and Bengio, 2010). We set the maximum sentence length to 160. That is, longer sentences were pruned, and shorter sentences were padded with zeros. We set the learning rate to be 0.0007 and the dropping probability 0.5. During the training, we ran 250 epochs of all the training examples. For each epoch, we randomized the training examples and conducted a mini-batch training with a batch size of 128 ($m = 128$).

In this paper, we experimented with three window sizes: 3, 5 and 7, each of which has 400 filters. Every filter performs convolution on the sentence matrix and generates variable-length feature maps. We got the best results using the single window of size 3 (see Section 4.2)

4 Results and Discussion

4.1 Data

We evaluated McDepCNN on two benchmarking PPI corpora, AIMed (Bunescu et al., 2005) and BioInfer (Pyysalo et al., 2007). These two corpora have different sizes (Table 1) and vary slightly in their definition of PPI (Pyysalo et al., 2008).

Tikk et al. (2010) conducted a comparison of a variety of PPI extraction systems on these two corpora[1]. In order to compare, we followed their

Table 1: Statistics of the corpora.

Corpus	Sentences	# Positives	# Negatives
AIMed	1,955	1,000	4,834
BioInfer	1,100	2,534	7,132

experimental setup to evaluate our methods: self-interactions were excluded from the corpora and 10-fold cross-validation (CV) was performed.

4.2 Results and discussion

Our system performance, as measured by Precision, Recall, and F1-score, is shown in Table 2. To compare, we also include the results published in (Tikk et al., 2010; Peng et al., 2015; Van Landeghem et al., 2008; Fundel et al., 2007). Row 2 reports the results of the previous best deep learning system on these two corpora. Rows 3 and 4 report the results of two previous best single kernel-based methods, an APG kernel (Airola et al., 2008; Tikk et al., 2010) and an edit kernel (Peng et al., 2015). Rows 5-6 report the results of two rule-based systems. As can be seen, McDepCNN achieved the highest results in both precision and overall F1-score on both datasets.

Note that we did not compare our results with two recent deep-learning approaches (Hua and Quan, 2016; Quan et al., 2016). This is because unlike other previous studies, they artificially removed sentences that cannot be parsed and discarded pairs which are in a coordinate structure. Thus, our results are not directly comparable with theirs. Neither did we compare our method with (Miwa et al., 2009b) because they combined, in a rich vector, analysis from different parsers and the output of multiple kernels.

To further test the generalizability of our method, we conducted the cross-corpus experiments where we trained the model on one corpus and tested it on the other (Table 3). Here we compared our results with the shallow linguistic model which is reported as the best kernel-based method in (Tikk et al., 2013).

The cross-corpus results show that McDepCNN achieved 24.4% improvement in F-score when trained on BioInfer and tested on AIMed, and 18.2% vice versa.

To better understand the advantages of McDepCNN over kernel-based methods, we followed the lead of (Tikk et al., 2013) to compare the method performance on some known "difficult"

Table 2: Evaluation results. Performance is reported in terms of Precision, Recall, and F1-score.

	Method	AIMed			BioInfer		
		P	R	F	P	R	F
1	McDepCNN	67.3	60.1	**63.5**	62.7	68.2	**65.3**
2	Deep neutral network (Zhao et al., 2016a)	51.5	63.4	56.1	53.9	72.9	61.6
3	All-path graph kernel (Tikk et al., 2010)	49.2	64.6	55.3	53.3	70.1	60.0
4	Edit kernel (Peng et al., 2015)	65.3	57.3	61.1	59.9	57.6	58.7
5	Rich-feature (Van Landeghem et al., 2008)	49.0	44.0	46.0	–	–	–
6	RelEx (Fundel et al., 2007)	40.0	50.0	44.0	39.0	45.0	41.0

Table 3: Cross-corpus results. Performance is reported in terms of Precision, Recall, and F1-score.

Method	Training corpus	AIMed			BioInfer		
		P	R	F	P	R	F
McDepCNN	AIMed	–	–	–	39.5	61.4	**48.0**
	BioInfer	40.1	65.9	**49.9**	–	–	–
Shallow linguistic (Tikk et al., 2010)	AIMed	–	–	–	29.2	66.8	40.6
	BioInfer	76.8	27.2	41.5	–	–	–

Table 4: Instances that are the most difficult to classify correctly by the collection of kernels using cross-validation (Tikk et al., 2013).

Corpus	Positive difficult	Negative difficult
AIMed	61	184
BioInfer	111	295

instances in AIMed and BioInfer. This subset of difficult instances is defined as 10% of all pairs with the least number of 14 kernels being able to classify correctly (Table 4).

Table 5 shows the comparisons between McDepCNN and kernel-based methods on difficult instances. The results of McDepCNN were obtained from the difficult instances combined from AIMed and BioInfer (172 positives and 479 negatives). And the results of APG, Edit, and SL were obtained from AIMed, BioInfer, HPRD50, IEPA, and LLL (190 positives and 521 negatives) (Tikk et al., 2013). While the input datasets are different, our outcomes are remarkably higher than the prior studies. The results show that McDepCNN achieves 17.3% in F1-score on difficult instances which is more than three times better than other kernels. Since there are no examples of difficult instances that could not be classified correctly by at least one of the 14 kernel methods, below, we only list some examples that McDepCNN can classify correctly.

1. Immunoprecipitation experiments further re-

veal that the fully assembled receptor complex is composed of two **IL-6**$_{\text{PROT1}}$, two **IL-6R alpha**$_{\text{PROT2}}$, and two gp130 molecules.

2. The phagocyte NADPH oxidase is a complex of membrane cytochrome b558 (comprised of subunits p22-phox and gp91-phox) and three cytosol proteins (**p47-phox**$_{\text{PROT1}}$, p67-phox, and p21rac) that translocate to membrane and bind to **cytochrome b558**$_{\text{PROT2}}$.

Together with the conclusions in (Tikk et al., 2013), "positive pairs are more difficult to classify in longer sentences" and "most of the analyzed classifiers fail to capture the characteristics of rare positive pairs in longer sentences", this comparison suggests that McDepCNN is probably capable of better capturing long distance features from the sentence and are more generalizable than kernel methods.

Finally, Table 6 compares the effects of different parts in McDepCNN. Here we tested McDepCNN using 10-fold of AIMed. Row 1 used a single window with the length of 3, row 2 used two windows, and row 3 used three windows. The reduced performance indicate that adding more windows did not improve the model. This is partially because the multichannel in McDepCNN has captured good context features for PPI. Second, we used the single channel and retrained the model with window size 3. The performance then dropped 1.1%. The results underscore the effectiveness of using the head word as a separate chan-

Table 5: Comparisons on the difficult instances with CV evaluation. Performance is reported in terms of Precision, Recall, and F1-score*.

Method	P	R	F
McDepCNN	14.0	22.7	**17.3**
All-path graph kernel	4.3	7.9	5.5
Edit kernel	4.8	5.8	5.3
Shallow linguistic	3.6	7.9	4.9

* The results of McDepCNN were obtained on the difficult instances combined from AIMed and BioInfer (172 positives and 479 negatives). The results of others (Tikk et al., 2013) were obtained from AIMed, BioInfer, HPRD50, IEPA, and LLL (190 positives and 521 negatives).

Table 6: Contributions of different parts in McDepCNN. Performance is reported in terms of Precision, Recall, and F1-score.

Method	P	R	F	Δ
window = 3	67.3	60.1	63.5	
window = [3,5]	60.9	62.4	61.6	(1.9)
window = [3,5,7]	61.7	61.9	61.8	(1.7)
Single channel	62.8	62.3	62.6	(1.1)

nel in CNN.

5 Conclusion

In this paper, we describe a multichannel dependency-based convolutional neural network for the sentence-based PPI task. Experiments on two benchmarking corpora demonstrate that the proposed model outperformed the current deep learning model and single feature-based or kernel-based models. Further analysis suggests that our model is substantially more generalizable across different datasets. Utilizing the dependency structure of sentences as a separated channel also enables the model to capture global information more effectively.

In the future, we would like to investigate how to assemble different resources into our model, similar to what has been done to rich-feature-based methods (Miwa et al., 2009b) where the current best performance was reported (F-score of 64.0% (AIMed) and 66.7% (BioInfer)). We are also interested in extending the method to PPIs beyond the sentence boundary. Finally, we would like to test and generalize this approach to other biomedical relations such as chemical-disease relations (Wei et al., 2016).

Acknowledgments

This work was supported by the Intramural Research Programs of the National Institutes of Health, National Library of Medicine. We are also grateful to Robert Leaman for the helpful discussion.

References

Martín Abadi, Ashish Agarwal, Paul Barham, Eugene Brevdo, Zhifeng Chen, Craig Citro, Greg S Corrado, Andy Davis, Jeffrey Dean, Matthieu Devin, et al. 2016. Tensorflow: Large-scale machine learning on heterogeneous distributed systems. *arXiv preprint arXiv:1603.04467* https://arxiv.org/abs/1603.04467.

Antti Airola, Sampo Pyysalo, Jari Björne, Tapio Pahikkala, Filip Ginter, and Tapio Salakoski. 2008. All-paths graph kernel for protein-protein interaction extraction with evaluation of cross-corpus learning. *BMC Bioinformatics* 9(Suppl 11):1–12. https://doi.org/10.1186/1471-2105-9-s11-s2.

Quoc-Chinh Bui, Sophia Katrenko, and Peter M. A. Sloot. 2011. A hybrid approach to extract protein-protein interactions. *Bioinformatics* 27(2):259–265. https://doi.org/10.1186/1471-2105-9-s11-s2.

Razvan C. Bunescu, Ruifang Ge, Rohit J. Kate, Edward M. Marcotte, Raymond J. Mooney, Arun K. Ramani, and Yuk Wah Wong. 2005. Comparative experiments on learning information extractors for proteins and their interactions. *Artificial Intelligence in Medicine* 33(2):139–155. https://doi.org/10.1016/j.artmed.2004.07.016.

Eugene Charniak. 2000. A maximum-entropy-inspired parser. In *Annual Conference of the North American Chapter of the Association for Computational Linguistics (NAACL)*. pages 132–139. http://dl.acm.org/citation.cfm?id=974323.

Faisal Md. Chowdhury, Alberto Lavelli, and Alessandro Moschitti. 2011. A study on dependency tree kernels for automatic extraction of protein-protein interaction. In *Proceedings of BioNLP 2011 Workshop*. pages 124–133. http://aclweb.org/anthology/W11-0216.

Michael Collins and Nigel Duffy. 2002. New ranking algorithms for parsing and tagging: Kernels over discrete structures, and the voted perceptron. In *Proceedings of the 40th annual meeting on association for computational linguistics (ACL)*. Association for Computational Linguistics, pages 263–270. http://aclweb.org/anthology/P02-1034.

Ronan Collobert, Jason Weston, Léon Bottou, Michael Karlen, Koray Kavukcuoglu, and Pavel Kuksa. 2011. Natural language processing (almost) from scratch. *Journal of Machine Learning Research* 12(Aug):2493–2537.

Marie-Catherine De Marneffe, Timothy Dozat, Natalia Silveira, Katri Haverinen, Filip Ginter, Joakim Nivre, and Christopher D Manning. 2014. Universal Stanford dependencies: A cross-linguistic typology. In *Proceedings of 9th International Conference on Language Resources and Evaluation (LREC)*. volume 14, pages 4585–4592.

Günes Erkan, Arzucan Özgür, and Dragomir R Radev. 2007. Semi-supervised classification for extracting protein interaction sentences using dependency parsing. In *Proceedings of the 2007 Joint Conference on Empirical Methods in Natural Language Processing and Computational Natural Language Learning (EMNLP-CoNLL)*. volume 7, pages 228–237. http://aclweb.org/anthology/D07-1024.

Katrin Fundel, Robert Küffner, and Ralf Zimmer. 2007. RelEx-relation extraction using dependency parse trees. *Bioinformatics* 23(3):365–371. https://doi.org/10.1093/bioinformatics/btl616.

Xavier Glorot and Yoshua Bengio. 2010. Understanding the difficulty of training deep feedforward neural networks. In *Proceedings of the 13th International Conference on Artificial Intelligence and Statistics (AISTATS)*. volume 9, pages 249–256.

Lei Hua and Chanqin Quan. 2016. A shortest dependency path based convolutional neural network for protein-protein relation extraction. *BioMed Research International* 2016. https://doi.org/10.1155/2016/8479587.

Chung-Chi Huang and Zhiyong Lu. 2015. Community challenges in biomedical text mining over 10 years: success, failure and the future. *Briefings in Bioinformatics* 17(1):132–144. https://doi.org/10.1093/bib/bbv024.

Seonho Kim, Juntae Yoon, Jihoon Yang, and Seog Park. 2010. Walk-weighted subsequence kernels for protein-protein interaction extraction. *BMC Bioinformatics* 11(107):1–21. https://doi.org/10.1186/1471-2105-11-107.

Yoon Kim. 2014. Convolutional neural networks for sentence classification. In *the 2014 Conference on Empirical Methods in Natural Language Processing (EMNLP)*. pages 1746–1751. http://aclweb.org/anthology/D14-1181.

Diederik Kingma and Jimmy Ba. 2015. Adam: A method for stochastic optimization. In *International Conference on Learning Representations (ICLR)*. pages 1–15. https://arxiv.org/abs/1412.6980.

Martin Krallinger, Florian Leitner, Carlos Rodriguez-Penagos, Alfonso Valencia, et al. 2008. Overview of the protein-protein interaction annotation extraction task of BioCreative II. *Genome Biology* 9(Suppl 2):S4. https://doi.org/10.1186/gb-2008-9-s2-s4.

Alex Krizhevsky, Ilya Sutskever, and Geoffrey E Hinton. 2012. ImageNet classification with deep convolutional neural networks. In *Advances in neural information processing systems*. pages 1097–1105. http://dl.acm.org/citation.cfm?id=2999257.

John Lafferty, Andrew McCallum, and Fernando Pereira. 2001. Conditional random fields: probabilistic models for segmenting and labeling sequence data. In *Proceedings of the International Conference on Machine Learning (ICML 01)*. pages 282–289. http://dl.acm.org/citation.cfm?id=655813.

Robert Leaman and Zhiyong Lu. 2016. TaggerOne: joint named entity recognition and normalization with semi-markov models. *Bioinformatics* 32(18):2839–2846. https://doi.org/10.1093/bioinformatics/btw343.

Yann LeCun, Léon Bottou, Yoshua Bengio, and Patrick Haffner. 1998. Gradient-based learning applied to document recognition. *Proceedings of the IEEE* 86(11):2278–2324. https://doi.org/10.1109/5.726791.

Christopher D Manning. 2015. Computational linguistics and deep learning. *Computational Linguistics* 41(4):701–707. https://doi.org/10.1162/coli_a_00239.

David McClosky. 2009. *Any domain parsing: automatic domain adaptation for natural language parsing*. Thesis, Department of Computer Science, Brown University.

Tomas Mikolov, Ilya Sutskever, Kai Chen, Greg S Corrado, and Jeff Dean. 2013. Distributed representations of words and phrases and their compositionality. In *Advances in neural information processing systems*. pages 3111–3119. http://dl.acm.org/citation.cfm?id=2999959.

Makoto Miwa, Rune Sætre, Yusuke Miyao, and Juníchi Tsujii. 2009a. Protein-protein interaction extraction by leveraging multiple kernels and parsers. *International Journal of Medical Informatics* 78(12):e39–46. https://doi.org/10.1016/j.ijmedinf.2009.04.010.

Makoto Miwa, Rune Sætre, Yusuke Miyao, and Jun'ichi Tsujii. 2009b. A rich feature vector for protein-protein interaction extraction from multiple corpora. In *Proceedings of the 2009 Conference on Empirical Methods in Natural Language Processing*. volume 1, pages 121–130. http://aclweb.org/anthology/D09-1013.

Yifan Peng, Samir Gupta, Cathy H Wu, and K Vijay-Shanker. 2015. An extended dependency graph for relation extraction in biomedical texts. In *Proceedings of BioNLP 2015 Workshop*. pages 21–30. http://aclweb.org/anthology/W15-3803.

Yifan Peng, Chih-Hsuan Wei, and Zhiyong Lu. 2016. Improving chemical disease relation extraction with rich features and weakly labeled data. *Journal of Cheminformatics* 8(53):1–12. https://doi.org/10.1186/s13321-016-0165-z.

Sampo Pyysalo, Antti Airola, Juho Heimonen, Jari Björne, Filip Ginter, and Tapio Salakoski. 2008. Comparative analysis of five protein-protein interaction corpora. *BMC Bioinformatics* 9(Suppl 3):1–11. https://doi.org/10.1186/1471-2105-9-S3-S6.

Sampo Pyysalo, Filip Ginter, Juho Heimonen, Jari Björne, Jorma Boberg, Jouni Järvinen, and Tapio Salakoski. 2007. BioInfer: a corpus for information extraction in the biomedical domain. *BMC Bioinformatics* 8(50):1–24. https://doi.org/10.1186/1471-2105-8-50.

Sampo Pyysalo, Filip Ginter, Hans Moen, Tapio Salakoski, and Sophia Ananiadou. 2013. Distributional semantics resources for biomedical text processing. In *International Symposium on Languages in Biology and Medicine (LBM)*. pages 39–44.

Chanqin Quan, Lei Hua, Xiao Sun, and Wenjun Bai. 2016. Multichannel convolutional neural network for biological relation extraction. *BioMed Research International* 2016:1–10. https://doi.org/10.1155/2016/1850404.

Sunil Kumar Sahu, Ashish Anand, Krishnadev Oruganty, and Mahanandeeshwar Gattu. 2016. Relation extraction from clinical texts using domain invariant convolutional neural network. In *Proceedings of the 15th Workshop on Biomedical Natural Language Processing*. pages 206–215. https://aclweb.org/anthology/W16-2928.

Hongjie Shi, Takashi Ushio, Mitsuru Endo, Katsuyoshi Yamagami, and Noriaki Horii. 2016. A multichannel convolutional neural network for cross-language dialog state tracking. In *2016 IEEE Workshop on Spoken Language Technology Workshop (SLT)*. IEEE, pages 559–564. https://arxiv.org/abs/1701.06247.

Sonse Shimaoka, Pontus Stenetorp, Kentaro Inui, and Sebastian Riedel. 2017. Neural architectures for fine-grained entity type classification. In *Proceedings of the 15th Conference of the European Chapter of the Association for Computational Linguistics (EACL)*. pages 1271–1280. http://aclweb.org/anthology/E17-1119.

Ayush Singhal, Michael Simmons, and Zhiyong Lu. 2016. Text mining genotype-phenotype relationships from biomedical literature for database curation and precision medicine. *PLOS Computational Biology* 12(11):e1005017. https://doi.org/10.1371/journal.pcbi.1005017.

Nitish Srivastava, Geoffrey E Hinton, Alex Krizhevsky, Ilya Sutskever, and Ruslan Salakhutdinov. 2014. Dropout: a simple way to prevent neural networks from overfitting. *Journal of Machine Learning Research* 15(1):1929–1958.

Domonkos Tikk, Illés Solt, Philippe Thomas, and Ulf Leser. 2013. A detailed error analysis of 13 kernel methods for protein-protein interaction extraction. *BMC Bioinformatics* 14(1):12. https://doi.org/10.1186/1471-2105-14-12.

Domonkos Tikk, Philippe Thomas, Peter Palaga, Jörg Hakenberg, and Ulf Leser. 2010. A comprehensive benchmark of kernel methods to extract protein-protein interactions from literature. *PLoS Computational Biology* 6(7):1–19. https://doi.org/10.1371/journal.pcbi.1000837.

Yoshimasa Tsuruoka and Jun'ichi Tsujii. 2005. Bidirectional inference with the easiest-first strategy for tagging sequence data. In *Proceedings of the Conference on Human Language Technology and Empirical Methods in Natural Language Processing (HLT-EMNLP)*. HLT '05, pages 467–474. https://doi.org/10.3115/1220575.1220634.

Sofie Van Landeghem, Yvan Saeys, Bernard De Baets, and Yves Van de Peer. 2008. Extracting protein-protein interactions from text using rich feature vectors and feature selection. In *Proceedings of the 3rd International Symposium on Semantic Mining in Biomedicine (SMBM)*. pages 77–84.

Vladimir N. Vapnik. 1995. *The nature of statistical learning theory*. Springer-Verlag New York, Inc.

Chih-Hsuan Wei, Yifan Peng, Robert Leaman, Allan Peter Davis, Carolyn J. Mattingly, Jiao Li, Thomas C. Wiegers, and Zhiyong Lu. 2016. Assessing the state of the art in biomedical relation extraction: overview of the BioCreative V chemical-disease relation (CDR) task. *Database* pages 1–8. https://doi.org/10.1093/database/baw032.

Wenpeng Yin and Hinrich Schütze. 2015. Multichannel variable-size convolution for sentence classification. In *Proceedings of the 19th Conference on Computational Language Learning (CoNLL)*. Association for Computational Linguistics, pages 204–214. http://aclweb.org/anthology/K15-1021.

Daojian Zeng, Kang Liu, Siwei Lai, Guangyou Zhou, Jun Zhao, et al. 2014. Relation classification via Convolutional Deep Neural Network. In *Proceedings of the 25th International Conference on Computational Linguistics (COLING)*. pages 2335–2344. http://aclweb.org/anthology/C14-1220.

Zhehuan Zhao, Zhihao Yang, Hongfei Lin, Jian Wang, and Song Gao. 2016a. A protein-protein interaction extraction approach based on deep neural network. *International Journal of Data Mining and Bioinformatics* 15(2):145–164. https://doi.org/10.1504/ijdmb.2016.076534.

Zhehuan Zhao, Zhihao Yang, Ling Luo, Hongfei Lin, and Jian Wang. 2016b. Drug drug interaction extraction from biomedical literature using syntax convolutional neural network. *Bioinformatics* 32(22):3444–3453. https://doi.org/10.1093/bioinformatics/btw486.

Guodong Zhou, Min Zhang, Dong Hong, and
Ji Qiaoming Zhu. 2007. Tree kernel-based relation
extraction with context-sensitive structured parse
tree information. In *Proceedings of the 2007 Joint
Conference on Empirical Methods in Natural Language Processing and Computational Natural Language Learning (EMNLP-CoNLL)*. pages 728–736.
http://aclweb.org/anthology/D07-1076.

Stacking with Auxiliary Features for Entity Linking in the Medical Domain

Nazneen Fatema Rajani
Department of Computer Science
University of Texas at Austin
nrajani@cs.utexas.edu

Mihaela Bornea   **Ken Barker**
T.J. Watson Research Center
IBM Research
{mabornea, kjbarker}@us.ibm.com

Abstract

Linking spans of natural language text to concepts in a structured source is an important task for many problems. It allows intelligent systems to leverage rich knowledge available in those sources (such as concept properties and relations) to enhance the semantics of the mentions of these concepts in text. In the medical domain, it is common to link text spans to medical concepts in large, curated knowledge repositories such as the *Unified Medical Language System*. Different approaches have different strengths: some are precision-oriented, some recall-oriented; some better at considering context but more prone to hallucination. The variety of techniques suggests that *ensembling* could outperform component technologies at this task. In this paper, we describe our process for building a *Stacking* ensemble using additional, auxiliary features for Entity Linking in the medical domain. Our best model beats several baselines and produces state-of-the-art results on several medical datasets.

1 Introduction

Entity Linking is the task of mapping phrases in text (*mention spans*) to concepts in a structured source, such as a knowledge base. The mention span is usually a word or short phrase describing a single, coherent concept. For example, "back pain" may be a mention span for a *Dorsalgia* concept in a knowledge base. The *span context* is a window of text surrounding the mention span that may be useful for disambiguating it. For example, the sentence "The patient reports suffering from back pain for several years prior to treatment" may

be useful for determining that "back pain" refers to the concept *Chronic Dorsalgia* in this context. In the medical domain, it is common to map mention spans to concepts in the *Unified Medical Language System (UMLS)*[1]. Concepts in UMLS have unique identifiers called *CUIs* (Concept Unique Identifiers). For example, the CUI for the concept *Dorsalgia* is C0004604.

The concepts in UMLS come from merging concepts from many disparate contributing vocabularies. Since automatic merging is imperfect, *UMLS* often contains multiple distinct CUIs for what amounts to the same semantic concept. For example, the three distinct CUIs C0425687, C1167958 and C3263244 are all *Jugular Venous Distension*. An Entity Linking system attempting to link a span such as "engorgement of the jugular vein" should be required to return all three CUIs. A ground truth dataset should include all the three mappings as well. *UMLS* also contains multiple textual labels for each CUI (called "variants") and semantic relations between CUIs, such as *Acetaminophen may_treat: Pain*.

Ensembling multiple systems is a well known standard approach to improving accuracy in machine learning (Dietterich, 2000). Ensembles have been applied to a wide variety of problems in all domains of artificial intelligence including natural language processing (NLP). However, these techniques do not learn to discriminate adequately across the component systems and thus are unable to integrate them optimally. Combining systems intelligently is crucial for improving the overall performance. In this paper, we use an approach called Stacking with Auxiliary Features (SWAF) (Rajani and Mooney, 2017) for combining multiple diverse models. Stacking (Wolpert, 1992) uses supervised learning to train a meta-classifier to

---

[1]UMLS: http://www.nlm.nih.gov/research/umls/

39

Proceedings of the BioNLP 2017 workshop, pages 39–47,
Vancouver, Canada, August 4, 2017. ©2017 Association for Computational Linguistics

combine multiple system outputs. SWAF enables the stacker to fuse additional relevant knowledge from multiple systems and thus leverage them to improve prediction. The idea behind using auxiliary features is that an output is more reliable if not just multiple systems produce it but also agree on its provenance and there is sufficient supporting evidence. We are the first to use ensembling for entity linking in the medical domain that lacks labeled data. All the publicly available datasets are very small and thus learning is a problem. Our approach is designed to overcome these challenges in the medical domain by using auxiliary features that are precision-focused and can be used to form a classification boundary from small amounts of data.

2   Component Entity Linking Systems

The entity linking ensemble we have built includes eight component systems. Given a span of text, each component links the entities in text to zero or more matching concepts in UMLS. The ensemble examines all concepts produced by each component system for the given span and determines the final entity linking outcome. All the component systems use traditional rule-based methods and thus only perform well on certain types of concepts. The errors produced by these base systems are de-correlated and our goal is to leverage the systems to the fullest by using carefully designed auxiliary features. We used the following component systems in our ensemble.

Medical Concept Resolution: Three of the components systems are variations of the Medical Concept Resolution (MCR) approach introduced in (Aggarwal et al., 2015). The MCR systems find UMLS concepts that best capture the meaning of the input span as expressed in the textual context where the span appears. The algorithms consist of two main steps: candidate overgeneration and candidate ranking. Candidate overgeneration finds all concepts having any variant containing any of the tokens in the mention text. This step results in a large number of candidate concepts, many of them irrelevant. In the second step, the candidate concepts are ranked by measuring the similarity between mention context and candidate context. The mention context is a window of text surrounding the span. The candidate context is generated differently by each of the three MCR systems. Both the span context and the candidate con-

text are treated as IDF-weighted bags-of-words for computing their cosine similarity. The higher the cosine similarity, the higher the rank of the candidate concept for the given span. The three variations of the MCR systems used are:

- *Gloss-Based MCR (GBMCR)*: generates the candidate context from the concept definitions in UMLS. In GBMCR, candidates are ranked according to the similarity between the words in the span mention (and its context) and the words in the UMLS definitions of the candidate.

- *Neighbor-Based MCR (NBMCR)*: generates the candidate context from the set of variants of the candidate's neighbors in UMLS. Neighbors are CUIs related to the candiate CUI by any of a select set of UMLS semantic relations. In NBMCR, candidates are ranked according to the similarity between the words in the span+context and the words in the variants of the candidate's neighbors.

- *Variants-Based MCR (VBMCR)*: generates the candidate context from the candidate's variants in UMLS. In VBMCR, candidates are ranked according to the similarity between the words in the span+context and the words in the candidate's variants.

Concept Mapper: Apache Concept Mapper matches text to dictionary entries. The dictionary contains surface forms and the concept identifiers those surface forms map to. The system included in the ensemble is based on a dictionary derived from the complete set of UMLS variants. Preprocessing of UMLS variants removes some supreflu-ous acronyms (e.g. "nos" = "not otherwise specified"; "nec" = "not elsewhere classified"). The dictionary is also expanded beyond the UMLS variants by including adjective-to-noun and plural-to-singular transformations, as well as additional spelling variants and synonymous phrases derived from wikipedia redirect pages.

CUI Finder Verbatim (CFV): CFV (Aggarwal et al., 2015) is a dictionary-based system similar to ConceptMapper with advanced matching algorithms and synonym expansion. If no concept is found when matching the dictionary using the entire span, CFV attempts to find concepts for smaller windows by removing words from the span iteratively. The algorithm considers both left-to-right and right-to-left shrinking of the span. If no concepts are found, it reduces the window size further. As soon as any concept is found, the algorithm stops, returning all concepts found for subspans of the given window size at any position within the original span.

MetaMap: This system is provided by the National Library of Medicine for detecting UMLS concepts in medical text.[2] It is NLP-based and uses domain-specific knowledge to map text to concepts. The ensemble includes MetaMap configured with the default settings.

cTAKES: Apache cTAKES[3] is an open source entity recognition system, originally developed at Mayo Clinic for identifying UMLS concepts in electronic medical records. cTAKES implements a terminology-agnostic dictionary lookup algorithm. Through the dictionary lookup, each named entity is mapped to a concept from the terminology. The dictionary lookup includes permutation of words in the spans, exact matches of the span and canonical forms of the words.

Structured Term Recognizer (STR): This system takes a span of text as input and produces a list of possible UMLS concepts for that span, as well as semantic types, if desired. Concept recognition proceeds in two phases: UMLS candidate generation and scoring of the candidate concepts. The candidate UMLS concepts are found by an inverted index, mapping tokens in the concepts to the concepts themselves. Once the candidate UMLS concepts are found, they are scored for similarity with the input span based on shared tokens and shared stems.

3 Stacking With Auxiliary Features

In this section we describe our algorithm and the auxiliary features used for classification. Figure 1 shows an overview of our ensembling approach.

3.1 Stacking

Stacking uses a meta-classifier to combine the outputs of multiple underlying systems. The stacker learns a classification boundary based on the confidence scores provided by individual systems for each possible output. Stacking has been shown to improve performance on tasks such as slot filling and tri-lingual entity linking (Viswanathan et al., 2015; Rajani and Mooney, 2016).

---

[2]MetaMap: http://metamap.nlm.nih.gov/
[3]cTAKES: https://ctakes.apache.org/

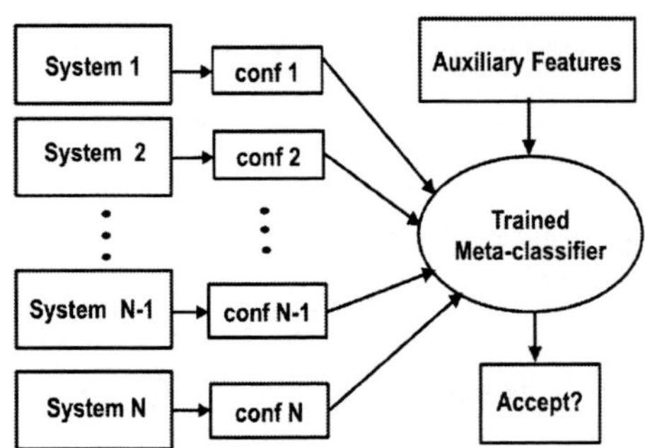

Figure 1: Ensemble Architecture using Stacking with Auxiliary Features. Given an input span, the ensemble judges every possible concept produced by the component systems and determines the final entity linking output.

3.2 Auxiliary Features

Stacking relies on systems producing a confidence score for every output. However, many times systems do not produce confidence scores or the scores produced are not probabilities or well calibrated and cannot be meaningfully compared. In such circumstances, it is beneficial to have other reliable auxiliary features. Auxiliary features enable the stacker to learn to rely on systems that not just agree on an output but also the provenance or the source of the output and other supporting evidence. We used four types of auxiliary features as part of our ensembling approach, described below.

3.2.1 CUI type

Every CUI in UMLS is associated with one or more semantic types (out of roughly 130 types). For example, the types associated with the CUI C0000970 (acetaminophen) are T109 (Organic chemical) and T121 (Pharmacologic substance).

The CUI type is represented by a binary vector of size 130. The CUI type vector has ones for each associated semantic type of the CUI under consideration and zeros elsewhere. This CUI type vector is used as an auxiliary feature for ensembling. The CUI type enables the stacker to learn to rely on systems that perform better for certain CUI types.

3.2.2 Span-CUI document similarity

The second auxiliary feature is the cosine similarity between the *tf-idf* vectors of the words in the mention span and the words in the candidate *CUI documents*. For each CUI in UMLS, we created a *pseudo document* which we call the CUI document. The CUI document is a concatenation of the following information from UMLS:

1. CUI ID and label; for example, *C0000970 (acetaminophen)*

2. Names of the types of the CUI; e.g., *Organic Chemical; Pharmacologic Substance*

3. Definition text for the CUI; e.g., *analgesic antipyretic derivative of acetanilide; weak antiinflammatory properties and is used as a common analgesic, but may cause liver, blood cell, and kidney damage.*

4. All variants for the CUI; e.g., *Acetaminophen, Paracetamol*

5. Select semantic relations between the CUI under consideration and other CUIs; for example, *(may_treat: fever), (may_treat: pain).*

The intuition behind using this feature is that the span would have a greater lexical overlap with a CUI document that it links to and thus have a higher similarity score.

3.2.3 Context-CUI document similarity

This auxiliary feature is very much like the span-CUI document similarity feature. For this feature as well, we use the *pseudo* CUI documents created using UMLS. However, instead of using the span for calculating the similarity we use the entire context surrounding the span. In the earlier example, the entire sentence "The patient reports suffering from back pain for several years prior to treatment" is the context. We note that for short documents, the context may be the entire document that contains the span to be linked. This means that some unique spans could have the same context. The context-CUI document similarity is the cosine similarity between the *tf-idf* vectors of words in the context and words in the CUI document.

3.2.4 Word embeddings

The auxiliary features discussed so far only capture the superficial lexical aspects of the data used for ensembling. The word embeddings features capture the semantic dimension of the data. We trained the continuous bag of words model (Mikolov et al., 2013) on the entire UMLS knowledge base with word vector dimension of 200 and window-size of 10. Ling et al. (2015) show that these parameters enable capturing long range dependencies. In this way we obtain a vector representation for every word in UMLS. We note that

we chose the UMLS corpus as opposed to medical documents so as to have better CUI coverage.

We used these word vectors to create the CUI document vector representation in the following way. Recall that the CUI document is a pseudo document made up of information about the CUI in UMLS. In order to obtain the embedding for a context, span or document, we use the technique described in (Le and Mikolov, 2014). We add up all the embedding vectors representing the words in the CUI document and normalize the sum by the number of words. The resultant vector represents the CUI document embedding. Similarly, we also obtain the span and the context embeddings by adding and normalizing the vectors representing the words in the span and context respectively. Note that if a word in the span or context does not have a vector representation then we just ignore it. Finally, we measure the cosine similarity between the span-CUI document and context-CUI document embedding vectors and use it as a feature for our classifier. Representing the concepts in vector space enables the stacker to learn deep semantic patterns for cases where just lexical information is not sufficient.

4 Experimental Results

4.1 Baselines

We compare our approach to several supervised and unsupervised baselines. The first is *Union* which accepts all predictions for all systems to maximize recall. It classifies all span-CUI links as correct and always includes them.

The second baseline is *Voting*. For this approach, we vary the threshold on the number of systems that must agree on a span-CUI link from one to all. This gradually changes the system behavior from union to intersection of the links. We identify the threshold that results in the highest F1 score on the training dataset. We use this threshold for the voting baseline on the test dataset.

The third baseline is an *oracle threshold* version of Voting. Since the best threshold on the training data may not necessarily be the best threshold for the test data, we identify the best threshold for the test data by plotting a precision-recall curve and finding the best F1 score for the voting baseline. Note that this gives an upper bound on the best results that can be achieved with voting, assuming an optimal threshold is chosen. Since the upper bound can not be predicted without using the test dataset, this baseline has an unfair advantage.

In addition to the above common baselines, we also compare our approach to a state-of-the-art ensembling system, Bipartite Graph based Consensus Maximization (BGCM) (Gao et al. (2009)). In addition to the output of supervised models, this ensembling technique uses unsupervised models to provide additional constraints and evidence to the classification algorithm. The rationale behind this approach is that objects that are in the same cluster should be more likely to receive the same class label compared to the objects in different clusters. The objective is to predict the class label of an instance in a way that favors agreement between supervised components and at the same time satisfies the constraints enforced by the clustering models. BGCM ensembles multiple models by performing an optimization over a bipartite graph of systems and outputs.

4.2 Dataset Description

All systems and baselines were evaluated on three datasets. Scores reflect the quality of concepts assigned to text spans, as decided by human judges. Detecting span boundaries is not part of this evaluation – all systems are given the same span as input. Annotations were performed by several human judges. For scoring, each text span was paired with a list of concepts produced by all component systems. Annotators marked each span-concept pair correct or incorrect.

The *MCR* dataset (Aggarwal et al., 2015) resulted from running a CRF-based entity recognition system that extracted 1,570 clinical factors from 100 short descriptions (averaging 8 sentences, 100 words) of patient scenarios. The annotated dataset contains a subset of 400 spans resulting in 6,139 annotated span-CUI pairs. The average of the pairwise kappa scores for annotator agreement on the *MCR* dataset was 0.56.

The *i2b2* dataset (Uzuner et al., 2011) is based on the annotated patient discharge summaries released with the 2010 i2b2/VA challenge. The concept extraction task was to identify and extract the text span corresponding to patient medical problems, treatments and tests in unannotated patient record text. We created an entity linking dataset from a random subset of 100 annotated text spans. We ran all available entity linking systems and produced 2,224 annotated span-CUI pairs. The average pairwise kappa score for annotator agree-ment on the *i2b2* dataset was 0.52.

The Electronic Medical Record dataset (*EMR*) is a private dataset containing spans of medical terms identified in doctors' notes within patient medical records. This dataset has 350 text spans with 3,991 annotated span-CUI pairs. Annotators for the *EMR* dataset reconciled their annotations to build the ground truth.

4.3 Evaluation Metrics

As noted in section 1, *UMLS* often has multiple distinct CUIs for the same semantic concept. So for a given span from a dataset, there may be many true positive concepts in the ground truth. This leads to two possible scoring schemes: *CUI level* and *Span level*. For CUI level scoring, every CUI in the ground truth is a ground truth positive instance. A CUI produced by the Entity Linking system for a given span is a true positive if it is in the ground truth for that span and a false positive if it is not. CUIs in the ground truth for the span that are not produced by the system are counted as false negatives. Spans that have many CUIs in the ground truth, therefore, will have more weight in the precision and recall than spans with fewer CUIs. But since the number of appropriate CUIs for a span is often a side effect of the imperfect automatic merging of concepts in building *UMLS*, the bias is unnatural.

An alternative scoring scheme awards only one true positive, false positive or false negative for each span, not each CUI. For this span level scoring, we report two versions of the metrics. The first version, which we call "Factor Level" in the reported results, aggregates CUI scores using *MAX*. The system scores a true positive if *any* of the CUIs it produces are in the ground truth for the span. It scores a false positive if *none* of its CUIs are in the ground truth. It scores a false negative if it produces no CUIs and there is at least one CUI in the ground truth.

The second version of span level scoring accounts for the fact that the system may produce a mixture of correct and incorrect CUIs for the same span. Each span still has a weight of one in the overall precision and recall, but the system's score for "true positiveness" and "false positiveness" can be a real number between 0 and 1. We call this scoring scheme "Quantum". The quantum true positive score for a span is the number of CUIs produced by the system that are in the

Approach	CUI Level			Factor Level			Quantum		
	P	**R**	**F1**	**P**	**R**	**F1**	**P**	**R**	**F1**
GBMCR	0.349	0.242	0.286	0.395	0.437	0.415	0.357	0.268	0.306
NBMCR	0.414	0.179	0.250	0.463	0.511	0.486	0.423	0.163	0.236
VBMCR	0.496	0.215	0.300	0.548	0.605	0.575	0.513	0.198	0.285
CFV	0.587	0.405	**0.479**	**0.903**	0.461	0.611	**0.716**	0.188	0.298
CTakes	0.384	0.245	0.299	0.711	0.577	0.637	0.498	0.202	0.287
MetaMap	0.447	0.219	0.293	0.623	0.652	0.637	0.535	0.215	0.306
CMap	0.179	**0.549**	0.270	0.802	**0.870**	**0.834**	0.305	**0.461**	**0.367**
STR	**0.623**	0.217	0.322	0.623	0.688	0.654	0.623	0.217	0.322
Union	0.207	**0.797**	0.329	**0.888**	**0.981**	**0.932**	0.278	**0.765**	0.408
Majority Voting	**0.746**	0.182	0.293	0.768	0.522	0.622	**0.745**	0.169	0.275
Oracle Voting	0.626	0.290	0.396	0.723	0.707	0.715	0.629	0.251	0.359
BGCM	0.481	0.430	**0.454**	0.753	0.822	0.786	0.525	0.368	**0.433**
Stacking	0.481	0.508	0.494	0.785	0.848	0.815	**0.501**	0.412	0.452
+ CUI Type	0.474	0.573	0.519	0.816	0.889	0.851	0.484	0.502	0.493
+ Span & Context Similarity	0.472	**0.575**	0.519	0.811	0.886	0.847	0.485	**0.508**	0.496
+ CBOW embedding	**0.567**	0.500	**0.532**	0.824	0.892	0.857	0.491	0.507	**0.499**

Table 1: Results on the MCR dataset.

ground truth for the span divided by the total number of CUIs produced by the system (*i.e.*, the span-level Precision). Quantum false positive score is the number of incorrect CUIs produced by the system divided by the total number of CUIs produced.

4.4 Results

We present results for entity linking in the medical domain on the three datasets described in section 4.2 using the evaluation metrics defined in section 4.3. The results include the performance of the individual models, several baselines and various ablations of the auxiliary features using stacking. Tables 1, 2 and 3 show performance on the *MCR*, *i2b2* and *EMR* datasets respectively.

Although we observe similar trends across all the datasets, no single individual model performs better than others across all the evaluation metrics. This led us to conclude that each individual model is optimized for a particular type of entity or data. For example, a model that is good at linking medical drugs might not perform as well on linking medical diseases. In order to leverage the strengths of each individual model, we ensemble them into one powerful model that works across all datasets as well as different evaluation metrics.

As expected, the *Union* baseline obtains the best recall and *Majority Voting* has the highest precision across all datasets. *Oracle Voting* is optimized for F1 and thus obtains an F1 higher than *Majority Voting*. Vanilla stacking beats the best component and baseline systems' F1 scores for CUI level and quantum metrics on all datasets. Adding each aux-

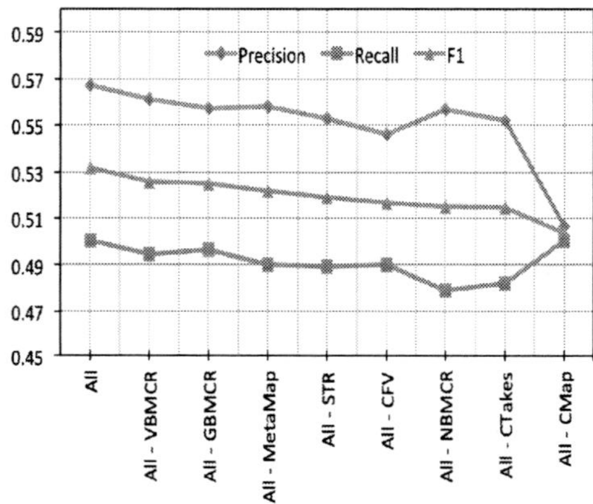

Figure 2: Ablation on the component systems in the ensemble for the *MCR* dataset using the CUI level metric. The systems are arranged in decreasing order of F1 score.

iliary feature further boosts the performance and we obtain the highest F1 for all datasets using all the features combined. Stacking outperforms the *BGCM* ensembling baseline on all datasets.

For a deeper understanding of the results, we performed ablation tests on the systems used in the final ensemble. Figure 2 shows the performance of the ensemble with each component ablated in turn. This experiment shows that every component system contributes to the ensemble in either precision, recall or both. While each component contributes to the overall performance, the strength of the ensemble is determined by the combination of

Approach	CUI Level			Factor Level			Quantum		
	P	**R**	**F1**	**P**	**R**	**F1**	**P**	**R**	**F1**
GBMCR	0.507	0.375	0.431	0.790	0.807	0.798	0.515	0.427	0.467
NBMCR	0.478	0.356	0.408	0.779	0.796	0.787	0.486	0.403	0.441
VBMCR	0.554	0.404	**0.467**	0.800	0.817	0.809	0.564	**0.468**	**0.511**
CFV	0.173	**0.457**	0.251	0.884	0.903	0.894	0.577	0.327	0.417
CTakes	0.564	0.213	0.309	0.861	0.731	0.791	**0.677**	0.195	0.303
MetaMap	0.565	0.154	0.242	0.750	0.742	0.746	0.647	0.153	0.248
CMap	0.216	0.360	0.270	**0.894**	**0.903**	**0.898**	0.410	0.260	0.318
STR	**0.825**	0.176	0.290	0.833	0.860	0.847	0.566	0.236	0.333
Union	0.191	**0.855**	0.312	0.969	1.000	0.984	0.352	**0.849**	**0.498**
Majority Voting	**0.705**	0.189	0.298	0.846	0.828	0.837	**0.766**	0.176	0.286
Oracle Voting	0.624	0.270	0.373	0.874	0.893	0.883	0.709	0.227	0.344
BGCM	0.469	0.406	**0.435**	0.938	0.968	0.952	0.509	0.386	0.439
Stacking	0.434	0.697	0.535	**0.958**	**0.989**	**0.974**	0.481	0.655	0.555
+ CUI Type	0.525	0.730	0.611	0.927	0.957	0.942	**0.547**	0.563	0.555
+ Span & Context Similarity	0.528	0.756	0.622	0.927	0.957	0.942	0.544	0.639	0.588
+ CBOW embedding	**0.528**	**0.756**	**0.622**	0.938	0.968	0.952	0.546	**0.700**	**0.607**

Table 2: Results on the *i2b2* dataset.

Approach	CUI Level			Factor Level			Quantum		
	P	**R**	**F1**	**P**	**R**	**F1**	**P**	**R**	**F1**
GBMCR	0.338	0.134	0.192	0.369	0.351	0.36	0.360	0.315	0.196
NBMCR	0.381	0.151	0.217	0.410	0.390	0.400	0.396	0.148	0.216
VBMCR	0.564	0.224	0.321	0.618	0.589	0.603	0.600	0.225	0.327
CFV	0.510	0.353	**0.417**	0.914	0.607	0.729	0.692	0.249	0.366
CTakes	0.403	0.321	0.357	0.706	0.628	0.665	0.527	0.268	0.355
MetaMap	0.460	0.220	0.298	0.575	0.568	0.571	0.527	0.223	0.313
CMap	0.205	**0.597**	0.305	0.761	**0.766**	**0.763**	0.334	**0.597**	**0.428**
STR	**0.714**	0.284	0.406	**0.714**	0.748	0.730	0.714	0.284	0.406
Union	0.187	**0.739**	0.299	0.857	**0.852**	**0.854**	0.272	**0.676**	0.388
Majority Voting	**0.879**	0.225	0.359	**0.912**	0.561	0.695	**0.894**	0.220	0.353
Oracle Voting	0.668	0.297	0.412	0.820	0.661	0.732	0.719	0.276	0.399
BGCM	0.453	0.419	**0.435**	0.801	0.809	0.805	0.482	0.409	**0.442**
Stacking	0.443	0.517	0.477	0.794	**0.832**	**0.812**	0.488	**0.463**	0.475
+ CUI Type	0.559	0.548	0.554	0.807	0.778	0.792	0.571	0.436	0.495
+ Span & Context Similarity	0.593	**0.554**	0.573	0.820	0.781	0.800	0.616	0.443	0.515
+ CBOW embedding	**0.667**	0.549	**0.602**	**0.830**	0.775	0.801	**0.669**	0.439	**0.530**

Table 3: Results on the *EMR* dataset.

the component systems. The ablation of the CMap system has the highest impact on the ensemble, reducing the F1 score by 5.2%. We obtained similar plots for the factor level and quantum metrics and we expect to see similar trends for the *i2b2* and the *EMR* datasets as well.

5 Discussion

The experimental results presented in section 4.4 confirm that the different component systems show significantly different behavior on different metrics for different datasets. No individual system was universally the best. *CMap* had consistently good Recall but low Precision. *CFV* scored well in certain circumstances on precision, recall and F1 score, but this varied from dataset to dataset and metric to metric. *STR* usually had relatively high precision, but low recall, and *VBMCR* had very good F1 scores on *i2b2*, but was less impressive on the other datasets.

These observations imply good conditions for ensembling to make a difference. Even so, the best baseline ensemble only outperforms the best component system on F1 in four of the nine experiments (metric-dataset combinations). Stacking outperforms the best component system in all nine, and outperforms the best ensembling baseline for six of the nine – all of the CUI level metrics and quantum, but never at the factor level. The factor level scoring is much more generous, but it is not immediately clear why this would benefit naïve ensembling over stacking.

Auxiliary features almost always improve stacking. Again the exception is with factor level scoring. Interestingly, auxiliary features almost universally improve precision significantly without too damaging an effect on recall. This result suggests that it would be worthwhile experimenting with the precision-vs-recall bias of component systems to see if Stacking with auxiliary features could be used, for example, to recover precision with recall-biased components.

6 Related Work

The problem of entity linking has received considerable attention in the research community. Several community tasks are focused specifically on the medical domain and are addressing the problem of linking disease/disorder entities to *SNOMED CT*.[4] *SNOMED CT* concepts are also included in *UMLS*.

The ShARe/CLEF eHealth Evaluation Lab 2013 (Suominen et al., 2013) consists of a collection of tasks focused on facilitating patients' understanding of their medical discharge summaries. The assumption is that an improved understanding of medical concepts in such documents can be achieved by normalizing all health conditions to standardized *SNOMED CT* concepts. Using these concepts, the medical documents can further be connected to other patient friendly sources.

The Open Biomedical Annotator (*OBA*) (Jonquet et al., 2009) is an ontology-based Web service that annotates public datasets with biomedical ontology concepts, including concepts from *UMLS*. The *OBA* is based on dictionary matching. The dictionary is a list of strings that identify ontology concepts. The dictionary is constructed by accessing biomedical ontologies and extracting all concept names, their synonyms or labels. The web service takes as input the user's free text. The tool recognizes concepts using string matching on the dictionary and outputs the concept annotations.

There are several notable approaches to perform entity linking in the open domain. These open domain approaches often deal with named entities. The linking targets in this case are often single, unambiguous, specific concepts. The problem of finding domain-specific concepts, on the other hand, can be more challenging as there may be appropriate concepts at different levels of specificity, and concepts are more compositional and contextual. Approaches such as DBPedia Spotlight (Mendes et al., 2011) and AIDA (Hoffart et al., 2011) use Wikipedia to find the links of recognized entity mentions.

To overcome challenges of obtaining labeled medical datasets, Zheng et al. (2015) proposed an unsupervised approach for entity linking. More traditional sieve-based techniques have been used for this task recently (D'Souza and Ng, 2015).

Using ensembling techniques for open domain entity linking has shown good performance in the past (Rajani and Mooney, 2017) on the Trilingual Entity Discovery and Linking (TEDL) task. TEDL is an entity linking task conducted by NIST. The goal of this task is to discover entities in the three included languages (English, Spanish and Chinese) from a supplied text corpus and link these entities to an existing English knowledge base (a reduced version of FreeBase).

Rajani and Mooney (2016) proposed an approach for combining multiple supervised and unsupervised models for entity linking. Their technique improves the previous result on the TEDL task. Another ensembling approach is Mixtures of Experts (Jacobs et al., 1991) which employs divide-and-conquer principle to soft switch between learners covering different sub-spaces of the input using Expectation-Maximization (EM). Our work is the first we know of to use ensembling for entity linking in the medical domain.

7 Conclusion

We have identified an entity linking task in the medical domain for which existing technologies perform differently on different metrics for different datasets. Such an environment presents an obvious opportunity for ensembling techniques.

We have built a stacking ensembler using multiple diverse entity linking systems. The auxiliary features further boost the stacker's performance. Experiments confirm that naïve ensembling does not always outperform component entity linking systems, but that vanilla stacking does. Adding auxiliary features to the stacker almost universally improves its precision without harming recall, giving it generally the best F1 scores overall.

Our model is able to fuse additional relevant knowledge from multiple systems and leverage them to improve prediction.

---

[4]SNOMED CT: http://www.snomed.org/

References

Nitish Aggarwal, Ken Barker, and Chris Welty. 2015. Medical concept resolution. In *Proceedings of the ISWC 2015 Posters & Demonstrations Track co-located with the 14th International Semantic Web Conference (ISWC-2015), Bethlehem, PA, USA, October 11, 2015.*.

T. Dietterich. 2000. Ensemble methods in machine learning. In J. Kittler and F. Roli, editors, *First International Workshop on Multiple Classifier Systems, Lecture Notes in Computer Science*. Springer-Verlag, pages 1–15.

Jennifer D'Souza and Vincent Ng. 2015. Sieve-based entity linking for the biomedical domain. In *ACL (2)*. pages 297–302.

Jing Gao, Feng Liang, Wei Fan, Yizhou Sun, and Jiawei Han. 2009. Graph-based consensus maximization among multiple supervised and unsupervised models. In *NIPS2009*. pages 585–593.

Johannes Hoffart, Mohamed Amir Yosef, Ilaria Bordino, Hagen Fürstenau, Manfred Pinkal, Marc Spaniol, Bilyana Taneva, Stefan Thater, and Gerhard Weikum. 2011. Robust disambiguation of named entities in text. In *Proceedings of the Seventh International Conference on Semantic Systems*. pages 1–8.

Robert A Jacobs, Michael I Jordan, Steven J Nowlan, and Geoffrey E Hinton. 1991. Adaptive mixtures of local experts. *Neural computation* 3(1):79–87.

Clement Jonquet, Nigam Shah, and Mark Musen. 2009. The open biomedical annotator. pages 56–60.

Quoc V Le and Tomas Mikolov. 2014. Distributed representations of sentences and documents. In *Proceedings of the International Conference on Machine Learning*. volume 14, pages 1188–1196.

Wang Ling, Lin Chu-Cheng, Yulia Tsvetkov, and Silvio Amir. 2015. Not all contexts are created equal: Better word representations with variable attention. In *Proceedings of the 2015 Conference on Empirical Methods in Natural Language Processing (EMNLP-15)*. Citeseer.

Pablo N. Mendes, Max Jakob, Andrés Garcia-Silva, and Christian Bizer. 2011. DBpedia spotlight: shedding light on the web of documents. In *Proceedings of the Seventh International Conference on Semantic Systems*. pages 1–8.

Tomas Mikolov, Ilya Sutskever, Kai Chen, Greg S Corrado, and Jeff Dean. 2013. Distributed representations of words and phrases and their compositionality. In *Advances in neural information processing systems*. pages 3111–3119.

Nazneen Fatema Rajani and Raymond J. Mooney. 2016. Combining Supervised and Unsupervised Ensembles for Knowledge Base Population. In *Proceedings of the 2016 Conference on Empirical Methods in Natural Language Processing (EMNLP-16)*.

Nazneen Fatema Rajani and Raymond J. Mooney. 2017. Stacking With Auxiliary Features. In *Proceedings of the Twenty-Sixth International Joint Conference on Artificial Intelligence (IJCAI2017)*. Melbourne, Australia.

Hanna Suominen, Sanna Salanterä, Sumithra Velupillai, Wendy W. Chapman, Guergana Savova, Noemie Elhadad, Sameer Pradhan, Brett R. South, Danielle L. Mowery, Gareth J. F. Jones, Johannes Leveling, Liadh Kelly, Lorraine Goeuriot, David Martinez, and Guido Zuccon. 2013. *Overview of the ShARe/CLEF eHealth Evaluation Lab 2013*, pages 212–231.

Ozlem Uzuner, Brett R. South, Shuying Shen, and Scott L. DuVall. 2011. 2010 i2b2/va challenge on concepts, assertions, and relations in clinical text. *JAMIA* 18(5):552–556.

Vidhoon Viswanathan, Nazneen Fatema Rajani, Yinon Bentor, and Raymond J. Mooney. 2015. Stacked Ensembles of Information Extractors for Knowledge-Base Population. In *Association for Computational Linguistics (ACL2015)*. Beijing, China, pages 177–187.

David H. Wolpert. 1992. Stacked Generalization. *Neural Networks* 5:241–259.

Jin G Zheng, Daniel Howsmon, Boliang Zhang, Juergen Hahn, Deborah McGuinness, James Hendler, and Heng Ji. 2015. Entity linking for biomedical literature. *BMC medical informatics and decision making* 15(1):S4.

Results of the fifth edition of the BioASQ Challenge

Anastasios Nentidis[1], Konstantinos Bougiatiotis[1], Anastasia Krithara[1],
Georgios Paliouras[1] and Ioannis Kakadiaris[2]
[1]National Center for Scientific Research "Demokritos", Athens, Greece
[2]University of Houston, Texas, USA

Abstract

The goal of the BioASQ challenge is to engage researchers into creating cutting-edge biomedical information systems. Specifically, it aims at the promotion of systems and methodologies that are able to deal with a plethora of different tasks in the biomedical domain. This is achieved through the organization of challenges. The fifth challenge consisted of three tasks: semantic indexing, question answering and a new task on information extraction. In total, 29 teams with more than 95 systems participated in the challenge. Overall, as in previous years, the best systems were able to outperform the strong baselines. This suggests that state-of-the art systems are continuously improving, pushing the frontier of research.

1 Introduction

The aim of this paper is twofold. First, we aim to give an overview of the data issued during the BioASQ challenge in 2017. In addition, we aim to present the systems that participated in the challenge and evaluate their performance. To achieve these goals, we begin by giving a brief overview of the tasks, which took place from February to May 2017, and the challenge's data. Thereafter, we provide an overview of the systems that participated in the challenge. Detailed descriptions of some of the systems are given in workshop proceedings. The evaluation of the systems, which was carried out using state-of-the-art measures or manual assessment, is the last focal point of this paper, with remarks regarding the results of each task. The conclusions sum up this year's challenge.

2 Overview of the Tasks

The challenge comprised three tasks: (1) a large-scale semantic indexing task (Task 5a), (2) a ques-

tion answering task (Task 5b) and (3) a funding information extraction task (Task 5c), described in more detail in the following sections.

2.1 Large-scale semantic indexing - 5a

In Task 5a the goal is to classify documents from the PubMed digital library into concepts of the MeSH hierarchy. Here, new PubMed articles that are not yet annotated by MEDLINE indexers are collected and used as test sets for the evaluation of the participating systems. In contrast to previous years, articles from all journals were included in the test data sets of task 5a. As soon as the annotations are available from the MEDLINE indexers, the performance of each system is calculated using standard flat information retrieval measures, as well as, hierarchical ones. As in previous years, an on-line and large-scale scenario was provided, dividing the task into three independent batches of 5 weekly test sets each. Participants had 21 hours to provide their answers for each test set. Table 1 shows the number of articles in each test set of each batch of the challenge. 12,834,585 articles with 27,773 labels were provided as training data to the participants.

2.2 Biomedical semantic QA - 5b

The goal of Task 5b was to provide a large-scale question answering challenge where the systems had to cope with all the stages of a question answering task for four types of biomedical questions: yes/no, factoid, list and summary questions (Balikas et al., 2013). As in previous years, the task comprised two phases: In phase A, BioASQ released 100 questions and participants were asked to respond with relevant elements from specific resources, including relevant MEDLINE articles, relevant snippets extracted from the articles, relevant concepts and relevant RDF triples. In phase B, the released questions were enhanced with relevant articles and snippets selected manu-

Proceedings of the BioNLP 2017 workshop, pages 48–57,
Vancouver, Canada, August 4, 2017. ©2017 Association for Computational Linguistics

Batch	Articles	Annotated Articles	Labels per Article
	6,880	6,661	12.49
	7,457	6,599	12.49
1	10,319	9,656	12.49
	7,523	4,697	11.78
	7,940	6,659	12.50
Total	40,119	34,272	12.39
	7,431	7,080	12.40
	6,746	6,357	12.62
2	5,944	5,479	12.87
	6,986	6,526	12.65
	6,055	5,492	12.41
Total	33,162	30,934	12.58
	9,233	5,341	12.78
	7,816	2,911	12.58
3	7,206	4,110	12.70
	7,955	3,569	12.17
	10,225	984	13.72
Total	42,435	21,323	12.68

Table 1: Statistics on test datasets for Task 5a.

ally and the participants had to respond with *exact answers*, as well as with summaries in natural language (dubbed *ideal answers*). The task was split into five independent batches and the two phases for each batch were run with a time gap of 24 hours. In each phase, the participants received 100 questions and had 24 hours to submit their answers. Table 2 presents the statistics of the training and test data provided to the participants. The evaluation included five test batches.

Batch	Size	Documents	Snippets
Train	1,799	11.86	20.38
Test 1	100	4.87	6.03
Test 2	100	3.93	5.13
Test 3	100	4.03	5.47
Test 4	100	3.23	4.52
Test 5	100	3.61	5.01
Total	2,299	10.14	17.09

Table 2: Statistics on the training and test datasets of Task 5b. All the numbers for the documents and snippets refer to averages.

2.3 Funding information extraction - 5c

Task 5c was introduced for the first time this year and the challenge at hand was to extract grant information from Biomedical articles. Funding information can be very useful; in order to estimate, for example, the impact of an agency's funding in the biomedical scientific literature or to identify agencies actively supporting specific directions in research. MEDLINE citations are annotated with information about funding from specified agencies[1]. This funding information is either provided by the author manuscript submission systems or extracted manually from the full text of articles during the indexing process. In particular, NLM human indexers identify the grant ID and the funding agencies can be extracted from the string of the grant ID[2]. In some cases, only the funding agency is mentioned in the article, without the grant ID.

In this task funding information from MEDLINE was used, as golden data, in order to train and evaluate systems. The systems were asked to extract grant information mentioned in the full text, but author-provided information is not necessarily mentioned in the article. Therefore, grant IDs not mentioned in the article were filtered out. This filtering also excluded grant IDs deviating from NLM's general policy of storing grant IDs as published, without any normalization. When an agency was mentioned in the text without a grant ID, it was kept only if it appeared in the list of agencies and abbreviations provided by NLM. Cases of misspellings or alternative naming of agencies were removed. In addition, information for funding agencies that are no longer indexed by NLM was omitted. Consequently, the golden data used in the task consisted of a subset of all funding information mentioned in the articles.

During the challenge, a training and a test dataset were prepared. The test set of MEDLINE documents with their full-text available in PubMed Central was released and the participants were asked to extract grant IDs and grant agencies mentioned in each test article. The participating systems were evaluated on (a) the extraction of grant IDs, (b) the extraction of grant agencies and (c) full-grant extraction, i.e. the combination of grant ID and the corresponding funding agency. Table 3 contains details regarding the datasets for training and test.

---

[1]https://www.nlm.nih.gov/bsd/grant_acronym.html
[2]https://www.nlm.nih.gov/bsd/mms/medlineelements.html#gr

Dataset	Articles	Grant IDs	Agencies	Time Period
Training	62,952	111,528	128,329	2005-13
Test	22,610	42,711	47,266	2015-17

Table 3: Dataset overview for Task 5c.

3 Overview of Participants

3.1 Task 5a

For this task, 10 teams participated and results from 31 different systems were submitted. In the following paragraphs we describe those systems for which a description was obtained, stressing their key characteristics. An overview of the systems and their approaches can be seen in Table 4.

System	Approach
Search system	search engine, UIMA ConceptMapper
MZ	tf-idf, LDA, BR classification
Sequencer	recurrent neural networks
DeepMesh	d2v, tf-idf, MESHlabeler
AUTH	d2v, tf-idf, LLDA, SVM, ensembles
Iria	bigrams, Luchene Index, k-NN, ensembles, UIMA ConceptMapper

Table 4: Systems and approaches for Task 5a. Systems for which no description was available at the time of writing are omitted.

The "*Search system*" and its variants were developed as a UIMA-based text and data mining workflow, where different search strategies were adopted to automatically annotate documents with MeSH terms. On the other hand, the "*MZ*" systems applied Binary Relevance (BR) classification, using TF-IDF features, and Latent Dirichlet allocation (LDA) models with label frequencies per journal as prior frequencies, using regression for threshold prediction. A different approach is adopted by the "*Sequencer*" systems, developed by the team from the Technical University of Darmstadt, that considers the task as a sequence-to-sequence prediction problem and use recurrent neural networks based algorithm to cope with it.

The "*DeepMeSH*" systems implement document to vector (*d2v*) and tf-idf feature embeddings

(Peng et al., 2016), alongside the MESHLabeler system (Liu et al., 2015) that achieved the best scores overall, integrating multiple evidence using learning to rank (LTR). A similar approach, with regards to the d2v and tf-idf representations of the text, is followed by the "*AUTH*" team. Regarding the learning algorithms they've extended their previous system (Papagiannopoulou et al., 2016), improving the Labeled LDA and SVM base models, as well as introducing a new ensemble methodology based on label frequencies and multi-label stacking. Last but not least, the team from the University of Vigo developed the "*Iria*" systems. Building upon their previous approach (Ribadas et al., 2014) that uses an Apache Lucene Index to provide most similar citations, they developed two systems that follow a multilabel k-NN approach. They also incorporated token bigrams and PMI scores to capture relevant multiword terms through a voting ensemble scheme and the ConceptMapper annotator tool, from the Apache UIMA project (Tanenblatt et al., 2010), to match subject headings with the citation's abstract text.

Baselines: During the challenge, two systems served as baselines. The first baseline is a state-of-the-art method called Medical Text Indexer (MTI) (Mork et al., 2014) with recent improvements incorporated as described in (Zavorin et al., 2016). MTI is developed by the National Library of Medicine (NLM) and serves as a classification system for articles of MEDLINE, assisting the indexers in the annotation process. The second baseline is an extension of the system MTI, incorporating features of the winning system of the first BioASQ challenge (Tsoumakas et al., 2013).

3.2 Task 5b

The question answering task was tackled by 51 different systems, developed by 17 teams. In the first phase, which concerns the retrieval of information required to answer a question, 9 teams with 25 systems participated. In the second phase, where teams are requested to submit exact and ideal answers, 10 teams with 29 different systems participated. Two of the teams participated in both phases. An overview of the technologies employed by each team can be seen in Table 5.

The "*Basic QA pipeline*" approach is one of the two that participated in both Phases. It uses MetaMap for query expansion, taking into account

Systems	Phase	Approach
Basic QA pipeline	A, B	MetaMap, BM25
Olelo	A, B	NER, UMLS, SAP HANA, SRL
USTB	A	sequential dependence models, ensembles
fdu	A	MESHLabeler, Language model, word similarity
UNCC	A	Stanford Parser, Semantic Indexing
MQU	B	deep learning, neural nets, regression
Oaqa	B	agglomerative clustering, tf-idf, word embeddings, maximum margin relevance
LabZhu	B	PubTator, Standford POS tool, ranking
DeepQA	B	FastQA, SQuAD
sarrouti	B	UMLS, BM25, dictionaries

Table 5: Systems and approaches for Task 5b. Systems for which no information was available at the time of writing are omitted.

the text and the title of each article, and the BM25 probabilistic model (Robertson et al., 1995) in order to match questions with documents, snippets etc. The same goes for phase B, except for the exact answers, where stop words were removed and the top-k most frequent words were selected. "*Olelo*" is the second approach that tackles both phases of task B. It is built on top of the SAP HANA database and uses various NLP components, such as question processing, document and passage retrieval, answer processing and multi-document summarization based on previous approaches (Schulze et al., 2016) to develop a comprehensive system that retrieves relevant information and provides both exact and ideal answers for biomedical questions. Semantic role labeling (SRL) based extensions were also investigated.

One of the teams that participated only in phase A, is "*USTB*" who combined different strategies to enrich query terms. Specifically, sequential dependence models (Metzler and Croft, 2005), pseudo-relevance feedback models, fielded sequential dependence models and divergence from random-

ness models are used on the training data to create better search queries. The "*fdu*" systems, as in previous years (Peng et al., 2015), use a language model in order to retrieve relevant documents and keyword scoring with word similarity for snippet extraction. The "*UNCC*" team on the other hand, focused mainly on the retrieval of relevant concepts and articles using the Stanford Parser (Chen and Manning, 2014) and semantic indexing.

In Phase B, the Macquarie University (MQU) team focused on ideal answers (Molla, 2017), submitting different models ranging from a "trivial baseline" of relevant snippets to deep learning under regression settings (Malakasiotis et al., 2015) and neural networks with word embeddings. The Carnegie Mellon University team ("*OAQA*"), focused also on ideal answer generation, building upon previous versions of the "*OAQA*" system. They used extractive summarization techniques and experimented with different biomedical ontologies and algorithms including agglomerative clustering, Maximum Marginal Relevance and sentence compression. They also introduced a novel similarity metric that incorporates both semantic information (using word embeddings) and tf-idf statistics for each sentence/question.

Many systems used a modular approach breaking the problem down to question analysis, candidate answer generation and answer ranking. The "*LabZhu*" systems, followed this approach, based on previous years' methodologies (Peng et al., 2015). In particular, they applied rule-based question type analysis and used Standford POS tool and PubTator for candidate answer generation. They also used word frequencies for candidate answer ranking. The "*DeepQA*" systems focused on factoid and list questions, using an extractive QA model, restricting the system to output substrings of the provided text snippets. At the core of their system stands a state-of-the-art neural QA system, namely FastQA (Weissenborn et al., 2017), extended with biomedical word embeddings. The model was pre-trained on a large-scale open-domain QA dataset, SQuAD (Rajpurkar et al., 2016), and then the parameters were fine-tuned on the BioASQ training set. Finally, the "*sarrouti*" system, from Morocco's USMBA, uses among others a dictionary approach, term frequencies of UMLS metathesaurus' concepts and the BM25 model.

Baselines: For this challenge the open source

OAQA system proposed by (Yang et al., 2016) for BioASQ4 was used as a strong baseline. This system, as well as its previous version (Yang et al., 2015) for BioASQ3, had achieved top performance in producing exact answers. The system uses an UIMA based framework to combine different components. Question and snippet parsing is based on ClearNLP. MetaMap, TmTool, C-Value and LingPipe are used for concept identification and UMLS Terminology Services (UTS) for concept retrieval. In addition, identification of concept, document and snippet relevance is based on classifier components and scoring, ranking and reranking techniques are also applied in the final steps.

3.3 Task 5c

In this inaugural year for task c, 3 teams participated with a total of 11 systems. A brief outline of the techniques used by the participating systems is provided in table 6.

Systems	Approach
Simple	regions of interest, SVM, regular expressions, hand-made rules, char-distances, ensemble
DZG	regions of interest, SVM, tf-idf of bigrams, HMMs, MaxEnt, CRFs, ensemble
AUTH	regions of interest, regular expressions

Table 6: Overview of the methodologies used by the participating systems in Task 5c.

The Fudan University team, participated with a series of similar systems ("*Simple*" systems) as well as their ensemble. The general approach included the following steps: First, the articles were parsed and some sections, such as affiliation or references, were removed. Then, using NLP techniques, alongside pre-defined rules, each paragraph was split into sentences. These sentences were classified as *positive* (*i.e.* containing grant information) or not, using a linear SVM. The positive sentences were scanned for grant IDs and agencies through the use of regular expressions and hand-made rules. Finally, multiple classifiers were trained in order to merge grant IDs and agencies into suitable pairs, based on a wide range of features, such as character-level features of the grant ID, the agency in the sentence and the distance between the grant ID and the agency in the sentence.

The "*DZG*" systems followed a similar methodology, in order to classify snippets of text as possible grant information sources, implementing a linear SVM with tf-idf vectors of bigrams as input features. However, their methodology differed from that of Fudan in two ways. Firstly, they used an in-house-created dataset consisting of more than 1,600 articles with grant information in order to train their systems. Secondly, the systems deployed were based on a variety of sequential learning models namely conditional random fields (Finkel et al., 2005), hidden markov models (Collins, 2002) and maximum entropy models (Ratnaparkhi, 1998). The final system deployed was a pooling ensemble of these three approaches, in order to maximize recall and exploit complementarity between predictions of different models. Likewise, the AUTH team, with systems "*Asclepius*", "*Gallen*" and "*Hippocrates*" emphasized on specific sections of the text that could contain grant support information and extracted grant IDs and agencies using regular expressions.

Baselines: For this challenge a baseline was provided by NLM ("*BioASQ Filtering*") which is based on a two-step procedure. First, the system classifies snippets from the full-text, as possible grant support "zones" based on the average probability ratio, generated separately by Naive Bayes (Zhang et al., 2009) and SVM (Kim et al., 2009). Then, the system identified grant IDs and agencies in these selected grant support "zones", using mainly heuristic rules, such as regular expressions, especially for detecting uncommon and irregularly formatted grant IDs.

4 Results

4.1 Task 5a

Each of the three batches of task 5a was evaluated independently. The classification performance of the systems was measured using flat and hierarchical evaluation measures (Balikas et al., 2013). The micro F-measure (MiF) and the Lowest Common Ancestor F-measure (LCA-F) were used to choose the winners for each batch (Kosmopoulos et al., 2013).

According to (Demsar, 2006) the appropriate way to compare multiple classification systems over multiple datasets is based on their average rank across all the datasets. On each dataset the system with the best performance gets rank 1.0,

System	Batch 1		Batch 2		Batch 3	
	MiF	LCA-F	MiF	LCA-F	MiF	LCA-F
auth1	8.88	8.25	10.50	9.75	10.25	9.75
auth2	7.25	6.50	7.63	7.50	8.88	9.75
auth3	6.75	8.25	7.50	10.25	6.50	7.00
auth4	-	-	7.38	8.25	9.63	9.75
auth5	-	-	7.50	7.00	8.50	7.50
DeepMeSH1	1.88	1.88	**1.00**	2.00	**1.00**	1.50
DeepMeSH2	**1.00**	**1.00**	3.00	3.00	2.50	2.75
DeepMeSH3	4.00	4.63	4.00	4.00	4.00	4.13
DeepMeSH4	5.00	4.38	5.00	5.50	4.88	5.63
DeepMeSH5	2.63	2.63	1.75	**1.00**	2.25	**1.25**
iria-1	-	-	13.75	13.75	12.75	12.75
iria-2	-	-	-	-	11.75	11.75
MZ1	10.75	10.75	-	-	-	-
Optimize Macro AUC	-	-	-	-	19.25	19.25
Optimize Micro AUC	-	-	-	-	15.75	18.25
Search system-1	12.25	12.25	-	-	13.75	13.25
Search system-2	13.25	13.25	-	-	14.75	14.25
Search system-3	16.25	16.25	-	-	18.50	17.50
Search system-4	15.25	15.25	-	-	16.75	16.25
Search system-5	14.25	14.25	-	-	15.75	15.25
Default MTI	7.50	6.25	8.75	6.00	7.50	6.75
MTI First Line Index	9.13	9.25	11.50	11.50	9.50	8.75

Table 7: Average system ranks across the batches of the Task 5a. A hyphenation symbol (-) is used whenever the system participated in fewer than 4 tests in the batch. Systems with fewer than 4 participations in all batches are omitted.

the second best rank 2.0 and so on. In case two or more systems tie, they all receive the average rank. Table 7 presents the average rank (according to MiF and LCA-F) of each system over all the test sets for the corresponding batches. Note, that the average ranks are calculated for the 4 best results of each system in the batch according to the rules of the challenge.

On both test batches and for both flat and hierarchical measures, the *DeepMeSH* systems (Peng et al., 2016) and the AUTH systems outperform the strong baselines, indicating the importance of the methodologies proposed, including d2v and tf-idf transformations to generate feature embeddings, for semantic indexing. More detailed results can be found in the online results page [3].

4.2 Task 5b

Phase A: For phase A and for each of the four types of annotations: documents, concepts, snippets and RDF triples, we rank the systems according to the Mean Average Precision (MAP) measure. The final ranking for each batch is calculated as the average of the individual rankings in the different categories. In tables 8 and 9 some indicative results from batch 3 are presented. Full results are available in the online results page of task 5b, phase A [4].

It is worth noting that document and snippet retrieval for the given questions were the most popular part of the task. Moreover, for different evaluation metrics, there are different systems performing best, indicating that different approaches to the task may be preferable depending on the target

[3] http://participants-area.bioasq.org/results/5a/

[4] http://participants-area.bioasq.org/results/5b/phaseA/

System	Mean Precision	Mean Recall	Mean F-measure	MAP	GMAP
testtext	0.1255	0.1789	0.1331	0.0931	**0.0017**
ustb-prir1	0.1306	0.1838	0.1372	0.0935	0.0016
ustb-prir4	0.1323	**0.2003**	**0.1412**	**0.1027**	0.0016
ustb-prir3	0.1307	0.1846	0.1376	0.0982	0.0015
ustb-prir2	0.1270	0.1832	0.1340	0.0975	0.0013
fdu	0.1551	0.1401	0.1286	0.0650	0.0005
fdu2	**0.1611**	0.1296	0.1185	0.0653	0.0005
Olelo	0.0702	0.1135	0.0764	0.0386	0.0003
HPI-S1	0.0475	0.1032	0.0593	0.0367	0.0003
KNU-SG	0.0678	0.0980	0.0702	0.0465	0.0003
c-e-50	0.0493	0.0662	0.0488	0.0345	0.0001
c-50	0.0520	0.0772	0.0530	0.0360	0.0001
c-idf-qe-1	0.0414	0.0574	0.0427	0.0326	0.0001
c-f-200	0.0485	0.0685	0.0484	0.0299	0.0001

Table 8: Results for snippet retrieval in batch 3 of phase A of Task 5b.

System	Mean Precision	Mean Recall	Mean F-measure	MAP	GMAP
ustb-prir4	0.1707	0.4787	0.2200	0.1143	**0.0066**
ustb-prir1	0.1680	0.4750	0.2155	0.1108	0.0060
fdu2	0.1645	0.4628	0.2135	0.0976	0.0059
ustb-prir2	0.1737	0.4754	0.2220	0.1134	0.0059
ustb-prir3	0.1620	**0.4803**	0.2111	**0.1157**	0.0050
fdu	0.1615	0.4475	0.2120	0.1021	0.0049
testtext	0.1610	0.4690	0.2087	0.1138	0.0048
fdu4	0.1420	0.4310	0.1856	0.0926	0.0044
fdu3	0.1390	0.4098	0.1809	0.0976	0.0031
UNCC System 1	**0.2317**	0.3340	**0.2322**	0.0825	0.0009
fdu5	0.1060	0.2461	0.1298	0.0737	0.0007
Olelo	0.1327	0.2444	0.1481	0.0658	0.0005
HPI-S1	0.0823	0.2152	0.0997	0.0464	0.0005
KNU-SG	0.0730	0.2149	0.0967	0.0521	0.0005
c-e-50	0.0720	0.1921	0.0861	0.0547	0.0003
c-50	0.0720	0.1921	0.0861	0.0547	0.0003
c-idf-qe-1	0.0720	0.1921	0.0861	0.0547	0.0003
c-f-200	0.0720	0.1921	0.0861	0.0547	0.0003

Table 9: Results for document retrieval in batch 3 of phase A of Task 5b.

outcome. For example, one can see that the *UNCC System 1* performed the best on some unordered measures, namely mean precision and f-measure, however using MAP or GMAP to consider the order of retrieved elements, it is out preformed by other systems, such as the *ustb-prir*. Additionally, the combination of some of these approaches seem like a promising direction for future research.

Phase B: In phase B of Task 5b the systems were asked to produce exact and ideal answers. For ideal answers, the systems will eventually be ranked according to manual evaluation by the BioASQ experts (Balikas et al., 2013). Regarding exact answers[5], the systems were ranked according to accuracy for the yes/no questions, mean reciprocal rank (MRR) for the factoids and mean

---

[5]For summary questions, no exact answers are required

System	Yes/No	Factoid			List		
	Accuracy	Strict Acc.	Lenient Acc.	MRR	Precision	Recall	F-measure
Lab Zhu,Fudan Univer	0.5517	0.1818	0.3030	0.2298	**0.3608**	0.4231	**0.3752**
LabZhu,FDU	0.5517	0.2424	0.3636	0.2904	0.3608	0.4231	**0.3752**
LabZhu-FDU	0.5517	0.2727	0.3939	0.3207	0.3608	0.4231	**0.3752**
Deep QA (ensemble)	0.5517	**0.3030**	**0.4545**	**0.3606**	0.2833	0.3436	0.2927
Deep QA (single)	0.5517	0.2424	0.3939	0.2965	0.2254	0.3564	0.2419
Oaqa-5b	**0.6552**	0.1515	0.1818	0.1667	0.1252	**0.5353**	0.1909
Oaqa 5b	0.6207	0.0909	0.1212	0.1061	0.1165	0.4615	0.1792
Oaqa5b-tfidf	0.6207	0.0909	0.1212	0.1061	0.1165	0.4615	0.1792
LabZhu-FDU	0.5517	0.0909	0.1818	0.1313	0.1239	0.3077	0.1692
Lab Zhu ,Fdan Univer	0.5517	0.1212	0.2121	0.1591	0.1143	0.3077	0.1599
sarrouti	0.6207	0.0909	0.1212	0.0970	0.1077	0.2013	0.1369
Basic QA pipline	0.5517	0.0606	0.1818	0.1035	0.0769	0.1462	0.0967
SemanticRole Labeling	0.5517	0.0303	0.0606	0.0379	0.0846	0.1122	0.0943
fa1	0.5517	0.0909	0.1818	0.1187	0.0564	0.1333	0.0718
Olelo	0.5517	0.0000	0.0606	0.0253	0.0513	0.0513	0.0513
Olelo-GS	0.5172	-	-	-	0.0513	0.0513	0.0513
L2PS - Relations	0.5172	0.0303	0.0303	0.0303	0.0371	0.1667	0.0504
L2PS - DeepQA	0.5172	0.0000	0.0303	0.0061	0.0207	0.2423	0.0338
L2PS	0.5172	-	-	-	0.0192	0.0513	0.0280
Simple system	0.5517	-	-	-	-	-	-
fa2	0.5517	0.0303	0.0606	0.0404	-	-	-
fa3	0.5517	0.0303	0.0909	0.0465	-	-	-
Using NNR	0.5517	-	-	-	-	-	-
Using regression	0.5517	-	-	-	-	-	-
Trivial baseline	0.5517	-	-	-	-	-	-
BioASQ-Baseline	0.4828	0.0303	0.1212	0.0682	0.1624	0.4276	0.2180

Table 10: Results for batch 4 for exact answers in phase B of Task 5b.

F-measure for the list questions. Table 10 shows the results for exact answers for the fourth batch of task 5b. The symbol (-) is used when systems don't provide exact answers for a particular type of question. The full results of phase B of task 5b are available online[6].

From the results presented in Table 10, it can be seen that systems achieve high scores in the yes/no questions. This was especially in the first batches, where a high imbalance in yes-no classes leaded to trivial baseline solutions being very strong. This was amended in the later batches, as shown in the table for batch 4, where the best systems outper-

form baseline approaches.

On the other hand, the performance in factoid and list questions indicates that there is more room for improvement in these types of answer.

4.3 Task 5c

Regarding the evaluation of Task 5c and taking into account the fact that only a subset of grant IDs and agencies mentioned in the full text were included in the ground truth data sets, both for training and testing, micro-recall was the evaluation measure used for all three sub-tasks. This means that each system was assigned a micro-recall score for grant IDs, agencies and full-grants independently and the top-two contenders for each sub-

---

[6] http://participants-area.bioasq.org/results/5b/phaseB/

System	Grant ID MR	Grant Agency MR	Full-Grant MR
Simple-ML2	**0.9750**	0.9900	**0.9526**
Simple-ML	0.9702	**0.9907**	0.9523
simpleSystem	0.9684	0.9890	0.9505
Simple-Regex2	0.9550	0.9847	0.9416
Gallen	0.9498	0.9862	0.9412
Hippocrates	0.9491	0.9859	0.9409
Simple-Regex	0.9530	0.9844	0.9397
Asclepius	0.9472	0.9859	0.9390
DZG1	0.9232	0.9122	0.8443
DZG-agency	0.0000	0.8829	0.0000
DZG-grants	0.9235	0.0000	0.0000
BIOASQ Filtering	0.8167	0.8312	0.7174

Table 11: Micro Recall (MR) results on the test set of Task 5c.

task were selected as winners.

The results of the participating systems can be seen in Table 11. Firstly, it can be seen that the grant ID extraction task is harder compared to the agency extraction. Moreover, the overall performance of the participants was very good, and certainly better than the baseline system. This indicates that the currently deployed techniques can be improved and as discussed in section 3.3, this can be done through the use of multiple methodologies. Finally, these results, despite being obtained on a filtered subset of the data available, could serve as a springboard to enhance and redeploy the currently implemented systems.

5 Conclusion

In this paper, an overview of the fifth BioASQ challenge is presented. The challenge consisted of three tasks: semantic indexing, question answering and funding information extraction. Overall, as in previous years, the best systems were able to outperform the strong baselines provided by the organizers. This suggests that advances over the state of the art were achieved through the BioASQ challenge but also that the benchmark in itself is challenging. Consequently, we believe that the challenge is successfully towards pushing the research frontier in on biomedical information systems.

In future editions of the challenge, we aim to provide even more benchmark data derived from a community-driven acquisition process and design a multi-batch scenario for Task 5c similar to the other tasks. Finally, as a concluding remark, it is worth mentioning that the increase in challenge participation this year[7] highlights the healthy growth of the BioASQ community, gathering attention from different teams around the globe and constituting a reference point for biomedical semantic indexing and question answering.

Acknowledgments

The fifth edition of BioASQ is supported by a conference grant from the NIH/NLM (number 1R13LM012214-01) and sponsored by the Atypon Systems inc. BioASQ is grateful to NLM for providing baselines for tasks 5a and 5c and the CMU team for providing the baselines for task 5b. Finally, we would also like to thank all teams for their participation.

References

Georgios Balikas, Ioannis Partalas, Aris Kosmopoulos, Sergios Petridis, Prodromos Malakasiotis, Ioannis Pavlopoulos, Ion Androutsopoulos, Nicolas Baskiotis, Eric Gaussier, Thierry Artieres, and Patrick Gallinari. 2013. Evaluation framework specifications. Project deliverable D4.1, UPMC.

Danqi Chen and Christopher D Manning. 2014. A fast and accurate dependency parser using neural networks. In *EMNLP*. pages 740–750.

Michael Collins. 2002. Discriminative training methods for hidden markov models: Theory and experiments with perceptron algorithms. In *Proceedings of the ACL-02 conference on Empirical methods in natural language processing-Volume 10*. Association for Computational Linguistics, pages 1–8.

[7]In BioASQ4, 6 teams participated in task 4a with 16 Systems and 11 teams in task 4b with 25 systems.

Janez Demsar. 2006. Statistical comparisons of classifiers over multiple data sets. *Journal of Machine Learning Research* 7:1–30.

Jenny Rose Finkel, Trond Grenager, and Christopher Manning. 2005. Incorporating non-local information into information extraction systems by gibbs sampling. In *Proceedings of the 43rd annual meeting on association for computational linguistics*. Association for Computational Linguistics, pages 363–370.

Jongwoo Kim, Daniel X Le, and George R Thoma. 2009. Inferring grant support types from online biomedical articles. In *Computer-Based Medical Systems, 2009. CBMS 2009. 22nd IEEE International Symposium on*. IEEE, pages 1–6.

Aris Kosmopoulos, Ioannis Partalas, Eric Gaussier, Georgios Paliouras, and Ion Androutsopoulos. 2013. Evaluation Measures for Hierarchical Classification: a unified view and novel approaches. *CoRR* abs/1306.6802. http://arxiv.org/pdf/1306.6802v2.

Ke Liu, Shengwen Peng, Junqiu Wu, Chengxiang Zhai, Hiroshi Mamitsuka, and Shanfeng Zhu. 2015. Meshlabeler: improving the accuracy of large-scale mesh indexing by integrating diverse evidence. *Bioinformatics* 31(12):i339–i347.

Prodromos Malakasiotis, Emmanouil Archontakis, Ion Androutsopoulos, Dimitrios Galanis, and Harris Papageorgiou. 2015. Biomedical question-focused multi-document summarization: Ilsp and aueb at bioasq3. In *CLEF (Working Notes)*.

Donald Metzler and W Bruce Croft. 2005. A markov random field model for term dependencies. In *Proceedings of the 28th annual international ACM SIGIR conference on Research and development in information retrieval*. ACM, pages 472–479.

Diego Molla. 2017. Macquarie university at bioasq 5b query-based summarisation techniques for selecting the ideal answers. In *Proceedings BioNLP 2017*.

James G. Mork, Dina Demner-Fushman, Susan C. Schmidt, and Alan R. Aronson. 2014. Recent enhancements to the nlm medical text indexer. In *Proceedings of Question Answering Lab at CLEF*.

E Papagiannopoulou, Y Papanikolaou, D Dimitriadis, S Lagopoulos, G Tsoumakas, M Laliotis, N Markantonatos, and I Vlahavas. 2016. Large-scale semantic indexing and question answering in biomedicine. *ACL 2016* page 50.

Shengwen Peng, Ronghui You, Hongning Wang, Chengxiang Zhai, Hiroshi Mamitsuka, and Shanfeng Zhu. 2016. Deepmesh: deep semantic representation for improving large-scale mesh indexing. *Bioinformatics* 32(12):i70–i79.

Shengwen Peng, Ronghui You, Zhikai Xie, Yanchun Zhang, and Shanfeng Zhu. 2015. The fudan participation in the 2015 bioasq challenge: Large-scale biomedical semantic indexing and question answering. In *CEUR Workshop Proceedings*. CEUR Workshop Proceedings, volume 1391.

Pranav Rajpurkar, Jian Zhang, Konstantin Lopyrev, and Percy Liang. 2016. Squad: 100, 000+ questions for machine comprehension of text. *CoRR* abs/1606.05250. http://arxiv.org/abs/1606.05250.

Adwait Ratnaparkhi. 1998. *Maximum entropy models for natural language ambiguity resolution*. Ph.D. thesis, University of Pennsylvania.

Francisco J Ribadas, Luis M De Campos, Víctor M Darriba, and Alfonso E Romero. 2014. Cole and utai participation at the 2014 bioasq semantic indexing challenge. In *Proceedings of the CLEF BioASQ Workshop*. Citeseer, pages 1361–1374.

Stephen E Robertson, Steve Walker, Susan Jones, Micheline M Hancock-Beaulieu, Mike Gatford, et al. 1995. Okapi at trec-3. *Nist Special Publication Sp* 109:109.

Frederik Schulze, Ricarda Schüler, Tim Draeger, Daniel Dummer, Alexander Ernst, Pedro Flemming, Cindy Perscheid, and Mariana Neves. 2016. Hpi question answering system in bioasq 2016. In *Proceedings of the Fourth BioASQ workshop at the Conference of the Association for Computational Linguistics*. pages 38–44.

Michael A Tanenblatt, Anni Coden, and Igor L Sominsky. 2010. The conceptmapper approach to named entity recognition. In *LREC*.

Grigorios Tsoumakas, Manos Laliotis, Nikos Markontanatos, and Ioannis Vlahavas. 2013. Large-Scale Semantic Indexing of Biomedical Publications. In *1st BioASQ Workshop: A challenge on large-scale biomedical semantic indexing and question answering*.

Dirk Weissenborn, Georg Wiese, and Laura Seiffe. 2017. Fastqa: A simple and efficient neural architecture for question answering. *arXiv preprint arXiv:1703.04816* .

Zi Yang, Niloy Gupta, Xiangyu Sun, Di Xu, Chi Zhang, and Eric Nyberg. 2015. Learning to answer biomedical factoid & list questions: Oaqa at bioasq 3b. In *CLEF (Working Notes)*.

Zi Yang, Yue Zhou, and Nyberg Eric. 2016. Learning to answer biomedical questions: Oaqa at bioasq 4b. *ACL 2016* page 23.

Ilya Zavorin, James G Mork, and Dina Demner-Fushman. 2016. Using learning-to-rank to enhance nlm medical text indexer results. *ACL 2016* page 8.

Xiaoli Zhang, Jie Zou, Daniel X Le, and George Thoma. 2009. A semi-supervised learning method to classify grant support zone in web-based medical articles. In *IS&T/SPIE Electronic Imaging*. International Society for Optics and Photonics, pages 72470W–72470W.

Tackling Biomedical Text Summarization: OAQA at BioASQ 5B

Khyathi Raghavi Chandu[1] Aakanksha Naik[1] Aditya Chandrasekar[1] Zi Yang[1]
Niloy Gupta[2] Eric Nyberg[1]

Language Technologies Institute, Carnegie Mellon University
[1]{kchandu,anaik,adityac,ziy,ehn}@cs.cmu.edu
[2]niloygupta@gmail.com

Abstract

In this paper, we describe our participation in phase B of task 5b of the fifth edition of the annual BioASQ challenge, which includes answering factoid, list, yes-no and summary questions from biomedical data. We describe our techniques with an emphasis on ideal answer generation, where the goal is to produce a relevant, precise, non-redundant, query-oriented summary from multiple relevant documents. We make use of extractive summarization techniques to address this task and experiment with different biomedical ontologies and various algorithms including agglomerative clustering, Maximum Marginal Relevance (MMR) and sentence compression. We propose a novel word embedding based tf-idf similarity metric and a soft positional constraint which improve our system performance. We evaluate our techniques on test batch 4 from the fourth edition of the challenge. Our best system achieves a ROUGE-2 score of 0.6534 and ROUGE-SU4 score of 0.6536.

1 Introduction

In recent years, there has been a huge surge in the number of biomedical articles being deposited online. The National Library of Medicine (NLM) provides MEDLINE, a gigantic database of 23 million references to biomedical journal papers. Approximately 200,000 articles [1] from this database have been cited since 2015. The rapid growth of information in this centralized repository makes it difficult for medical researchers to manually find an *exact answer* for a question

or to summarize the enormous content to answer a query. The problem of extracting *exact answers* for factoid questions from this data is being studied extensively, resulting in the development of several techniques including inferencing (Moldovan et al., 2002), noisy-channel transformation (Echihabi and Marcu, 2003) and exploitation of resources like WordNet (Lin and Hovy, 2003). However, recent times have also seen an interest in developing *ideal answer* generation systems which can produce relevant, precise, non-repetitive and readable summaries for biomedical questions (Tsatsaronis et al., 2015). A query based summarization system called "BioSQUASH" (Shi et al., 2007) uses domain specific ontologies like the Unified Medical Language System (UMLS) (Schuyler et al., 1993) to create a conceptual model for sentence ranking. Experiments with biomedical ontology based concept expansion and weighting techniques were conducted, where the strength of the semantic relationships between concepts was used as a similarity metric for sentence ranking (Chen and Verma, 2006). Similar methods (Yenala et al., 2015; Weissenborn et al., 2013) are used for this task where the difference lies in query similarity ranking methods.

This paper describes our efforts in creating a system that can provide ideal answers for biomedical questions. More specifically, we develop a system which can answer the kinds of biomedical questions present in the dataset for the BioASQ challenge (Tsatsaronis et al., 2015), which is a challenge on large-scale biomedical semantic indexing and question answering. We participate in Phase B of Task 5b (biomedical question-answering) for the 2016 edition of this challenge comprising of factoid, yes/no, list and summary type questions. We develop a system for biomedical summarization using MMR and clustering based techniques. To answer factoid, list and

---

[1]https://www.nlm.nih.gov/bsd/medline_lang_distr.html

58

Proceedings of the BioNLP 2017 workshop, pages 58–66,
Vancouver, Canada, August 4, 2017. ©2017 Association for Computational Linguistics

yes/no questions, we use one of the winning systems (Yang et al., 2016) from the 2015 edition of the BioASQ challenge, open-sourced after the conclusion of the challenge [2].

We build on standard techniques such as Maximal Marginal Relevance (Carbonell and Goldstein, 1998) and Sentence Compression (Filippova et al., 2015) and incorporate domain-specific knowledge using biomedical ontologies such as the UMLS metathesaurus and SNOMEDCT (Stearns et al., 2001) to build an ideal answer generator for biomedical questions. We also experiment with several similarity metrics such as jaccard similarity and a novel word embedding based tf-idf (w2v tf-idf) similarity metric within our system. We evaluate the performance of our system on the dataset for test batch 4 of the fourth edition of the challenge and report our system performance on ROUGE-2 and ROUGE-SU4 (Lin and Hovy, 2003), which are the standard metrics used for official evaluation in the BioASQ challenge. Our best system achieves ROUGE-2 and ROUGE-SU4 scores of 0.6534 and 0.6536 respectively on test batch 4 for task 4b when evaluated on *BioASQ Oracle*[3]. Various configurations and similarity metrics, granularity and algorithms selection enabled us to secure top 1,2,3 in test batch 4 and top 1,2,3,4 in test batch 5 on automatic evaluation metrics of ROUGE-2 and ROUGE-SU4, from our participation in Task 5b of ideal answer generation.

The rest of the paper is organized as follows: Section 2 describes the datasets used. In section 3, we describe our summarization pipeline, while section 4 gives a brief overview of the system used for factoid, list and yes-no questions. Section 5 presents the evaluation results of our summarization system and our observations about various system configurations. Section 6 presents a comparative qualitative error analysis of some of our system configurations. Section 7 concludes and describes future work in this area.

2 Dataset

The training data for Phase B of task 5b provides biomedical questions, where each question is associated with question type, urls of relevant PubMed articles and relevant snippets from those articles. This dataset consists of 1,799 questions.

Though our ideal answer generation system is unsupervised, we use a brief manual inspection of the training data for this edition of the challenge to make an informed choice of hyperparameters for the algorithms used by our system.

To develop an ideal answer generator which can produce query-oriented summaries for each question, we can adopt one of two popular approaches: extractive or abstractive. Extractive summarization techniques choose sentences from relevant documents and combine them to form a summary. Abstractive summarization methods use relevant documents to create a semantic representation of the knowledge from these documents and then generate a summary using reasoning and natural language generation techniques. Brief analysis on a randomly sampled subset from the training data shows us that most of the sentences in the gold ideal answers are present either in the relevant snippets or relevant abstracts of PubMed articles. Hence we perform extractive summarization. We also observe an interesting ordering trend among relevant snippets which is used to develop a positional constraint. Adding this positional constraint to our similarity metrics gives us a slight boost in performance. We explain the intuition behind this idea in more detail in section 3.1.2.

For evaluation, we use the dataset from test batch 4 of the fourth edition of the BioASQ challenge which consists of 100 questions.

3 Summarization Pipeline

In this section, we describe our system pipeline for the ideal answer generation task which mainly comprises of three stages: *question-sentence relevance ranker*, *sentence selection* and *sentence tiling*. Each stage has multiple configurations depending upon various choices for algorithms, concept expansion and similarity metrics. Figure 1 shows the overall architecture of our system and also briefly mentions various algorithms used in each stage. We describe these stages and choices in more detail in subsequent sections.

3.1 Question-Sentence Relevance ranker:

In this phase, we retrieve a list of candidate sentences from gold abstracts and snippets provided for each question and compute relevance scores with respect to the question for these sentences. We can choose from several similarity metrics, biomedical ontologies and different granularities

---

[2]https://github.com/oaqa/bioasq
[3]http://participants-area.bioasq.org/oracle/

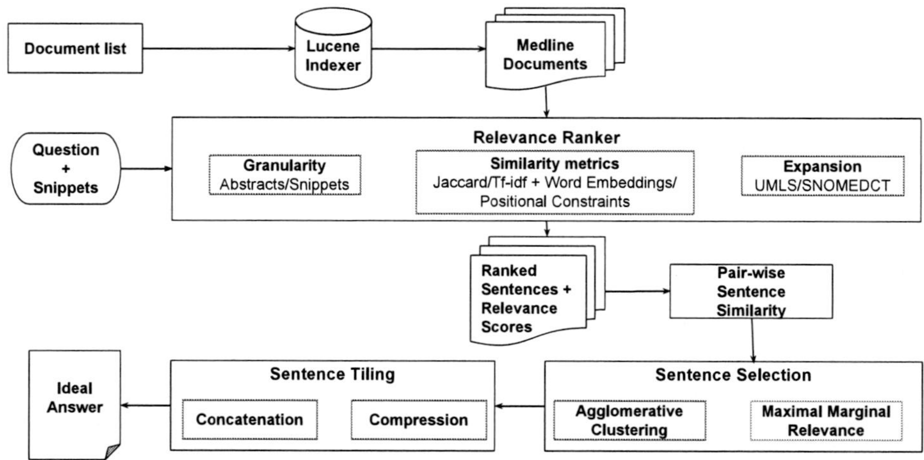

Figure 1: System pipeline for Ideal Answer Generation (with configuration choices)

for sentence scoring in this stage.

3.1.1 Granularity for Candidate Sentence Extraction

The training data provided for the BioASQ task contains a list of PubMed IDs of gold relevant documents from NLM, along with gold relevant snippets from these documents, for each question. Since, the training data only contains PubMed IDs of relevant documents, we extract complete abstract text for these documents by first indexing all Medline abstracts [4] via Lucene and then retrieving relevant documents based on PubMed IDs.

We now have two choices of granularity for candidate sentence extraction: using entire abstract texts from relevant documents or using only relevant snippets. We experiment with both possibilities. However, since relevant snippets for each question are a subset of abstract texts, which are highly relevant to the question, leveraging this insight and using only snippets for candidate sentence extraction gives us better performance, as we see from the results in Section 5.

3.1.2 Similarity metrics

The performance of both, the relevance ranker and the sentence selection phase (which is the following phase in the pipeline), depends on the similarity metrics used to capture question-sentence relevance and sentence-sentence similarity. In

this section, we describe various similarity metrics which we experiment with.

Jaccard similarity: For each sentence, its relevance with respect to the question is computed as the Jaccard index between the sets containing all words occurring in the question and the sentence. This is the simplest metric which captures surface (word-level) similarity between the question and the sentence. Including related concepts obtained by concept expansion in these word sets provides some measure of semantic overlap, but this technique is not very effective as we show in section 5.

Tf-idf based similarity with word embeddings: Using ontologies such as WordNet (for general English) and UMLS/ SNOMEDCT (for biomedical domain) for concept expansion to incorporate some semantics while computing sentence similarity, is not sufficient due to the unbounded nature of such ontologies. Hence, to assimilate semantic information in a more controlled manner, we use a novel similarity metric inspired by the widely-used tf-idf cosine similarity metric which incorporates semantic information by making use of word embeddings (Mikolov et al., 2013).

Let $\mathbf{W}$ represent the symmetric word-to-word similarity matrix and $\vec{a}$, $\vec{b}$ represent tf-idf vectors for the sentences. The similarity metric is defined as:

$$sim(\vec{a}, \vec{b}) = \frac{\vec{a}^T \mathbf{W} \vec{b}}{\sqrt{\vec{a}^T \mathbf{W} \vec{a}} \sqrt{\vec{b}^T \mathbf{W} \vec{b}}} \quad (1)$$

---

[4] https://www.nlm.nih.gov/databases/download/pubmed_medline.html

The word-to-word similarity matrix $\mathbf{W}$ is computed using cosine similarity between word embeddings for each word. We use word embeddings which have been pre-trained on PubMed, PMC and Wikipedia articles to incorporate domain knowledge [5].

Similarity function with positional constraints: As described in section 2, the data provided for each question contains a list of relevant abstracts of PubMed articles, as well as a list of relevant snippets extracted from these abstracts. The abstracts are ordered by relevance. Snippets on the other hand, are not ordered by relevance, but are ordered according to the abstracts that they are extracted from. Since the abstracts themselves are ordered by relevance, this gives an inherent discourse structure to the snippets. This observation motivates us to incorporate information about a snippet's position in the list into the similarity function to improve the summaries generated by our system. We first test this hypothesis using a simple baseline which gives the first snippet in the list as the summary for every question. This simple baseline is able to achieve good ROUGE scores as shown in Table 1. We experiment with two different ways of incorporating this constraint:

• **Hard positional constraint:** In this method, we enforce snippet position as a hard constraint. We achieve this by restricting the algorithm to select the first sentence of the summary from the first snippet (most relevant snippet) in the list. Remaining sentences can be selected from any snippet. This method does not have much improvement on our ROUGE scores as explained in section 5.

• **Soft positional constraint:** This method incorporates snippet position as a soft constraint by adding it to the similarity function. The augmented similarity function after incorporating snippet position is presented below:

$$positionalSim(q, s) = \alpha * sim(q, s) + \\ (1 - \alpha) * rank(s) \tag{2}$$

Here, q and s denote the question and sentence respectively; $sim(q, s)$ denotes a function which computes similarity between question and sentence (we experiment with Jaccard and tf-idf based similarities); $rank(s)$ denotes the boost

given to the sentence based on the position of the snippet to which it belongs and α is a weighting parameter. The value of $rank(s)$ for a sentence is computed as follows:

$$rank(s) = 1 - pos(s) \\ pos(s) = snippetPos(s)/\#snippets$$

Here, $snippetPos(s)$ denotes the position (index) of the snippet, to which the sentence belongs, in the list of relevant snippets. If a sentence belongs to multiple snippets, we consider the lowest index. $\#snippets$ denotes the number of relevant snippets for the current question. This positional boost gives higher weight to sentences with lower position values (since they occur earlier in the list) and returns a normalized value in the range 0-1, to ensure that it is comparable to the range of values produced by the similarity function. Adding this constraint boosts our ROUGE scores.

3.1.3 Biomedical Tools and Ontologies

We experiment with various biomedical tools and ontologies for concept expansion, in order to incorporate relations between concepts while computing similarity. To perform concept expansion, the first step is to identify biomedical concepts from a sentence. We choose the MetaMap concept identification tool and use a python wrapper, pymetamap[6] for this purpose. This API identifies biomedical concepts from a sentence and returns a Concept Unique Identification (CUI) for each concept. This CUI acts as a unique identifier for the concept which is shared across ontologies, i.e it can be used as an ID to retrieve the same concept from the UMLS ontology. After biomedical concepts are identified, we experiment with two ontologies for concept expansion: UMLS Metathesaurus and SNOMEDCT.

• **UMLS Metathesaurus:** The UMLS Metathesarus contains many types of relations for each biomedical concept. For our task, three relation types are of interest to us: 'RB' (broader relationship), 'RL' (similar or alike relationship) and 'RQ' (related and possibly synonymous relationship). However, none of the biomedical concepts identified from questions and sentences in

---

[5] These pre-trained word vectors are provided by `http://evexdb.org/pmresources/vec-space-models/`

[6] `https://github.com/AnthonyMRios/pymetamap`

our training dataset contained relations of the type 'RL' or 'RQ'. Hence we perform expansion for each biomedical concept by collecting all concepts linked to it by the 'RB' relation.

- **SNOMEDCT:** The SNOMEDCT ontology does not contain CUIs for biomedical concepts. Hence, we need to use a different technique to locate concepts in this ontology. In addition to CUI, pymetamap also provides a "preferred name" for each concept. We use this preferred name to perform a full-text search in the SNOMEDCT ontology. All concepts returned by this search are then considered to be related concepts and used for expansion. Using this ontology for concept expansion returns a much larger number of related concepts, due to the nature of our search (using fuzzy text search instead of precise identifiers).

We use these techniques to perform concept expansion on both questions and sentences from relevant snippets. In Section 6, we present the results of various system configurations with and without domain specific concept expansion.

3.2 Sentence Selection

In this stage, we want to select sentences for the final summary from candidate sentences extracted by the previous stage. Since the BioASQ task has a word limit of 200, we limit the number of sentences selected for the final summary to five. This sentence limit gives us good ROUGE scores across multiple system configurations.

The simplest way of performing sentence selection is to continue selecting the sentence with the highest relevance score with respect to the question, till the sentence limit is reached. However, sentences having high relevance with respect to the question may be semantically similar, thus introducing redundancy in the generated summary. We use two algorithms to combat this issue: agglomerative clustering based on sentence similarity and Maximum Marginal Relevance (MMR) (Carbonell and Goldstein, 1998). Both algorithms require effective similarity metrics to compute semantic similarity between sentences. We experiment with various similarity metrics described in section 3.1.2. We also experiment with concept expansion using multiple biomedical ontologies.

3.2.1 Agglomerative Clustering

Redundancy reduction via clustering is one of the techniques that was proposed for biomedical query-oriented summarization (Chen and Verma, 2006). In this technique, we create all possible sentence pairs from our set of candidate sentences and compute pair-wise similarities. We then perform agglomerative clustering on the sentences using these pair-wise similarity scores. Finally, we select one sentence from each cluster to generate the final summary, in such a way that the sentence having maximum question relevance score is selected from every cluster. The number of clusters is set to the maximum number of sentences we need in the final summary (five in this case). The intuition behind this technique is that agglomerative clustering forces semantically similar sentences to fall into the same cluster. Since we only select one sentence from each cluster in the end, we discard sentences which are highly similar to the selected ones.

3.2.2 Maximal Marginal Relevance

Maximal Marginal Relevance (Carbonell and Goldstein, 1998) is a widely-used summarization algorithm which was proposed to tackle the issue of redundancy while maintaining query relevance in summarization. This algorithm selects new sentences based on a combination of relevance score with respect to the question as well as similarity score with respect to the sentences which have already been selected for the final summary. Thus, this algorithm incorporates sentence similarity as a constraint, instead of explicitly clustering sentences.

3.3 Sentence Tiling

In the final stage, we combine all selected sentences to produce the final summary. The simplest way is to append all selected sentences while constraining summary length (because of the word-limit constraint for this task). We also experiment with an LSTM-based sentence compression method. We train a neural network based on a work done previously (Filippova et al., 2015) for sentence compression. We generate training data for this network by pairing sentences from abstract texts with their full text versions. Given that this dataset is too small to train the neural network, we add in training instances from existing sentence compression data-sets. Input to this model includes the word vector representation for a word

	Experiment	ROUGE-2	ROUGE-SU4
1	Clustering + Abstract texts (with average constraint)	0.2906	0.3138
2	Clustering + Snippets (with average constraint)	0.4314	0.4347
3	Clustering + Snippets (without average constraint)	0.5609	0.5632
4	Clustering + UMLS expansion	0.5488	0.5521
5	Clustering + SNOMEDCT expansion	0.5514	0.5586
6	Clustering + UMLS expansion + weighting	0.5402	0.5431
7	Clustering + SNOMEDCT expansion + weighting	0.5530	0.5588
8	Clustering + UMLS expansion + weighted normalization	0.5592	0.5632
9	Clustering + SNOMEDCT expansion + weighted normalization	0.5585	0.5650
10	MMR	0.6338	0.6296
11	MMR + w2v tf-idf similarity	0.6168	0.6126
12	First snippet baseline	0.3363	0.3308
13	MMR + Hard positional constraint + Jaccard similarity	0.6338	0.6296
14	MMR + Soft positional constraint + Jaccard similarity	0.6419	0.6410
15	Hard positional constraint + Jaccard similarity	0.6328	0.6254
16	Soft positional constraint + Jaccard similarity	0.6433	0.6429
17	**Soft positional constraint + w2v tf-idf similarity**	**0.6534**	**0.6536**
18	MMR + tf-idf similarity + LSTM compression	0.5689	0.5723

Table 1: ROUGE scores with different algorithms, ontologies and similarity metrics

and a binary value to indicate whether the previous word was included in the output sentence. Based on these inputs, the output of the model predicts whether the word should be deleted or not. Sentences generated after word deletion are concatenated together to generate the final summary. It is to be noted that this model does not require any linguistic features.

4 Overview of system for exact answer generation

To answer factoid, list and yes/no questions, we use the publicly available system (Yang et al., 2016), which builds on participation in 2015 (Yang et al., 2015). This system uses TmTool in place of UTS (unlike (Yang et al., 2015)) for concept identification as some of the constituent parsers of TmTool identify concepts based on morphological features instead of previously coded ontologies. Also, the c-value method is used to mine frequent multi-word concepts that might not have been identified by tools such as TmTool, MetaMap and LingPipe. The idea of reranking a candidate answer based on its similarity to other candidate answers is introduced in this system for list type questions. The intuition behind this approach is that all answers to a list type question should have the same semantic type and therefore, it is useful to increase the score of a low-ranked

candidate answer that has the same semantic type, and vice-versa.

Yes/No questions are answered using the technique of question inversion. The last biomedical concept present in the question is considered to be the *expected answer*. The concept mentions and tokens of the expected answer are removed from the question, which is then converted to a factoid type question. Candidate answers are generated for this factoid question using the snippets for the original question. The expected answer is then compared to the ranked list of candidate answers retrieved. The answer to the yes/no type question will be *yes* if the expected answer is among the top ranked candidate answers and *no* otherwise.

5 Evaluation and Discussion

We experiment with ideal answer generation using various system configurations which differ in similarity metrics, biomedical ontologies, sentence selection algorithms(clustering/MMR) and tiling algorithms used. The official evaluation for *ideal answers* includes manual evaluation by biomedical experts in the BioASQ team as well as automatic evaluation via ROUGE scores. To present comparable and standardized results, we run our system on the batch 4 dataset for Phase B of task 4b and get our results evaluated via the *BioASQ Oracle*. These results are shown in Table 1. We

Category	Question/Summary
Question	What is the effect that EZH2 has on chromatin?
Gold Ideal Answer	**Ez that catalyzes di- and trimethylation of histone H3 lysine 27 (H3K37me2/3), marks repressive to transcription.[1] The mammalian homologs Ezh1 and Ezh2 form similar PRC2 complexes but exhibit contrasting repressive roles.[2] PRC2-Ezh2 catalyzes H3K27me2/3 and its knockdown affects global H3K27me2/3 levels.[3]** EZH2 thus maintains chromatin in a repressive state.
Clustering + Snippets	During progenitor cell differentiation and ageing, PcG silencer EZH2 attenuates, causing loss of PRC binding and transcriptional activation of INK4b and INK4a Treg-specific ablation of Ezh2 resulted in spontaneous autoimmunity with reduced Foxp3(+) cells in non-lymphoid tissues and impaired resolution of experimental autoimmune encephalomyelitis We report that **the mammalian homologs Ezh1 and Ezh2 form similar PRC2 complexes but exhibit contrasting repressive roles.[2]** In this study, we found the inverse correlation between FOXP3 and Ezh2, an enzyme for histone H3K27 trimethylation (**H3K27me3**) and a central epigenetic regulator in cancer the PRC2-Ezh2 complex, which is bound to the myogenin (MyoG) promoter and muscle creatine kinase (mCK) enhancer in proliferating myoblasts, and the PRC2-Ezh1 complex, which replaces PRC2-Ezh2 on MyoG promoter in post-mitotic myotubes.
Clustering + UMLS expansion + weighted normalization	During progenitor cell differentiation and ageing, PcG silencer EZH2 attenuates, causing loss of PRC binding and transcriptional activation of INK4b and INK4a. **We report that the mammalian homologs Ezh1 and Ezh2 form similar PRC2 complexes but exhibit contrasting repressive roles.** [2] Treg-specific ablation of Ezh2 resulted in spontaneous autoimmunity with reduced Foxp3(+) cells in non-lymphoid tissues and impaired resolution of experimental **autoimmune encephalomyelitis Ez that catalyzes di- and trimethylation of histone H3 lysine 27 (H3K37me2/3) [3], marks repressive to transcription.[1]** the PRC2-Ezh2 complex, which is bound to the myogenin (MyoG) promoter and muscle creatine kinase (mCK) enhancer in proliferating myoblasts, and the PRC2-Ezh1 complex, which replaces PRC2-Ezh2 on MyoG promoter in post-mitotic myotubes.
Clustering + SNOMEDCT expansion + weighted normalization	During progenitor cell differentiation and ageing, PcG silencer EZH2 attenuates, causing loss of PRC binding and transcriptional activation of INK4b and INK4a Treg-specific ablation of Ezh2 resulted in spontaneous autoimmunity with reduced Foxp3(+) cells in non-lymphoid tissues and impaired resolution of experimental autoimmune encephalomyelitis We report that **the mammalian homologs Ezh1 and Ezh2 form similar PRC2 complexes but exhibit contrasting repressive roles.[2] Ez that catalyzes di- and trimethylation of histone H3 lysine 27 (H3K37me2/3), marks repressive to transcription.[1]** the PRC2-Ezh2 complex, which is bound to the myogenin (MyoG) promoter and muscle creatine kinase (mCK) enhancer in proliferating myoblasts, and the PRC2-Ezh1 complex, which replaces PRC2-Ezh2 on MyoG promoter in post-mitotic myotubes.
MMR	**Ezh1 and Ezh2 maintain repressive chromatin through different mechanisms.[1]** The chromatin-modifying enzyme Ezh2 is critical for the maintenance of regulatory T cell identity after activation. Treg-specific ablation of Ezh2 resulted in spontaneous autoimmunity with reduced Foxp3(+) cells in non-lymphoid tissues and impaired resolution of experimental autoimmune encephalomyelitis. the PRC2-Ezh2 complex, which is bound to the myogenin (MyoG) promoter and muscle creatine kinase (mCK) enhancer in proliferating myoblasts, and the PRC2-Ezh1 complex, which replaces PRC2-Ezh2 on MyoG promoter in post-mitotic myotubes. In this study, we found the inverse correlation between FOXP3 and Ezh2, an enzyme for histone H3K27 trimethylation (H3K27me3) and a central epigenetic regulator in cancer.
MMR + w2v tf-idf	**Ezh1 and Ezh2 maintain repressive chromatin through different mechanisms.[1]** In this study, we found the inverse correlation between FOXP3 and Ezh2, an enzyme for histone H3K27 trimethylation (H3K27me3) and a central epigenetic regulator in cancer. These studies reveal a critical role for Ezh2 in the maintenance of Treg cell identity during cellular activation. We report that **the mammalian homologs Ezh1 and Ezh2 form similar PRC2 complexes but exhibit contrasting repressive roles.[2]** The chromatin-modifying enzyme Ezh2 is critical for the maintenance of regulatory T cell identity after activation.
Soft constraint + w2v tf-idf	**Ezh1 and Ezh2 maintain repressive chromatin through different mechanisms.**[1] We report that **the mammalian homologs Ezh1 and Ezh2 form similar PRC2 complexes but exhibit contrasting repressive roles.[2] Ez that catalyzes di- and trimethylation of histone H3 lysine 27 (H3K37me2/3), marks repressive to transcription.** During progenitor cell differentiation and ageing, PcG silencer EZH2 attenuates, causing loss of PRC binding and transcriptional activation of INK4b and INK4a. the **PRC2-Ezh2** complex, which is bound to the myogenin (MyoG) promoter and muscle creatine kinase (mCK) enhancer in proliferating myoblasts, and the PRC2-Ezh1 complex, which replaces **PRC2-Ezh2** on MyoG promoter in post-mitotic myotubes.
MMR + w2v tf-idf + LSTM sentence compression	and **ezh2 maintain repressive chromatin through different mechanisms.**[1] this study , found the inverse correlation between foxp3 and ezh2 , an enzyme for histone h3k27 trimethylation (h3k27me3) and a central epigenetic regulator in cancer . prc2-ezh2 complex , which is bound to the myogenin (myog) promoter and muscle creatine kinase (mck) enhancer in proliferating myoblasts , and the prc2-ezh1 complex , which replaces prc2-ezh2 on myog promoter in post-mitotic myotubes .

Figure 2: Summaries generated with different techniques

obtain the best results among these configurations by using soft positional constraint with tf-idf based similarity on snippets.

The first three rows in Table 1 show our experiments with different granularities for sentence extraction. While using abstract texts for sentence selection, we observe that our clustering technique frequently puts sentences with low query relevance into the same clusters. Since our selection method picks one sentence from each cluster, some sentences with low query relevance from these "bad" clusters are also selected for the final summary. To solve this issue, we imposed a constraint which filtered out sentences having a lower-than-average relevance score with respect to the question before clustering. We also tried adding this constraint while using relevant snippets, but this reduced our scores, because sentences from snippets are already relevant to the question and we end up discarding important information by fil-

tering. We also observed that *switching granularity from abstract texts to relevant snippets significantly boosted the ROUGE scores*. Hence all subsequent experiments (rows 4-18) use snippets for sentence extraction.

Rows 4-9 show our experiments with concept expansion using various biomedical ontologies and weighting techniques. We use the following weighting technique: while calculating similarity, words from the original question and sentences carry a weight of 1, while words obtained added after concept expansion carry a weight of 0.5. *We do not observe significant gains using concept expansion.* The unbounded nature of concept expansion hurts our performance and so we refrain from using this technique in further experiments. Row 10 shows our experiment using MMR for sentence selection instead of clustering. *MMR provides a significant boost in ROUGE score.* Row 11 shows our experiment with the *w2v tf-idf based similar-*

ity metric instead of Jaccard similarity, which *decreases our ROUGE scores slightly, but is still better than previous system configurations.* Row 12 shows the scores of a baseline system which returns the first snippet from the list, which is quite high, *validating our assumption that snippet position is an important factor.* Rows 13-17 shows our experiments with different ways of adding positional constraints described in section 3.1.2. While *using a hard constraint does not show much improvement, soft positional constraint gives a slight boost.* Results with and without MMR for this metric are nearly comparable. *Soft constraint gives a huge boost when used with w2v tf-idf based similarity.* Row 18 shows our experiment *adding LSTM-based compression* on top of MMR with w2v tf-idf based similarity, which *reduces our scores.* Row 17 is the system configuration with the highest ROUGE score on our dataset, which uses soft positional constraint with w2v tf-idf similarity.

6 Comparative Qualitative Error Analysis

Figure 2 presents ideal answers generated by some of our system configurations for a randomly selected summary question from Task 4b Phase B data to provide a comparative qualitative error analysis. Each sentence in the ideal gold answer is indexed with a number as shown in the figure. We perform a relative analysis of the extent of information captured by a selected subset of system configurations from Table 1.

The sentence indexed [1] in the gold ideal answer is present word-for-word in summaries created by two configurations: Clustering + SNOMEDCT expansion + weighted normalization and Soft constraint + w2v tf-idf. Clustering + UMLS expansion + weighted normalization contains a longer version of this sentence. We also observe that this sentence does not contain any of the terms from the original question. Hence, summaries generated by all configurations using only Jaccard similarity (Clustering + Snippets, MMR) do not contain this sentence since there is no surface-level similarity. However, methods which incorporate some semantic information via word embeddings (w2v tf-idf similarity) or concept expansion (UMLS/ SNOMEDCT) include this sentence in the final summary, which shows that incorporating semantic information is important to

bridge the vocabulary gap in some situations.

The sentence indexed [2] in the gold answer is present in summaries generated by most of the configurations as shown but with extra phrases such as 'We report that' at the beginning of the sentence. Though the presence of such words does not have a major impact on automatic scores like ROUGE, it influences the manual evaluation which also judges summary readability. However, the LSTM-based compression method removes these words via deletion. We observe that this sentence contains the concept "Ezh2" which is also present in the question. Hence, some configurations which use surface-level similarity (Clustering+Snippets) also pick this sentence for the final summary. But this sentence is not present in the summary generated by the MMR + snippets configuration. This happens because many sentences selected by the algorithm already contain the concept "Ezh2" and so this sentence is excluded due to its similarity to already selected sentences.

7 Conclusion and Future Work

In this paper, we present a system for query-oriented summary generation. Our comparison of MMR and agglomerative clustering-based techniques shows that while clustering selects distinct sentences, it is unable to select sentences with high query relevance. This can be improved by learning hyperparameters like number of clusters and number of sentences to be selected from each cluster based on the type of question. We plan to investigate this in the future. We find that unbounded concept expansion hurts our system scores. LSTM-based compression also hurts our system scores and we need to investigate upon this in the future to select the optimal parameters for compression ratio in order to maximize recall and precision. We also find that incorporating word embedding based tf-idf similarity along with soft positional constraints outperforms surface level word similarity with soft positional constraints. This is because the former captures both semantic information of the content as well as relevance to query based on sentence position.

Acknowledgments

This research was supported in parts by grants from Accenture PLC (PI: Anatole Gershman), NSF IIS 1546393 and NHLBI R01 HL122639.

References

Jaime Carbonell and Jade Goldstein. 1998. The use of mmr, diversity-based reranking for reordering documents and producing summaries. In *Proceedings of the 21st annual international ACM SIGIR conference on Research and development in information retrieval*. ACM, pages 335–336.

Ping Chen and Rakesh Verma. 2006. A query-based medical information summarization system using ontology knowledge. In *Computer-Based Medical Systems, 2006. CBMS 2006. 19th IEEE International Symposium on*. IEEE, pages 37–42.

Abdessamad Echihabi and Daniel Marcu. 2003. A noisy-channel approach to question answering. In *Proceedings of the 41st Annual Meeting on Association for Computational Linguistics-Volume 1*. Association for Computational Linguistics, pages 16–23.

Katja Filippova, Enrique Alfonseca, Carlos A Colmenares, Lukasz Kaiser, and Oriol Vinyals. 2015. Sentence compression by deletion with lstms. In *EMNLP*. pages 360–368.

Chin-Yew Lin and Eduard Hovy. 2003. Automatic evaluation of summaries using n-gram co-occurrence statistics. In *Proceedings of the 2003 Conference of the North American Chapter of the Association for Computational Linguistics on Human Language Technology-Volume 1*. Association for Computational Linguistics, pages 71–78.

Tomas Mikolov, Ilya Sutskever, Kai Chen, Greg S Corrado, and Jeff Dean. 2013. Distributed representations of words and phrases and their compositionality. In *Advances in neural information processing systems*. pages 3111–3119.

Dan I Moldovan, Sanda M Harabagiu, Roxana Girju, Paul Morarescu, V Finley Lacatusu, Adrian Novischi, Adriana Badulescu, and Orest Bolohan. 2002. Lcc tools for question answering. In *TREC*.

Peri L Schuyler, William T Hole, Mark S Tuttle, and David D Sherertz. 1993. The umls metathesaurus: representing different views of biomedical concepts. *Bulletin of the Medical Library Association* 81(2):217.

Zhongmin Shi, Gabor Melli, Yang Wang, Yudong Liu, Baohua Gu, Mehdi M Kashani, Anoop Sarkar, and Fred Popowich. 2007. Question answering summarization of multiple biomedical documents. In *Advances in Artificial Intelligence*, Springer, pages 284–295.

Michael Q Stearns, Colin Price, Kent A Spackman, and Amy Y Wang. 2001. Snomed clinical terms: overview of the development process and project status. In *Proceedings of the AMIA Symposium*. American Medical Informatics Association, page 662.

George Tsatsaronis, Georgios Balikas, Prodromos Malakasiotis, Ioannis Partalas, Matthias Zschunke, Michael R Alvers, Dirk Weissenborn, Anastasia Krithara, Sergios Petridis, Dimitris Polychronopoulos, et al. 2015. An overview of the bioasq large-scale biomedical semantic indexing and question answering competition. *BMC bioinformatics* 16(1):138.

Dirk Weissenborn, George Tsatsaronis, and Michael Schroeder. 2013. Answering factoid questions in the biomedical domain. *BioASQ@ CLEF* 1094.

Zi Yang, Niloy Gupta, Xiangyu Sun, Di Xu, Chi Zhang, and Eric Nyberg. 2015. Learning to answer biomedical factoid & list questions: Oaqa at bioasq 3b. In *CLEF (Working Notes)*.

Zi Yang, Yue Zhou, and Eric Nyberg. 2016. Learning to answer biomedical questions: Oaqa at bioasq 4b. *ACL 2016* page 23.

Harish Yenala, Avinash Kamineni, Manish Shrivastava, and Manoj Kumar Chinnakotla. 2015. Iiith at bioasq challange 2015 task 3b: Bio-medical question answering system. In *CLEF (Working Notes)*.

Macquarie University at BioASQ 5b – Query-based Summarisation Techniques for Selecting the Ideal Answers

Diego Mollá

Department of Computing
Macquarie University
Sydney, Australia
`diego.molla-aliod@mq.edu.au`

Abstract

Macquarie University's contribution to the BioASQ challenge (Task 5b Phase B) focused on the use of query-based extractive summarisation techniques for the generation of the ideal answers. Four runs were submitted, with approaches ranging from a trivial system that selected the first n snippets, to the use of deep learning approaches under a regression framework. Our experiments and the ROUGE results of the five test batches of BioASQ indicate surprisingly good results for the trivial approach. Overall, most of our runs on the first three test batches achieved the best ROUGE-SU4 results in the challenge.

1 Introduction

The main goal of query-focused multi-document summarisation is to summarise a collection of documents from the point of view of a particular query. In this paper we compare the use of various techniques for query-focused summarisation within the context of the BioASQ challenge. The BioASQ challenge (Tsatsaronis et al., 2015) started in 2013 and it comprises various tasks centred on biomedical semantic indexing and question answering. The fifth run of the BioASQ challenge (Nentidis et al., 2017), in particular, had three tasks:

- BioASQ 5a: Large-scale online biomedical semantic indexing.

- BioASQ 5b: Biomedical semantic question answering. This task had two phases:

  - Phase A: Identification of relevant information.
  - Phase B: Question answering.

- BioASQ 5c: Funding information extraction from biomedical literature.

The questions used in BioASQ 5b were of three types: yes/no, factoid, list, and summary. Submissions to the challenge needed to provide an exact answer and an ideal answer. Figure 1 shows examples of exact and ideal answers for each type of question. We can see that the ideal answers are full sentences that expand the information provided by the exact answers. These ideal answers could be seen as the result of query-focused multi-document summarisation. We therefore focused on Task 5b Phase B, and in that phase we did not attempt to provide exact answers. Instead, our runs provided the ideal answers only.

In this paper we will describe the techniques and experiment results that were most relevant to our final system runs. Some of our runs were very simple, yet our preliminary experiments revealed that they were very effective and, as expected, the simpler approaches were much faster than the more complex approaches.

Each of the questions in the BioASQ test sets contained the text of the question, the question type, a list of source documents, and a list of relevant snippets from the source documents. We used this information, plus the source documents which are PubMed abstracts accessible using the URL provided in the test sets.

Overall, the summarisation process of our runs consisted of the following two steps:

1. Split the input text (source documents or snippets) into candidate sentences and score each candidate sentence.

2. Return the n sentences with highest score.

The value of n was determined empirically and it depended on the question type, as shown in Table 1.

Proceedings of the BioNLP 2017 workshop, pages 67–75,
Vancouver, Canada, August 4, 2017. ©2017 Association for Computational Linguistics

yes/no Does Apolipoprotein E (ApoE) have anti-inflammatory activity?

- **Exact answer**: yes
- **Ideal answer**: Yes. ApoE has anti-inflammatory activity

factoid Which type of lung cancer is afatinib used for?

- **Exact answer**: EGFR-mutant non small cell lung carcinoma
- **Ideal answer**: Afatinib is a small molecule covalently binding and inhibiting the EGFR, HER2 and HER4 receptor tyrosine kinases. Trials showed promising efficacy in patients with EGFR-mutant NSCLC or enriched for clinical benefit from EGFR tyrosine kinase inhibitors gefitinib or erlotinib.

list Which are the Yamanaka factors?

- **Exact answer**: [OCT4, SOX2, MYC, KLF4]
- **Ideal answer**: The Yamanaka factors are the OCT4, SOX2, MYC, and KLF4 transcription factors

summary What is the role of brain natriuretic peptide in traumatic brain injury patients ?

- **Exact answer**: N/A
- **Ideal answer**: Brain natriuretic peptide concentrations are elevated in patients with traumatic brain during the acute phase and correlate with poor outcomes. In traumatic brain injury patients higher brain natriuretic peptide concentrations are associated with more extensive SAH, elevated ICP and hyponatremia. Brain natriuretic peptide may play an adaptive role in recovery through augmentation of cerebral blood flow.

Figure 1: Examples of questions with their exact and ideal answers in BioASQ 5b.

	Summary	Factoid	Yesno	List
n	6	2	2	3

Table 1: Value of n (the number of sentences returned as the ideal answer) for each question type.

2   Simple Runs

As a first baseline, we submitted a run labelled **trivial** that simply returned the first n snippets of each question. The reason for this choice was that, in some of our initial experiments, we incorporated the position of the snippet as a feature for a machine learning system. In those experiments, the resulting system did not learn anything and simply returned the input snippets verbatim. Subsequent experiments revealed that a trivial baseline that returned the first snippets of the question was very hard to beat. In fact, for the task of summarisation of other domains such as news, it has been observed that a baseline that returns the first sentences often outperformed other methods (Brandow et al., 1995).

As a second baseline, we submitted a run labelled **simple** that selected the n snippets what were most similar to the question. We used cosine similarity, and we tried two alternatives for computing the question and snippet vectors:

tfidf-svd: First, generate the $tf.idf$ vector of the question and the snippets. We followed the usual procedure, and the $tf.idf$ vectors of these sentences are bag-of-word vectors where each dimension represents the $tf.idf$ of a word. Then, reduce the dimensionality of the vectors by selecting the first 200 components after applying Singular Value Decomposition. In contrast with a traditional approach to generate the $tf.idf$ (and SVD) vectors where the statistics are based on the input text solely (question and snippets in our case), we used the text of the question and the text of the ideal answers of the training data.[1] The reason for using this variant was based on empirical results during our preliminary experiments.

word2vec: Train Word2Vec (Mikolov et al.,

	trivial	simple	
		tfidf-svd	word2vec
Mean F1	0.2157	0.1643	0.1715
Stdev F1	0.0209	0.0097	0.0128

Table 2: ROUGE-SU4 of the simple runs.

2013) using a set of over 10 million PubMed abstracts provided by the organisers of BioASQ. Using these pre-trained word embeddings, look up the word embeddings of each word in the question and the snippet. The vector representing a question (or snippet) is the sum of embeddings of each word in the question (or snippet). The dimension of the word embeddings was set to 200.

Table 2 shows the F1 values of ROUGE-SU4 of the resulting summaries. The table shows the mean and the standard deviation of the evaluation results after splitting the training data set for BioASQ 5b into 10 folds (for comparison with the approaches presented in the following sections).

We observe that the trivial run has the best results, and that the run that uses word2vec is second best. Our run labelled "simple" therefore used cosine similarity of the sum of word embeddings returned by word2vec.

3   Regression Approaches

For our run labelled **regression**, we experimented with the use of Support Vector Regression (SVR). The regression setup and features are based on the work by Malakasiotis et al. (2015), who reported the best results in BioASQ 3b (2015).

The target scores used to train the SVR system were the F1 ROUGE-SU4 score of each individual candidate sentence.

In contrast with the simple approaches described in Section 2, which used the snippets as the input data, this time we used all the sentences of the source abstracts. We also incorporated information about whether the sentence was in fact a snippet as described below.

As features, we used:

- $tf.idf$ vector of the candidate sentence. In contrast with the approach described in Section 2, The statistics used to determine the $tf.idf$ vectors were based on the text of the question, the text of the ideal answers, and the text of the snippets.

---

[1] In particular, we used the "TfidfVectorizer" module of the sklearn toolkit (`http://scikit-learn.org`) and fitted it with the list of questions and ideal answers. We then used the "TruncatedSVD" module and fitted it with the tf.idf vectors of the list of questions and ideal answers.

- Cosine similarity between the $tf.idf$ vector of the question and the $tf.idf$ vector of the candidate sentence.

- The smallest cosine similarity between the $tf.idf$ vector of candidate sentence and the $tf.idf$ vector of each of the snippets related to the question. Note that this feature was not used by Malakasiotis et al. (2015).

- Cosine similarity between the sum of word2vec embeddings of the words in the question and the word2vec embeddings of the words in the candidate sentence. As in our run labelled "simple", we used vectors of dimension 200.

- Pairwise cosine similarities between the words of the question and the words of the candidate sentence. As in the work by Malakasiotis et al. (2015), we used word2vec to compute the word vectors. These word vectors were the same as used in Section 2. We then computed the pairwise cosine similarities and selected the following features:

    - The mean, median, maximum, and minimum of all pairwise cosine similarities.
    - The mean of the 2 highest, mean of the 3 highest, mean of the 2 lowest, and mean of the 3 lowest.

- Weighted pairwise cosine similarities, also based in the work by Malakasiotis et al. (2015). In particular, now each word vector was multiplied by the $tf.idf$ of the word, we computed the pairwise cosine similarities, and we used the mean, median, maximum, minimum, mean of 2 highest, mean of 3 highest, mean of 2 lowest, and mean of 3 lowest.

Figure 2 shows the result of grid search by varying the $gamma$ parameter of SVR, fixing C to 1.0, and using the RBF kernel.[2] The figure shows the result of an extrinsic evaluation that reports the F1 ROUGE-SU4 of the final summary, and the result of an intrinsic evaluation that reports the Mean Square Error (MSE) between the target and the predicted SU4 of each individual candidate sentence.

We can observe discrepancy between the results of the intrinsic and the extrinsic evaluations. This

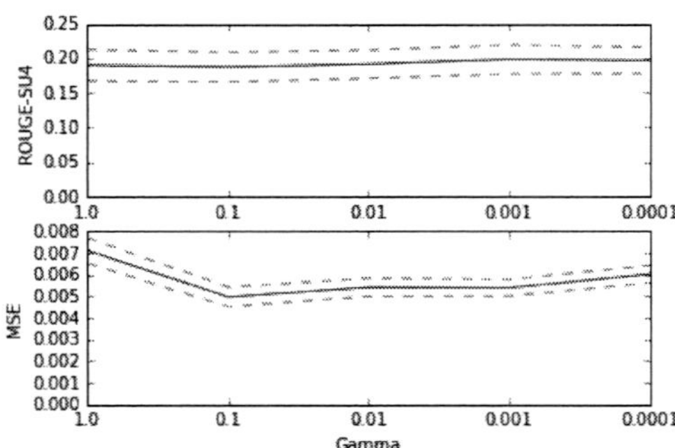

Figure 2: Grid search of the Gamma parameter for the experiments using Support Vector Regression. The continuous lines indicate the mean of 10-fold cross-validation over the training data set of BioASQ 5b. The dashed lines indicate 2 × the standard deviation.

discrepancy could be due to the fact that the data are highly imbalanced in the sense that most annotated SU4 scores in the training data have low values. Consequently, the regressor would attempt to minimise the errors in the low values of the training data at the expense of errors in the high values. But the few sentences with high SU4 scores are most important for the final summary, and these have higher prediction error. This can be observed in the scatter plot of Figure 3, which plots the target against the predicted SU4 in the SVR experiments for each value of $gamma$. The SVR system has learnt to predict the low SU4 scores to some degree, but it does not appear to have learnt to discriminate among SU4 scores over a value of 0.4.

Our run labelled "regression" used $gamma = 0.1$ since it gave the best MSE in our intrinsic evaluation, and Figure 3 appeared to indicate that the system learnt best.

4 Deep Learning Approaches

For our run labelled **nnr** we experimented with the use of deep learning approaches to predict the candidate sentence scores under a regression setup. The regression setup is the same as in Section 3.

Figure 4 shows the general architecture of the deep learning systems explored in our experiments. In a pre-processing stage, and not shown in the figure, the main text of the source PubMed abstracts is split into sentences by using the default NLTK[3] sentence segmenter. The candidate sen-

---

[2] We used the Scikit-learn Python package.

[3] http://www.nltk.org

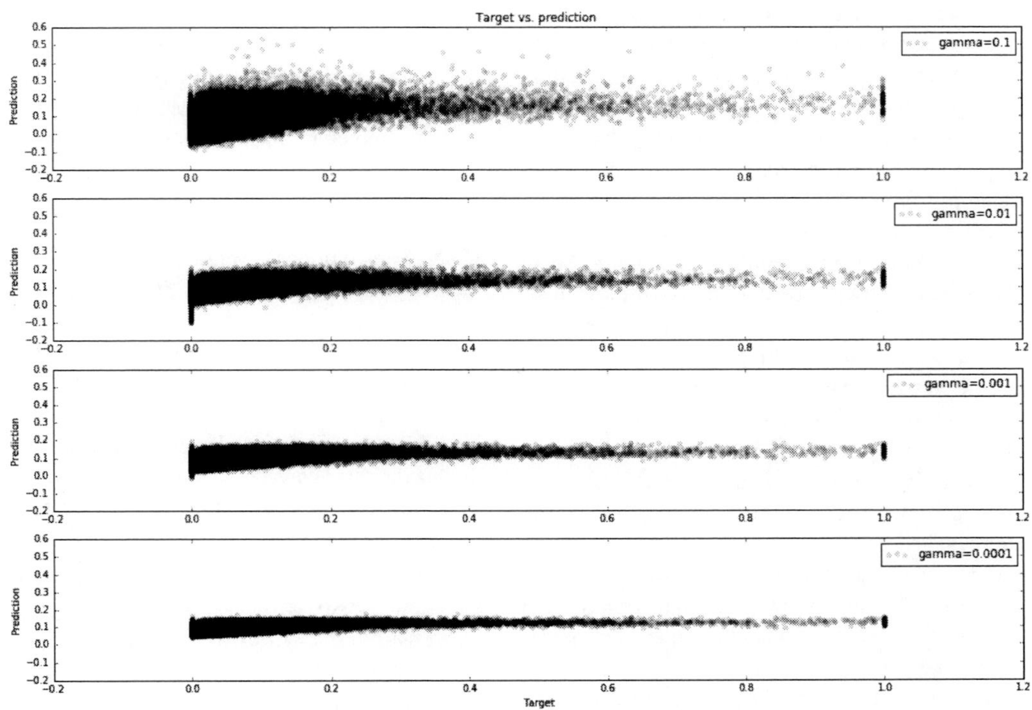

Figure 3: Target vs. predicted SU4 in the SVR experiments for various values of $gamma$.

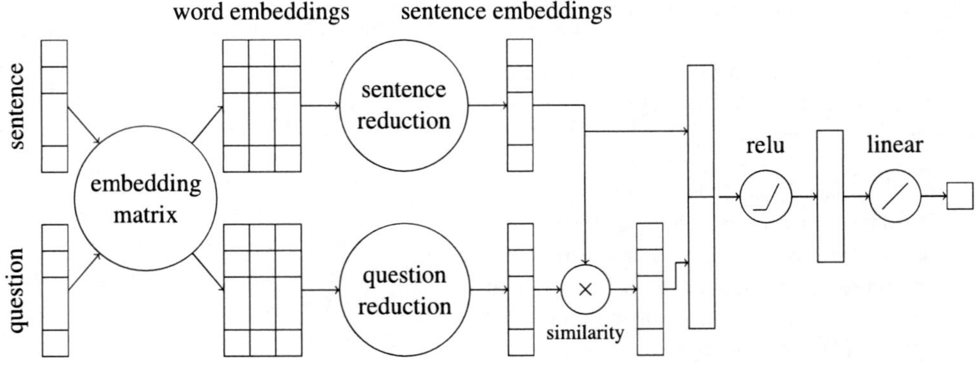

Figure 4: Architecture of the regression system.

tences and questions undergo a simple preprocessing stage that removes punctuation characters, and lowercases the string and splits on blank spaces. Then, these are fed to the system as a sequence of token identifiers. Figure 4 shows that the input to the system is a candidate sentence and the question (as sequences of token IDs). The input is first converted to sequences of word embeddings by applying an embedding matrix. The word embedding stage is followed by a sentence and question reduction stage that combines the word embeddings of each sentence into a sentence embedding. Then, the sentence embedding and the question embedding are compared by applying a similarity operation, and the vector resulting from the comparison is concatenated to the sentence embedding for a final regression comprising of a hidden layer of rectilinear units (relu) and a final linear combination.

The weights of all stages are optimised by backpropagation in order to minimise the MSE of the predicted score at training time. Our experiments varied on the approach for sentence and question reduction, and the approach to incorporate the similarity between sentence and question, as described below.

To produce word embeddings we use word2vec, trained on a collection of over 10 million PubMed abstracts as described in previous sections. The resulting word embeddings are encoded in the embedding matrix of Figure 4. We experimented with the possibility of adjusting the weights of the embedding matrix by backpropagation, but the results did not improve. The results reported in this paper, therefore, used a constant embedding matrix. We experimented with various sizes of word embeddings and chose 100 for the experiments in this paper.

After obtaining the word embeddings, we experimented with the following approaches to produce the sentence vectors:

Mean: The word embeddings provided by word2vec map words into a dimensional space that roughly represents the word meanings, such that words that are similar in meaning are also near in the embedded space. This embedding space has the property that some semantic relations between words are also mapped in the embedded space (Mikolov et al., 2013). It is therefore natural to apply vector arithmetics such as the sum or the mean of word embeddings of a sentence in order to obtain the sentence embedding. In fact, this approach has been used in a range of applications, on its own, or as a baseline against which to compare other more sophisticated approaches to obtain word embeddings, e.g. work by Yu et al. (2014) and Kageback et al. (2014). To accommodate for different sentence lengths, in our experiments we use the mean of word embeddings instead of the sum.

CNN: Convolutional Neural Nets (CNN) were originally developed for image processing, for tasks where the important information may appear on arbitrary fragments of the image (Fukushima, 1980). By applying a convolutional layer, the image is scanned for salient information. When the convolutional layer is followed by a maxpool layer, the most salient information is kept for further processing.

We follow the usual approach for the application of CNN for word sequences, e.g. as described by Kim (2014). In particular, the embeddings of the words in a sentence (or question) are arranged in a matrix where each row represents a word embedding. Then, a set of convolutional filters are applied. Each convolutional filter uses a window of width the total number of columns (that is, the entire word embedding). Each convolutional filter has a fixed height, ranging from 2 to 4 rows in our experiments. These filters aim to capture salient ngrams. The convolutional filters are then followed by a maxpool layer.

Our final sentence embedding concatenates the output of 32 different convolutional filters, each at filter heights 2, 3, and 4. The sentence embedding, therefore, has a size of $32 \times 3 = 96$.

LSTM: The third approach that we have used to obtain the sentence embeddings is recurrent networks, and in particular Long Short Term Memory (LSTM). LSTM has been applied successfully to applications that process sequences of samples (Hochreiter et al., 1997). Our experiments use TensorFlow's implementation of LSTM cells as described by Pham et al. (2013).

In order to incorporate the context on the left

and right of each word we have used the bidirectional variant that concatenates the output of a forward and a backward LSTM chain. As is usual practice, all the LSTM cells in the forward chain share a set of weights, and all the LSTM cells in the backward chain share a different set of weights. This way the network can generalise to an arbitrary position of a word in the sentence. However, we expect that the words of the question behave differently from the words of the candidate sentence. He have therefore used four distinct sets of weights, two for the forward and backward chains of the candidate sentences, and two for the question sentences.

In our experiments, the size of the output of a chain of LSTM cells is the same as the number of features in the input data, that is, the size of the word embeddings. Accounting for forward and backward chains, and given word embeddings of size 100, the size of the final sentence embedding is 200.

Figure 4 shows how we incorporated the similarity between the question and the candidate sentence. In particular, we calculated a weighted dot product, where the weights w_i can be learnt by backpropagation:

$$sim(q, s) = \sum_i w_i q_i s_i$$

Since the sum will be performed by the subsequent relu layer, our comparison between the sentence and the question is implemented as a simple element-wise product between the weights, sentence embeddings, and question embeddings.

An alternative similarity metric that we have also tried is as proposed by Yu et al. (2014). Their similarity metric allows for interactions between different components of the sentence vectors, by applying a $d \times d$ weight matrix W, where d is the sentence embedding size, and adding a bias term:

$$simYu(q, s) = q^T W s + b$$

In both cases, the optimal weights and bias are learnt by backpropagation as part of the complete neural network model of the system.

Table 3 shows the average MSE of 10-fold cross-validation over the training data of BioASQ 5b. "Tf.idf" is a neural network with a hidden layer of 50 relu cells, followed by a

Method	Plain	Sim	SimYu
Tf.idf	0.00354		
SVD	0.00345	0.00334	0.00342
Mean	0.00341	0.00330	0.00331
CNN	0.00350	0.00348	0.00349
LSTM	0.00344	0.00335	0.00336

Table 3: Average MSE of 10-fold cross-validation.

linear cell, where the inputs are the *tf.idf* of the words. "SVD" computes the sentence vectors as described in Section 2, with the only difference being that now we chose 100 SVD components (instead of 200) for comparison with the other approaches shown in Table 3.

We observe that all experiments perform better than the Tf.idf baseline, but there are no major differences between the use of SVD and the three approaches based on word embeddings. The systems which integrated a sentence similarity performed better than those not using it, though the differences when using CNN are negligible. Each cell in Table 3 shows the best results after grid searches varying the dropout rate and the number of epochs during training.

For the "nnr" run, we chose the combination "Mean" and "Sim" of Table 3, since they produced the best results in our experiments (although only marginally better than some of the other approaches shown in the table).

5 Submission Results

At the time of writing, the human evaluations had not been released, and only the ROUGE results of all 5 batches were available. Table 4 shows the F1 score of ROUGE-SU4.

Figure 5 shows the same information as a plot that includes our runs and all runs of other participating systems with higher ROUGE scores. The figure shows that, in the first three batches, only one run by another participant was among our results (shown as a dashed line in the figure). Batches 4 and 5 show consistent results by our runs, and improved results of runs of other entrants.

The results are consistent with our experiments, though the absolute values are higher than those in our experiments. This is probably because we used the entire training set of BioASQ 5b for our cross-validation results, and this data is the aggre-

73

System	Batch 1	Batch 2	Batch 3	Batch 4	Batch 5
trivial	**0.5498**	0.4901	0.5832	0.5431	0.4950
simple	0.5068	**0.5182**	**0.6186**	**0.5769**	**0.5840**
regression	0.5186	0.4795	0.5785	0.5436	0.4784
nnr	0.4192	0.3920	0.5196	0.4445	0.4000

Table 4: ROUGE-SU4 of the 5 batches of BioASQ 2017.

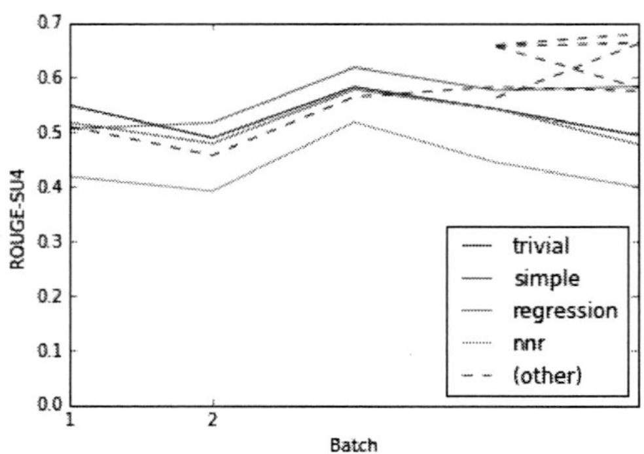

Figure 5: Top ROUGE-SU4 scores of the 5 batches of BioASQ 2017.

gation of the training sets of the BioASQ tasks of previous years. It is possible that the data of latter years are of higher quality, and it might be useful to devise learning approaches that would account for this possibility.

6 Conclusions

At the time of writing, only the ROUGE scores of BioASQ 5b were available. The conclusions presented here, therefore, do not incorporate any insights of the human judgements that are also part of the final evaluation of BioASQ.

Our experiments show that a trivial baseline system that returned the first n snippets appears to be hard to beat. This implies that the order of the snippets matters. Even though the judges were not given specific instructions about the order of the snippets, it would be interesting to study what criteria they used to present the snippets.

Our runs using regression were not significantly better than simpler approaches, and the runs using deep learning reported the lowest results. Note, however, that the input features used in the runs using deep learning did not incorporate information about the snippets. Table 3 shows that the results using deep learning are comparable to results using tf.idf and using SVD, so it is possible that an extension of the system that incorporates information from the snippets would equal or better the other systems.

Note that none of the experiments described in this paper used information specific to the biomedical domain and therefore the methods described here could be applied to any other domain.

Acknowledgments

Some of the experiments in this research were carried out in cloud machines under a Microsoft Azure for Research Award.

References

Ronald Brandow, Karl Mitze, and Lisa F. Rau. 1995. Automatic condensation of electronic publications by sentence selection. *Information Processing and Management* 31(5):675–685.

Kunihiko Fukushima. 1980. Neocognitron: A self-organizing neural network model for a mechanism of pattern recognition unaffected by shift in position. *Biological Cybernetics* 36(4):193–202.

Sepp Hochreiter, Jürgen Schmidhuber, Sepp Hochreiter, Jürgen Schmidhuber, and Jürgen Schmidhuber. 1997. Long short-term memory. *Neural Computation* 9(8):1735–80.

Mikael Kageback, Olof Mogren, Nina Tahmasebi, and Devdatt Dubhashi. 2014. Extractive summarization using continuous vector space models. In *Proceedings of the 2nd Workshop on Continuous Vector Space Models and their Compositionality (CVSC)*. pages 31–39.

Yoon Kim. 2014. Convolutional neural networks for sentence classification. *Proceedings of the 2014 Conference on Empirical Methods in Natural Language Processing (EMNLP 2014)* pages 1746–1751.

Prodromos Malakasiotis, Emmanouil Archontakis, and Ion Androutsopoulos. 2015. Biomedical question-focused multi-document summarization: ILSP and AUEB at BioASQ3. In *CLEF 2015 Working Notes*.

Tomas Mikolov, Kai Chen, Greg Corrado, and Jeffrey Dean. 2013. Efficient estimation of word representations in vector space. In *Proceedings of Workshop at ICLR*. pages 1–12.

Anastasios Nentidis, Konstantinos Bougiatiotis, Anastasia Krithara, Georgios Paliouras, and Ioannis Kakadiaris. 2017. Results of the fifth edition of the BioASQ Challenge. In *Proceedings BioNLP 2017*.

Vu Pham, Théodore Bluche, Christopher Kermorvant, and Jérôme Louradour. 2013. Dropout improves recurrent neural networks for handwriting recognition. Technical report.

George Tsatsaronis, Georgios Balikas, Prodromos Malakasiotis, Ioannis Partalas, Matthias Zschunke, Michael R Alvers, Dirk Weissenborn, Anastasia Krithara, Sergios Petridis, Dimitris Polychronopoulos, Yannis Almirantis, John Pavlopoulos, Nicolas Baskiotis, Patrick Gallinari, Thierry Artiéres, Axel-Cyrille Ngonga Ngomo, Norman Heino, Eric Gaussier, Liliana Barrio-Alvers, Michael Schroeder, Ion Androutsopoulos, and Georgios Paliouras. 2015. An overview of the BIOASQ large-scale biomedical semantic indexing and question answering competition. *BMC Bioinformatics* 16(1):138.

Lei Yu, Karl Moritz Hermann, Phil Blunsom, and Stephen Pulman. 2014. Deep learning for answer sentence selection. In *NIPS Deep Learning Workshop*. page 9.

Neural Question Answering at BioASQ 5B

Georg Wiese[1,2], Dirk Weissenborn[2] and Mariana Neves[1]
[1] Hasso Plattner Institute, August Bebel Strasse 88, Potsdam 14482 Germany
[2] Language Technology Lab, DFKI, Alt-Moabit 91c, Berlin, Germany
`georg.wiese@student.hpi.de,`
`dewe01@dfki.de, mariana.neves@hpi.de`

Abstract

This paper describes our submission to the 2017 BioASQ challenge. We participated in Task B, Phase B which is concerned with biomedical question answering (QA). We focus on factoid and list question, using an *extractive* QA model, that is, we restrict our system to output substrings of the provided text snippets. At the core of our system, we use FastQA, a state-of-the-art neural QA system. We extended it with biomedical word embeddings and changed its answer layer to be able to answer list questions in addition to factoid questions. We pre-trained the model on a large-scale open-domain QA dataset, SQuAD, and then fine-tuned the parameters on the BioASQ training set. With our approach, we achieve state-of-the-art results on factoid questions and competitive results on list questions.

1 Introduction

BioASQ is a semantic indexing, question answering (QA) and information extraction challenge (Tsatsaronis et al., 2015). We participated in Task B of the challenge which is concerned with biomedical QA. More specifically, our system participated in Task B, Phase B: Given a *question* and gold-standard *snippets* (i.e., pieces of text that contain the answer(s) to the question), the system is asked to return a list of answer candidates.

The fifth BioASQ challenge is taking place at the time of writing. Five batches of 100 questions each were released every two weeks. Participating systems have 24 hours to submit their results. At the time of writing, all batches had been released.

The questions are categorized into different question types: factoid, list, summary and yes/no.

Our work concentrates on answering *factoid* and *list* questions. For factoid questions, the system's responses are interpreted as a ranked list of answer candidates. They are evaluated using mean-reciprocal rank (MRR). For list questions, the system's responses are interpreted as a set of answers to the list question. Precision and recall are computed by comparing the given answers to the gold-standard answers. F1 score, i.e., the harmonic mean of precision and recall, is used as the official evaluation measure [1].

Most existing biomedical QA systems employ a traditional QA pipeline, similar in structure to the baseline system by Weissenborn et al. (2013). They consist of several discrete steps, e.g., named-entity recognition, question classification, and candidate answer scoring. These systems require a large amount of resources and feature engineering that is specific to the biomedical domain. For example, OAQA (Zi et al., 2016), which has been very successful in last year's challenge, uses a biomedical parser, entity tagger and a thesaurus to retrieve synonyms.

Our system, on the other hand, is based on a neural network QA architecture that is trained end-to-end on the target task. We build upon FastQA (Weissenborn et al., 2017), an extractive factoid QA system which achieves state-of-the-art results on QA benchmarks that provide large amounts of training data. For example, SQuAD (Rajpurkar et al., 2016) provides a dataset of $\approx 100,000$ questions on Wikipedia articles. Our approach is to train FastQA (with some extensions) on the SQuAD dataset and then fine-tune the model parameters on the BioASQ training set.

Note that by using an extractive QA network as our central component, we restrict our system's

---

[1]The details of the evaluation can be found at `http://participants-area.bioasq.org/ Tasks/b/eval_meas/`

Proceedings of the BioNLP 2017 workshop, pages 76–79,
Vancouver, Canada, August 4, 2017. ©2017 Association for Computational Linguistics

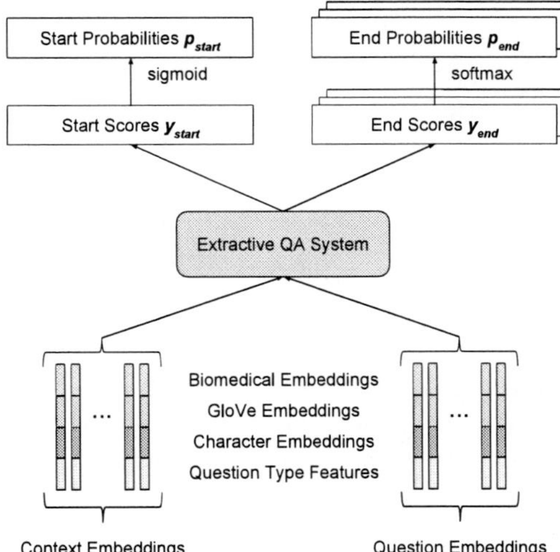

Figure 1: Neural architecture of our system. Question and context (i.e., the *snippets*) are mapped directly to start and end probabilities for each context token. We use FastQA (Weissenborn et al., 2017) with modified input vectors and an output layer that supports list answers in addition to factoid answers.

responses to substrings in the provided snippets. This also implies that the network will not be able to answer yes/no questions. We do, however, generalize the FastQA output layer in order to be able to answer list questions in addition to factoid questions.

2 Model

Our system is a neural network which takes as input a question and a context (i.e., the snippets) and outputs start and end pointers to tokens in the context. At its core, we use FastQA (Weissenborn et al., 2017), a state-of-the-art neural QA system. In the following, we describe our changes to the architecture and how the network is trained.

2.1 Network architecture

In the input layer, the context and question tokens are mapped to high-dimensional word vectors. Our word vectors consists of three components, which are concatenated to form a single vector:

- **GloVe embedding:** We use 300-dimensional GloVe embeddings [2] (Pennington et al.,

---

[2]We use the 840B embeddings available here: https://nlp.stanford.edu/projects/glove/

2014) which have been trained on a large collection of web documents.

- **Character embedding:** This embedding is computed by a 1-dimensional convolutional neural network from the characters of the words, as introduced by Seo et al. (2016).

- **Biomedical Word2Vec embeddings:** We use the biomedical word embeddings provided by Pavlopoulos et al. (2014). These are 200-dimensional Word2Vec embeddings (Mikolov et al., 2013) which were trained on ≈ 10 million PubMed abstracts.

To the embedding vectors, we concatenate a one-hot encoding of the question type (*list* or *factoid*). Note that these features are identical for all tokens.

Following our embedding layer, we invoke FastQA in order to compute start and end scores for all context tokens. Because end scores are conditioned on the chosen start, there are $O(n^2)$ end scores where n is the number of context tokens. We denote the start index by $i \in [1, n]$, the end index by $j \in [i, n]$, the start scores by y_{start}^i, and end scores by $y_{end}^{i,j}$.

In our output layer, the start, end, and span probabilities are computed as:

$$p_{start}^i = \sigma(y_{start}^i) \tag{1}$$

$$p_{end}^{i,\cdot} = softmax(y_{end}^{i,\cdot}) \tag{2}$$

$$p_{span}^{i,j} = p_{start}^i \cdot p_{end}^{i,j} \tag{3}$$

where σ denotes the sigmoid function. By computing the start probability via the sigmoid rather than softmax function (as used in FastQA), we enable the model to output multiple spans as likely answer spans. This generalizes the factoid QA network to list questions.

2.2 Training & decoding

Loss We define our loss as the cross-entropy of the correct start and end indices. In the case of multiple occurrences of the same answer, we only minimize the span of the lowest loss.

	Factoid MRR		List F1	
Batch	**Single**	**Ensemble**	**Single**	**Ensemble**
1	52.0% (2/10)	57.1% (1/10)	33.6% (1/11)	33.5% (2/11)
2	38.3% (3/15)	42.6% (2/15)	29.0% (8/15)	26.2% (9/15)
3	43.1% (1/16)	42.1% (2/16)	41.5% (2/17)	49.5% (1/17)
4	30.0% (3/20)	36.1% (1/20)	24.2% (5/20)	29.3% (4/20)
5	39.2% (3/17)	35.1% (4/17)	36.1% (4/20)	39.1% (2/20)
Average	40.5%	42.6%	32.9%	35.1%

Table 1: Preliminary results for factoid and list questions for all five batches and for our single and ensemble systems. We report MRR and F1 scores for factoid and list questions, respectively. In parentheses, we report the rank of the respective systems relative to all other systems in the challenge. The last row averages the performance numbers of the respective system and question type across the five batches.

Optimization We train the network in two steps: First, the network is trained on SQuAD, following the procedure by Weissenborn et al. (2017) (*pre-training phase*). Second, we fine-tune the network parameters on BioASQ (*fine-tuning phase*). For both phases, we use the Adam optimizer (Kingma and Ba, 2014) with an exponentially decaying learning rate. We start with learning rates of 10^{-3} and 10^{-4} for the pre-training and fine-tuning phases, respectively.

BioASQ dataset preparation During fine-tuning, we extract answer spans from the BioASQ training data by looking for occurrences of the gold standard answer in the provided snippets. Note that this approach is not perfect as it can produce false positives (e.g., the answer is mentioned in a sentence which does not answer the question) and false negatives (e.g., a sentence answers the question, but the exact string used is not in the synonym list).

Because BioASQ usually contains multiple snippets for a given question, we process all snippets independently and then aggregate the answer spans, sorting globally according to their probability $p_{span}^{i,j}$.

Decoding During the inference phase, we retrieve the top 20 answers span via beam search with beam size 20. From this sorted list of answer strings, we remove all duplicate strings. For factoid questions, we output the top five answer strings as our ranked list of answer candidates. For list questions, we use a *probability cutoff threshold* t, such that $\{(i, j)|p_{span}^{i,j} \geq t\}$ is the set of answers. We set t to be the threshold for which the list F1 score on the development set is optimized.

Ensemble In order to further tweak the performance of our systems, we built a model ensemble. For this, we trained five single models using 5-fold cross-validation on the entire training set. These models are combined by averaging their start and end scores before computing the span probabilities (Equations 1-3). As a result, we submit two systems to the challenge: The best single model (according to its development set) and the model ensemble.

Implementation We implemented our system using TensorFlow (Abadi et al., 2016). It was trained on an NVidia GForce Titan X GPU.

3   Results & discussion

We report the results for all five test batches of BioASQ 5 (Task 5b, Phase B) in Table 1. Note that the performance numbers are not final, as the provided synonyms in the gold-standard answers will be updated as a manual step, in order to reflect valid responses by the participating systems. This has not been done by the time of writing[3]. Note also that – in contrast to previous BioASQ challenges – systems are no longer allowed to provide an own list of synonyms in this year's challenge.

In general, the single and ensemble system are performing very similar relative to the rest of field: Their ranks are almost always right next to each other. Between the two, the ensemble model performed slightly better on average.

On factoid questions, our system has been very successful, winning three out of five batches. On

---

[3]The final results will be published at `http://participants-area.bioasq.org/results/5b/phaseB/`

list questions, however, the relative performance varies significantly. We expect our system to perform better on factoid questions than list questions, because our pre-training dataset (SQuAD) does not contain any list questions.

Starting with batch 3, we also submitted responses to yes/no questions by always answering *yes*. Because of a very skewed class distribution in the BioASQ dataset, this is a strong baseline. Because this is done merely to have baseline performance for this question type and because of the naivety of the method, we do not list or discuss the results here.

4 Conclusion

In this paper, we summarized the system design of our BioASQ 5B submission for factoid and list questions. We use a neural architecture which is trained end-to-end on the QA task. This approach has not been applied to BioASQ questions in previous challenges. Our results show that our approach achieves state-of-the art results on factoid questions and competitive results on list questions.

References

Martín Abadi, Ashish Agarwal, Paul Barham, Eugene Brevdo, Zhifeng Chen, Craig Citro, Greg S Corrado, Andy Davis, Jeffrey Dean, Matthieu Devin, et al. 2016. Tensorflow: Large-scale machine learning on heterogeneous distributed systems. *arXiv preprint arXiv:1603.04467* .

Diederik Kingma and Jimmy Ba. 2014. Adam: A method for stochastic optimization. *arXiv preprint arXiv:1412.6980* .

Tomas Mikolov, Ilya Sutskever, Kai Chen, Greg S Corrado, and Jeff Dean. 2013. Distributed representations of words and phrases and their compositionality. In *Advances in neural information processing systems*. pages 3111–3119.

Ioannis Pavlopoulos, Aris Kosmopoulos, and Ion Androutsopoulos. 2014. Continuous space word vectors obtained by applying word2vec to abstracts of biomedical articles http://bioasq.lip6.fr/info/BioASQword2vec/.

Jeffrey Pennington, Richard Socher, and Christopher D. Manning. 2014. Glove: Global vectors for word representation. In *Empirical Methods in Natural Language Processing (EMNLP)*. pages 1532–1543. http://www.aclweb.org/anthology/D14-1162.

Pranav Rajpurkar, Jian Zhang, Konstantin Lopyrev, and Percy Liang. 2016. Squad: 100,000+ questions for machine comprehension of text. *arXiv preprint arXiv:1606.05250* .

Minjoon Seo, Aniruddha Kembhavi, Ali Farhadi, and Hannaneh Hajishirzi. 2016. Bidirectional attention flow for machine comprehension. *arXiv preprint arXiv:1611.01603* .

George Tsatsaronis, Georgios Balikas, Prodromos Malakasiotis, Ioannis Partalas, Matthias Zschunke, Michael R Alvers, Dirk Weissenborn, Anastasia Krithara, Sergios Petridis, Dimitris Polychronopoulos, et al. 2015. An overview of the bioasq large-scale biomedical semantic indexing and question answering competition. *BMC bioinformatics* 16(1):1.

Dirk Weissenborn, George Tsatsaronis, and Michael Schroeder. 2013. Answering factoid questions in the biomedical domain. *BioASQ@ CLEF* 1094.

Dirk Weissenborn, Georg Wiese, and Laura Seiffe. 2017. Making neural qa as simple as possible but not simpler. *arXiv preprint arXiv:1703.04816* .

Yang Zi, Zhou Yue, and Eric Nyberg. 2016. Learning to answer biomedical questions: Oaqa at bioasq 4b. *ACL 2016* page 23.

End-to-End System for Bacteria Habitat Extraction

Farrokh Mehryary[1,2]*, Kai Hakala[1,2]*, Suwisa Kaewphan[1,2,3]*,
Jari Björne[1], Tapio Salakoski[1,3] and Filip Ginter[1]
1. Turku NLP Group, Department of FT, University of Turku, Finland
2. The University of Turku Graduate School (UTUGS), University of Turku, Finland
3. Turku Centre for Computer Science (TUCS), Finland
`firstname.lastname@utu.fi`

Abstract

We introduce an end-to-end system capable of named-entity detection, normalization and relation extraction for extracting information about bacteria and their habitats from biomedical literature. Our system is based on deep learning, CRF classifiers and vector space models. We train and evaluate the system on the BioNLP 2016 Shared Task Bacteria Biotope data. The official evaluation shows that the joint performance of our entity detection and relation extraction models outperforms the winning team of the Shared Task by 19pp on F-score, establishing a new top score for the task. We also achieve state-of-the-art results in the normalization task. Our system is open source and freely available at `https://github.com/TurkuNLP/BHE`.

1 Introduction

Knowledge about habitats of bacteria is crucial for the study of microbial communities, e.g. metagenomics, as well as for various applications such as food processing and health sciences. Although this type of information is available in the biomedical literature, comprehensive resources accumulating the knowledge do not exist (Deléger et al., 2016).

The BioNLP Bacteria Biotope (BB) Shared Tasks are organized to provide a common evaluation platform for language technology researchers interested in developing information extraction methods adapted for the detection of bacteria and their physical locations mentioned in the literature. So far three BB shared tasks have been organized, the latest in 2016 (BB3) consisting of three main

subtasks: named entity recognition and categorization (BB3-cat and BB3-cat+ner), event extraction (BB3-event and BB3-event+ner) and knowledge base extraction. The NER task includes three relevant entity types: HABITAT, BACTERIA and GEOGRAPHICAL, the categorization task focuses on normalizing the mentions to established ontology concepts, although GEOGRAPHICAL entities are excluded from this task, whereas the event extraction aims at finding the relations between these entities, i.e. extracting in which locations certain bacteria live in. The knowledge base extraction task is centered upon aggregating this type of information from a large text corpus.

In this paper we revisit the BB3 subtasks of NER, categorization and event extraction, all of which are essential for building a real-world information extraction pipeline. As a result, we present a text mining pipeline which achieves state-of-the-art results for the joint evaluation of NER and event extraction as well as for the categorization task using the official BB3 shared task datasets and evaluation tools. Building such end-to-end system is important for bringing the results from the shared tasks to the actual intended users. To our best knowledge, no such system is openly available for bacteria habitat extraction.

The pipeline utilizes deep neural networks, conditional random field classifiers and vector space models to solve the various subtasks and the code is freely available at `https://github.com/TurkuNLP/BHE`. In the following sections we discuss our system, divided into three modules: entity recognition, categorization and event extraction. We then analyze the results and finally discuss the potential future research directions.

---

*These authors contributed equally.

80

Proceedings of the BioNLP 2017 workshop, pages 80–90,
Vancouver, Canada, August 4, 2017. ©2017 Association for Computational Linguistics

2 Method

2.1 Named entity detection

Detecting the BB3 HABITAT, BACTERIA and GE-
OGRAPHICAL mentions is a standard named entity
recognition task, evaluated based on the correctness of the type and character offsets of the discovered text spans. In our NER pipeline, all documents are preprocessed following the approach
of Hakala et al. (2016). In brief, we first convert all documents and annotation files from UTF-
8 to ASCII encoding using a modified version
of publicly available tool designed for parsing
PubMed documents (Pyysalo et al., 2013) [1]. Next
we split documents into sentences using the Genia Sentence Splitter (Sætre et al., 2007) and the
sentences are subsequently tokenized and part-of-
speech tagged using the tokenization and POS-
tagging modules in NERsuite [2], respectively.

To detect the entity mentions we use NERsuite,
a named entity recognition toolkit, as it is relatively easy to train on new corpora, yet supports
adding novel user-defined features. In biomedical
NER, NERsuite has been a versatile tool achieving excellent performance for various entity types
(Ohta et al., 2012; Kaewphan et al., 2014, 2016),
however, it is not capable of dealing with overlapping entities. Therefore, we only use the longest
spans of overlapping annotated entities as our
training data, ignoring embedded entities which
are substrings of the longest spans.

In biomedical NER, domain knowledge such
as controlled vocabularies has been crucial for
achieving high performance. In this work we prepare 3 dictionaries, specific for each entity type.
For BACTERIA, we compile a dictionary of names
exclusively from the NCBI Taxonomy database[3]
by including all names under bacteria superkingdom (NCBI taxonomy identifier 2). The *scientific names* are expanded to include abbreviations
whose genus names are conventionally abbreviated with the first and/or second alphabet, whereas
the rest of the names, such as species epithet and
strains, remains unchanged. For HABITAT, we
combine all symbols from the OntoBiotope ontology [4] and use them without any further modifications. Similar to HABITAT, we also prepare dictionary for GEOGRAPHICAL by taking all

strings under the semantic type *geographical area*
from UMLS database (version 2016AA) (Bodenreider, 2004). All dictionaries prepared in this
step are directly provided to NERsuite through the
dictionary-tagging module without any normalization. The tagging provides additional features describing whether the tokens are present in some
semantic categories, such as bacteria names or geographical places. For GEOGRAPHICAL model,
we also add token-level tagging results for *location* from Stanford NER (SNER) (Finkel et al.,
2005) as binary values to NERsuite; 1 and 0 for
location and non-location, respectively.

Although utilizing dictionary features is beneficial for NER, strict string matching tends to lead
to low coverage, an issue which is also common in
the categorization task. To remedy this problem,
we also generate fuzzy matching features based
on our categorization system (see Section 2.2) by
measuring the maximum similarity of each token
against the NCBI Taxonomy and OntoBiotope ontologies for BACTERIA and HABITAT respectively.
Thus, instead of a binary feature denoting whether
a token is present in the ontology or not, a similarity score ranging from 0 to 1 is assigned for
each token. This approach is similar to (Kaewphan
et al., 2014), but instead of using word embedding
similarities, our fuzzy matching relies on character
ngrams. We do not use these features for the GEO-
GRAPHICALentities, which are not categorized by
our system.

In the official BB3 evaluation, NER is jointly
evaluated with either categorization or event extraction system. In BB3-cat+ner task, SER (Slot
Error Rate) is used as the main scoring metric,
whereas in BB3-event+ner, participating teams
are ranked based on F-score of extracted relations. Due to the lack of an official evaluation
on NER for all entities in BB3-event+ner and
for GEOGRAPHICAL in BB3-cat+ner, we use our
own implementation by calculating the F-score
using exact string matching criteria as our main
scoring metric. In this study, we consider BB3-
event+ner as our primary subtask and thus all
hyper-parameters in model selection are optimized
against F-score instead of SER.

2.2 Named entity categorization

In the BB3 categorization subtask each BACTE-
RIA and HABITAT mention has to be assigned to
the corresponding ontology concepts, specifically

---

[1] https://github.com/spyysalo/nxml2txt
[2] http://nersuite.nlplab.org/
[3] https://www.ncbi.nlm.nih.gov/taxonomy
[4] http://agroportal.lirmm.fr/ontologies/ONTOBIOTOPE

to NCBI Taxonomy and OntoBiotope identifiers respectively. This task is commonly known as named entity normalization or entity linking and various approaches ranging from Levenshtein edit distances to recurrent neural networks have been suggested as the plausible solutions (Tiftikci et al., 2016; Limsopatham and Collier, 2016).

Our categorization method is based on the common approach of TFIDF weighted sparse vector space representations (Salton and Buckley, 1988; Leaman et al., 2013; Hakala, 2015), i.e. the problem is seen as an information retrieval task where each concept name in the ontology is considered a document and the IDF weights are based on these names. Consequently, each concept name and each entity mention is represented with a TFIDF weighted vector and the concept with the highest cosine similarity is assigned for a given entity. Whereas these representations are commonly formed in a bag-of-words fashion, in our experiments using character-level ngrams resulted in better outcome. In the final system we use ngrams of length 1, 2 and 3 characters. These ngram lengths produced the highest accuracy on the official development set for both BACTERIA and HABITAT entities, each entity type evaluated separately. The TFIDF vectorization was implemented using the scikit-learn library (Pedregosa et al., 2011) and default parameter values except for using the character level ngrams instead of words.

For both included ontologies we use the preferred names as well as the listed synonyms to represent the concepts. Since the task is restricted to bacteria mentions instead of all organisms, we also narrow down the NCBI Taxonomy ontology to cover only the Bacteria superkingdom, i.e. the categorization system is not allowed to assign taxonomy identifiers which do not belong to this superkingdom. Otherwise all concepts from the used ontologies are included.

As preprocessing steps we use three main approaches: abbreviation expansion, acronym expansion and stemming. For stemming we use the Porter stemmer (Porter, 1980) and stem each token in the entities and concept names. According to our evaluation this is not beneficial for the BACTERIA entities and is thus included only for the HABITAT entities.

In biomedical literature the genus names in BACTERIA mentions are commonly shortened af-

ter the first mention, e.g. *Staphylococcus aureus* is abbreviated as *S. aureus*, but the NCBI Taxonomy ontology does not include these abbreviated forms as synonyms for the corresponding concepts. Thus, if an entity mention includes a token with a period in it, we expand the given token by finding the most common token with the same initial from all previously mentioned entities of the same type within the same document.

Another commonly used naming convention for BACTERIA mentions is forming acronyms, e.g. *lactic acid bacteria* is often referred to as *LAB*. Consequently, when we detect a BACTERIA mention with less than five characters or written in uppercase, we try to find the corresponding full form by generating acronyms for all previously mentioned BACTERIA entities by simply concatenating their initials. However, many BACTERIA acronyms do not follow this format strictly, e.g *Lactobacillus casei strain Shirota* should be shortened to *LcS* instead of *LCSS* and *Francisella tularensis Live Vaccine Strain* as *LVS* instead of *FTLVS*. Thus, instead of using strict matching to find the corresponding full form, we utilize the same character-level TFIDF representations as used for the actual categorization for these acronyms to find the most similar full form. In our evaluation, using the same approach for HABITAT entities dramatically decreased the performance hence was thus not used for this entity type (see Section 3.2).

Both of these expansion methods have similar intentions as the preprocessing steps utilized by the winning system in BB3 (BOUN) by Tiftikci et al. (2016), but our system uses more relaxed criteria for finding the full forms and should thus result in better recall at the expense of precision.

2.3 Event extraction

The BB3-event and BB3-event+ner tasks demand extraction of undirected binary associations of two named entities: a BACTERIA entity and either a HABITAT or a GEOGRAPHICAL entity; and these relations represent the locations in which bacteria live. We thus formulate this task as a binary classification task and assign the label *positive* if such relation holds for a given entity pair and *negative* otherwise.

To address this task, we present a deep learning-based relation extraction system that generates features along the *shortest dependency path (SDP)*

	Train	Devel	Test
Total sentences	527	319	508
Sentences w/examples	158	117	158
Sentences w/o examples	369	202	350
Total examples	524	506	534
Positive examples	251	177	-
Negative examples	273	329	-

Table 1: BB3-event data statistics.

which connects the two candidate entities in the syntactic parse graph. Many successful relation extraction systems have been built utilizing SDP (Cai et al., 2016; Mehryary et al., 2016; Xu et al., 2015; Björne and Salakoski, 2013; Björne et al., 2012; Bunescu and Mooney, 2005) since it is known to contain most of the relevant words for expressing the relation between the two entities while excluding less relevant and uninformative words. Since this approach focuses on a *single* sentence parse graph at a time, it is unable to detect plausible cross-sentence relations, i.e, the cases in which the two candidate entities belong to different sentences. As discussed by Kim et al. (2011), detecting such relations is a major challenge for relation extraction systems. We simply exclude any cross-sentence relations from training, development and test sets.[5] Table 1 summarizes the statistics of the data that is used for building our relation extraction system after removing cross-sentence relations.

2.3.1 Preprocessing

For preprocessing, we use the preprocessing pipeline of the TEES system (Björne and Salakoski, 2013) which automates tokenization, part-of-speech tagging and sentence parsing. TEES runs the BLLIP parser (Charniak and Johnson, 2005) with the biomedical domain model created by McClosky (2010). The resulting phrase structure trees are then converted to dependency graphs (*nonCollapsed* variant of Stanford Dependency) using the Stanford conversion tool (version 2.0.1) (de Marneffe et al., 2006).

2.3.2 Relation extraction system architecture

The architecture of our deep learning-based relation extraction system is centered around utilizing three separate convolutional neural networks (CNN): for the sequence of *words*, the sequence of

POS tags, and the sequence of *dependency types* (the edges of the parse graph), along the SDP connecting the two candidate entities (see Figure 1). Even though the parse graph is directed, we regard it as an undirected graph and always traverse the SDP by starting the path from the BACTERIA entity mention to the HABITAT/GEOGRAPHICAL, regardless of the order of their occurrence in the sentence. Evaluation against the development set showed that this approach leads to better generalization in comparison with simply traversing the path from the first occurring entity mention to the second (with/without considering the direction of the edges).

The structure of each CNN is similar: the words (or POS tags or dependency types) in the sequence are mapped into their corresponding vector representations using an embedding lookup layer. The resulting sequence of vectors is then forwarded into a convolutional layer which creates a convolution kernel that is applied on the layer input over a single spatial dimension to produce a tensor of outputs. These outputs are then forwarded to a max-pooling layer that gathers information from local features of the SDP. Hence, the three CNNs produce three vector representations.

Subsequently, the output vectors of the CNNs and two 1-hot-encoded entity-type vectors are concatenated. The first entity-type vector represents the type of the first occurring entity in the sentence (BACTERIA, HABITAT or GEOGRAPHICAL), and the other is used for the second one. The resulting vector is then forwarded into a fully connected hidden layer and finally, the hidden layer connects to a single-node binary classification layer.

For the word features, we use a vector space model with 200-dimensional word embeddings pre-trained by Pyysalo et al. (2013). These are fine-tuned during the training while the POS-tag and dependency type embeddings are learned from scratch after being randomly initialized.

Based on experiments on the development set, we have set the dimensionality of the POS tag embeddings to 200, and for dependency types to 300. For all convolutional layers, the number of filters has been set to 100 and the window size (filter length) to 4. Finally, dimensionality of the hidden layer has been set to 100.The *ReLU* activation function is applied on the output of the convolutional layers while we apply *sigmoid* activation to

---

[5]Official evaluation results on the development and test data are of course comparable to those of other systems: any cross-sentence relations in the development/test data count against our submissions as false negatives.

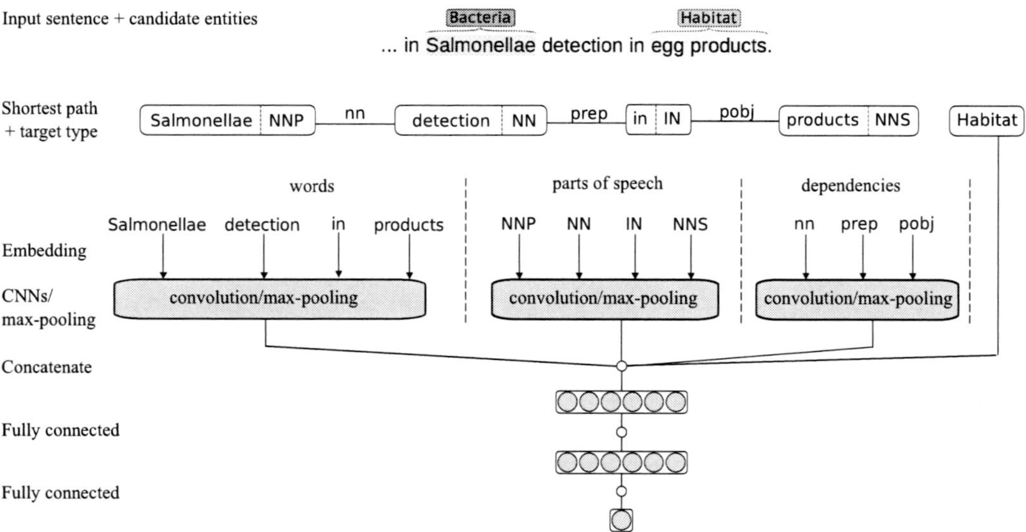

Figure 1: Proposed network architecture.

the output of the hidden layer.

2.3.3 Training and optimization

We use *binary cross-entropy* as the objective function and the *Adam* optimization algorithm (Kingma and Ba, 2014) for training the network. Applying the *dropout* (Srivastava et al., 2014) with rate of 50% on the output of the hidden layer is the only network regularization method used to avoid overfitting.

When the number of weights in a neural network is high and the training set is very small (e.g., there are only 524 examples in the BB3-event training set), the initial random state of the model can have a significant impact on the final model and its generalization performance. Mehryary et al. (2016) have reported that the F-score on the development set of BB3-event task can vary up to 9 percentage points based on the different initial random state of the network.

To overcome this problem, we implement the simple but effective strategy proposed by them, which consists of training the neural network model 15 times with different initial random states, predicting the development/test set examples and aggregating the 15 classifiers' predictions using a simple voting algorithm.

For each development/test example, the voting algorithm combines the predictions based on a given threshold parameter t: the relation is voted to be positive if at least t classifiers have predicted it to be positive, otherwise, it will be considered as a negative. Obviously, the lowest threshold value ($t = 1$) produces the highest recall and lowest precision and the highest threshold ($t = 15$) produces the highest precision and lowest recall and the aim is to find be best threshold value which maximizes the F-score.

Our experiments on the development set (using the proposed network architecture) showed that for the BB3-event task the optimal results are achieved when we train the networks for 2 epochs and set the threshold value to 4, and for the BB3-event+ner task, when we train the networks for 2 epochs and set the threshold value to 3.

3 Results and discussion

3.1 Named entity detection

For the named entity detection task, we obtain the baseline performance by training NERsuite for each entity-type independently. As shown in Table 2, the F-scores for BACTERIA, GEOGRAPHICAL and HABITAT are 0.713, 0.516 and 0.482 respectively. The baseline performance of HABITAT and GEOGRAPHICAL models is significantly lower than BACTERIA.

For all entities, adding dictionary features improves the performance of the model. A substantial improvement in F-score is found for GEOGRAPHICAL where the performance is increased

Entity/Experiment	Precision	Recall	F-score
Bacteria			
BB3	0.787	0.652	0.713
BB3 + dict	0.833	0.697	0.759
BB3 + tfidf	0.793	0.660	0.720
BB3 + tfidf + dict	0.822	0.717	0.766
BB3 + BB2 + dict	0.902	0.713	0.796
BB3 + BB1 + dict	0.893	0.721	**0.798**
Habitat			
BB3	0.589	0.407	0.482
BB3 + dict	0.649	0.465	0.541
BB3 + tfidf	0.697	0.482	0.570
BB3 + tfidf + dict	0.715	0.520	**0.602**
BB3 + BB2 + dict	0.560	0.500	0.529
Geographical			
BB3	0.667	0.421	0.516
BB3 + dict	0.719	0.605	0.657
BB3 + SNER	0.694	0.658	0.676
BB3 + dict + SNER	0.788	0.684	0.732
BB3 + BB2 + dict	0.903	0.737	**0.812**

Table 2: The performance of our named entity detection system on BACTERIA, HABITAT and GEOGRAPHICAL mentions using internal evaluation system. The models are evaluated on the BB3 development data.

by more than 14pp compared to 6pp and 5pp for HABITAT and BACTERIA, respectively. Adding fuzzy matching features further improves the F-score for HABITAT by more than 12pp compared to 8pp for BACTERIA. This result shows that having both domain knowledge and relaxed matching criteria can significantly enhance the model performance.

We improve equally the baseline performance for GEOGRAPHICAL by adding features from SNER tagging. The increase in F-score, 0.657 versus 0.676, is about the same as independently adding *UMLS-geographical area* dictionary features. Further increase in F-score is achieved by combining both features, likely due to the expanded coverage of geographical names.

The BB3 corpus is relatively small in terms of entity frequency and the number of unique entities. We explore the possibility of increasing model performance through adding additional training data from previously organized BB Shared Tasks (i.e, BB1 (Bossy et al., 2011) and BB2 (Bossy et al., 2013)). Annotations for BACTERIA mentions are available in both BB1 and BB2 Shared Tasks and we thus train NERsuite models by adding these annotations to the training data. The results show that the models, trained with additional datasets, achieve higher performance. BB1 provides a slightly better F-score than BB2, 0.798 vs 0.796.

For GEOGRAPHICAL and HABITAT entities, compatible annotations are only available from BB2 (Bossy et al., 2013), subtask 2. We thus train NERsuite for both HABITAT and GEOGRAPHICAL by using combined BB3 and BB2 data. The result for GEOGRAPHICAL is similar to the one observed with BACTERIA and additional data can increase the model F-score by more than 15pp. However, the result for HABITAT is different as F-score slightly drops from 0541 to 0.529. The best NER model for HABITAT thus remains unchanged.

Finally, we train our final model by combining training and development datasets and use hyperparameters obtained from the best performing system on development dataset. The official evaluation of the NER task jointly with either categorization or event extraction system is discussed in Section 3.2 and Section 3.3, respectively.

3.2 Categorization

To analyze our categorization approaches, we evaluate their performance on the official development set. During the development we used accuracy for evaluating the effects of different hyperparameters and preprocessing steps. To get comparable results to previous systems we, however, report the results in this paper using the precision scores from the official evaluation service. As the used ontologies form hierarchical structures, the official evaluation penalizes the incorrect predictions based on the distance from the gold standard annotations, whereas our internal accuracy evaluation measures exact matches. Our accuracy scores and the official evaluation seem to correlate to the level that all improvements validated using the accuracy score also improved the performance according to the official evaluation.

The performance of our system and various preprocessing steps are shown incrementally in Table 3. As a baseline system we use TFIDF bag-of-words representations without any of our preprocessing steps. By simply switching to character level representations the precision is increased by 1.3pp for HABITAT and 14.1pp for BACTERIA mentions.

Adding the abbreviation expansion step further improves precision for BACTERIA by 14.1pp, but does not influence HABITAT entities as most likely there are no abbreviated mentions in this category. The acronym expansion has a lesser, but still noticeable impact and improves precision for BAC-TERIA by 4.9pp. However, applying this method to HABITAT entities decreases the performance by 4.5pp and is thus left out in the final system for this entity type. This is probably due to the fact that we consider all tokens with less than 5 characters to be acronyms, which seems to hold for BACTERIA mentions, but is a bad assumption for HABITAT entities. The final preprocessing step, stemming, improves the performance on HABITAT entities by mere 1.3pp, but has a negative impact on BACTERIA and is left out for this entity type in the final system.

The results on the official test set are consistently lower than on the development set for both entity types (see Table 4), suggesting that the hyperparameters selected based on the development set might have been slightly overfit on this data. However, our system is able to outperform BOUN (Tiftikci et al., 2016), the winning system from the BioNLP'16 BB3 Shared Task, by 1pp, 1.5pp and 1.2pp on HABITAT, BACTERIA and all entities respectively.

Since the BB3 tasks do not evaluate named entity recognition independently, but only in conjunction with either categorization or event extraction, we also report the official numbers for the BB-cat+ner task in Table 5. In this combined evaluation our system is not able to reach the performance level of the state-of-the-art system TagIt (Cook et al., 2016), but does outperform the other systems which participated in the given subtask.

Our combined system is also performing clearly worse on the test set than on the development set. Unfortunately, due to the test set being blinded, we are unable to specify the exact cause for this. However, the official evaluation service does provide relaxed evaluation modes where e.g. entity boundaries are ignored, i.e. the evaluation focuses on the categorization task. Based on these evaluations our categorization system seems to perform on the same level on both development and test sets, but the performance of our NER model drops, especially for the BACTERIA mentions. This might be simply due to overfitting on the development set, but requires further investigation.

	Habitat	Bacteria	Overall
BOW TFIDF	0.634	0.531	0.568
Char TFIDF	0.647	0.672	0.656
+ abbreviations	0.647	0.813	0.705
+ acronyms	0.602	**0.862**	0.693
+ stemming	**0.660**	0.858	0.729
Final system	**0.660**	**0.862**	**0.731**

Table 3: Evaluation of our categorization system with different preprocessing steps compared to a baseline system with TFIDF weighted bag-of-words (unigrams) representations. The scoring is produced by the official evaluation service. Any added processing step, which decreases the performance is left out for the given entity type for the following experiments.

	Habitat	Bacteria	Overall
Our system	**0.630**	**0.816**	**0.691**
BOUN	0.620	0.801	0.679

Table 4: Comparison of our entity categorization system and the best performing system in BioNLP'16 BB3 Shared Task on the test set using the official evaluation service.

	Habitat	Bacteria	Overall
	Development set		
Our system	0.645	0.377	0.553
TagIt	**0.511**	**0.303**	**0.439**
	Test set		
Our system	0.804	0.706	0.766
TagIt	**0.775**	**0.399**	**0.628**

Table 5: Official results for the combined evaluation of named entity recognition and categorization compared against the state-of-the-art system. The results are evaluated in slot error rate (SER), i.e. a smaller value is better. The scores for the TagIt system are as reported in their paper.

3.3 Event extraction

As discussed earlier, there are two tasks in the BB3 which involve extracting the relations between BACTERIA and HABITAT/GEOGRAPHICAL entities: (1) The BB3-event task, for which all manually annotated entities are given (even for the test set). This task aims to assess the performance of relation extraction systems; (2) The BB3-event+ner task, for which, entities for the test set are hidden and the aim is assessing the joint performance of the NER and the relation extraction systems.

It should be highlighted that the performance of the NER system has a direct impact on the relation extraction system and subsequently on the performance of an end-to-end system for the

BB3-event+ner task. On one hand, if the NER system produces extremely low recall outputs, the relation extraction system will fail to extract some of the valid relations, simply because it only investigates the existence of possible relations among the *given* entities. On the other hand, if the NER system provides high recall but very low precision predictions, this means that words mistakenly detected as valid entities are given to the relation extraction system. For each given entity, the relation extraction system pairs it with other provided entities in the sentence and tries to classify all candidate pairs. Hence, invalid entities will lead to generation of candidate pairs in which one or even both of the entities are actually invalid. Since the relation extraction system is trained on valid entity pairs, i.e., (BACTERIA,HABITAT) or (BACTERIA,GEOGRAPHICAL), it can easily produce a plethora of false-positives and hence, its precision will dramatically drop.

To summarize, if the NER system performance is low (low precision and/or low recall), even a very high-performance relation extraction system will not be able to compensate. Thus, when building an end-to-end system, the joint performance of NER and relation extraction should be assessed since individual performances do not reflect how efficiently the system will work in real-world applications.

The official performance of our relation extraction system alone when evaluated against the test set of the BB3-event task is 0.512 measured in F-score (0.444 recall and 0.605 precision), achieving the third place among Shared Task participants for this task.

Dataset	Overall	Habitat	Geography
Development set			
With sub-optimal entities	0.423	0.390	0.576
With optimal entities	0.429	0.395	0.604
Test set			
With sub-optimal entities	0.372	0.388	0.207
With optimal entities	0.381	0.386	0.319

Table 6: Combined performance of our named entity recognition and event extraction systems on the event+ner task reported in F-score as measured by the official evaluation service.

For the BB3-event+ner task, the official results on the development and the test set are given in Table 6. As discussed earlier, to increase the performance of the NER system, we combine the BB3 with older BB datasets. This leads to the best prediction performance (denoted as *optimal*). Thus, we report and compare the overall performance of the end-to-end system when we use these entities. To establish a fair comparison with previously published systems we also report results for models trained only on the BB3 (denoted as *sub-optimal*). As Table 6 shows, using previous BB-ST data for training the NER leads to 3pp increase in F-score of (BACTERIA,GEOGRAPHICAL) relations on the development set and about 11pp for the test set, probably due to the drastically increased performance for GEOGRAPHICAL entity detection. Unfortunately, since there are much less (BACTERIA,GEOGRAPHICAL) relations than (BACTERIA,HABITAT) relations in the data, our approach increases the overall F-score only by 1pp for the test set.

Table 7 compares the performance of our end-to-end system with the winning team in the BB3-event+ner task (LIMSI, developed by Grouin (2016)). As it can be seen in the table, our system outperforms the winning team by 19pp in F-score, achieving the new state-of-the-art score for the task. Even if we solely rely on BB3 data for the NER system, the improvement is 18pp in F-score. We emphasize that no other data than BB3 is used for training/optimization of our relation extraction system in any way.

Teams	F-score	Recall	Precision	SER
LIMSI	0.192	0.191	0.193	1.558
Our system	**0.381**	**0.292**	**0.548**	**0.891**

Table 7: Official evaluation results for BB3-event+ner test data of our system compared to LIMSI, the winning team in the Shared Task.

4 Conclusions and future work

In this work, we introduced an open-source end-to-end system, capable of named-entity detection/normalization and relation extraction to extract information about bacteria and their habitats from text. Our system is trained and evaluated on the BioNLP Shared Task 2016 Bacteria Biotope data.

According to the official evaluation, our entity detection and categorization system would have achieved the second place in BB3. Compared to the best performing system on cat+ner, TagIt, we consider that our approach on NER can still be improved, especially for HABITAT entities. First, we consider employing a *post-processing* step in order to recover embedded entities which are not currently handled by NERsuite. An effective post-processing step should have a substantial impact on our NER system as the embedded entities accounted for over 10% of all HABITAT mentions.

Our categorization system outperforms the best performing system of BB3 by 1.2pp in the official evaluation, constituting the new state-of-the-art for this task. Our system also relies less on rule-based or heuristic preprocessing steps and uses the same general approach for both BACTERIA and HABITAT mentions suggesting that it will be more adaptable for new entity types.

As 9.6% of the HABITAT entities in the official training set have more than one gold standard ontology annotation whereas our current system is only assigning a single concept for each entity, one future work direction is to assess different ways of associating entities with multiple concepts. In the simplest form this could be implemented by defining a similarity threshold instead of selecting only the best matching concept.

Since the character level ngrams resulted in significantly better performance than our word level baseline, the exploration of character level neural approaches is also warranted for the categorization task and will be tested in the future.

Official evaluation shows that the joint performance of entity detection and relation extraction of our end-to-end system outperforms the winning team by 19pp on F-score, establishing a new top score for the event+ner task. In this work we did not use previous BB Shared Task data for training the relation extraction system. However, as a future work we would like to investigate the effect of utilizing previous BB Shared Task data.

As a future work, we would like to run our system on large-scale, on all PubMed abstracts and PubMed Central Open Access full articles to form a publicly available knowledge base.

We highlight that the methods discussed and used in this work are not only applicable for BB3 tasks and can be beneficial for other entity detection/normalization and relation extraction projects as well.

5   Acknowledgements

This work was supported by ATT Tieto käyttöön grant. Computational resources were provided by CSC - IT Center For Science Ltd., Espoo, Finland.

References

Jari Björne, Filip Ginter, and Tapio Salakoski. 2012. University of Turku in the BioNLP'11 Shared Task. *BMC bioinformatics* 13(11):S4.

Jari Björne and Tapio Salakoski. 2013. TEES 2.1: Automated annotation scheme learning in the BioNLP 2013 Shared Task. In *Proceedings of the BioNLP Shared Task 2013 Workshop*. pages 16–25.

Olivier Bodenreider. 2004. The unified medical language system (UMLS): integrating biomedical terminology. *Nucleic acids research* 32(suppl 1):D267–D270.

Robert Bossy, Wiktoria Golik, Zorana Ratkovic, Philippe Bessières, and Claire Nédellec. 2013. BioNLP Shared Task 2013–an overview of the Bacteria Biotope task. In *Proceedings of the BioNLP Shared Task 2013 Workshop*. pages 161–169.

Robert Bossy, Julien Jourde, Philippe Bessieres, Maarten Van De Guchte, and Claire Nédellec. 2011. BioNLP Shared Task 2011: Bacteria Biotope. In *Proceedings of the BioNLP Shared Task 2011 Workshop*. Association for Computational Linguistics, pages 56–64.

Razvan C. Bunescu and Raymond J. Mooney. 2005. A shortest path dependency kernel for relation extraction. In *Proceedings of Human Language Technology Conference and Conference on Empirical Methods in Natural Language Processing (HLT/EMNLP)*. pages 724–731.

Rui Cai, Xiaodong Zhang, and Houfeng Wang. 2016. Bidirectional recurrent convolutional neural network for relation classification. In *Proceedings of the 54th Annual Meeting of the Association for Computational Linguistics, ACL 2016, August 7-12, 2016, Berlin, Germany, Volume 1: Long Papers*.

Eugene Charniak and Mark Johnson. 2005. Coarse-to-fine n-best parsing and maxent discriminative reranking. In *Proceedings of the 43rd annual meeting on association for computational linguistics*. Association for Computational Linguistics, pages 173–180.

Helen V Cook, Evangelos Pafilis, and Lars Juhl Jensen. 2016. A dictionary-and rule-based system for identification of bacteria and habitats in text. In *Proceedings of the 4th BioNLP Shared Task 2016 Workshop*. Association for Computational Linguistics, Berlin, Germany.

Marie-Catherine de Marneffe, Bill MacCartney, and Christopher D. Manning. 2006. Generating typed dependency parses from phrase structure parses. In *Proceedings of the 5th International Conference on Language Resources and Evaluation (LREC 2006)*. pages 449–454.

Louise Deléger, Robert Bossy, Estelle Chaix, Mouhamadou Ba, Arnaud Ferré, Philippe Bessieres, and Claire Nédellec. 2016. Overview of the Bacteria Biotope task at BioNLP Shared Task 2016. In *Proceedings of the 4th BioNLP Shared Task Workshop. Berlin: Association for Computational Linguistics*. Association for Computational Linguistics, Berlin, Germany.

Jenny Rose Finkel, Trond Grenager, and Christopher Manning. 2005. Incorporating non-local information into information extraction systems by Gibbs sampling. In *Proceedings of the 43rd annual meeting on association for computational linguistics*. Association for Computational Linguistics, pages 363–370.

Cyril Grouin. 2016. Identification of mentions and relations between bacteria and biotope from PubMed abstracts. In *Proceedings of the 4th BioNLP Shared Task Workshop*. Association for Computational Linguistics, Berlin, Germany, pages 64–72.

Kai Hakala. 2015. UTU: Adapting biomedical event extraction system to disorder attribute detection. In *Proceedings of the 9th International Workshop on Semantic Evaluation (SemEval 2015)*. Association for Computational Linguistics, pages 375–379.

Kai Hakala, Suwisa Kaewphan, Tapio Salakoski, and Filip Ginter. 2016. Syntactic analyses and named entity recognition for PubMed and PubMed Central–up-to-the-minute. In *Proceedings of the 4th BioNLP Shared Task Workshop*. Association for Computational Linguistics, Berlin, Germany, pages 102–107.

Suwisa Kaewphan, Kai Hakaka, and Filip Ginter. 2014. UTU: Disease mention recognition and normalization with CRFs and vector space representations. *Proceedings of the 8th International Workshop on Semantic Evaluation (SemEval 2014)* pages 807–11.

Suwisa Kaewphan, Sofie Van Landeghem, Tomoko Ohta, Yves Van de Peer, Filip Ginter, and Sampo Pyysalo. 2016. Cell line name recognition in support of the identification of synthetic lethality in cancer from text. *Bioinformatics* 32(2):276–282.

Jin-Dong Kim, Tomoko Ohta, Sampo Pyysalo, Yoshinobu Kano, and Jun'ichi Tsujii. 2011. Extracting bio-molecular events from literature – the BioNLP'09 shared task. *Computational Intelligence* 27(4):513–540.

Diederik P. Kingma and Jimmy Ba. 2014. Adam: A method for stochastic optimization. *CoRR* abs/1412.6980. http://arxiv.org/abs/1412.6980.

Robert Leaman, Rezarta Islamaj Doğan, and Zhiyong Lu. 2013. DNorm: disease name normalization with pairwise learning to rank. *Bioinformatics* page btt474.

Nut Limsopatham and Nigel Collier. 2016. Normalising medical concepts in social media texts by learning semantic representation. In *Proceedings of the 54th Annual Meeting of the Association for Computational Linguistics (Volume 1: Long Papers)*. Association for Computational Linguistics, Berlin, Germany, pages 1014–1023.

David McClosky. 2010. *Any Domain Parsing: Automatic Domain Adaptation for Natural Language Parsing*. Ph.D. thesis.

Farrokh Mehryary, Jari Bjrne, Sampo Pyysalo, Tapio Salakoski, and Filip Ginter. 2016. Deep learning with minimal training data: TurkuNLP Entry in the BioNLP Shared Task 2016. In *Proceedings of the 4th BioNLP Shared Task Workshop*. Association for Computational Linguistics, Berlin, Germany, pages 73–81.

Tomoko Ohta, Sampo Pyysalo, Jun'ichi Tsujii, and Sophia Ananiadou. 2012. Open-domain anatomical entity mention detection. In *Proceedings of the Workshop on Detecting Structure in Scholarly Discourse*. Association for Computational Linguistics, pages 27–36.

Fabian Pedregosa, Gaël Varoquaux, Alexandre Gramfort, Vincent Michel, Bertrand Thirion, Olivier Grisel, Mathieu Blondel, Peter Prettenhofer, Ron Weiss, Vincent Dubourg, Jake Vanderplas, Alexandre Passos, David Cournapeau, Matthieu Brucher, Matthieu Perrot, and Édouard Duchesnay. 2011. Scikit-learn: Machine learning in Python. *Journal of Machine Learning Research* 12:2825–2830.

Martin F Porter. 1980. An algorithm for suffix stripping. *Program* 14(3):130–137.

Sampo Pyysalo, Filip Ginter, Hans Moen, Tapio Salakoski, and Sophia Ananiadou. 2013. Distributional semantics resources for biomedical text processing. In *Proceedings of the 5th International Symposium on Languages in Biology and Medicine (LBM 2013)*. pages 39–44.

Rune Sætre, Kazuhiro Yoshida, Akane Yakushiji, Yusuke Miyao, Yuichiro Matsubayashi, and Tomoko Ohta. 2007. AKANE system: protein-protein interaction pairs in BioCreAtIvE2 challenge, PPI-IPS subtask. In *Proceedings of the Second BioCreative Challenge Workshop*. pages 209–212.

Gerard Salton and Christopher Buckley. 1988. Termweighting approaches in automatic text retrieval. *Information processing & management* 24(5):513–523.

Nitish Srivastava, Geoffrey Hinton, Alex Krizhevsky, Ilya Sutskever, and Ruslan Salakhutdinov. 2014.

Dropout: A simple way to prevent neural networks from overfitting. *Journal of Machine Learning Research* 15:1929–1958.

Mert Tiftikci, Hakan Şahin, Berfu Büyüköz, Alper Yayıkçı, and Arzucan Özgür. 2016. Ontology-based categorization of bacteria and habitat entities using information retrieval techniques. In *Proceedings of the 4th BioNLP Shared Task 2016 Workshop*. Association for Computational Linguistics, Berlin, Germany.

Yan Xu, Lili Mou, Ge Li, Yunchuan Chen, Hao Peng, and Zhi Jin. 2015. Classifying relations via long short term memory networks along shortest dependency paths. In *In Proceedings of the 2015 Conference on Empirical Methods in Natural Language Processing*. pages 1785–1794.

Creation and evaluation of a dictionary-based tagger
for virus species and proteins

Helen Victoria Cook
Rūdolfs Bērziņš
Cristina Leal Rodríguez
Novo Nordisk Foundation
Center for Protein Research,
Faculty of Health and Medical Sciences,
University of Copenhagen, Denmark
`helen.cook@cpr.ku.dk`

Juan Miguel Cejuela
TUM, Department of Informatics,
Bioinformatics & Computational Biology,
i12, Boltzmannstr. 3, 85748
Garching/Munich, Germany

Lars Juhl Jensen
Novo Nordisk Foundation
Center for Protein Research,
Faculty of Health and Medical Sciences,
University of Copenhagen, Denmark
`lars.juhl.jensen@cpr.ku.dk`

Abstract

Text mining automatically extracts information from the literature with the goal of making it available for further analysis, for example by incorporating it into biomedical databases. A key first step towards this goal is to identify and normalize the named entities, such as proteins and species, which are mentioned in text. Despite the large detrimental impact that viruses have on human and agricultural health, very little previous text-mining work has focused on identifying virus species and proteins in the literature. Here, we present an improved dictionary-based system for viral species and the first dictionary for viral proteins, which we benchmark on a new corpus of 300 manually annotated abstracts. We achieve 81.0% precision and 72.7% recall at the task of recognizing and normalizing viral species and 76.2% precision and 34.9% recall on viral proteins. These results are achieved despite the many challenges involved with the names of viral species and, especially, proteins. This work provides a foundation that can be used to extract more complicated relations about viruses from the literature.

1 Introduction

Viruses are major human and agricultural pathogens. Influenza A in the US alone costs billions of dollars each year in lost wages and medical expenses (Molinari et al., 2007). Worldwide, Influenza, Human papilloma virus and Hepatitis C virus are each responsible for at least a quarter of a million deaths each year (WHO, 2014). At the same time, viruses such as Zika virus are emerging as global health threats as the habitats of their vectors are expanding due to climate change (Mills et al., 2010; Fauci and Morens, 2016). Such arboviruses are previously neglected diseases, and as such vaccines and antiviral drugs are not available for them, posing a large health risk.

The impact of outbreaks in livestock can also be immense. The 2001 Foot and mouth disease virus outbreak in the UK cost an estimated 8 billion (Knight-Jones and Rushton, 2013) and still today much remains unknown about the virus, including the mechanisms for persistent infection (Paul et al., 2010), and the virus' interactions with the immune system that may aid cross serotype vaccine production (Paton and Taylor, 2011).

The study of viruses is aided by bioinformatics resources such as protein–protein interaction databases. Having a comprehensive picture of a virus protein's interaction partners is crucial to the understanding of the viral lifecycle and aids in the search for vaccines and antiviral drugs (Shah et al., 2015). However, manually creating and maintaining such resources is a cost, time and labour intensive endeavour (Attwood et al., 2015). Text mining provides a means to automatically identify relevant publications and the entities of interest that are mentioned in them quickly and at low cost. A first step towards building these resources for viruses is the identification of viral species and their proteins in text.

1.1 Background

Text mining for viruses presents several challenges over text mining for species of cellular organisms. Viruses are often known by many different names, either because a virus was identified in different countries and given different names (e.g. Bovine pestivirus and Bovine viral diarrhoea virus), or because the taxonomy has changed (e.g.

Proceedings of the BioNLP 2017 workshop, pages 91–98,
Vancouver, Canada, August 4, 2017. ©2017 Association for Computational Linguistics

polyomaviruses). Another source of synonyms is the use of the disease that the virus causes in place of the virus name.

Viral proteins are even more challenging to text mine, as they are often referred to by one-letter names such as E or M. Further, even if their names are longer, they can be written in many different orthographic variants e.g. U(S)11, Us11, or US11. Viral proteins may also have many synonyms related to the gene name, position on a segment, or may be referred to by their function e.g. "the polymerase". Some RNA viruses have polyproteins which complicates their analysis. Their viral mRNA codes for a single open reading frame that is translated to a polypeptide product, which is then post translationally cleaved into functional protein products. Bioinformatics databases such as UniProt (The UniProt Consortium, 2014) often give first class identifiers to the polyprotein but not to the cleavage products, thus complicating the process of referring to the functional protein product.

These challenges can be mitigated by using a dictionary approach to text mining. In such an approach, comprehensive dictionaries are created to contain all the alternative names that are likely to be referred to in a corpus. In this work, we have chosen to use a dictionary based method based on the success of this approach to identify bacteria species and biotopes (Cook et al., 2016). We have chosen to use curated databases (NCBI taxonomy and UniProt) to populate the dictionary instead of other approaches such as unsupervised methods to learn which items are named entities (Neelakantan and Collins, 2014), as the data available in these databases is high quality and openly available. Furthermore, starting with a resource dramatically reduces the difficulty of normalization of recognized entities.

Previous work in this field includes LINNAEUS (Gerner et al., 2010), a dictionary-based system that is also designed to recognize species in abstracts. The SPECIES tagger (Pafilis et al., 2013) is a newer and faster dictionary system that aims to identify names of any species in biomedical abstracts. SPECIES has achieved good performance when tagging names of viruses species in abstracts from virology journals. A more recent and specialized effort used the dictionary and template-based ANDSystem to text mine the HCV interactome (Saik et al., 2015).

Here, we improve on the SPECIES dictionary for all virus species, and additionally tag names of virus proteins for those proteins that have reference proteomes in UniProt. We have created viral species and protein dictionaries, and a gold-standard corpus that has been annotated by 4 human annotators.

2   Availability

The version of the dictionaries used in this publication are available at `http://figshare.com/articles/ virus_entities_tsv/4721287` and the most recent version will be available at `http://download.jensenlab.org/`. The V300 corpus and annotator guidelines is publicly available at `http: //www.tagtog.net/-corpora`. The evaluation code is available at `http://github. com/bitmask/textmining-stats`. The tagger software used for this work is available at `http://bitbucket.org/ larsjuhljensen/tagger`.

3   Methods

3.1   Dictionary creation and tagging

Virus names were taken from NCBI Taxonomy (Sayers et al., 2009) and included all synonyms at all taxonomic levels under the viruses superkingdom. Disease names were taken from the Disease Ontology (Kibbe et al., 2015) and were manually mapped onto the correct virus taxid, giving an additional 387 names for 102 species that are human pathogens. This resulted in a total of 173,367 names for 150,885 virus tax levels.

Virus species name acronyms were taken from the ninth ICTV report on virus taxonomy (King et al., 2012) by text mining the document and extracting any text in parentheses that appears to be an acronym and that follows a match for a virus name. This way we found 778 acronyms, more than 500 of which were not found in the previous sources, for 662 virus species.

Virus protein names were taken from UniProt reference proteomes (The UniProt Consortium, 2014) as of Aug 31, 2015. Viruses that did not have complete proteomes were not included in the protein dictionary, although they are included in the species dictionary. Protein names and synonyms were taken from all fields in the UniProt record, including the protein name, short name, gene, and chain entries if the protein is a polyprotein. Additionally, many variants of the protein names were generated following a set of rules to cover orthographic variation, such as "X protein" is expanded to "protein X" and "X". For a complete list of rules, refer to the code. This resulted in 16,580 proteins with 112,013 names from 397 virus species.

Stopwords were adapted from the text mining done for the text-mining channel of the STRING database (Szklarczyk et al., 2015). Additional stopwords were found by running the dictionary over all documents in PubMed and inspecting the 100 most frequent matches to determine if they should be stopworded. Although normally considered to be stopwords by the tagging system, specific one and two letter names from the dictionary were permitted to be matches to enable finding very short protein names.

Automated tagging used the dictionaries described above and the tagger text-mining system developed for the SPECIES resource (Pafilis et al., 2013).

3.2 Corpus creation and gold standard creation

300 abstracts were selected randomly by filtering abstracts mentioned in reviewed UniProt entries for viral proteins for top virology journals as determined by impact factor. Documents were divided among four annotators such that each pair of annotators shared 10 documents, implying that 20% of the documents were annotated by two annotators. These overlapping documents were used to calculate inter-annotator agreement (IAA), and the annotators were blind to which documents were in this set throughout the project.

Annotation guidelines were agreed upon following the annotation of 10 documents in a pilot set, which were not used in the evaluation of IAA or to assess the performance of the tagger. All abstracts were manually annotated using tagtog (Cejuela et al., 2014), an online system for text mining. Species names were normalized to NCBI taxonomic identifiers. Protein names were normalized to UniProt entry names, unless they were the cleavage product of a polyprotein, in which case they were normalized to their chain name.

3.3 Evaluation

The IAA among the human annotators was determined separately for viral species and proteins by determining the number of annotations that overlap and contain the same normalization. Boundaries of annotations were considered to match if the annotations overlapped.

Species normalizations were considered to match if one was a parent of the other and if both were at or below species level, or if both were below species level and had a common parent. For example, both of the following pairs were considered matches: "Influenza A" and "Influenza A H1N1", and "Influenza A H1N1" and "Influenza A H7N9". This allowed for an annotation to not be penalized if the strain was annotated instead of the species, or if two different strains of the same species were annotated. Protein normalizations were considered to match if they were within 90% identity according to BLAST (Zhang et al., 2000).

IAA was measured by F-score, however since we allow boundaries to overlap, this measure may not be symmetric. If one annotator has annotated "long form (short form)" as one annotation, and another annotator has annotated it as two annotations, then this will count as one true positive when comparing the first annotator to the second, but as two true positives when comparing the second annotator to the first. To avoid this asymmetry, we counted all the true positives, false negatives and false positives across both annotators.

The guidelines specify that if a span refers to multiple entities, then it should be normalized to each of them. Each normalization was treated as contributing separately to the number of true or false positives. A special case was established for Adenovirus, which is a large genus containing very many species of viruses that have a highly conserved set of proteins. Adenovirus proteins are often referred to in general in the literature, without specifying a specific species. Manual annotation of Adenovirus proteins required that only one representative protein from one species be tagged, thus effectively treating this genus as a single species.

The recall and precision of the tagger was calculated against the consensus of the human annotations. The consensus was determined as follows. If only one annotator annotated the document, their annotations were taken as the gold standard. The annotations were similarly accepted as the gold standard if two annotators agreed on position and normalization. However, if there was a disagreement, then a third annotator was asked to resolve it. For positions that overlapped, the union of the spans was used as the consensus.

The precision and recall were calculated in three different ways. The first method required that the boundaries and normalizations of the consensus and tagger annotations match. The second method, "boundaries only", required only the boundaries of the annotations to match. The last method, "document level normalization", compared the lists of unique normalizations found in the document, regardless of position and number of occurrences.

4 Results and Discussion

4.1 Corpus and Inter-annotator agreement

The corpus consisted of 300 documents with 1,826 species and 2,540 protein annotations. There was overall good agreement between annotators for both species and proteins. The mean IAA F-score for species was 87.3%, and considering boundaries only was 90.0%. For proteins, the mean IAA F-score was 76.5%, which rose to 86.9% when considering boundaries only. Detailed results are shown in figure 1.

There was substantial agreement between annotators regarding the location of species and protein annotations, and there was also good agreement on the normalization of species. However, there was less agreement among protein normalizations than those for species. 26% of these disagreements involve one annotator normalizing a protein name to a UniProt entry, and the second annotator reporting the normalization as unknown. An additional 20% of the disagreement is due to an annotator normalizing a span to multiple entities and another annotator normalizing it to fewer entities. Such cases, in which an abstract discusses a protein in one virus and compares it to a closely related protein, can be ambiguous and refer to the protein without being completely clear about which species is being referred to.

However, the largest part this disagreement comes from instances in which annotators have normalized to different proteins that are different enough to not pass the 90% identity BLAST criterion. Manual inspection of these proteins indicate that the majority are correct, but that fast viral evolution has caused the protein sequences of similar isolates to diverge. The set of documents randomly chosen to calculate IAA was unlucky to contain a few documents containing proteins that

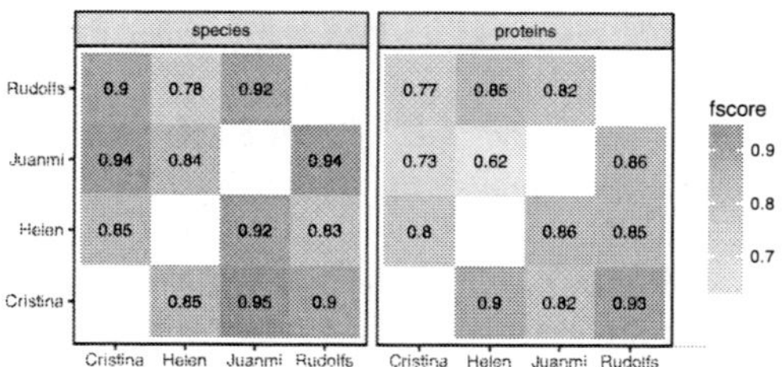

Figure 1: Inter-annotator agreement for viral proteins and species. Above the diagonal both normalization and boundaries are required to be correct, below the diagonal only identification of boundaries are required to be correct.

are quite divergent, but this is not representative of the whole corpus. This can be seen by dropping the BLAST identity criterion to 50%, which then accounts for 29% of the difference between annotators, but increases the tagger precision and recall by only 1%.

4.2 Tagger performance for species

The automatic tagger achieved 81.5% precision and 73.3% recall for the combined task of recognizing and normalizing viral species. When requiring only the boundaries to be correct, i.e. recognition but not normalization, the precision and recall were 93.1% and 79.8% respectively. At the document level, the normalization precision was 74.9% and the recall was 85.4%. Results are summarized in table 1. Combined, this shows that if the tagger identifies a viral species, it is very likely that a viral species is mentioned at the reported position, and it is also likely that the tagger has normalized it correctly. Also, the tagger correctly identifies most of the species that are mentioned in a document.

In 43% of the cases of incorrect species normalization, the tagger has identified both the correct species normalization and additional normalizations with the same abbreviation. For example, the tagger normalized SV40 to "Simian virus 40", which is correct, but also to "Polyomavirus sp." under unclassified Polyomaviridae because both taxa have SV40 as an abbreviation in the NCBI taxonomy. The abbreviation SV40 will thus count as both a true positive and a false positive with respect to normalization. If instead such partially correct normalizations were counted only as true positives, the precision would rise from 81.5% to 85.8%.

The tagger does not attempt to correctly identify all referenced entities in sentence constructs such as "HSV types 1 and 2" although such normalizations are obvious to human annotators. More ambiguously, papers that discuss Influenza proteins or Adenovirus proteins, without specifying the species (such as Influenza A, or Adenovirus type 1) are not clear about what exactly is being referred to.

In an additional 32% of the cases of incorrect species normalization, an annotator identified the virus as unclassified in which case it and the taxa identified by the tagger joined the taxonomic tree above the species level, and so was not considered to be a match by the matching code. If the match is relaxed to genus level, then the precision will rise from 81.5% to 85.0% and to 86.3% if accepting also partially correct normalizations as described above.

Despite efforts to be comprehensive, some abbreviations are missing from the virus dictionary, for example the abbreviations Ad2 and Ad5 for Adenovirus type 2 and 5 respectively were not included in the dictionary. The tagger does contain logic to identify and expand acronyms on the fly, but has very strict matching criteria to prevent false positives (Pafilis et al., 2013). Further, synonyms that are not present in NCBI taxonomy will not be identified. For example "Blackberry yellow vein disease" was not identified as as synonym for "Blackberry yellow vein virus" and so was not found by the tagger. This could be improved with more comprehensive synonym generation.

The tagger will tag all instances of entries in its dictionary, even in contexts that are not appropriate. The annotation guidelines state that viruses that are used as vectors should not be tagged, since the scientific work they are mentioned in is not primarily about the virus. However, this is a matter of opinion and the opposite case could also be argued. Regardless, the tagger cannot distinguish the context in which viruses are mentioned, and will blindly tag all occurrences of the virus name.

4.3 Tagger performance for proteins

For combined recognition and normalization of viral proteins, the precision and recall of the tagger were 76.2% and 34.9% respectively. Observing boundaries only, the precision and recall rose to 87.4% and 40.0% respectively. At the document level, the normalization precision was 76.2% and

	Precision	Recall
Normalisation	81.5%	73.3%
Boundaries only	93.1%	79.8%
Doc level normalisation	74.9%	85.4%
Partially correct norm	85.8%	73.3%
Match at genus level	85.0%	73.5%
Previous two criteria	86.3%	73.5%

Table 1: Summary of species precision and recall for different evaluation criteria: Normalization and recognition, recognition of boundaries only, normalization at the document level, treating entities that have been normalized to multiple entities as correct if one of the normalizations is correct, relaxing the matching criterion to the genus level, and finally allowing both of the previous two criteria.

the recall was 38.1%. Results are summarized in table 2.

Since viral protein names are so short and not unique to one species, the tagger will only tag protein names for species that have already been identified. This means that the theoretical upper bound for tagging proteins is equivalent to the species document level normalization recall (85.4%) assuming that all the proteins are present in the dictionary. However, the dictionary only contains protein names for species that are contained in reviewed UniProt proteomes, a total of 348 species and 88.1% of the proteins mentioned in the corpus. This gives a maximum possible recall of 75.2% for proteins. Conversely, since the tagger detects proteins only after the species has been detected, the normalization of the viral proteins that are found is quite accurate.

Considering only annotation of the proteins in the dictionary, the precision was 86.0% and the recall 35.5%. Recall does not change significantly from considering all proteins because there are 10 times more false negatives due to not locating the protein compared to false negatives due to incorrectly normalizing the protein. At the document level, the normalization precision of proteins that were present in the dictionary is 77.1% and the recall is 50.7%.

Viral proteins are very hard for the tagger to identify due to the diversity of names that are used to refer to them. For example, the tagger has missed 97% of names in which the protein is referred to by its molecular weight (e.g. "the 33K protein"). Including these synonyms would increase the recall by 4 percentage points. Similarly, the tagger has tagged only 10% of the cases in which the viral protein is referred to by its function (e.g. "the helicase"). Including these synonyms would increase the recall by 6 percentage points. As observed for species, the tagger does not recognize novel abbreviations, such as "sGP" for the Ebola virus nonstructural small glycoprotein, and such constructs are used quite frequently in the literature. Better on-the-fly acronym identification in the tagger may help increase this recall rate.

Another source of error is the ambiguity of terms used in the text to refer to parts of the virus that are also names of proteins such as "capsid". Although the frequently-named capsid protein is the main constituent of the viral capsid, references in the text to "capsid" are often ambiguous as to whether they refer to the protein or to the assembled virus part. The annotation guidelines state that such terms should only be tagged if they refer to the protein and should not be tagged if they refer to part of the virus, but these cases are often difficult to distinguish in practice.

The tagger identifies false positives at a much lower rate than false negatives. Since very short protein names are present in the dictionary, it is much more likely for these names to appear in places that are not in the context of a protein. For example, Coronavirus infectious bronchitis virus has a spike protein abbreviated S, however discussion of the polyprotein cleavage site before a serine residue will be false positively tagged as serine is also abbreviated S.

Normalization of protein names to multiple entities can also be incorrect in instances where an abstract discusses both a specific protein in one species, and the same protein in many species. The tagger will tag all instances of the protein name with all species and will not be able to distinguish the instances that refer only to the protein in a specific species, whereas human annotators are more easily able to distinguish these cases.

4.4 Results in other corpora

Compared to the S800 virus corpus (Pafilis et al., 2013), the improved dictionary finds over 100 more mentions, including new abbreviations, but does not tag more general terms such as "infectious virus" and "avian viruses" which refer to

	Precision	Recall
Normalisation	77.5%	35.5%
Boundaries only	87.4%	40.0%
Doc level normalisation	77.1%	50.7%
Theoretical max recall	-	75.2%

Table 2: Summary of results for protein detection for different evaluation criteria: normalization and recognition, recognition of boundaries only and document level normalization. The theoretical maximum recall based on requiring the species to be recognized and present in the dictionary is also listed.

more than one species. Measured against the S800 gold standard for only virus annotations in the virus subset of the corpus, the improved tagger has a precision and recall of 63.3% and 57.0% respectively, compared to the initial results from SPECIES of 63.2% and 53.0% respectively.

Running the tagger over all of Medline finds over 53 million mentions of 8063 viral species in more than 1.5M articles. Of these, we have protein level detail for 348 species, and find over 10M mentions of 4668 unique proteins. The most commonly mentioned species is HIV-1, making up over 3% of species mentions.

5   Conclusions and Perspectives

As the biomedical literature continues to grow at an exponential rate (Lu, 2011), automated tools, such as text mining, are necessary to enable extracting information from the literature in a timely and efficient manner. Text mining is a means to automatically extract information from the literature without requiring manual curation of a large number of documents. It can be used successfully to extract virus species and proteins from abstracts that pertain to viruses with good precision and also, in the case of species, good recall. There is still much room to improve the recall of proteins due to the abundance of alternative names that are used to refer to them. Further, the tagger does not recognize disjoint entities, and since there has recently been progress in this field (Tang et al., 2013), this could also be an area for future improvement of the tagger.

These results can be used in future work to extract co-occurrences of virus and host proteins, which could imply an interaction between these proteins. Integrating virus-host protein-protein interactions into the larger host interaction network may provide insight into viral mechanisms of disease. Work done specifically on EBV, HPV, and Hepatitis C virus (Gulbahce et al., 2012; Mosca et al., 2014) revealed that host proteins local to viral targets form network modules that are related to the diseases caused by these viruses. With the virus-agnostic tools presented here, such work can be scaled up to easily enable investigation of all viruses for which there is sufficient data.

The work presented here could also be used as a foundation to identify viruses that are understudied compared to their impact, and may reveal future directions that are promising to study. The interrelationship of proteins and diseases has been explored recently using text mining to assess both the strength of an interaction between a protein and a disease, and also the scarceness of publications about a given protein target (Cannon et al., 2017). This gives researchers an overview of understudied proteins that could be relevant for disease etiology. A similar approach could be taken to reveal new directions in virus research.

Acknowledgements

The authors would like to thank Jorge Campos for his work on the interface of tagtog to support this project.

References

Teresa Attwood, Bora Agit, and Lynda Ellis. 2015. Longevity of Biological Databases. *EMBnet.journal* 21(0).

Daniel C Cannon, Jeremy J Yang, Stephen L Mathias, Oleg Ursu, Subramani Mani, Anna Waller, Stephan C Schurer, Lars Juhl Jensen, Larry A Sklar, Cristian G Bologa, and Tudor I Oprea. 2017. TIN-X: Target Importance and Novelty Explorer. *Bioinformatics* pages 1–3.

Juan Miguel Cejuela, Peter McQuilton, Laura Ponting, S. J. Marygold, Raymund Stefancsik, Gillian H. Millburn, and Burkhard Rost. 2014. Tagtog: Interactive and text-mining-assisted annotation of gene mentions in PLOS full-text articles. *Database* 2014:1–8. https://doi.org/10.1093/database/bau033.

Helen Cook, Evangelos Pafilis, and Lars Jensen. 2016. A dictionary- and rule-based system for identification of bacteria and habitats in text. In *Proceedings of the 4th BioNLP Shared Task Workshop*. pages 50–55. http://www.aclweb.org/anthology/W/W16/W16-30.pdf#page=60.

Anthony S Fauci and David M Morens. 2016. Zika Virus in the Americas Yet Another Arbovirus Threat. *The New England journal of medicine* 374:601–604. https://doi.org/10.1056/NEJMp1600297.

Martin Gerner, Goran Nenadic, and Casey M Bergman. 2010. LINNAEUS: a species name identification system for biomedical literature. *BMC bioinformatics* 11(1):85. https://doi.org/10.1186/1471-2105-11-85.

Natali Gulbahce, Han Yan, Amélie Dricot, Megha Padi, Danielle Byrdsong, Rachel Franchi, Deok-Sun Lee, Orit Rozenblatt-Rosen, Jessica C. Mar, Michael A. Calderwood, Amy Baldwin, Bo Zhao, Balaji Santhanam, Pascal Braun, Nicolas Simonis, Kyung-Won Huh, Karin Hellner, Miranda Grace, Alyce Chen, Renee Rubio, Jarrod A. Marto, Nicholas A. Christakis, Elliott Kieff, Frederick P. Roth, Jennifer Roecklein-Canfield, James A. DeCaprio, Michael E. Cusick, John Quackenbush, David E. Hill, Karl Münger, Marc Vidal, and Albert-László Barabási. 2012. Viral Perturbations of Host Networks Reflect Disease Etiology. *PLoS Computational Biology* 8(6):e1002531. https://doi.org/10.1371/journal.pcbi.1002531.

Warren A. Kibbe, Cesar Arze, Victor Felix, Elvira Mitraka, Evan Bolton, Gang Fu, Christopher J. Mungall, Janos X. Binder, James Malone, Drashtti Vasant, Helen Parkinson, and Lynn M. Schriml. 2015. Disease Ontology 2015 update: An expanded and updated database of Human diseases for linking biomedical knowledge through disease data. *Nucleic Acids Research* 43(D1):D1071–D1078. https://doi.org/10.1093/nar/gku1011.

Andrew M.Q. King, Michael J. Adams, Eric B. Carstens, and Elliot J. Lefkowitz, editors. 2012. *Virus Taxonomy: Classification and Nomenclature of Viruses: Ninth Report of the International Committee on Taxonomy of Viruses*. Elsevier Academic Press. https://doi.org/10.1016/B978-0-12-384684-6.X0001-8.

T. J D Knight-Jones and J. Rushton. 2013. The economic impacts of foot and mouth disease - What are they, how big are they and where do they occur? *Preventive Veterinary Medicine* 112(3-4):162–173. https://doi.org/10.1016/j.prevetmed.2013.07.013.

Zhiyong Lu. 2011. PubMed and beyond: A survey of web tools for searching biomedical literature. *Database* 2011:1–13. https://doi.org/10.1093/database/baq036.

James N. Mills, Kenneth L. Gage, and Ali S. Khan. 2010. Potential influence of climate change on vector-borne and zoonotic diseases: A review and proposed research plan. *Environmental Health Perspectives* 118(11):1507–1514. https://doi.org/10.1289/ehp.0901389.

Noelle Angelique M. Molinari, Ismael R. Ortega-Sanchez, Mark L. Messonnier, William W. Thompson, Pascale M. Wortley, Eric Weintraub, and Carolyn B. Bridges. 2007. The annual impact of seasonal influenza in the US: Measuring disease burden and costs. *Vaccine* 25(27):5086–5096. https://doi.org/10.1016/j.vaccine.2007.03.046.

Ettore Mosca, Roberta Alfieri, and Luciano Milanesi. 2014. Diffusion of Information throughout the Host Interactome Reveals Gene Expression Variations in Network Proximity to Target Proteins of Hepatitis C Virus. *PLoS ONE* 9(12):e113660. https://doi.org/10.1371/journal.pone.0113660.

Arvind Neelakantan and Michael Collins. 2014. Learning Dictionaries for Named Entity Recognition using Minimal Supervision. *Proceedings of the 14th Conference of the European Chapter of the Association for Computational Linguistics* pages 452–461. http://www.aclweb.org/anthology/E14-1048.

Evangelos Pafilis, Sune P. Frankild, Lucia Fanini, Sarah Faulwetter, Christina Pavloudi, Aikaterini Vasileiadou, Christos Arvanitidis, and Lars Juhl Jensen. 2013. The SPECIES and ORGANISMS Resources for Fast and Accurate Identification of Taxonomic Names in Text. *PLoS ONE* 8(6):2–7. https://doi.org/10.1371/journal.pone.0065390.

David J Paton and Geraldine Taylor. 2011. Developing vaccines against foot-and-mouth disease and some other exotic viral diseases of livestock. *Philosophical transactions of the Royal Society of London. Series B, Biological sciences* 366(1579):2774–81. https://doi.org/10.1098/rstb.2011.0107.

William E Paul, Michael Mcheyzer-williams, and David Barthold, Stephen. Bowen, R. Hedrick, Ronald. Knowles, Donald. Lairmore, Michael. Parrish, Colin. Saif, Linda. Swayne. 2010. Fenner'S Veterinary Virology. *Elsevier* 5th editio(August):43–51. https://doi.org/10.1016/B978-0-12-375158-4.X0001-6.

Olga V. Saik, Timofey V. Ivanisenko, Pavel S. Demenkov, and Vladimir A. Ivanisenko. 2015. Interactome of the hepatitis C virus: Literature mining with ANDSystem. *Virus Research* 218:40–48. https://doi.org/10.1016/j.virusres.2015.12.003.

Eric W Sayers, Tanya Barrett, Dennis A Benson, Stephen H Bryant, Kathi Canese, Vyacheslav Chetvernin, Deanna M Church, Michael DiCuccio, Ron Edgar, Scott Federhen, Michael Feolo, Lewis Y Geer, Wolfgang Helmberg, Yuri Kapustin, David Landsman, David J Lipman, Thomas L Madden, Donna R Maglott, Vadim Miller, Ilene Mizrachi, James Ostell, Kim D Pruitt, Gregory D Schuler, Edwin Sequeira, Stephen T Sherry, Martin Shumway, Karl Sirotkin, Alexandre Souvorov, Grigory Starchenko, Tatiana A Tatusova, Lukas Wagner, Eugene Yaschenko, and Jian Ye. 2009. Database resources of the National Center for Biotechnology Information. *Nucleic Acids Research* 37:D5–15. https://doi.org/10.1093/nar/gkn741.

Priya S. Shah, Jason A. Wojcechowskyj, Manon Eckhardt, and Nevan J. Krogan. 2015. Comparative mapping of host-pathogen protein-protein interactions. *Current Opinion in Microbiology* 27:62–68. https://doi.org/10.1016/j.mib.2015.07.008.

Damian Szklarczyk, Andrea Franceschini, Stefan Wyder, Kristoffer Forslund, Davide Heller, Jaime Huerta-Cepas, Milan Simonovic, Alexander Roth, Alberto Santos, Kalliopi P. Tsafou, Michael Kuhn, Peer Bork, Lars J. Jensen, and Christian Von Mering. 2015. STRING v10: Protein-protein interaction networks, integrated over the tree of life. *Nucleic Acids Research* 43(D1):D447–D452. https://doi.org/10.1093/nar/gku1003.

Buzhou Tang, Yonghui Wu, Min Jiang, Joshua C. Denny, and Hua Xu. 2013. Recognizing and Encoding Disorder Concepts in Clinical Text using Machine Learning and Vector Space. *Proceedings of the ShARe/CLEF Evaluation Lab* http://www.clef-initiative.eu/documents/71612/d596ae25-c4b3-4a9a-be4a-648a77712aaf.

The UniProt Consortium. 2014. UniProt: a hub for protein information. *Nucleic Acids Research* 43(D1):D204–212. https://doi.org/10.1093/nar/gku989.

WHO. 2014. WHO Fact Sheets: Influenza, HCV, HPV. http://www.who.int/mediacentre/factsheets/.

Zheng Zhang, Scott Schwartz, Lukas Wagner, and Webb Miller. 2000. A Greedy Algorithm for Aligning DNA Sequences. *Journal of Computational Biology* 7(12):203–214. https://doi.org/10.1089/10665270050081478.

Representation of complex terms in a vector space structured by an ontology for a normalization task

Arnaud Ferré [1,2], Pierre Zweigenbaum[2], Claire Nédellec[1]
[1]MaIAGE, INRA, Université Paris-Saclay, 78350 Jouy-en-Josas, France
[2]LIMSI, CNRS, Université Paris-Saclay, 91405 Orsay, France
`arnaud.ferre@universite-paris-saclay.fr`

Abstract

We propose in this paper a semi-supervised method for labeling terms of texts with concepts of a domain ontology. The method generates continuous vector representations of complex terms in a semantic space structured by the ontology. The proposed method relies on a distributional semantics approach, which generates initial vectors for each of the extracted terms. Then these vectors are embedded in the vector space constructed from the structure of the ontology. This embedding is carried out by training a linear model. Finally, we apply a cosine similarity to determine the proximity between vectors of terms and vectors of concepts and thus to assign ontology labels to terms. We have evaluated the quality of these representations for a normalization task by using the concepts of an ontology as semantic labels. Normalization of terms is an important step to extract a part of the information contained in texts, but the vector space generated might find other applications. The performance of this method is comparable to that of the state of the art for this task of standardization, opening up encouraging prospects.

1 Introduction

A lot of biomedical or biological knowledge is in a non-structured form, such as that expressed in scientific articles (Kang et al., 2013). For experts from these fields, the substantial increase in the specialized literature has created a significant need for automatic methods of information extraction (Ananiadou and McNaught, 2006). The task of normalization is one of the main tasks to respond to this need.

Normalization consists in standardizing terms (single- or multi-word) extracted from texts by linking them to non-ambiguous references, such as entries from existing knowledge bases. Concepts from an ontology can be used to represent these references in a formal and structured way. Term and their relationships carry a lot of the knowledge contained in texts, thus successful term identification is a key to getting access to the information (Krauthammer and Nenadic, 2004).

Standardization encounters several difficulties, such as the significant variability of the form of the terms, whether they are represented by one word (e.g. "child" / "kid" or "accommodation" / "home", etc.) or by several (e.g. "child" / "little boy" or "accommodation" / "dwelling place", etc.) (Nazarenko et al., 2006). Multiword terms, which have varied morphosyntactic structures and complex imbrications (mainly complex noun phrases), are particularly difficult to normalize (e.g. only with a different syntactic organization: "breast cancer" / "cancer of the breast"). In the literature, such as scientific articles in life sciences, complex noun groups are abundant (Maniez, 2007). An approach based on the similarity of form between term and semantic label appears limited to perform this task (Golik et al., 2011), because the form of the labels of the concepts is not necessarily close to the form of the terms to be annotated. Another difficulty arises from the large number of ontology concepts, making a supervised classification approach costly in manual annotation (e.g. over 2,000 categories for example in the ontology of bacterial habitats OntoBiotope (Bossy et al., 2015)).

An alternative approach is to calculate the semantic proximity between terms by distributional semantics. It is an approach based on the correlation between the similarity of meaning and the distribution similarity of semantic units (word, combination of words, sentence, documents, ...) (Firth, 1957; Harris, 1954). A semantic unit can then be represented by a vector: it is constructed from the context information in which the semantic unit is found. The proximity of vectors in

Proceedings of the BioNLP 2017 workshop, pages 99–106,
Vancouver, Canada, August 4, 2017. ©2017 Association for Computational Linguistics

this space can be transposed to a semantic proximity (Fabre and Lenci, 2015). Today, there are many methods for generating such vector spaces, such as Word2Vec (Mikolov et al., 2013), but they usually focus on massive data sets (Fabre et al., 2014) in which information is often repeated.

The question is: how to use distributional semantics to normalize terms by an ontology? In other words how to relate distributional information to the categories of ontology? In the context of specialized literature, we often deal with relatively small corpora and a large number of semantic categories.

We propose an original method in which we represent complex terms based on word embedding, embed the ontology in a vector space, and learn a transformation from term vectors to concept vectors. Then, this transformation is used to determine the most suitable concept for an input term.

2   Material

The data used are those of the Bacteria Biotope categorization task (Task 3) of the 2016 BioNLP Shared Task (Deléger et al., 2016). The documents are references from MEDLINE, composed of titles and abstracts of scientific articles in the field of biology. The task consists in assigning a category from the OntoBiotope ontology to given corpus terms related to bacterial habitats. The corpus is divided into three subparts: the training corpus, the development corpus and the test corpus. In the training and development corpus, the categories of terms are given: they have been used to train our method. The terms from the test corpus are those which categories have to be predicted: it is the corpus used to evaluate our method for the task of normalization. The entities of each of these corpora have been manually annotated. Table 1 provides a summary of their characteristics:

	Train	Dev.	Test	Total
Documents	71	36	54	**161**
Words	16,295	8,890	13,797	**38,982**
Entities	747	454	720	**1,921**
Distinct entities	476	267	478	**1,125**
Semantic cat.	825	535	861	**2,221**
Distinct cat.	210	122	177	**329**

Table 1: Descriptive statistics for the Bacteria Biotope corpus ("cat." = categories, "Dev." = development corpus)

In addition to this corpus, an extended corpus of the same domain is used to generate vector representations of each word. It is composed of approximately 100,000 sentences (4,800,000 words) from titles and abstracts of scientific articles in the field of biology available on PubMed. This represents a relatively small size corpus, which contains a majority of words with a low frequency of occurrence (cf. Table 2). Other corpus, larger and/or more general could be used, also direct words embedding as the one released by BioASQ (Pavlopoulos et al., 2014). Nevertheless, the very accurate domain of the used extended corpus and its desired small size seemed to be more adapted.

Repeated >2	72,412	35%
Repeated 2 times	31,569	15%
Not repeated	105,364	50%
Words (without stopwords)	209,345	100%

Table 2: Descriptive statistics of extended corpus

3   Method

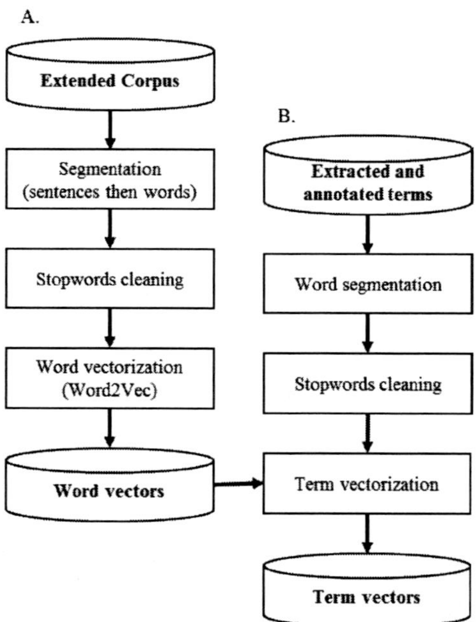

Figure 1: A. Process to create word vectors. B. Process to create term vectors.

3.1   Word vectors

The vector space of the terms (VST) is obtained by generating a vector for each word of the extended corpus and the Bacteria Biotope corpus. For this, we used the Word2Vec tool

(Mikolov et al., 2013), taking as context of a word, a list containing all the words of their sentence. To have enough training data for the generation of meaningful word vectors, and also to avoid taking into account typos or errors, it is usually advisable to use Word2Vec without the infrequent words appearing only once or twice throughout the corpus. But our corpus contains many words of interest with a low frequency, so we choose not to apply this frequency threshold. After some performance tests, the dimension 200 was selected for the output vectors (cf. Figure 1A), which is of the same order of magnitude as what is usually advised (Mikolov et al., 2013).

3.2 Term vectors

To compute the vector representations of the multiword terms (cf. Figure 1B), segmenting them into words is the first step. For each word, which is not a stopword, the vector calculated by Word2Vec is used. Then the vector of the multiword term is obtained by averaging the vectors of the words which compose it:

$$v_{t_k} = \sum_{i=1}^{n_k} v_{m_i^k}/n_k \qquad (1)$$

where v_{t_k} is the associated vector of the term t_k, n_k is the number of words (without stopwords) of the term t_k, $v_{m_i^k}$ is the vector of the word m_i^k from our Word2Vec computation, and the term t_k is such that :

$$\forall i \in [1, n_k], m_i^k \in t_k \qquad (2)$$

Even if it is not the aim of this paper, future works could test other methods.

3.3 Concept vectors

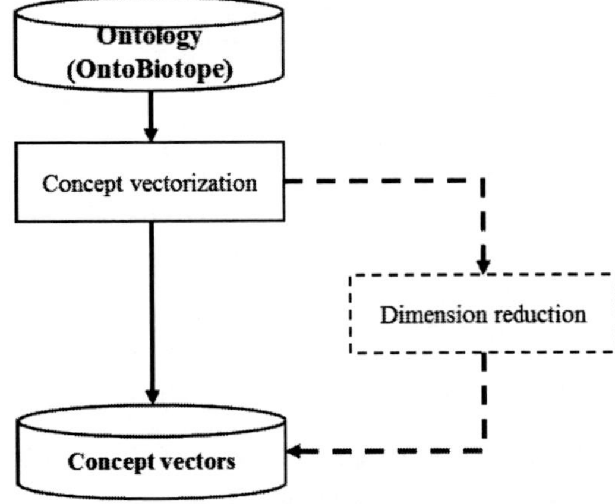

Figure 2: Process to create concept vector

To construct the concept vectors and thus a vector space of an ontology (VSO), null vectors with as many dimensions as the number of concepts in the ontology are initialized. Each value of the vector is thus related to one of the concepts of the ontology, which is set to 1 for the considered concept. The value is also 1 if the current axis is related to a concept which is an ancestor of the considered concept, and 0 otherwise:

$$v_{c_k} = \left(w_{c_k}^0, \dots, w_{c_k}^i, \dots, w_{c_k}^n \right) \qquad (3)$$

where v_{c_k} is the vector related to the concept c_k, n is the number of concepts in the ontology and $w_{c_k}^i$ is the value of vector v_{c_k} for the axis i, such as:

$$w_{c_k}^i = \begin{cases} 1, if\ i = k \\ 1, if\ c_i\ parent\ of\ c_k \\ 0, otherwise \end{cases} \qquad (4)$$

This representation has the advantage of preserving the similarity arrangement (with cosine distance) expected between the concepts (cf. Figure 3 and Table 3): a concept is more similar to his children and his parents.

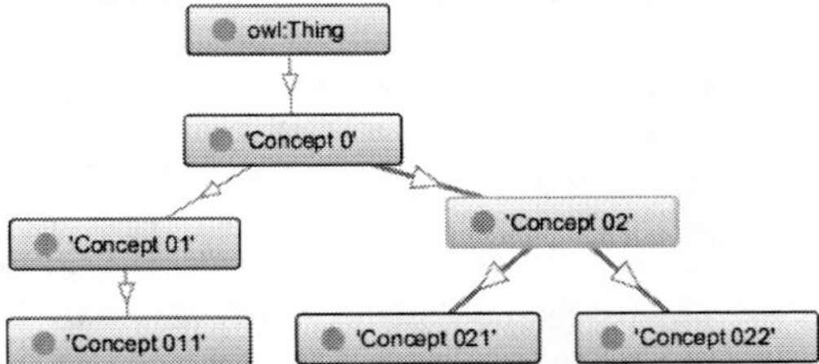

Figure 3: Abstract ontology representation (displayed by Protégé)

Concept 02	Similarity
Concept 02	1,0000
Concept 021	0,8165
Concept 022	0,8165
Concept 0	0,7071
Concept 01	0,5000
Concept 011	0,4082

Table 3: Cosine distances between concepts of an abstract ontology (cf. Figure 3)

We can notice that the dimension of the generated VSO is the number of concepts of the ontology (e.g. more than 2,000 for the OntoBiotope ontology). It is a high dimension in comparison to the VST but concept vectors are very sparse (with a maximum of 13 non-zero values in a vector) and they only contain binary values. Therefore, to make them more comparable to term vectors, we experimented with reducing the VSO to denser representations in a lower-dimension space (cf. Figure 2). Two methods have been tested: Principal Component Analysis (PCA) and Multi-Dimensional Scaling (MDS).

3.4 Training with general linear model

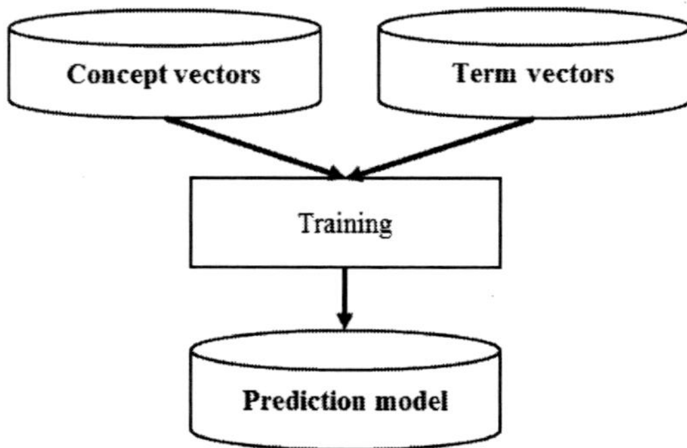

Figure 4: Training process to determine a transformation VST to VSO

The objective of the training step is to determine a transformation from VST to VSO, which minimizes all the distances between the vectors of terms resulting from this transformation and the vectors of the associated concepts. In this paper, a linear transformation is studied with the aim of keeping a strong similarity between the distribution of term vectors in the VST and the distribution of the projections in the VSO. Indeed, a non-linear transformation could strongly distort the resulting distribution to fit better to training data.

This training aims to obtain the best parameters to approximate this matrix equation:

$$Y = X.B + U \qquad (5)$$

where Y is a matrix resulting in a series of concept vectors, X is a matrix resulting in a series of term vectors (where the ith line of X is the vector of a term which has for category a concept which has for vector the ith line of Y), B is a matrix containing parameters that are usually to be estimated and U is a matrix containing noise following a multivariate Gaussian distribution. This training is performed on the training and development corpora (cf. Figure 4).

The obtained matrix enables us to design a linear transformation function then make it possible to predict new vectors associated with the terms of the test corpus expressed in the VSO:

$$f: \left(\begin{array}{c} VST \rightarrow VSO \\ v_{\text{term}} \rightarrow v'_{\text{term}} = f(v_{\text{term}}) \end{array} \right) \qquad (6)$$

where v_{term} is a vector of term in the VST and v'_{term} is the resulting vector of the same term projected in the VSO. To satisfy the requirements of the evaluation task, the concept vector nearest to v'_{term} (as determined by cosine distance) is chosen as category for the annotated term (cf. Figure 5).

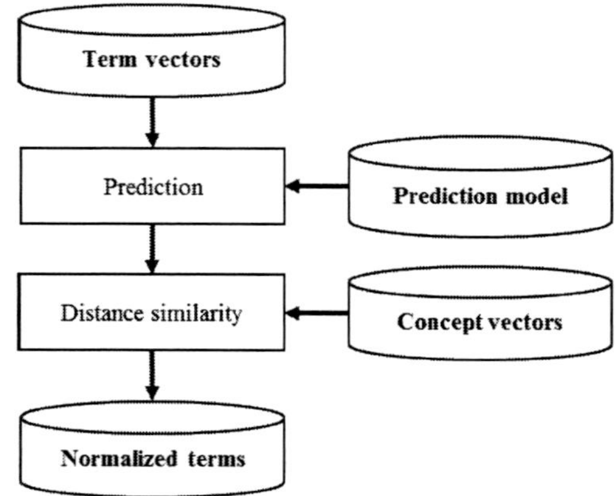

Figure 5: Process of predicting semantic categories associated with extracted terms

3.5 Evaluation

We evaluate the performance of our normalization method on the Bacteria Biotope normalization task of the BioNLP Shared Task 2016. The dataset was presented in Section 2. The predicted concepts identifiers are compared to the gold standard concepts according to the similarity measure of (Wang et al., 2007), with the weight parameter set to 0.65. The evaluation was performed by submitting our results to the evaluation server run at the BioNLP-ST 2016 challenge site.

4 Results

4.1 Normalization

Team	Similarity score
BOUN	0.6200
CONTES	**0.5968**
LIMSI	0.4380
Baseline	0.3217

Table 4: Results on the normalization task of BioNLP-ST 2016

We applied our concept normalization method to the test dataset of the Bacteria Biotope 2017 Task 3. We computed baseline results by assigning all terms to the concept "bacteria habitat", which is the root of the OntoBiotope ontology hierarchy. We also compared these results to those of the two teams who participated in this task of BioNLP-ST 2016. We report all results in Table 4. The baseline obtains a score of 0.3217. Our method

(CONTES - CONcept-TErm System) obtained a score of 59.68%, much higher than the baseline, and close to that of the top team (Tiftikci et al., 2016). This score is also significantly above the method of LIMSI (Grouin, 2016), which is based on a morphological approach.

4.2 Term vectors

In spite of the low frequency of occurrence of the words of the extended corpus (cf. Table 2), the resulting word vectors seem to have relatively satisfactory proximities, from the point of view of the semantic similarity of the associated terms. Moreover, the method used to compute vectors for complex terms also seems satisfactory, as illustrated Table 5.

cell	Similarity
HCE cell	0.9999
13C-labeled cell	0.9998
parietal cell	0.9989
Schwann cell	0.9965
CD8+ T cell	0.9770
PMN cell	0.9669
macrophage cell	0.9473

Table 5: Terms nearest to the term "cell"

It also appears that lexical variation can be overcome (cf. Table 6 and Table 7), which was one of the desired properties. Although more generally, it seems that terms with similar lexical forms are closer (Table 5).

Nevertheless, the co-occurrence of some words seems to cluster certain terms from different categories: two words appearing frequently in common contexts are then found close. This similarity persists when calculating multiword term vectors. This applies, for example, to the terms relating to fish and those relating to fish farms (cf. Table 8). These cases are less satisfactory because they do not differentiate between terms which should be annotated with different semantic categories (e.g. "fish" and "healthy fish" should be annotated by <OBT:001902: fish>, "fish farm" and "disease-free fish farm" by <OBT:000295: fish farm> and "fish farm sediments" by <OBT:000704: sediment>).

younger ones	Similarity
children less than five years of age	0.8087
children less than 2 years of age	0.8060
children less than two years of age	0.7995

Table 6: Terms nearest to the term 'younger ones'

seawater	Similarity
sediments	0.7696
sediment sample from a disease-free fish farm	0.7499
fish farm sediments	0.7342
subterranean brine	0.7320
lagoon on the outskirts of the city of Cagliari	0.7128
petroleum reservoir	0.7095
marine environments	0.7077
marine bivalves	0.6896
sediment samples from diseased farms	0.6870
urine sediments	0.6819
petroleum	0.6576
subterranean environment	0.6497
fresh water	0.6494
fresh water supply	0.6395
Seafood	0.6390
marine	0.6366

Table 7: Terms nearest to the term 'seawater'

fish	Similarity
fish farming	0.9875
fish farm	0.9170
disease-free fish farm	0.9124
fish farm sediments	0.8683
healthy fish	0.8145

Table 8: Terms nearest to the term 'fish'

4.3 Concept vectors

<OBT:001922: algae> sans ACP	Similarity
<OBT:001777: aquatic plant>	0.9258
<OBT:001895: submersed aquatic plant>	0.8571
<OBT:001967: seaweed>	0.8018

Table 9: Concepts nearest to the concept <OBT:001922: algae>

We can estimate the quality of the created concept vectors by observing the consistency between the proximity of two vectors and the similarity of their meanings. Table 9 and Figure 6 show the example of the 'algae' concept: the nearest neighbors of its vector are its father in the ontology, its sibling and the immediate descendant of its sibling.

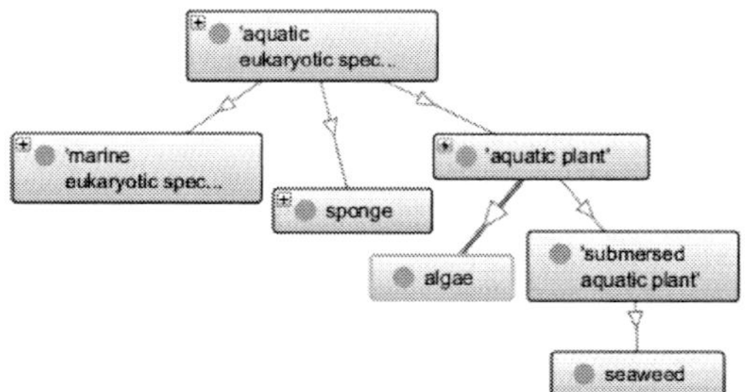

Figure 6: Taxonomy of concepts around concept
"algae" (displayed by Protégé)

By comparing several examples, it seems that
PCA does not modify the order of proximity of
the concepts, but an increase in vector density can
be observed (cf. comparison between Table 9 and
Table 10).

<OBT:001922: algae> avec ACP	Similarity
<OBT:001777: aquatic plant>	0.9990
<OBT:001895: submersed aquatic plant>	0.9982
<OBT:001967: seaweed>	0.9943
<OBT:000372: sponge>	0.9303
<OBT:000269: marine eukaryotic species>	0.9303

Table 10: Concepts nearest to the concept
<OBT:001922: algae> after a PCA with a final
dimension of 100

4.4 Impact of the size of the VST

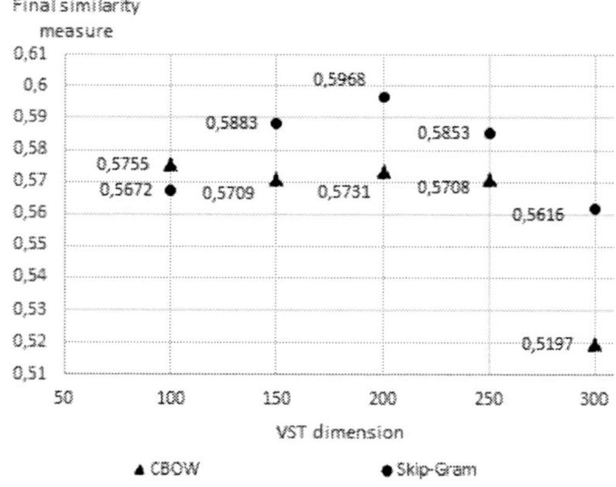

Figure 7: Comparison between CBOW and
Skip-Gram architectures for the VST

Word2Vec allows the use of 2 different architectures to generate word vectors from a corpus: Continuous Bag Of Words (CBOW) and Skip-Gram. We tested the 2 architectures on different output vector sizes (cf. Figure 7). For vector spaces generated with a dimension between 100 and 250, the final scores appear to be relatively stable, especially with CBOW. Similarly, the score difference between the two architectures remains below 3%. Above a dimension of 250, there is a decrease in the score for the 2 architectures, with a greater slope for CBOW.

4.5 Impact of a dimension reduction on the VSO

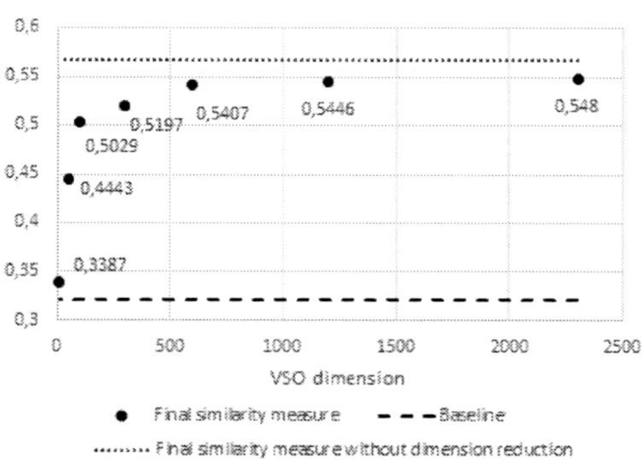

Figure 8: Evolution of performance depending
on the final size of the VSO after reduction
(here with a VST with 100 dimensions)

The VSO has a large dimension compared to the specific information that it contains (i.e. the ontology structure). This may present combinatorial but also theoretical difficulties: a linear projection of the VST on the VSO (with a higher dimension than the VST) should then only be performed on a subspace of the VSO. Thus, it theoretically limits the results. It was therefore interesting to study the impact of a reduction of the VSO size on the final score. We can then observe that a reduction PCA (with similar results with MDS) systematically decreases the score obtained when using a non-reduced VSO (cf. Figure 8).

Nevertheless, there is a level with relatively high performance (less than 3% below the score without reduction) which collapses below a certain dimension. This threshold might have a link with the number of concepts that have at least 2 distinct parents.

5 Discussion

To extend the interpretations derived from examples, it would be interesting to evaluate the overall quality of the generated vector spaces: vector spaces of words, terms, concepts as well as the final space containing the transformations of the vectors of the terms. We plan to perform this in further work.

One of the difficulties of the task is that in this normalization task, a term can be annotated by several distinct concepts of the ontology (e.g. "school age children with wheezing illness" should be annotated by the concept <OBT: 002307: pediatric patient> as well as the concept <OBT: 002187: patient with disease>). This difficulty is linked to the ontology of interest. In 2016, all participating systems of the task skip this difficulty, which is not anecdotal among the extracted terms.

6 Future work

For future work, it would be relevant to apply methods of global evaluation of the quality of the generated vector spaces. In particular, this would make it possible to evaluate the intermediate processes more thoroughly and to observe the impact of the modifications on their internal parameters more precisely. New methods could then be considered to improve outcomes. For example, it would certainly be positive to use a method of vector representation of an ontology that would generate a space with a smaller dimension while retaining the possibility of discerning the initial structure of the ontology. Similarly, the method used here to generate the VST vectors could be improved to take into account the syntactic context of the terms. This could solve the semantic similarity problems between "fish" and "fish farm" (cf. Table 8).

In the Bacteria Biotope normalization task, terms often have to be annotated with several concepts of the target ontology (for example, "children greater than 9 years of age who had lower respiratory illness" should be annotated by the concept <OBT: 002307: pediatric patient> and by the concept <OBT: 002187: patient with disease>). Having a completely defined ontology (i.e. containing all the concepts sufficient to annotate uniquely each possible extracted term - for example, a concept 'pediatric patient with disease' which is a subset of <OBT: 002307: pediatric patient> and of <OBT: 002187: patient with disease>) should improve the results. If such ontologies seem to be relatively rare in the biological domain, it might be interesting to start by automatically generating all the concepts equivalent to the intersection of the non-disjoint concepts to answer this problem. Nevertheless, if the concepts share many intersections between them or the disjoint property has not been formalized, the size of the generated ontology may pose combinatorial difficulties.

We addressed a task in which entities have already been detected in text. Since entity detection and terminology extraction methods have relatively acceptable performance, it would be useful to use them to extend the current task to an end-to-end concept detection and normalization system.

Finally, despite the inherent limitation of normalization methods based on word form similarity, these could nevertheless be used to carry out a pre-normalization of the corpus. As a result, one might consider using these annotations to drive the training part of the method (cf. 3.4 Training with general linear model) instead of using a manual annotation (i.e. a test corpora). Thus, this would transform this method into a fully unsupervised method.

7 Conclusion

The aim of this article was to propose an approach for the creation of vector representations for (complex or non-complex) terms in a semantic space. In addition, it aimed to propose a method capable of adapting to a small specialized corpus where the interest terms appear with a relatively low frequency. The most widely used methods currently generate vector spaces which meaning is difficult to interpret other than in terms of spatial proximity / semantic similarity. Our method seems to show that by combining relatively classical approaches, it is possible to use an ontology to generate vectors in a more interpretable vector space. The results are comparable to those of the state of the art, which seems to open up encouraging prospects. Beyond the standardization task, new efficient methods of generating interpretable vector spaces could apply to a number of further tasks.

Acknowledgments

This work is supported by the "IDI 2015" project funded by the IDEX Paris-Saclay, ANR-11-IDEX-0003-02.

References

Sophia Ananiadou and John McNaught, editors. 2006. *Text mining for biology and biomedicine*. Artech House, Boston.

Robert Bossy, Wiktoria Golik, Zorana Ratkovic, Dialekti Valsamou, Philippe Bessières, and Claire Nédellec. 2015. Overview of the gene regulation network and the bacteria biotope tasks in BioNLP'13 shared task. *BMC bioinformatics*, 16(10):S1.

Louise Deléger, Estelle Chaix, Mouhamadou Ba, Arnaud Ferré, Philippe Bessières, and Claire Nédellec. 2016. Overview of the Bacteria Biotope Task at BioNLP Shared Task 2016.

Cécile Fabre, Nabil Hathout, Franck Sajous, and Ludovic Tanguy. 2014. Ajuster l'analyse distributionnelle à un corpus spécialisé de petite taille. In *21e Conférence sur le Traitement Automatique des Langues Naturelles (TALN 2014)*, pages 266–279.

Cécile Fabre and Alessandro Lenci. 2015. Distributional Semantics Today Introduction to the special issue. *Traitement Automatique des Langues*, 56(2):7–20.

J. R. Firth. 1957. The technique of semantics.

Wiktoria Golik, Pierre Warnier, and Claire Nédellec. 2011. Corpus-based extension of termino-ontology by linguistic analysis: a use case in biomedical event extraction. In *WS 2 Workshop Extended Abstracts, 9th International Conference on Terminology and Artificial Intelligence*, pages 37–39.

Cyril Grouin. 2016. Identification of Mentions and Relations between Bacteria and Biotope from PubMed Abstracts. *ACL 2016*:64.

Zellig S. Harris. 1954. Distributional Structure. WORD, 10(2–3):146–162, August.

Ning Kang, Bharat Singh, Zubair Afzal, Erik M van Mulligen, and Jan A Kors. 2013. Using rule-based natural language processing to improve disease normalization in biomedical text. *Journal of the American Medical Informatics Association*, 20(5):876–881, September.

Michael Krauthammer and Goran Nenadic. 2004. Term identification in the biomedical literature. *Journal of Biomedical Informatics*, 37(6):512–526, December.

François Maniez. 2007. Prémodification et coordination : quelques problèmes de traduction des groupes nominaux complexes en anglais médical. *ASp*(51–52):71–94, December.

Tomas Mikolov, Kai Chen, Greg Corrado, and Jeffrey Dean. 2013. Efficient estimation of word representations in vector space. *arXiv preprint arXiv:1301.3781*.

Adeline Nazarenko, Claire Nédellec, Erick Alphonse, Sophie Aubin, Thierry Hamon, and Alain-Pierre Manine. 2006. Semantic annotation in the alvis project. In *International Workshop on Intelligent Information Access (IIIA)*, page 5–pages.

Ioannis Pavlopoulos, Aris Kosmopoulos, and Ion Androutsopoulos. 2014. Continuous Space Word Vectors Obtained by Applying Word2Vec to Abstracts of Biomedical Articles. March.

Mert Tiftikci, Hakan Sahin, Berfu Büyüköz, Alper Yayıkçı, and Arzucan Ozgür. 2016. Ontology-based Categorization of Bacteria and Habitat Entities using Information Retrieval Techniques. *ACL 2016*:56.

J. Z. Wang, Z. Du, R. Payattakool, P. S. Yu, and C.-F. Chen. 2007. A new method to measure the semantic similarity of GO terms. *Bioinformatics*, 23(10):1274–1281, May.

Improving Correlation with Human Judgments
by Integrating Semantic Similarity with Second–Order Vectors

Bridget T. McInnes
Department of Computer Science
Virginia Commonwealth University
Richmond, VA 23284 USA
btmcinnes@vcu.edu

Ted Pedersen
Department of Computer Science
University of Minnesota
Duluth, MN 55812 USA
tpederse@d.umn.edu

Abstract

Vector space methods that measure semantic similarity and relatedness often rely on distributional information such as co–occurrence frequencies or statistical measures of association to weight the importance of particular co–occurrences. In this paper, we extend these methods by incorporating a measure of semantic similarity based on a human curated taxonomy into a second–order vector representation. This results in a measure of semantic relatedness that combines both the contextual information available in a corpus–based vector space representation with the semantic knowledge found in a biomedical ontology. Our results show that incorporating semantic similarity into a second order co–occurrence matrices improves correlation with human judgments for both similarity and relatedness, and that our method compares favorably to various different word embedding methods that have recently been evaluated on the same reference standards we have used.

1 Introduction

Measures of semantic similarity and relatedness quantify the degree to which two concepts are similar (e.g., *lung–heart*) or related (e.g., *lung–bronchitis*). Semantic similarity can be viewed as a special case of semantic relatedness – to be similar is one of many ways that a pair of concepts may be related. The automated discovery of groups of semantically similar or related terms is critical to improving the retrieval (Rada et al., 1989) and clustering (Lin et al., 2007) of biomedical and clinical documents, and the development

of biomedical terminologies and ontologies (Bodenreider and Burgun, 2004).

There is a long history in using distributional methods to discover semantic similarity and relatedness (e.g., (Lin and Pantel, 2002; Reisinger and Mooney, 2010; Radinsky et al., 2011; Yih and Qazvinian, 2012)). These methods are all based on the distributional hypothesis, which holds that two terms that are distributionally similar (i.e., used in the same context) will also be semantically similar (Harris, 1954; Weeds et al., 2004). Recently word embedding techniques such as word2vec (Mikolov et al., 2013) have become very popular. Despite the prominent role that neural networks play in many of these approaches, at their core they remain distributional techniques that typically start with a word by word co–occurrence matrix, much like many of the more traditional approaches.

However, despite these successes distributional methods do not perform well when data is very sparse (which is common). One possible solution is to use second–order co–occurrence vectors (Schütze, 1992; Schütze, 1998). In this approach the similarity between two words is not strictly based on their co–occurrence frequencies, but rather on the frequencies of the other words which occur with both of them (i.e., second order co–occurrences). This approach has been shown to be successful in quantifying semantic relatedness (Islam and Inkpen, 2006; Pedersen et al., 2007). However, while more robust in the face of sparsity, second–order methods can result in significant amounts of noise, where contextual information that is overly general is included and does not contribute to quantifying the semantic relatedness between the two concepts.

Our goal then is to discover methods that automatically reduce the amount of noise in a second–order co–occurrence vector. We achieve this by incorporating pairwise semantic similarity scores

Proceedings of the BioNLP 2017 workshop, pages 107–116,
Vancouver, Canada, August 4, 2017. ©2017 Association for Computational Linguistics

derived from a taxonomy into our second–order vectors, and then using these scores to select only the most semantically similar co–occurrences (thereby reducing noise).

We evaluate our method on two datasets that have been annotated in multiple ways. One has been annotated for both similarity and relatedness, and the other has been annotated for relatedness by two different types of experts (medical doctors and medical coders). Our results show that integrating second order co–occurrences with measures of semantic similarity increases correlation with our human reference standards. We also compare our result to a number of other studies which have applied various word embedding methods to the same reference standards we have used. We find that our method often performs at a comparable or higher level than these approaches. These results suggest that our methods of integrating semantic similarity and relatedness values have the potential to improve performance of purely distributional methods.

2   Similarity and Relatedness Measures

This section describes the similarity and relatedness measures we integrate in our second–order co–occurrence vectors. We use two taxonomies in this study, SNOMED–CT and MeSH. SNOMED–CT (*Systematized Nomenclature of Medicine Clinical Terms*) is a comprehensive clinical terminology created for the electronic representation of clinical health information. MeSH (*Medical Subject Headings*) is a taxonomy of biomedical terms developed for indexing biomedical journal articles.

We obtain SNOMED–CT and MeSH via the Unified Medical Language System (UMLS) Metathesaurus (version 2016AA). The Metathesaurus contains approximately 2 million biomedical and clinical concepts from over 150 different terminologies that have been semi–automatically integrated into a single source. Concepts in the Metathesaurus are connected largely by two types of hierarchical relations: *parent/child* (PAR/CHD) and *broader/narrower* (RB/RN).

2.1 Similarity Measures

Measures of semantic similarity can be classified into three broad categories : path–based, feature–based and information content (IC). Path–based similarity measures use the structure of a taxon-omy to measure similarity – concepts positioned close to each other are more similar than those further apart. Feature–based methods rely on set theoretic measures of overlap between features (union and intersection). The information content measures quantify the amount of information that a concept provides – more specific concepts have a higher amount of information content.

2.1.1 Path–based Measures

Rada et al. (1989) introduce the *Conceptual Distance* measure. This measure is simply the length of the shortest path between two concepts ($c1$ and $c2$) in the MeSH hierarchy. Paths are based on *broader than* (RB) and *narrower than* (RN) relations. Caviedes and Cimino (2004) extends this measure to use *parent* (PAR) and *child* (CHD) relations. Our *path* measure is simply the reciprocal of this shortest path value (Equation 1), so that larger values (approaching 1) indicate a high degree of similarity.

$$path = \frac{1}{spath(c_1, c_2)} \qquad (1)$$

While the simplicity of *path* is appealing, it can be misleading when concepts are at different levels of specificity. Two very general concepts may have the same path length as two very specific concepts. Wu and Palmer (1994) introduce a correction to *path* that incorporates the depth of the concepts, and the depth of their Least Common Subsumer (LCS). This is the most specific ancestor two concepts share. In this measure, similarity is twice the depth of the two concept's LCS divided by the product of the depths of the individual concepts (Equation 2). Note that if there are multiple LCSs for a pair of concepts, the deepest of them is used in this measure.

$$wup = \frac{2 * depth(lcs(c_1, c_2))}{depth(c_1) + depth(c_2)} \qquad (2)$$

Zhong et al. (2002) take a very similar approach and again scale the depth of the LCS by the sum of the depths of the two concepts (Equation 3), where $m(c) = k^{-depth(c)}$. The value of k was set to 2 based on their recommendations.

$$zhong = \frac{2 * m(lcs(c_1, c_2))}{m(c_1) + m(c_2)} \qquad (3)$$

Pekar and Staab (2002) offer another variation on *path*, where the shortest path of the two concepts to the LCS is used, in addition to the shortest bath between the LCS and the root of the taxonomy (Equation 4).

$$pks = -\log \frac{spath(lcs(c_1, c_2), root)}{\sum_{x=c_1,c_2,root} spath(lcs(c_1, c_2), x)} \quad (4)$$

2.1.2 Feature–based Measures

Feature–based methods represent each concept as a set of features and then measure the overlap or sharing of features to measure similarity. In particular, each concept is represented as the set of their ancestors, and similarity is a ratio of the intersection and union of these features.

Maedche and Staab (2001) quantify the similarity between two concepts as the ratio of the intersection over their union as shown in Equation 5.

$$cmatch = \frac{|A(c_1) \cap A(c_2)|}{|A(c_1) \cup A(c_2)|} \quad (5)$$

Batet et al. (2011) extend this by excluding any shared features (in the numerator) as shown in Equation 6.

$$batet = -log_2(\frac{|A(c_1) \cup A(c_2)| - |A(c_1) \cap A(c_2)|}{|A(c_1) \cup A(c_2)|}) \quad (6)$$

2.1.3 Information Content Measures

Information content is formally defined as the negative log of the probability of a concept. The effect of this is to assign rare (low probability) concepts a high measure of information content, since the underlying assumption is that more specific concepts are less frequently used than more common ones.

Resnik (1995) modified this notion of information content in order to use it as a similarity measure. He defines the similarity of two concepts to be the information content of their LCS (Equation 7).

$$res = IC(lcs(c_1, c_2) = -\log(P(lcs(c_1, c_2))) \quad (7)$$

Jiang and Conrath (1997), Lin (1998), and Pirró and Euzenat (2010) extend res by incorporating the information content of the individual concepts in various different ways. Lin (1998) defines the similarity between two concepts as the ratio of information content of the LCS with the sum of the individual concept's information content (Equation 8). Note that lin has the same form as wup and $zhong$, and is in effect using information content as a measure of specificity (rather than depth). If there is more than one possible LCS, the LCS with the greatest IC is chosen.

$$lin = \frac{2 * IC(lcs(c_1, c_2))}{IC(c_1) + IC(c_2)} \quad (8)$$

Jiang and Conrath (1997) define the distance between two concepts to be the sum of the information content of the two concepts minus twice the information content of the concepts' LCS. We modify this from a distance to a similarity measure by taking the reciprocal of the distance (Equation 9). Note that the denominator of jcn is very similar to the numerator of $batet$.

$$jcn = \frac{1}{IC(c_1) + IC(c_2) - 2 * IC(lcs(c_1, c_2))} \quad (9)$$

Pirró and Euzenat (2010) define the similarity between two concepts as the information content of the two concept's LCS divided by the sum of their individual information content values minus the information content of their LCS (Equation 10). Note that $batet$ can be viewed as a set–theoretic version of $faith$.

$$faith = \frac{IC(lcs(c_1, c_2))}{IC(c_1) + IC(c_2) - IC(lcs(c_1, c_2))} \quad (10)$$

2.2 Information Content

The information content of a concept may be derived from a corpus (corpus–based) or directly from a taxonomy (intrinsic–based). In this work we focus on corpus–based techniques.

For corpus–based information content, we estimate the probability of a concept c by taking the sum of the probability of the concept $P(c)$ and the probability its descendants $P(d)$ (Equation 11).

$$P(c*) = P(c) + \sum_{d \in descendant(c)} P(d) \quad (11)$$

The initial probabilities of a concept ($P(c)$) and its descendants ($P(d)$) are obtained by dividing the number of times each concept and descendant occurs in the corpus, and dividing that by the total numbers of concepts (N).

Ideally the corpus from which we are estimating the probabilities of concepts will be sense–tagged. However, sense–tagging is a challenging problem in its own right, and it is not always possible to carry out reliably on larger amounts of text. In fact in this paper we did not use any sense–tagging of the corpus we derived information content from.

Instead, we estimated the probability of a concept by using the *UMLSonMedline* dataset. This was created by the National Library of Medicine and consists of concepts from the 2009AB UMLS and the counts of the number of times they occurred in a snapshot of Medline taken on 12 January, 2009. These counts were obtained by using the Essie Search Engine (Ide et al., 2007) which queried Medline with normalized strings from the 2009AB MRCONSO table in the UMLS. The frequency of a CUI was obtained by aggregating the frequency counts of the terms associated with the CUI to provide a rough estimate of its frequency. The information content measures then use this information to calculate the probability of a concept.

Another alternative is the use of *Intrinsic Information Content*. It assess the informativeness of concept based on its placement within a taxonomy by considering the number of incoming (ancestors) relative to outgoing (descendant) links (Sánchez et al., 2011) (Equation 12).

$$IC(c) = -log(\frac{\frac{|leaves(c)|}{|subsumers(c)|} + 1}{max\_leaves + 1}) \qquad (12)$$

where *leaves* are the number of descendants of concept c that are leaf nodes, *subsumers* are the number of concept c's ancestors and *max_leaves* are the total number of leaf nodes in the taxonomy.

2.3 Relatedness Measures

Lesk (1986) observed that concepts that are related should share more words in their respective definitions than concepts that are less connected. He was able to perform word sense disambiguation by identifying the senses of words in a sentence with the largest number of overlaps between their definitions. An overlap is the longest sequence of one or more consecutive words that occur in both definitions. Banerjee and Pedersen (2003) extended this idea to WordNet, but observed that WordNet glosses are often very short, and did not contain enough information to distinguish between multiple concepts. Therefore, they created a *super–gloss* for each concept by adding the glosses of related concepts to the gloss of the concept itself (and then finding overlaps).

Patwardhan and Pedersen (2006) adapted this measure to second–order co–occurrence vectors. In this approach, a vector is created for each word in a concept's definition that shows which words co–occur with it in a corpus. These word vectors are averaged to create a single co-occurrence vector for the concept. The similarity between the concepts is calculated by taking the cosine between the concepts second–order vectors. Liu et al. (2012) modified and extended this measure to be used to quantify the relatedness between biomedical and clinical terms in the UMLS. The work in this paper can be seen as a further extension of Patwardhan and Pedersen (2006) and Liu et al. (2012).

3 Method

In this section, we describe our *second–order similarity vector* measure. This incorporates both contextual information using the term pair's definition and their pairwise semantic similarity scores derived from a taxonomy. There are two stages to our approach. First, a co–occurrence matrix must be constructed. Second, this matrix is used to construct a second–order co–occurrence vector for each concept in a pair of concepts to be measured for relatedness.

3.1 Co–occurrence Matrix Construction

We build an $m \times n$ similarity matrix using an external corpus where the rows and columns represent words within the corpus and the element contains the similarity score between the row word and column word using the similarity measures discussed above. If a word maps to more than one possible sense, we use the sense that returns the highest similarity score.

For this paper our external corpus was the NLM 2015 Medline baseline. Medline is a bibliographic database containing over 23 million citations to journal articles in the biomedical domain and is maintained by National Library of Medicine. The 2015 Medline Baseline encompasses approximately 5,600 journals starting from 1948 and contains 23,343,329 citations, of which 2,579,239 contain abstracts. In this work, we use Medline titles and abstracts from 1975 to present day. Prior to 1975, only 2% of the citations contained an abstract. We then calculate the similarity for each bigram in this dataset and include those that have a similarity score greater than a specified threshold on these experiments.

3.2 Measure Term Pairs for Relatedness

We obtain definitions for each of the two terms we wish to measure. Due to the sparsity and inconsistencies of the definitions in the UMLS, we not only use the definition of the term (CUI) but also include the definition of its related concepts. This follows the method proposed by Patwardhan and Pedersen (2006) for general English and Word-Net, and which was adapted for the UMLS and the medical domain by Liu et al. (2012). In particular we add the definitions of any concepts connected via a parent (PAR), child (CHD), RB (broader than), RN (narrower than) or TERM (terms associated with CUI) relation. All of the definitions for a term are combined into a single *super–gloss*. At the end of this process we should have two super–glosses, one for each term to be measured for relatedness.

Next, we process each super–gloss as follows:

1. We extract a first–order co–occurrence vector for each term in the super–gloss from the co–occurrence matrix created previously.

2. We take the average of the first order co–occurrence vectors associated with the terms in a super–gloss and use that to represent the meaning of the term. This is a second–order co–occurrence vector.

3. After a second–order co–occurrence vector has been constructed for each term, then we calculate the cosine between these two vectors to measure the relatedness of the terms.

4 Data

We use two reference standards to evaluate the semantic similarity and relatedness measures [1]. UMNSRS was annotated for both similarity and relatedness by medical residents. MiniMayoSRS was annotated for relatedness by medical doctors (MD) and medical coders (coder). In this section, we describe these data sets and describe a few of their differences.

MiniMayoSRS: The MayoSRS, developed by Pakhomov et al. (2011), consists of 101 clinical term pairs whose relatedness was determined by nine medical coders and three physicians from the Mayo Clinic. The relatedness of each term pair was assessed based on a four point scale: (4.0) practically synonymous, (3.0) related, (2.0) marginally related and (1.0) unrelated. Mini-MayoSRS is a subset of the MayoSRS and consists of 30 term pairs on which a higher inter–annotator agreement was achieved. The average correlation between physicians is 0.68. The average correlation between medical coders is 0.78. We evaluate our method on the mean of the physician scores, and the mean of the coders scores in this subset in the same manner as reported by Pedersen et al. (2007).

UMNSRS: The University of Minnesota Semantic Relatedness Set (UMNSRS) was developed by Pakhomov et al. (2010), and consists of 725 clinical term pairs whose semantic similarity and relatedness was determined independently by four medical residents from the University of Minnesota Medical School. The similarity and relatedness of each term pair was annotated based on a continuous scale by having the resident touch a bar on a touch sensitive computer screen to indicate the degree of similarity or relatedness. The Intraclass Correlation Coefficient (ICC) for the reference standard tagged for similarity was 0.47, and 0.50 for relatedness. Therefore, as suggested by Pakhomov and colleagues,we use a subset of the ratings consisting of 401 pairs for the similarity set and 430 pairs for the relatedness set which each have an ICC of 0.73.

5 Experimental Framework

We conducted our experiments using the freely available open source software package UMLS::Similarity (McInnes et al., 2009) version 1.47[2]. This package takes as input two terms (or UMLS concepts) and returns their similarity or relatedness using the measures discussed in Section 2.

Correlation between the similarity measures and human judgments were estimated using Spearman's Rank Correlation (ρ). Spearman's measures the statistical dependence between two variables to assess how well the relationship between the rankings of the variables can be described using a monotonic function. We used Fisher's r-to-z transformation (Fisher, 1915) to calculate the significance between the correlation results.

---

[1] http://www.people.vcu.edu/ btmcinnes/downloads.html

[2] http://search.cpan.org/edist/UMLS-Similarity/

6 Results and Discussion

Table 1 shows the Spearman's Rank Correlation between the human scores from the four reference standards and the scores from the various measures of similarity introduced in Section 2. Each class of measure is followed by the scores obtained when integrating our second order vector approach with these measures of semantic similarity.

6.1 Results Comparison

The results for UMNSRS tagged for similarity (sim) and MiniMayoSRS tagged by coders show that all of the second-order similarity vector measures ($Integrated$) except for $vector\text{-}jcn$ obtain a higher correlation than the original measures. We found that $vector\text{-}res$ and $vector\text{-}faith$ obtain the highest correlations of all these results with human judgments.

For the UMNSRS dataset tagged for relatedness and MiniMayoSRS tagged by physicians (MD), the original $vector$ measure obtains a higher correlation than our measure ($Integrated$) although the difference is not statistically significant ($p \leq 0.2$).

In order to analyze and better understand these results, we filtered the bigram pairs used to create the initial similarity matrix based on the strength of their similarity using the $faith$ and the res measures. Note that the $faith$ measure holds to a 0 to 1 scale, while res ranges from 0 to an unspecified upper bound that is dependent on the size of the corpus from which information content is estimated. As such we use a different range of threshold values for each measure. We discuss the results of this filtering below.

6.2 Thresholding Experiments

Table 2 shows the results of applying the threshold parameter on each of the reference standards using the res measure. For example, a threshold of 0 indicates that all of the bigrams were included in the similarity matrix; and a threshold of 1 indicates that only the bigram pairs with a similarity score greater than one were included.

These results show that using a threshold cutoff of 2 obtains the highest correlation for the UMNSRS dataset, and that a threshold cutoff of 4 obtains the highest correlation for the MiniMayoSRS dataset. All of the results show an increase in correlation with human judgments when incorporating a threshold cutoff over all of the original

Table 1: Spearman's Correlation Results

	UMNSRS Resident		MiniMayoSRS MD Coder	
	sim	rel	relatedness	
Path				
path	0.52	0.28	0.35	0.45
wup	0.50	0.24	0.39	0.51
pks	0.49	0.25	0.38	0.50
zhong	0.50	0.25	0.42	0.50
Integrated				
vector-path	0.60	0.43	0.54	0.54
vector-wup	0.60	0.42	0.55	0.55
vector-pks	0.60	0.42	0.53	0.53
vector-zhong	0.58	0.41	0.54	0.53
Feature				
batet	0.16	0.33	0.16	0.15
cmatch	0.33	0.17	0.35	0.35
Integrated				
vector-batet	0.59	0.43	0.53	0.51
vector-cmatch	0.60	0.43	0.54	0.55
IC				
res	0.49	0.26	0.36	0.47
lin	0.51	0.29	0.44	0.54
jcn	0.52	0.33	0.42	0.52
faith	0.51	0.29	0.43	0.54
Integrated				
vector-res	0.57	0.41	0.58	**0.65**
vector-lin	0.57	0.41	0.59	0.64
vector-jcn	0.42	0.15	0.26	0.41
vector-faith	**0.59**	0.42	0.58	0.63
Intrinsic IC				
ires	0.49	0.26	0.40	0.50
ilin	0.50	0.28	0.41	0.50
ijcn	0.51	0.29	0.39	0.50
ifaith	0.50	0.28	0.41	0.50
Integrated				
vector-ires	0.57	0.41	0.50	0.52
vector-ilin	0.57	0.41	0.55	0.59
vector-ijcn	0.50	0.41	0.54	0.54
vector-ifaith	0.58	0.42	0.58	0.64
Relatedness				
lesk	0.49	0.33	0.52	0.56
o1vector	0.47	0.36	0.43	0.54
o2vector	0.54	**0.45**	**0.63**	0.59

Table 2: Threshold Correlation with *vector-res*

		UMNSRS		MiniMayoSRS	
T	# bigrams	sim	rel	MD	coder
0	850,959	0.58	0.41	0.58	0.65
1	166,003	0.56	0.39	0.60	0.67
2	65,502	**0.64**	**0.47**	0.56	0.62
3	27,744	0.60	0.46	0.62	0.71
4	10,991	0.56	0.43	**0.75**	**0.76**
5	3,305	0.26	0.16	0.36	0.36

Table 3: Threshold Correlation with *vector-faith*

	#	UMNSRS		MiniMayoSRS	
T	bigrams	sim	rel	MD	coder
0	838,353	0.59	0.42	**0.58**	**0.63**
0.1	197,189	0.58	0.41	0.57	**0.63**
0.2	121,839	0.58	0.41	**0.58**	**0.63**
0.3	71,353	0.63	0.46	0.54	0.55
0.4	45,335	0.64	0.48	0.50	0.51
0.5	29,734	**0.66**	**0.49**	0.49	0.53
0.6	19,347	0.65	0.49	0.52	0.56
0.7	11,946	0.64	0.48	0.53	0.55
0.8	7,349	0.64	0.49	0.53	0.56
0.9	4,731	0.62	0.49	0.53	0.57

measures. The increase in the correlation for the UMNSRS tagged for similarity is statistically significant ($p \leq 0.05$), however this is not the case for the UMNSRS tagged for relatedness nor for the MiniMayoSRS data.

Similarly, Table 3 shows the results of applying the threshold parameter (T) on each of the reference standards using the *faith* measure. Although, unlike *res* whose scores are greater than or equal to 0 without an upper limit, the *faith* measure returns scores between 0 and 1 (inclusive). Therefore, here a threshold of 0 indicates that all of the bigrams were included in the similarity matrix; and a threshold of 0.1 indicates that only the bigram pairs with a similarity score greater than 0.1 were included. The results show an increase in accuracy for all of the datasets except for the MiniMayoSRS tagged for physicians. The increase in the results for the UMNSRS tagged for similarity and the MayoSRS is statistically significant ($p \leq 0.05$). This is not the case for the UMNSRS tagged for relatedness nor the MiniMayoSRS.

Overall, these results indicate that including only those bigrams that have a sufficiently high similarity score increases the correlation results with human judgments, but what quantifies as sufficiently high varies depending on the dataset and measure.

6.3 Comparison with Previous Work

Recently, word embeddings (Mikolov et al., 2013) have become a popular method for measuring semantic relatedness in the biomedical domain. This is a neural network based approach that learns a representation of a word by word co–occurrence matrix. The basic idea is that the neural network learns a series of weights (the hidden layer within the neural network) that either maximizes the probability of a word given its context, referred to as the continuous bag of words (CBOW) approach, or that maximizes the probability of the context given a word, referred to as the Skip–gram approach. These approaches have been used in numerous recent papers.

Muneeb et al. (2015) trained both the Skip–gram and CBOW models over the PubMed Central Open Access (PMC) corpus of approximately 1.25 million articles. They evaluated the models on a subset of the UMNSRS data, removing word pairs that did not occur in their training corpus more than ten times. Chiu et al. (2016) evaluated both the the Skip–gram and CBOW models over the PMC corpus and PubMed. They also evaluated the models on a subset of the UMNSRS ignoring those words that did not appear in their training corpus. Pakhomov et al. (2016) trained CBOW model over three different types of corpora: clinical (clinical notes from the Fairview Health System), biomedical (PMC corpus), and general English (Wikipedia). They evaluated their method using a subset of the UMNSRS restricting to single word term pairs and removing those not found within their training corpus. Sajadi et al. (2015) trained the Skip–gram model over CUIs identified by MetaMap on the OHSUMED corpus, a collection of 348,566 biomedical research articles. They evaluated the method on the complete UMNSRS, MiniMayoSRS and the MayoSRS datasets; any subset information about the dataset was not explicitly stated therefore we believe a direct comparison may be possible.

In addition, a previous work very closely related to ours is a retrofitting vector method proposed by Yu et al. (2016) that incorporates ontological information into a vector representation by includ-

Table 4: Comparison with Previous Work

Method	UMNSRS				MayoSRS (N=101)	MiniMayoSRS (N=29)		
	Subsets		Full					
	sim	rel	sim (N=566)	rel (N=587)	rel	MD	coder	avg
vector–res (ours)	0.64 (N=401)	0.49 (N=430)	0.59	0.48	0.51	0.75	0.76	0.76
vector–faith (ours)	0.66 (N=401)	0.49 (N=430)	0.61	0.49	0.46	0.58	0.63	0.63
(Yu et al., 2016)						0.70	0.67	
(Sajadi et al., 2015)			0.39	0.39	0.63			0.8
(Pakhomov et al., 2016)	0.62 (N=449)	0.58 (N=458)						
(Muneeb et al., 2015)	0.52 (N=462)	0.45 (N=465)						
(Chiu et al., 2016)	0.65 (N=UK)	0.60 (N=UK)						

ing semantically related words. In their measure, they first map a biomedical term to MeSH terms, and second build a word vector based on the documents assigned to the respective MeSH term. They then retrofit the vector by including semantically related words found in the Unified Medical Language System. They evaluate their method on the MiniMayoSRS dataset.

Table 4 shows a comparison to the top correlation scores reported by each of these works on the respective datasets (or subsets) they evaluated their methods on. N refers to the number of term pairs in the dataset the authors report they evaluated their method. The table also includes our top scoring results: the *integrated vector-res* and *vector-faith*. The results show that integrating semantic similarity measures into second–order co–occurrence vectors obtains a higher or on–par correlation with human judgments as the previous works reported results with the exception of the UMNSRS rel dataset. The results reported by Pakhomov et al. (2016) and Chiu et al. (2016) obtain a higher correlation although the results can not be directly compared because both works used different subsets of the term pairs from the UMNSRS dataset.

7 Conclusion and Future Work

We have presented a method for quantifying the similarity and relatedness between two terms that integrates pair–wise similarity scores into second–order vectors. The goal of this approach is two–fold. First, we restrict the context used by the vector measure to words that exist in the biomedical domain, and second, we apply larger weights to those word pairs that are more similar to each other. Our hypothesis was that this combination would reduce the amount of noise in the vectors and therefore increase their correlation with human judgments. We evaluated our method on

datasets that have been manually annotated for relatedness and similarity and found evidence to support this hypothesis. In particular we discovered that guiding the creation of a second–order context vector by selecting term pairs from biomedical text based on their semantic similarity led to improved levels of correlation with human judgment.

We also explored using a threshold cutoff to include only those term pairs that obtained a sufficiently large level of similarity. We found that eliminating less similar pairs improved the overall results (to a point). In the future, we plan to explore metrics to automatically determine the threshold cutoff appropriate for a given dataset and measure. We also plan to explore additional features that can be integrated with a second–order vector measure that will reduce the noise but still provide sufficient information to quantify relatedness. We are particularly interested in approaches that learn word, phrase, and sentence embeddings from structured corpora such as literature (Hill et al., 2016a) and dictionary entries (Hill et al., 2016b). Such embeddings could be integrated into a second–order vector or be used on their own.

Finally, we compared our proposed method to other distributional approaches, focusing on those that used word embeddings. Our results showed that integrating semantic similarity measures into second–order co–occurrence vectors obtains the same or higher correlation with human judgments as do various different word embedding approaches. However, a direct comparison was not possible due to variations in the subsets of the UMNSRS evaluation dataset used. In the future, we would not only like to conduct a direct comparison but also explore integrating semantic similarity into various kinds of word embeddings by training on pair–wise values of semantic similarity as well as co–occurrence statistics.

References

S. Banerjee and T. Pedersen. 2003. Extended gloss overlaps as a measure of semantic relatedness. In *Proceedings of the Eighteenth International Joint Conference on Artificial Intelligence*. Acapulco, Mexico, pages 805–810.

M. Batet, D. Sánchez, and A. Valls. 2011. An ontology-based measure to compute semantic similarity in biomedicine. *Journal of biomedical informatics* 44(1):118–125.

O. Bodenreider and A. Burgun. 2004. Aligning knowledge sources in the UMLS: methods, quantitative results, and applications. In *Proceedings of the 11th World Congress on Medical Informatics (MEDINFO)*. San Fransico, CA, pages 327–331.

J.E. Caviedes and J.J. Cimino. 2004. Towards the development of a conceptual distance metric for the UMLS. *Journal of Biomedical Informatics* 37(2):77–85.

B. Chiu, G. Crichton, A. Korhonen, and S. Pyysalo. 2016. How to Train Good Word Embeddings for Biomedical NLP. In *Proceedings of the 15th Workshop on Biomedical Natural Language Processing*. pages 166–174.

R.A. Fisher. 1915. Frequency distribution of the values of the correlation coefficient in samples from an indefinitely large population. *Biometrika* pages 507–521.

Z. Harris. 1954. Distributional structure. *Word* 10(23):146–162.

F. Hill, K. Cho, and A. Korhonen. 2016a. Learning distributed representations of sentences from unlabelled data. In *Proceedings of the 2016 Conference of the North American Chapter of the Association for Computational Linguistics: Human Language Technologies*. Association for Computational Linguistics, San Diego, California, pages 1367–1377.

F. Hill, K. Cho, A. Korhonen, and Y. Bengio. 2016b. Learning to understand phrases by embedding the dictionary. *Transactions of the Association for Computational Linguistics* 4:17–30.

N.C. Ide, R.F. Loane, and D. Demner-Fushman. 2007. Essie: a concept-based search engine for structured biomedical text. *Journal of the American Medical Informatics Association* 14(3):253–263.

A. Islam and D. Inkpen. 2006. Second order co-occurrence pmi for determining the semantic similarity of words. In *Proceedings of the International Conference on Language Resources and Evaluation, Genoa, Italy*. pages 1033–1038.

J. Jiang and D. Conrath. 1997. Semantic similarity based on corpus statistics and lexical taxonomy. In *Proceedings on International Conference on Research in Computational Linguistics*. Tapei, Taiwan, pages 19–33.

M. Lesk. 1986. Automatic sense disambiguation using machine readable dictionaries: how to tell a pine cone from an ice cream cone. In *Proceedings of the 5th Annual International Conference on Systems Documentation*. Toronto, Canada, pages 24–26.

D. Lin. 1998. An information-theoretic definition of similarity. In *Intl Conf ML Proc.*. San Francisco, CA, pages 296–304.

D. Lin and P. Pantel. 2002. Concept discovery from text. In *Proceedings of the 19th International Conference on Computational Linguistics (COLING-02)*. Taipei, Taiwan, pages 577–583.

Y. Lin, W. Li, K. Chen, and Y. Liu. 2007. A document clustering and ranking system for exploring MEDLINE citations. *Journal of the American Medical Informatics Association* 14(5):651–661.

Y. Liu, B.T. McInnes, T. Pedersen, G. Melton-Meaux, and S. Pakhomov. 2012. Semantic relatedness study using second order co-occurrence vectors computed from biomedical corpora, UMLS and WordNet. In *Proceedings of the 2nd ACM SIGHIT symposium on International health informatics*. ACM, pages 363–372.

A. Maedche and S. Staab. 2001. *Comparing ontologies-similarity measures and a comparison study*. AIFB.

B.T. McInnes, T. Pedersen, and S.V. Pakhomov. 2009. UMLS-Interface and UMLS-Similarity: Open source software for measuring paths and semantic similarity. In *Proceedings of the American Medical Informatics Association (AMIA) Symposium*. San Fransico, CA, pages 431–435.

T. Mikolov, I. Sutskever, K. Chen, G.S. Corrado, and J. Dean. 2013. Distributed representations of words and phrases and their compositionality. In *Advances in neural information processing systems*. pages 3111–3119.

T.H. Muneeb, Sunil Sahu, and Ashish Anand. 2015. Evaluating distributed word representations for capturing semantics of biomedical concepts. In *Proceedings of BioNLP 15*. Association for Computational Linguistics, Beijing, China, pages 158–163.

S. Pakhomov, B.T. McInnes, T. Adam, Y. Liu, T. Pedersen, and G.B. Melton. 2010. Semantic similarity and relatedness between clinical terms: An experimental study. In *Proceedings of the American Medical Informatics Association (AMIA) Symposium*. Washington, DC, pages 572–576.

S.V. Pakhomov, G. Finley, R. McEwan, Y. Wang, and G.B. Melton. 2016. Corpus domain effects on distributional semantic modeling of medical terms — Bioinformatics — Oxford Academic. *Bioinformatics* 32:3635–3644.

S.V.S. Pakhomov, T. Pedersen, B. McInnes, G.B. Melton, A. Ruggieri, and C.G. Chute. 2011. Towards a framework for developing semantic relatedness reference standards. *Journal of Biomedical Informatics* 44(2):251–265.

S. Patwardhan and T. Pedersen. 2006. Using WordNet-based context vectors to estimate the semantic relatedness of concepts. In *Proceedings of the EACL 2006 Workshop Making Sense of Sense - Bringing Computational Linguistics and Psycholinguistics Together*. Trento, Italy, pages 1–8.

T. Pedersen, S.V.S. Pakhomov, S. Patwardhan, and C.G. Chute. 2007. Measures of semantic similarity and relatedness in the biomedical domain. *Journal of Biomedical Informatics* 40(3):288–299.

V. Pekar and S. Staab. 2002. Taxonomy learning: Factoring the structure of a taxonomy into a semantic classification decision. In *Proceedings of the 19th International Conference on Computational Linguistics - Volume 1*. Association for Computational Linguistics, Stroudsburg, PA, USA, COLING '02, pages 1–7.

G. Pirró and J. Euzenat. 2010. A feature and information theoretic framework for semantic similarity and relatedness. In *The Semantic Web–ISWC 2010*, Springer, pages 615–630.

R. Rada, H. Mili, E. Bicknell, and M. Blettner. 1989. Development and application of a metric on semantic nets. *IEEE Transactions on Systems, Man, and Cybernetics* 19(1):17–30.

K. Radinsky, E. Agichtein, E. Gabrilovich, and S. Markovitch. 2011. A word at a time: computing word relatedness using temporal semantic analysis. In *Proceedings of the 20th international conference on World wide web*. ACM, pages 337–346.

J. Reisinger and R.J. Mooney. 2010. Multi-prototype vector-space models of word meaning. In *Human Language Technologies: The 2010 Annual Conference of the North American Chapter of the Association for Computational Linguistics*. Association for Computational Linguistics, pages 109–117.

P. Resnik. 1995. Using information content to evaluate semantic similarity in a taxonomy. In *Proceedings of the 14th International Joint Conference on Artificial Intelligence*. Montreal, Canada, pages 448–453.

A. Sajadi, E.E. Milios, V. Kešelj, and J.C.M. Janssen. 2015. Domain-specific semantic relatedness from Wikipedia structure: A case study in biomedical text. In *Proceedings of the 16th International Conference on Computational Linguistics and Intelligent Text Processing (CICLing 2015))*. Cairo, Egypt, pages 347–360.

D. Sánchez, M. Batet, and D. Isern. 2011. Ontology-based information content computation. *Knowledge-Based Systems* 24(2):297–303.

H. Schütze. 1992. Dimensions of meaning. In *Proceedings of the ACM/IEEE Conference on Supercomputing*. Minneapolis, MN, pages 787–796.

H. Schütze. 1998. Automatic word sense discrimination. *Computational Linguistics* 24(1):97–123.

J. Weeds, D. Weir, and D. McCarthy. 2004. Characterising measures of lexical distributional similarity. In *Proceedings of the 20th international conference on Computational Linguistics*. Association for Computational Linguistics, page 1015.

Z. Wu and M. Palmer. 1994. Verbs semantics and lexical selection. In *Proceedings of the 32nd Meeting of Association of Computational Linguistics*. Las Cruces, NM, pages 133–138.

W. Yih and V. Qazvinian. 2012. Measuring word relatedness using heterogeneous vector space models. In *Proceedings of the 2012 Conference of the North American Chapter of the Association for Computational Linguistics: Human Language Technologies*. Association for Computational Linguistics, pages 616–620.

Z. Yu, T. Cohen, B. Wallace, E. Bernstam, and T. Johnson. 2016. Retrofitting word vectors of mesh terms to improve semantic similarity measures. In *Proceedings of the Seventh International Workshop on Health Text Mining and Information Analysis*. Association for Computational Linguistics, Auxtin, TX, pages 43–51.

J. Zhong, H. Zhu, J. Li, and Y. Yu. 2002. Conceptual graph matching for semantic search. In *Proceedings of the 10th International Conference on Conceptual Structures*. pages 92–106.

Proactive Learning for Named Entity Recognition

Maolin Li, Nhung T. H. Nguyen, Sophia Ananiadou
National Centre for Text Mining
School of Computer Science, The University of Manchester, United Kingdom
{maolin.li, nhung.nguyen, sophia.ananiadou}@manchester.ac.uk

Abstract

The goal of active learning is to minimise the cost of producing an annotated dataset, in which annotators are assumed to be perfect, i.e., they always choose the correct labels. However, in practice, annotators are not infallible, and they are likely to assign incorrect labels to some instances. Proactive learning is a generalisation of active learning that can model different kinds of annotators. Although proactive learning has been applied to certain labelling tasks, such as text classification, there is little work on its application to named entity (NE) tagging. In this paper, we propose a proactive learning method for producing NE annotated corpora, using two annotators with different levels of expertise, and who charge different amounts based on their levels of experience. To optimise both cost and annotation quality, we also propose a mechanism to present multiple sentences to annotators at each iteration. Experimental results for several corpora show that our method facilitates the construction of high-quality NE labelled datasets at minimal cost.

1 Introduction

Manually annotating a dataset with NEs is both time-consuming and costly. Active learning, a semi-supervised machine learning algorithm, aims to address such issues (Lewis, 1995; Settles, 2010). Instead of asking annotators to label the whole dataset, active learning methods present only representative and informative instances to annotators. Through iterative application of this process, a high-quality annotated corpus can be produced in less time and at lower cost than traditional annotation methods.

There are two strong assumptions in active learning: (1) instances are labelled by experts, who always produce correct annotations and are not affected by the tedious and repetitive nature of the task; (2) all annotators are paid equally, regardless of their annotation quality or level of expertise. However, in practice, it is highly unlikely that all annotators will assign accurate labels all of the time. For example, especially for complex annotation tasks, some labels are likely to be assigned incorrectly (Donmez and Carbonell, 2008, 2010; Settles, 2010). Furthermore, if annotation is carried out for long periods of time, tiredness and reduced concentration may ensue (Settles, 2010), which can lead to annotation errors. An additional issue is that different annotators may have varying levels of expertise, which could make them reluctant to annotate certain cases, and they may assign incorrect labels in other cases. It is also possible that an inexperienced annotator may assign random labels.

To address the above-mentioned assumptions, proactive learning has been proposed to model different types of experts (Donmez and Carbonell, 2008, 2010). Proactive learning assumes that (1) not all annotators are perfect, but that there is at least one "perfect" expert and one less experienced or "fallible" annotator; (2) as the perfect expert always provides correct answers, their time is more expensive than that of the fallible annotator. The annotation process in proactive learning is similar to traditional active learning. At each iteration, annotators will be asked to tag an unlabelled instance, the result of which will be added to the labelled dataset. However, the difference with proactive learning is that, in order to reduce annotation cost, an appropriate annotator is chosen to label each selected instance. For example, if

Proceedings of the BioNLP 2017 workshop, pages 117–125,
Vancouver, Canada, August 4, 2017. ©2017 Association for Computational Linguistics

there is a high probability that the fallible annotator will provide the correct label for an unlabelled instance, then proactive learning will send this instance to be annotated by fallible annotator. This aims to ensure a simultaneous saving of costs and maintenance of the quality of the data.

Proactive learning has been used for several annotation tasks, such as binary and multi-class text classification, and parsing (Donmez and Carbonell, 2008, 2010; Olsson, 2009). In contrast, this paper proposes a proactive learning method for NE tagging, i.e., a sequence labelling task.

Similarly to other efforts that have used proactive learning, our method models two annotators: a reliable one and a fallible one, who have different probabilities of providing correct labels. The reliable annotator is much more likely to produce correct annotations, but their time is expensive. In contrast, the fallible annotator is likely to assign incorrect annotations more often, but charges less for their services. It should be noted that the characteristics of our reliable expert are different from those proposed in previous work (Donmez and Carbonell, 2008, 2010). Specifically, in the conventional proactive learning, the reliable expert is assumed to be perfect, i.e., he/she always provides correct annotations. However, in practice, such an assumption is too strong, especially for NE annotation. Therefore, we assume that the reliable expert is not perfect, but that he/she has a higher expertise level in the target domain, and has a very low error rate. In order to determine an appropriate annotator for each sentence, we calculate the probability that an annotator will assign the correct sequence of labels in a selected unlabelled sentence. Furthermore, at each iteration, we use a batch sampling mechanism to select several sentences for annotators to label (instead of selecting only a single sentence), which optimises both cost and performance.

For evaluation purposes, we simulate the two annotators by using two machine-learning based NER methods, namely LSTM-CRF (Lample et al., 2016) as the reliable expert, and CRF (Lafferty et al., 2001) as the fallible expert. We then apply our method to three corpora from different domains: ACE2005 (Walker et al., 2006) for general language entities, COPIOUS—an in-house corpus of biodiversity entities[1], and GENIA (Kim et al., 2003)—a corpus of biomedical entities. Our ex-

---

[1] The corpus is available upon request.

perimental results demonstrate that by using the proposed method, we can obtain a high-quality labelled corpus at a lower cost than current baseline methods.

The contributions of our work are as follows. Firstly, we have modified the conventional proactive learning method to ensure its suitability for a sequence labelling task. Secondly, in contrast to previous work, which selects a single instance for each annotator at each iteration (Donmez and Carbonell, 2008, 2010; Moon and Carbonell, 2014), our method selects multiple sentences for presentation to annotators. Thirdly, by applying our method to a number of different corpora, we demonstrate that our method is generalisable to different domains.

2 Methodology

The proposed proactive learning for NE tagging is outlined in Algorithm 1. As an initial step, the performance of each expert is estimated based on a benchmark dataset (see Section 2.1). Subsequently, at each iteration, all sentences in the unlabelled dataset are sorted according to an active learning criterion. The top-N most informative sentences are then used as input to the batch sampling step. In this step, a batch of sentences is divided into two sets to be distributed to the reliable and fallible experts, respectively. Sentences distributed to the fallible experts are not only informative, but there is also a high probability that the expert will provide correct labels for them. Meanwhile, only those sentences that are estimated to be too difficult for the fallible expert to annotate will be sent to the reliable expert. By applying this process, annotation cost can be reduced. Further details about the batch sampling algorithm are presented in Section 2.2.

In Algorithm 1, UL_r is the set of selected unlabelled sentences assigned to the reliable expert and UL_f is the set assigned to the fallible expert. L_r, L_f are the annotated results of UL_r, UL_f.

2.1 Expert performance estimation

As mentioned above, our method assumes that there are two types of experts. One is reliable, who has a higher probability of assigning the correct sequence of labels for a sentence, and has a high cost for their time. The other expert is fallible, meaning that they may assign a higher proportion of incorrect labels for a sequence, but

---

Algorithm 1: Proactive Learning for NER

Input: a labelled dataset L, an unlabelled dataset UL, a test dataset T, a budget B, a reliable expert e_r with cost C_r for each sentence, a fallible expert e_f with cost C_f, the current cost C

Output: a labelled dataset L

1 Estimate the performance of each expert as described in Section 2.1;
2 **while** $C < B$ **do**
3   Train a named entity recognition model M on L;
4   Sort all sentences in the unlabelled dataset according to an active learning criterion;
5   Select the top N sentences;
6      $UL_r, UL_f = BatchSampling(M, top\ N\ \text{sentences})$;
7      $L_r, L_f \leftarrow e_r$ and e_f annotate UL_r and UL_f respectively;
8      $L = L \cup L_r \cup L_f$;
9      $UL = UL - UL_r - UL_f$;
10      $C = C + C_r * |L_r| + C_f * |L_f|$;
11 **end**

Algorithm 2: Batch Sampling

Input: a named entity recognition model M, top-N sentences selected according to an active learning criterion

Output: UL_r, UL_f

1 $UL_r = \emptyset$;
2 $UL_f = \emptyset$;
3 **while** *Batch Size* **do**
4   // Stage 1
4   **foreach** *sentence* x **do**
5   **if** $p(CorrectLabels|fallible, x) > \alpha$ **then**
6              $UL_f = UL_f \cup \{x\}$;
7   BatchSize = BatchSize - 1
8   **end**
9   **end**
     // Stage 2
10   **if** *Batch Size* $\neq 0$ **then**
11   Sort the remaining sentences according to a re-ranking criterion;
12   Calculate threshold β;
13   **foreach** *sentence* x **do**
14   **if** *Batch Size* $\neq 0$ **then**
15   **if** $diff(reliable, fallible, x) < \beta$ **then**
16                      $UL_f = UL_f \cup \{x\}$;
17   **else**
18                      $UL_r = UL_r \cup \{x\}$;
19   **end**
20   BatchSize = BatchSize - 1;
21   **end**
22   **end**
23   **end**
24 **end**

---

charges less for their time. The likely annotation quality of each expert is estimated based on two different probabilities: the class probability, $p(label|expert, c)$ and the sentence probability $p(CorrectLabels|expert, x)$.

2.1.1 Class probability

The class probability, $p(label|expert, c)$, is the probability that an *expert* provides a correct label when annotating a named entity of class c. This probability is obtained by asking both the reliable and fallible experts to annotate a benchmark dataset and calculating F_1 scores for each of them against the gold standard annotations.

2.1.2 Sentence probability

The sentence probability is the probability that an *expert* provides a sequence of correct labels for a sentence x.

We firstly compute the probability for each token in the sentence by combining the class probability and the likelihood that an *expert* provides a correct label for the token x_i, as shown in Equation 1. The equation is inspired by Moon and Carbonell (2014), who used it for a classification task.

$$p(CorrectLabel|expert, x_i) =$$
$$\sum_c^{|C|} p(c|x_i) * p(label|expert, c) \quad (1)$$

C is the set of all entity labels and the label O. $p(c|x_i)$ is the probability that a token x_i is an entity of class c, which is predicted by an NER model.

Given the probabilities that an expert will provide correct labels for each tokens in a sentence, the sentence probability is calculated by averaging all of these probabilities, as presented in Equation 2.

$$p(CorrectLabels|expert, x) =$$
$$\frac{\sum_i^{|x|} p(CorrectLabel|expert, x_i)}{|x|} \quad (2)$$

$|x|$ is the length of the sentence x.

2.2 Batch sampling

Instead of asking annotators to label only one sentence at each iteration, it is more efficient to ask them to annotate several sentences. To facilitate this, we propose a batch sampling algorithm that can select a set of sentences and assign them to appropriate annotators (see Algorithm 2).

The input of the algorithm is a set of sentences in the unlabelled dataset that are considered to be the most informative ones, based on an active learning criterion (as described in line 5 of Algorithm 1).

This batch sampling process is divided into two stages. In the first stage, unlabelled sentences for which the sentence probability for the fallible expert is higher than a threshold α, will be assigned to the fallible expert. Otherwise, the sentence will be passed to the second stage. In the second stage, we firstly reorder sentences according to a re-ranking criterion, as shown in Equation 3. The intuition behind this re-ranking step is that in order to save on annotation costs, we set a high priority for sentences to be assigned to the fallible expert in certain cases. Specifically, for sentences that are informative and for which there is a small difference between the sentence probabilities for the reliable and fallible experts, we favour the selection of the fallible one.

$$ReRankingCriterion = $$
$$\frac{ActiveLearningCriterion(\boldsymbol{x})}{diff(reliable, fallible, \boldsymbol{x})} \quad (3)$$

For an unlabelled sentence $\boldsymbol{x}$, the difference between the sentence probabilities for the two experts is calculated as shown in Equation 4.

$$diff(reliable, fallible, \boldsymbol{x})$$
$$= |p(CorrectLabels|reliable, \boldsymbol{x})$$
$$- p(CorrectLabels|fallible, \boldsymbol{x})| \quad (4)$$

If the above difference is not significant, i.e., it is less than a threshold β, $\boldsymbol{x}$ will be distributed to the fallible expert. Otherwise, $\boldsymbol{x}$ will be assigned to the reliable expert.

Equations (5) - (7) describe the estimation of the threshold β, in which $\boldsymbol{x}^i$ is the ith sentence in the top-N sentences selected by an active learning criterion. γ is a parameter that controls the value of the threshold β. γ ranges from 0 to 1. If $\gamma = 0$, no sentences will be given to the fallible expert to annotate. If $\gamma = 1$, the fallible expert will label all the $BatchSize$ sentences. It should be noted that β is a dynamic threshold, which is recalculated based on the difference between $diff_{max}$ and $diff_{min}$ at each iteration.

$$diff_{min} = min_i^N(diff(reliable, fallible, \boldsymbol{x}^i))$$
$$(5)$$

$$diff_{max} = max_i^N(diff(reliable, fallible, \boldsymbol{x}^i))$$
$$(6)$$

$$\beta = diff_{min} + \gamma(diff_{max} - diff_{min}) \quad (7)$$

3 Experiments

3.1 Dataset

We have applied our method to three different corpora: (1) ACE2005 (Walker et al., 2006) which includes named entities for the general domain, e.g., person, location, and organisation; (2) COPIOUS that includes five categories of biodiversity entities, such as taxon, habitat, and geographical location; (3) GENIA (Kim et al., 2003), a biomedical named entity corpus.

Table 1 shows the entity classes and the number of entities of each class that are annotated in the three corpora. As shown in the table, for the GENIA corpus, we combined the DNA and RNA entities into a single named entity class. Meanwhile, for ACE2005, although top-level entity classes are divided into a number of different subtypes, we only considered the top-level classes, as shown in the table.

For active and proactive learning experiments, 1% and 20% of sentences of each corpus were used as the initial labelled set and the test set, respectively. The remaining 79% of sentences were regarded as unlabelled data.

3.2 Expert simulation

We simulated the reliable and fallible experts by using two machine learning models: LSTM-CRF (Lample et al., 2016)—a neural network NER and CRF (Lafferty et al., 2001). To evaluate the performance of the two models, we conducted preliminary experiments, by firstly trained the two models on 80% of the labelled corpora and subsequently testing them on the remaining 20% of the data.

Word embeddings As the three corpora belong to three different domains, we used three corresponding pre-trained word embeddings as input to the LSTM-CRF model.

- ACE2005: GoogleNews vectors[2], which include approximately 100 billion words.

- COPIOUS: we applied word2vec to the English subset of the Biodiversity Heritage Library[3] to learn vectors for biodiversity entities. The set has approximately 26 million pages with more than 8 billion words.

---

[2]http://code.google.com/archive/p/word2vec/
[3]http://www.biodiversitylibrary.org/

Corpus	Entity	Labelled	Unlabelled	Test	Total
ACE2005	Person (PER)	291	22853	5179	28323
	Organization (ORG)	36	4554	690	5280
	Geo-Political Entity (GPE)	21	5813	1360	7194
	Location (LOC)	7	760	168	935
	Facility (FAC)	5	1136	227	1368
	Weapon (WEA)	8	609	178	795
	Vehicle (VEH)	7	640	123	770
COPIOUS	Habitat	23	619	366	1008
	Taxon	116	4485	1728	6329
	Person	24	768	258	1050
	Geographical Location (GeoLoc)	42	4373	1942	6357
	Temporal Expression (TempExp)	20	904	358	1282
GENIA	DNA&RNA	88	6592	1757	8437
	Cell	133	9623	2437	12193
	Protein	316	24940	6402	31658

Table 1: Statistic information of the three corpora

- GENIA: word vectors trained on a combination of PubMed, PMC and English Wikipedia texts (Pyysalo et al., 2013).

CRF features To train the CRF model, we used CRF++[4] and employed following features: word base, lemma, part-of-speech tag and chunk tag of a token. We also used unigram and bigram features that combine the features of the previous, current and following token.

As illustrated in Table 2, the LSTM-CRF model is mostly more precise and achieves wider coverage than CRF. We therefore selected LSTM-CRF to simulate the reliable expert and CRF to simulate the fallible expert.

Corpus	CRF			LSTM-CRF		
	Pre.	Re.	F1	Pre.	Re.	F1
ACE2005	73.89	65.07	69.20	75.69	74.11	74.89
COPIOUS	81.01	48.58	60.74	77.18	74.77	75.96
GENIA	73.90	64.52	68.89	75.41	73.91	74.66

Table 2: Performance of CRF and LSTM-CRF on the three corpora

The reliable expert (the LSTM-CRF model) was trained on 80% of the labelled data, while the fallible one (the CRF model) was trained on 60%. The F_1 scores of the reliable and fallible experts when applied to the test dataset are presented in Table 3.

Corpus	Fallible	Reliable
ACE2005	61.19	74.89
COPIOUS	50.92	75.96
GENIA	57.67	74.66

Table 3: F_1 scores of each expert on the three corpora

The class probability of each expert is precalculated based on the the the F_1 score of each class that an expert can achieve on the 1% initial labelled set. Meanwhile, the sentence probability of each expert is estimated at each iteration.

3.3 Active learning criteria

Various active learning criteria were investigated using the three corpora. We firstly estimated the performance (F_1 score) of a supervised NER model by using CRF++ and the above-mentioned features. We then compared the performance of each active learning criterion with that of the supervised model. If the performance of one criterion approximates that of the supervised with the least number of iterations, we consider the criterion as the best one for proactive learning experiments.

We experimented with the following criteria: least confidence (Culotta and McCallum, 2005), normalized entropy (Kim et al., 2006), MMR (Maximal Marginal Relevance) (Kim et al., 2006), density (Settles and Craven, 2008) when using feature vectors and word embeddings, and the combination of least confidence and density criterion. Equation 8 describes the combination criterion used in our experiments. In this equation, UL is the current unlabelled dataset, x^u is the uth unlabelled sentence in UL, the parameter $\lambda = 0.8$, and the similarity score (Settles and Craven, 2008) were calculated by using feature vectors.

$$x^* = \arg\max_{x}(\lambda * Least\_Confidence(x)$$

$$+ (1 - \lambda)\frac{1}{|UL|} \sum_{u=1}^{|UL|} similarity(x, x^u)) \quad (8)$$

[4]https://taku910.github.io/crfpp/

Corpus	Entity Class	Best Criterion
ACE2005	PER	Density (w2v)
	ORG	Density (f2v)
	GPE	Entropy
	LOC	Least Confidence
	FAC	Longest
	WEA	MMR
	VEH	Longest
		(Overall) Entropy
COPIOUS	Habitat	Density (f2v)
	Taxon	Entropy
	Person	Density (f2v)
	GeoLoc	Entropy
	TempExp	Least Confidence
		(Overall) Entropy
GENIA	Protein	Entropy
	Cell	LC+Density (f2v)
	DNA&RNA	Entropy
		(Overall) Entropy

Table 4: The best active learning criteria on the three corpora

We also implemented two baseline criteria. The first one is random selection, in which a batch of sentences is selected randomly at each iteration. The second one, namely *longest*, is a criterion that selects the longest sentences to be labelled.

Among these criteria, we selected the best criterion for further experiments. The best criterion is the one that produced competitive or better performance (F-score) than that of a supervised learning method with the least number of training instances. We report these criteria for each entity class as well as for the overall corpus in Table 4. In this table, Density (f2v) and Density (w2v) represent the density criteria when using feature and word vectors, respectively. Entropy is the normalized entropy. LC+Density is the combined criterion, described in Equation 8. As shown in the table, the best criteria at the level of individual classes are diverse. However, overall, normalized entropy is the best criterion for all three corpora. We therefore selected this criterion in our proactive learning experiments.

3.4 Proactive learning results

Our method was evaluated on the test datasets of the three corpora mentioned in Section 3.1. For all experiments with proactive learning, we used the following settings: $\alpha = 0.975$, $\gamma = 0.05$, $N = 200$, and the annotation costs are 3 and 1 per sentence for the reliable and fallible experts, respectively.

3.4.1 *BatchSize*

We investigated different values of *BatchSize* including 20, 10, 5, and 1. The results when *BatchSize* is 1 was not shown in Figure 1 as our method always selects the fallible expert at every iteration, which results in a performance that is inferior to the baselines. For the GENIA corpus, the F-scores are comparable, regardless of the *BatchSize* used. Meanwhile, for the ACE2005 corpus, the F-scores are the highest when the batch size is 20. In contrast, for the COPIOUS corpus, the best scores are obtained with a batch size of 10.

3.4.2 Comparison with baselines

Figure 2 compares the experimental results of the two baseline methods (*Reliable* and *Fallible*) and the best performance of the proposed proactive learning method (*PA*) with batch sizes of 20, 10, and 5, respectively, on the three corpora. *Reliable* refers to a baseline in which we only select the reliable expert at each iteration. Similarly, only the fallible expert was selected in the *Fallible* experiments.

It can be seen that the performance of the three models is comparable between ACE2005 and the COPIOUS corpus. For these two corpora, *PA* outperformed the two baselines. In most cases, by using *PA*, better F-scores are obtained at the same cost as the two baselines. Both *PA* and *Reliable* performance is increased when the total cost is increased. Meanwhile, for the *Fallible* model, the performance stabilises at a lower level than the other methods when cost rises above a certain level.

Regarding the GENIA corpus, *PA* acheived a higher performance than *Reliable*, but a lower performance than *Fallible* in the range of costs from 0 to approximately 3,500. This can be partly explained by the fact that there are only three NE classes in this corpus. Hence, the annotation task is simpler than for the the other corpora, even for the fallible expert. However, when the cost is greater than 3,500, the performance of *Fallible* becomes stable, while the performance of *PA* continues to increase.

We also investigated the number of times that each expert was selected during the iterative process of *PA*. The results are shown in Figure 3. *PA* (*Reliable*) and *PA* (*Fallible*) correspond to number of times that the reliable and fallible expert respectively, were selected in *PA*, while

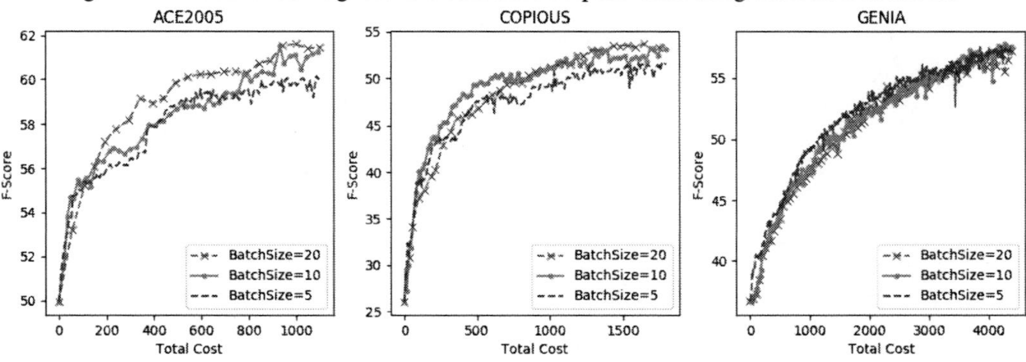

Figure 1: Pro-active learning results on the three corpora when using different $BatchSize$

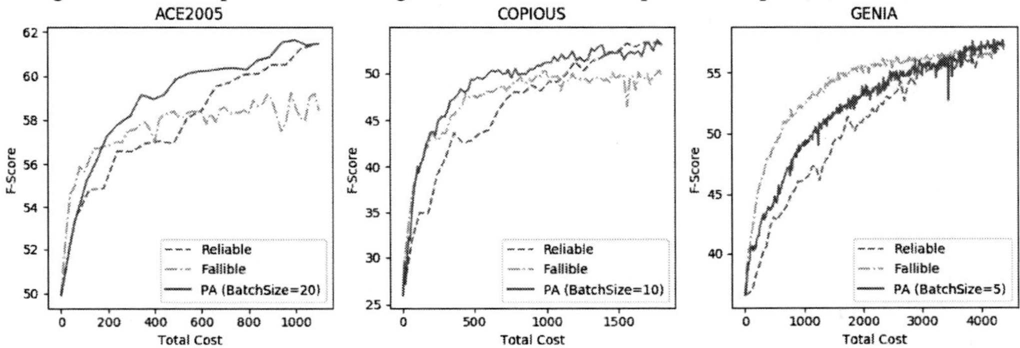

Figure 2: The best pro-active learning results on the three corpora in comparison to the baselines

Reliable corresponds to the number of times that the reliable expert was selected in *Reliable* baseline experiment. The figure illustrates that the number of times that the fallible expert is selected grows continually as the number of iterations increases. This shows that our method can effectively distribute appropriate unlabelled sentences to the fallible expert, in order to save on annotation costs.

4 Related work

4.1 Active learning for NER

Active learning aims to decrease annotation cost, whilst maintaining acceptable quality of annotated data. To achieve this, the method iteratively selects the most informative sentences to be annotated from an unlabelled data set.

One of the most common selection criteria used in applying active learning to the task of NE labelling is the uncertainty-based criterion. This criterion assumes that the most uncertain sentence is the most useful instance for learning an NER model. There are several ways to implement this, such as least confidence (Culotta and McCallum, 2005)–the lower the probability of a sequence of labels, the less confidence the model, and entropy (Kim et al., 2006) that can measure the uncertainty of a probability distribution. Some other criteria are a diversity measurement (Kim et al., 2006) and a density criterion (Settles and Craven, 2008).

4.2 Cost-sensitive active learning

Cost-sensitive active learning is a type of active learning method that considers the annotation cost, e.g., budget, time or effort required to complete the annotation process (Olsson, 2009). Since proactive learning also models the reliability or expertise of each annotator in addition to the annotation cost, it can be considered as another case of cost-sensitive active learning.

Donmez and Carbonell (2008, 2010) investigated proactive learning for binary classification.

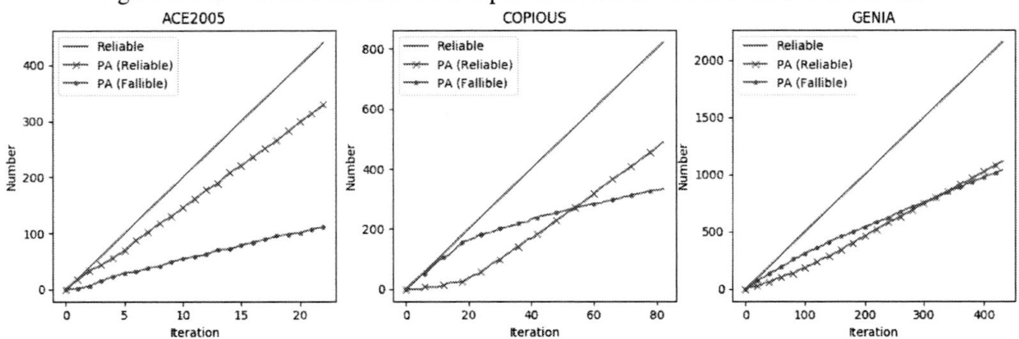

Figure 3: Number of times that each expert is selected in PA and *Reliable* models

They predicted the probability that a reluctant oracle refuses to annotate an instance and the probability that a fallible oracle assigns a random label to an instance. Each oracle charges a different amount for their efforts. They also proposed a model that assigns different costs to unlabelled instances according to their annotation difficulty. For the multi-class classification task, Moon and Carbonell (2014) used the same approach but they had multiple experts, each of whom is specialised for each class. Kapoor et al. (2007) proposed a decision-theoretic method for the task of voice mail classification. They defined a criterion named "expected value-of-information" that combines the misclassification risk with the labelling cost.

Cost-sensitive active learning was also applied to part-of-speech (POS) tagging (Haertel et al., 2008). In this work, an hourly cost measurement was determined and a linear regression model was trained to predict the annotation cost. Hwa (2000) aimed to reduce the manual effort for a parsing task by using tree entropy cost. Meanwhile, Baldridge and Osborne (2004) measured the total annotation cost to create a treebank by using unit cost and discriminant cost.

5 Conclusion and future work

Our work constitutes the first attempt to use proactive learning method for named entity labelling. We simulated the behaviour of reliable and fallible experts having different levels of expertise and different costs. To save annotation costs and to ensure acceptable quality of the resulting annotated data, the method favours the selection of the fallible expert. In order to increase efficiency, we also proposed a batch sampling algorithm to select more than one sentence in each iteration.

Experimental results for three corpora belonging to different domains demonstrate that the employment of non-perfect experts can help to build gold standard dataset at reasonable cost. Moreover, our method performed well across the three different corpora, demonstrating the generality of our approach.

A potential limitation of our approach is that the initial step is reliant on the availability of a gold standard corpus to estimate the experts' performance. However, for some domains, it may be difficult to obtain such a dataset. Therefore, as future work, we will explore how we can assess experts' performance without the need for gold-standard labelled data.

As a further extension to our work, we will explore the deployment of our method on crowd sourcing platforms, such as CrowdFlower[5] and Amazon Mechanical Turk[6]. These platforms allow annotations to be obtained from non-expert annotators in a rapid and cost-effective manner (Snow et al., 2008). These non-experts can be treated as non-perfect annotators in our proposed proactive learning method.

Acknowledgement

We would like to thank Paul Thompson for his valuable comments. This work is partially funded by the British Council (COPIOUS 172722806) and the Biotechnology and Biological Sciences Research Council, UK (EMPATHY, Grant No. BB/M006891/1).

---

[5]https://www.crowdflower.com/
[6]https://www.mturk.com/mturk/welcome

References

Jason Baldridge and Miles Osborne. 2004. Active Learning and the Total Cost of Annotation. In *EMNLP*. pages 9–16.

Aron Culotta and Andrew McCallum. 2005. Reducing labeling effort for structured prediction tasks. In *AAAI*. volume 5, pages 746–51.

Pinar Donmez and Jaime G Carbonell. 2008. Proactive learning: cost-sensitive active learning with multiple imperfect oracles. In *Proceedings of the 17th ACM conference on Information and knowledge management*. ACM, pages 619–628.

Pinar Donmez and Jaime G Carbonell. 2010. From active to proactive learning methods. In *Advances in Machine Learning I*, Springer, pages 97–120.

Robbie Haertel, Eric Ringger, Kevin Seppi, James Carroll, and Peter McClanahan. 2008. Assessing the costs of sampling methods in active learning for annotation. In *Proceedings of the 46th Annual Meeting of the Association for Computational Linguistics on Human Language Technologies: Short Papers*. Association for Computational Linguistics, pages 65–68.

Rebecca Hwa. 2000. Sample selection for statistical grammar induction. In *Proceedings of the 2000 Joint SIGDAT conference on Empirical methods in natural language processing and very large corpora: held in conjunction with the 38th Annual Meeting of the Association for Computational Linguistics-Volume 13*. Association for Computational Linguistics, pages 45–52.

Ashish Kapoor, Eric Horvitz, and Sumit Basu. 2007. Selective Supervision: Guiding Supervised Learning with Decision-Theoretic Active Learning. In *IJCAI*. volume 7, pages 877–882.

J-D Kim, Tomoko Ohta, Yuka Tateisi, and Junichi Tsujii. 2003. GENIA corpus–a semantically annotated corpus for bio-textmining. *Bioinformatics* 19(suppl 1):i180–i182.

Seokhwan Kim, Yu Song, Kyungduk Kim, Jeong-Won Cha, and Gary Geunbae Lee. 2006. Mmr-based active machine learning for bio named entity recognition. In *Proceedings of the Human Language Technology Conference of the NAACL, Companion Volume: Short Papers*. Association for Computational Linguistics, pages 69–72.

John Lafferty, Andrew McCallum, Fernando Pereira, et al. 2001. Conditional random fields: Probabilistic models for segmenting and labeling sequence data. In *Proceedings of the eighteenth international conference on machine learning, ICML*. volume 1, pages 282–289.

Guillaume Lample, Miguel Ballesteros, Sandeep Subramanian, Kazuya Kawakami, and Chris Dyer. 2016. Neural Architectures for Named Entity Recognition. In Kevin Knight, Ani Nenkova, and Owen Rambow, editors, *HLT-NAACL*. The Association for Computational Linguistics, pages 260–270.

David D Lewis. 1995. Evaluating and optimizing autonomous text classification systems. In *Proceedings of the 18th annual international ACM SIGIR conference on Research and development in information retrieval*. ACM, pages 246–254.

Seungwhan Moon and Jaime G Carbonell. 2014. Proactive learning with multiple class-sensitive labelers. In *Data Science and Advanced Analytics (DSAA), 2014 International Conference on*. IEEE, pages 32–38.

Fredrik Olsson. 2009. A literature survey of active machine learning in the context of natural language processing. Swedish Institute of Computer Science.

Sampo Pyysalo, Filip Ginter, Hans Moen, Tapio Salakoski, and Sophia Ananiadou. 2013. Distributional semantics resources for biomedical text processing. In *Proceedings of LBM 2013*. LBM, pages 39–44.

Burr Settles. 2010. Active learning literature survey. *University of Wisconsin, Madison* 52(55-66):11.

Burr Settles and Mark Craven. 2008. An analysis of active learning strategies for sequence labeling tasks. In *Proceedings of the conference on empirical methods in natural language processing*. Association for Computational Linguistics, pages 1070–1079.

Rion Snow, Brendan O'Connor, Daniel Jurafsky, and Andrew Y Ng. 2008. Cheap and fast—but is it good?: evaluating non-expert annotations for natural language tasks. In *Proceedings of the conference on empirical methods in natural language processing*. Association for Computational Linguistics, pages 254–263.

Christopher Walker, Stephanie Strassel, Julie Medero, and Kazuaki Maeda. 2006. ACE 2005 multilingual training corpus. *Linguistic Data Consortium, Philadelphia* 57.

Biomedical Event Extraction using Abstract Meaning Representation

Sudha Rao[1], Daniel Marcu[2], Kevin Knight[2], Hal Daumé III[1]
[1]Computer Science, University of Maryland, College Park,
[2]Information Sciences Institute, University of Southern California
raosudha@cs.umd.edu, marcu@isi.edu, knight@isi.edu, hal@cs.umd.edu

Abstract

We propose a novel, Abstract Meaning Representation (AMR) based approach to identifying molecular events/interactions in biomedical text. Our key contributions are: (1) an empirical validation of our hypothesis that an event is a subgraph of the AMR graph, (2) a neural network-based model that identifies such an event subgraph given an AMR, and (3) a distant supervision based approach to gather additional training data. We evaluate our approach on the 2013 Genia Event Extraction dataset[1] (Kim et al., 2013) and show promising results.

1 Introduction

For several years now, the biomedical community has been working towards the goal of creating a curated knowledge base of biomolecule entity interactions. The scientific literature in the biomedical domain runs to millions of articles and is an excellent source of such information. However, automatically extracting information from text is a challenge because natural language allows us to express the same information in several different ways. The series of Genia Event Extraction shared tasks (Kim et al., 2009, 2011, 2013, 2016) has resulted in various significant approaches to biomolecule event extraction spanning methods that use learnt patterns from annotated text (Bui et al., 2013) to machine learning methods (Björne and Salakoski, 2013) that use syntactic parses as features. In this work, we find that a semantic analysis of text that relies on Abstract Meaning Representations (Banarescu et al., 2013) is highly useful because it normalizes many lexical and syntactic variations in text.

---

[1]This dataset is different from BioNLP 2016 GE dataset

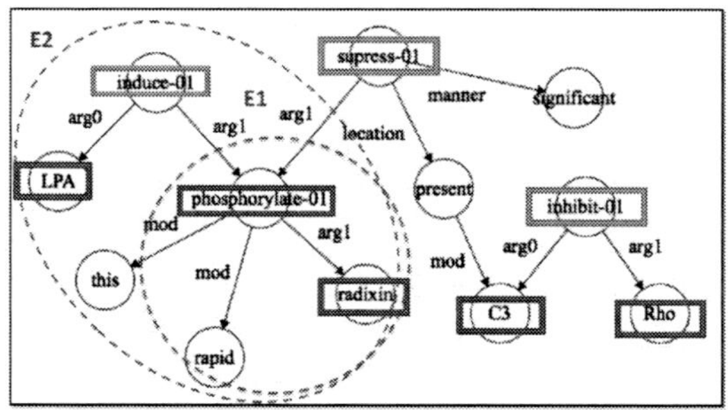

Figure 1: AMR with sample event annotations for sentence *"This LPA-induced rapid phosphorylation of radixin was significantly suppressed in the presence of C3 toxin, a potent inhibitor of Rho"*

AMR is a rooted, directed acyclic graph (DAG) that captures the notion of *who did what to whom* in text, in a way that sentences that have the same basic meaning often have the same AMR. The nodes in the graph (also called concepts) map to words in the sentence and the edges map to relations between the words. In the recent past, there have been several efforts towards parsing a sentence into its AMR (Flanigan et al., 2014; Wang et al., 2015; Pust et al., 2015; May, 2016). AMR naturally captures hierarchical relations between entities in text making it favorable for complex event detection. For example, consider the following sentence from the biomedical literature: *"This LPA-induced rapid phosphorylation of radixin was significantly suppressed in the presence of C3 toxin, a potent inhibitor of Rho"*. Figure 1 shows its Abstract Meaning Representation (AMR). The subgraph rooted at phosphorylate-01 identifies the event E_1 and the subgraph rooted at induce-01 identifies the event E_2 where

$$E_1 = phosphorylation\ of\ radixin;$$
$$E_2 = LPA\ induces\ E1.$$

We hypothesize that an event structure is a sub-

Proceedings of the BioNLP 2017 workshop, pages 126–135,
Vancouver, Canada, August 4, 2017. ©2017 Association for Computational Linguistics

Type	Primary Args.
Gene_expression	T(P)
Transcription	T(P)
Localization	T(P)
Protein_catabolism	T(P)
Binding	T(P)+
Phosphorylation	T(P/Ev), C(P/Ev)
Regulation	T(P/Ev), C(P/Ev)
Positive_regulation	T(P/Ev), C(P/Ev)
Negative_regulation	T(P/Ev), C(P/Ev)

Table 1: Event types and their arguments in the 2013 Genia Event Extraction task

graph of a DAG structure like AMR and under this assumption, we cast the event extraction task as a graph identification problem. Our **first contribution** is the testing of the above hypothesis that an event structure is a subgraph of an AMR graph. Given a sentence, we automatically obtain its AMR using an AMR parser (Pust et al., 2015) and explain how an event can be defined as a subgraph of the AMR graph. Under the assumption that we can correctly identify such an event subgraph from an AMR graph when it exists, we evaluate how good is our definition (Section 2).

Our **second contribution** is a supervised neural network-based model that is trained to identify an event subgraph given an AMR (Section 3). Our model is built on the intuition that the path between an interaction term and an entity term in an AMR graph contains important signal for identifying the relation between them. For e.g. in figure 1 the path { *'induce-01', 'arg0', 'LPA'* } suggests that *LPA* is the cause of *induce*. We encode this path using word embeddings pre-trained on millions of biomedical text and develop two pipelined neural network models: (a) to identify the *theme* of an *interaction*; and (b) to identify the *cause* of the *interaction*, if there exists one.

Experimental results show that our model, although achieves a reasonable precision, suffers from low recall. Our **third contribution** is a distant supervision (Mintz et al., 2009) based approach to collect additional annotated training data. Distant supervision works on the assumption that given a known relation between two entities, a sentence containing the two entities is likely to express this relation and hence can serve as training data for that relation. Data gathered using such a method can be noisy (Takamatsu et al., 2012). Roth et al. (2013) have discussed several prior work that address this issue. In our work, we introduce a method based on AMR path heuristic

This LPA-induced rapid phosphorylation of radixin was significantly suppressed in the presence of C3 toxin, a potent inhibitor of Rho

T1	(Protein, LPA)
T2	(Protein, radixin)
T2	(Protein, C3)
T4	(Protein, Rho)
T5	(Phosphorylation, phosphorylate)
T6	(Positive_regulation, induce)
T7	(Negative_regulation, suppress)
T8	(Negative_regulation, inhibit)
E1	(Type: T5, Theme: T2)
E2	(Type: T6, Theme: E1, Cause: T1)
E3	(Type: T7, Theme: E1)
E4	(Type: T8, Theme: T4, Cause: T3)

Table 2: Example event annotation. The protein annotations T1- T4 are given as starting points. The task is to identify the events E1-E4 with their interaction type and arguments.

to selectively sample the sentences we obtain using distant supervision (Section 3) and show its effectiveness over our vanilla neural network model.

We evaluate our event extraction model on the 2013 Genia Event Extraction dataset and show that our model achieves promising results when compared to the state-of-the-art system. Given that AMR parsing is still a young field, our model, which currently uses a parser of 67% accuracy, would perform better with improved AMR parsers.

2 AMR based event extraction model

2.1 Task description

The biomedical event extraction task in this work is adopted from the Genia Event Extraction subtask of the well-known BioNLP shared task ((Kim et al., 2009), (Kim et al., 2011), (Kim et al., 2013)). Table 2 shows a sample event annotation for the sentence in Figure 1. The protein annotations T1- T4 are given as starting points. The task is to identify the events E1-E4 with their interaction type and arguments. Table 1 describes the various event types and the arguments they accept. The first four event types require only unary theme argument. The binding event can take a variable number of theme arguments. The last four events take a theme argument and, when expressed, also a cause argument. Their theme or cause may in turn be another event, creating a nested event (For e.g. event E2 in Table 2).

2.2 Model description

We cast this event extraction problem as a subgraph identification problem. Given a sentence we first obtain its AMR graph automatically using an AMR parser (Pust et al., 2015). Next, we identify protein nodes and interaction nodes in the graph.

Protein Node Identification: In both the training and the test set, protein terms are pre-annotated (e.g. $T1$ to $T4$ in Table 2). We then use the AMR graph alignment information to identify nodes in the AMR graph aligned to these protein terms to get our protein nodes P.

Interaction Node Identification: In the training data, interaction terms are pre-annotated (e.g. $T5$ to $T8$ in Table 2). To identify the interaction terms in the test set we use the following heuristic: any term that was annotated as an interaction term more than once in the training data is considered as an interaction term in the test data as well. We then use the AMR graph alignment information to identify nodes in the AMR graph aligned to the interaction terms to get our interaction nodes T.

Given P and T, we identify an event sub-graph using the following two-step process:

a.   Theme Identification: Every pair (p_i, t_j) where $p_i \in P$ and $t_j \in T$, is a candidate for an event e_m defined as e_m: *(Type: t_j, Theme: p_i)* where *Type* is one of the nine event types in Table 1. If e_m can take other events as arguments (last four event types in Table 1) and if the shortest path between t_j and p_i includes an interaction term t_k, such that the pair (p_i, t_k) is an event e_n in itself, then we define the event e_m instead as e_m: *(Type: t_j, Theme: e_n)*. For e.g. in Figure 1, the path between *induce-01* and *radixin* includes *phosphorylate-01* which is an event in itself (E_1). Hence event E_2 is defined with E_1 as its theme (in Table 2).

b.   Cause Identification: For events e_m: *(Type: t_j: Theme: p_i)* that can take a cause argument, we identify possible candidates for their cause by again looking for all pairs (p_l, t_j) where $p_l \in P$ and $l \neq i$ and add cause to the event e_m as e_m: *(Type: t_j, Theme: p_i, Cause: p_l)*. Since these events can even take other events as their cause argument, we identify additional candidates for their cause by looking for all pairs (e_n, t_j) where $e_n \in E$ and $n \neq m$ and add cause to the event e_m as e_m: *(Type: t_j, Theme: p_i, Cause: e_n)*.

2.3   Upper bound using *"event is a subgraph of AMR"* hypothesis

Before we learn to identify event sub-graphs from an AMR graph, we first calculate the upper bound

Event Type	R	P	F1	F1 ()
Gene_expression	87.82	100.00	93.51	
Transcription	65.31	100.00	79.01	
Localization	86.80	100.00	92.93	
Protein_catabolism	90.00	100.00	94.74	
==[SVT-TOTAL]==	82.48	100.00	90.04	76.59
Binding	67.83	95.83	79.43	42.88
Phosphorylation	60.62	80.14	69.03	65.37
Regulation	42.61	61.73	50.42	
Positive_regulation	41.93	65.43	51.11	
Negative_regulation	50.94	65.85	57.45	
==[REG-TOTAL]==	45.16	64.33	53.00	38.41
==[ALL-TOTAL]==	65.98	85.44	74.18	50.97

Table 3:   Upper bound on the dev set using our *"event is a subgraph of AMR"* hypothesis

that we are setting for our model because we are using an AMR parser instead of obtaining gold AMRs. For calculating this upper bound, we first obtain the AMR graph of a sentence using the AMR parser and then assume that if an event is a sub-graph of this AMR graph then we can identify it correctly. Table 3 shows the upper bound we get on the dev set of the 2013 Genia Event Extraction dataset (described in Section 5.1). The last column in the table is the state-of-the-art F1 score obtained by the system EVEX (Hakala et al., 2013) on the test set of the dataset[2].

In case of simple events i.e. events that take only proteins as theme arguments, an event is always a subgraph of the AMR unless there is an alignment error causing the protein node or the interaction node to be missing. Hence the upper bound on our precision is 100% whereas the upper bound on our recall is 82.48% for these simple events. In case of the other event types where an event can take other events as arguments, an event is correctly identified only if the path between the pair (p_i, t_j) in the AMR graph includes all its subevents. Therefore we lose more on the precision and recall in these cases due to AMR parsing errors bringing our overall upper bound on precision down to 85.44% and our overall upper bound on recall down to 65.98%. These results give us following two important insights:

1. By using this hypothesis we have set an upper bound of 74.18% F1-score for our learning model.

2. As the accuracy of automatic AMR parsers improve, our model will perform better at the event extraction task.

---

[2]We compare our numbers on the dev set to the EVEX numbers on test set since gold annotations for the test set are not available for download

3 LSTM based learning model

In this section we will describe our model that learns to identify an event sub-graph from an AMR graph. The key idea is that the path between the interaction node and the entity node (where the term entity is used to denote both a protein and a sub-event) contains information about how the event is structured. We build on this idea to develop a supervised model using Long Short Term Memory (LSTM) (Hochreiter and Schmidhuber, 1997) architecture that can learn to identify events using the nodes and the edges in the AMR path between the interaction term and the entity term.

3.1 Motivation

The input to our problem is a sequence of words (w_i) interwound with edge labels (e_j) of the form: $w_1, e_1, w_2, e_2, ..., e_{n-1}, w_n$ that exists in the path between an interaction node and an entity node in an AMR graph. Due to large semantic variations that exist in naturally occurring texts, traditional feature based methods suffer from sparsity issues while learning from such a sequence. Neural network based models provide a framework for learning from non-sparse representations. Specifically, LSTM is known to handle sequences of variable length and capture long range dependencies well. Since the input sequence in our case falls into this category, we build our model using the LSTM framework.

3.2 Event identification

We model the event identification task as a two-step process: *Theme Identification* and *Cause Identification*. For simple events, this process includes only theme identification (since they don't have cause). We describe the two LSTM models corresponding to the two steps as follows:

3.2.1 Theme Identification

Given a pair of interaction node (t_j) and protein node (p_i), the task is to identify if there exists an event with t_j as the interaction and p_i as the theme; and if yes, what is the type of the event. We cast this problem as a multi-class classification task with label set as $L : \{NULL \cup Event\_types\}$ where *Event_types* correspond to the nine event types described in Table 1 and *NULL* corresponds to no event. We train an LSTM model for this task with the input layer as the embeddings corresponding to the sequence of words interwound

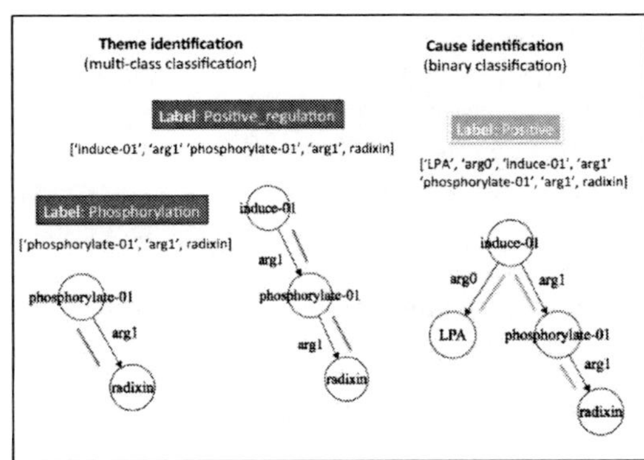

Figure 2: Theme identification and Cause identification stages

with edge labels in the shortest path between p_i and t_j in the AMR graph. We use a hidden layer of size 100 and an output layer of the size of our label set L. For e.g. in Figure 2, the sequence {*'phosphorylate-01', 'arg1', 'radixin'*} is the input sequence and the event type *Phosphorylation* is its label.

3.2.2 Cause Identification

The last four event types in Table 1 can take proteins or other events as cause argument. We cast this problem as a binary classification task where for an event we ask the question if a protein/event is its cause argument or not for every protein and every other event in that sentence. Let e_m be the event identified as $e_m : (Type : t_j, Theme : p_i)$ that can take a cause argument. Let $C = P \cup E$ where P is the set of all other proteins in the AMR graph (except p_i) and E is the set of all identified events (except e_m). For every $c_k \in C$, we get the shortest path between c_k and t_j and combine it with the shortest path between p_i and t_j and use the words and edges in this combined path as the input layer of our second LSTM model. We use a hidden layer of size 100 and an output layer of size one corresponding to the binary prediction of whether c_k is the cause of the event e_m or not.

3.3 Initialization of Embeddings

When initializing our model, we have two choices: we can initialize the embeddings in the input layer randomly or we can initialize them with values that reflect the meanings of the word types. It has been seen that using pre-defined word embeddings improves the performance of RNN models over random initializations (Collobert and Weston,

2008; Socher et al., 2011). We initialize the vectors corresponding to words in our input layer with 100-dimensional vectors generated by a word2vec (Mikolov et al., 2013) model trained on over one million words from the PubMed central article repository. Words not included in the pre-trained model and the edges are initialized randomly using uniform sampling from [-0.25, +0.25] to match the embedding standard deviation.

3.4 Event Construction

During test time, we first make predictions using our LSTM model for Theme identification. For every pair (p_i, t_j) with a non-zero label l, we construct events as follows: For label l corresponding to interaction types that take only proteins as theme arguments, we construct event as $e_m : (Type : t_j, Theme : p_i)$. For label l corresponding to interaction types that can take another event as its theme, we look at the path between t_j and p_i in the AMR. If this path includes a pair (t_k, p_i) that has a non-zero label, then we construct an event $e_n : (Type : t_j, Theme : e_p)$ where e_p is the event constructed from the pair (t_k, p_i). Otherwise, we construct the event as $e_n : (Type : t_j, Theme : p_i)$.

For each of the predicted event $e_m : (Type : t_j : Theme : p_i)$ that can take a cause argument, we run the second LSTM model for its Cause identification. If there is a pair (p_i, c_k) which has a positive label, then we assign c_k as the cause of the event e_m.

4 Distant Supervision

An empirical evaluation of our LSTM-based learning model (Section 5.4) shows that it can suffer from low recall. Obtaining additional human annotated data for our complex event extraction task can be very costly. This motivates us to develop an approach that can gather more training data with minimal supervision.

4.1 Motivation

Distant supervision as a learning paradigm was introduced by Mintz et al. (2009) for relation extraction in general domain. They use Freebase to get a set of relation instances and entity pairs participating in those relations, extract all sentences containing those two entity pairs from Wikipedia text and use these sentences as their training data. This work and many others show that distant supervi-

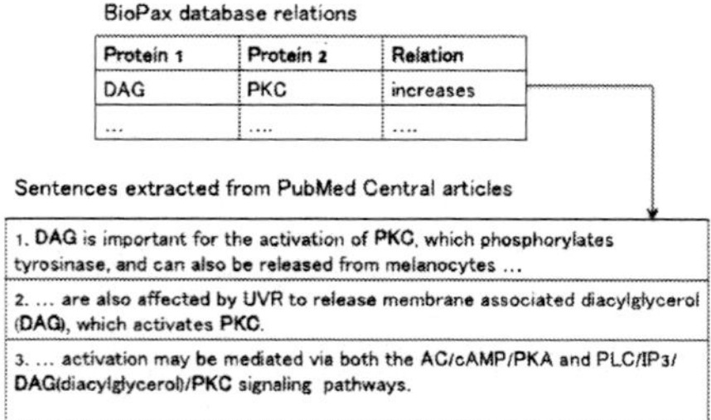

Figure 3: Distant Supervision: Sentences extracted from PubMed Central articles using BioPax database relations

sion technique yields significant improvements in relation extraction. Neural network models like LSTM need to be trained on substantial amounts of training data for them to be able to generalize well. However due to lack of labeled data in biomedical domain, most work in relation extraction in this domain has been restricted to purely supervised techniques. In this work we cope with this problem by gathering additional training data using distant supervision from a knowledge base.

4.2 Methodology

Relation extraction using distant supervision requires two things: 1) A knowledge base containing relations between proteins, and 2) A large corpus of unannotated text that contain protein mentions. We use the BioPax (Biological Pathway Exchange) database (Demir et al., 2010) as our knowledge base of protein relations and we use the PubMed central articles as our unannotated text corpus. Given a database entry of the form *('Protein1', 'Protein2', 'relation')*, we extract all sentences from the PubMed central articles in which the two proteins co-occur. For example, Figure 3 shows some sample sentences extracted for the database entry *('DAG', 'PKC', increases)*. The first two sentences in the figure indeed express the relation in the database but the third sentence just mentions the two proteins in a comma separated list. We observe that a lot of the extracted sentences fall into the category of the third sentence. Hence as a first step, we filter such instances by tagging the sentence with their parts-of-speech and removing those in which the two proteins are separated only by nouns (or punctuations).

4.3 AMR Path Based Selection

The traditional distant supervision approach says that all the sentences extracted using the method above can be used as additional training data un-

Event Type	Biopax relation
Gene_expression	adds_modification
Transcription	adds_modification
Localization	adds_modification
Protein_catabolism	adds_modification
Binding	binds
Phosphorylation	adds_modification
Regulation	increases, increases_activity
Positive_regulation	increases, increases_activity
Negative_regulation	-

Table 4: Mapping between event types and Biopax model relations

der the assumption that all sentences in which the proteins co-occur express the relation mentioned in the database. However Takamatsu et al. (2012) note that this approach can often lead to a lot of false positives. Roth et al. (2013) have discussed several prior work that try to reduce such noise in the data. In our work, we develop a novel selection technique for reducing such noise using AMR path heuristic. We make the observation that given two protein nodes in an AMR, if there is a relation r between the two then the shortest path between the two protein nodes in the AMR contains the interaction term expressing the relation r.

For e.g. Figure 4 shows the AMR for the sentence *"DAG is important for the activation of PKC, which phosphorylates tyrosinase, and can also be released..."* that was extracted using the database entry {*'DAG'*, *'PKC'*, *'increases'*}. The interaction term *'activate'* suggesting the relation 'increases' exists in the shortest path between the proteins *DAG* and *PKC*. Figure 5 shows AMR for the sentence *"The sun-network links TCF3 with ZYX and HOXA9 via NEDD9 and CREBBP, respectively."* extracted for the pair (*'TCF3'*, *'HOXA9'*, increases). There is no interaction term suggesting the relation 'increases' in the shortest path between the proteins *TCF3* and *HOXA9*.

Table 4 shows the mapping we define between the event types and the relations found in the entries (*'Protein1'*, *'Protein2'*, *'relation'*) that we extracted from the Biopax model. In each sentence extracted for the database entry (*'P_1'*, *'P_2'*, *'r'*) , we check if the shortest path between the two protein nodes P_1 and P_2 in the AMR of the sentence contains one of the interaction terms corresponding to the event type mapped to the relation r. We discard all those sentences that do not satisfy this constraint.

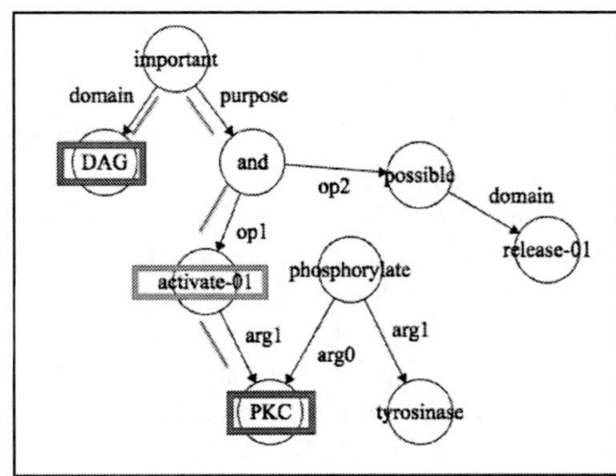

Figure 4: Interaction term *'activate'* corresponding to the relation 'increases' exists in the shortest path between *DAG* and *PKC*

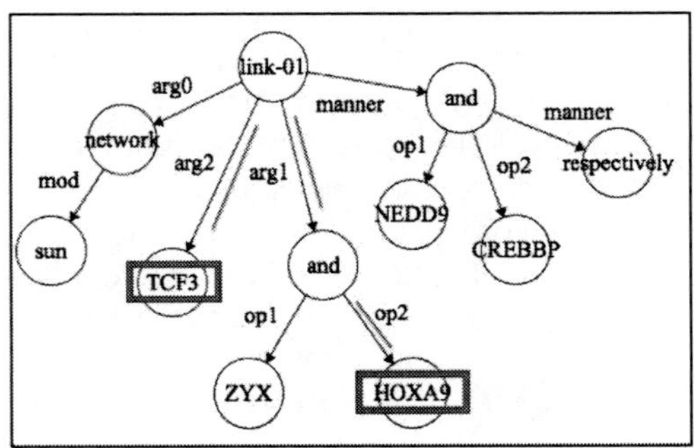

Figure 5: No interaction term corresponding to the relation 'increases' exists in the shortest path between *TCF3* and *HOXA9*

4.4 Using Data for LSTM Model

We use these selected sentences as additional training data for our two LSTM models as follows:
a. Theme identification: Let S be the sentence extracted for the database entry (*'DAG'*, *'PKC'*, *'increases'*) and let *'activates'* be the interaction term that exists in the shortest path between the protein nodes. Since the database entry refers to *'DAG'* as the cause and *'PKC'* as the theme, we assume these roles for the two proteins in the extracted sentence S as well. Therefore, we can now use the path between the interaction term *'activates'* and the theme *'PKC'* as an input sequence for our model with the label corresponding to the event type of the interaction term *'activates'*.
b. Cause identification: In case of cause identification instead of using the path between the interaction term and the theme entity, we use the shortest path between the cause entity and the theme entity via the interaction term and use this as an input sequence to our model with a positive label.

131

5 Experiments

5.1 Dataset and task setting

The event extraction task described in this work corresponds to the Task 1 of the Genia Event Extraction task described by the BioNLP Shared Task series (2009, 2011 and 2013). We train a model on a combination of *abstract collection* (from 2009 edition) and *full text collection* (from 2011 and 2013). We test our model on the dev set of the 2013 edition (since the gold annotation is publicly available only for the dev set and not the test set).

5.2 Data prepraration

The dataset made available for the Shared Task is in the form of sentences and event annotations as shown in Table 2. We convert these event annotations into input sequences and labels for our multi-class classification task (theme identification) and for our binary classification task (cause identification) as follows

a. Theme identification: Given a sentence, we define the set T as the set of interaction terms corresponding to all its event annotations. We define the set P as the set of all its protein mentions. For every pair (t_j, p_i) where $p_i \in P$ and $t_j \in T$, we create a training data of the form $\{w_1, e_1, w_2, e_2, ..., e_{n-1}, w_n, label\}$ where the input sequence corresponds to the words interwound with edge labels in shortest path between t_j and p_i; and the *label* is the event type of the event e_m if there exists an event $e_m : (Type : t_j, Theme : p_i)$, *NULL* otherwise. We create the test data similarly; except we do not use event annotations for creating the set T but instead identify terms in the sentence that was annotated as an interaction term in the training data more than once.

b. Cause identification: For every pair (t_j, p_k) where t_j is part of some event annotation $e_m : (Type : t_j, Theme : p_i)$ of event type that can take cause argument and $p_k \in P$, we create a training data of the form $\{w_1, e_1, w_2, e_2, ..., e_{n-1}, w_n, label\}$ where the input sequence corresponds to the shortest path between p_k and p_i via t_j; and the label is 1 if p_k is the cause of the event e_m, 0 otherwise.

5.3 LSTM model setup

We implement our LSTM model using the *lasagne* library. For the first LSTM model, we use softmax as our non-linear function and optimize the cat-egorical cross entropy loss using adam (Kingma and Ba, 2014). For the second LSTM model, we use a sigmoid non-linear function and optimize the binary loss using adam. We use a dropout of 0.5, batch size of 100 and a learning rate of 0.001.

5.4 Results and Discussion

Table 5 shows the results of our LSTM and distant supervision based event extraction model. We compare our results with the state-of-the-art event extraction system EVEX (Hakala et al., 2013). We report the Approximate Span/Approximate Recursive metric in all our tables (described in the Shared Task (Kim et al., 2013)). The columns to the left (with column heading LSTM) show the performance of our model trained only on the official training data. The columns to the right (with column heading LSTM+Distant Supervision) show the performance of our model trained on official training data plus the additional training data of 11792 sentences we gather using our distant supervision strategy.

The table highlights some of our results. Firstly, we note that, in cases where we obtain a large number of extra sentences using distant supervision (highlighted in the column "DS Sents"), we see a considerable gain in the recall values between "LSTM" and "LSTM+Distant Supervision" models. On the contrary, in cases where we extract only a small number, we see a small gain (or sometimes even a decrease in performance). This suggests we explore further ways of selecting our extra sentences. Secondly, although the overall performance of our model using the automatic AMR parser is lower than the current state-of-the-art system, the gap of 5% in the F1 score can hopefully be reduced with the ongoing improvements in AMR parsing.

6 Related work

The biomedical event extraction task described in this work was first introduced in the BioNLP Shared Task in 2009 (Kim et al., 2009). This task helped shift the focus of relation extraction efforts from identifying simple binary interactions to identifying complex nested events that better represent the biological interactions stated frequently in text. Existing approaches to this task include SVM (Björne and Salakoski, 2013) other ML based approaches (Riedel and McCallum, 2011; Miwa et al., 2010, 2012). Methods like

Event Type	LSTM			LSTM + Distant Supervision				EVEX		
	Recall	Precision	F1	Recall	Precision	F1	DS Sents	Recall	Precision	F1
Gene_expression	**66.33**	66.55	66.44	**76.98**	61.48	68.36	**868**			
Transcription	55.10	28.57	37.63	57.14	26.92	36.60	807			
Localization	36.55	63.72	46.45	38.07	85.06	52.60	96			
Protein_catabolism	73.33	84.62	78.57	60.00	94.74	73.47	7			
==[SVT-TOTAL]==	57.82	60.86	57.27	56.35	68.05	57.60		73.83	79.56	76.59
Binding	27.61	25.94	26.75	28.57	26.12	27.29	139	41.14	44.77	42.88
Phosphorylation	**49.21**	53.75	51.38	**73.45**	45.55	56.23	**3183**			
Regulation	**16.30**	29.18	20.92	**26.07**	21.00	23.26	**2131**			
Positive_regulation	**25.98**	35.16	29.88	**37.41**	29.17	32.78	**4561**			
Negative_regulation	23.17	30.50	26.33	22.97	29.44	25.81	0			
==[REG-TOTAL]==	21.81	31.61	25.71	28.81	26.53	27.28		32.41	47.16	38.41
==[ALL-TOTAL]==	44.42	51.01	46.37	46.73	46.60	**46.66**	11792	45.44	58.03	**50.97**

Table 5: Evaluation results (Recall/Precision/F1) on the 2013 Genia Event Extraction dev set. LSTM and LSTM + Distant Supervision are our models. The last column corresponds to the results of EVEX (Hakala et al., 2013) model on the 2013 test set. Certain notable numbers are emphasized and discussed under results 5.4.

(Liu et al., 2013; MacKinlay et al., 2013) learn subgraph patterns from the event annotations in the training data and cast the event detection as subgraph matching problem. Non-feature based approaches like graph kernels compare syntactic structures directly (Airola et al., 2008; Bunescu et al., 2005). Rule based methods that either use manually crafted rules or generate rules from training data (Cohen et al., 2009; Kaljurand et al., 2009; Kilicoglu and Bergler, 2011; Bui et al., 2013) have obtained high precision on these tasks.

In our work, we take inspiration from the Turk Event Extraction System (TEES) (Björne and Salakoski, 2013) (the event extraction system for EVEX) that has consistently been the top performer in these series of tasks. They represent events using a graph format and break the event extraction task into separate multi-class classification tasks using SVM as their classifier. In our work we take a step further by making use of a deeper semantic representation as a starting point and identifying subgraphs in the AMR graph.

AMR has been successfully used for deeper semantic tasks like entity linking (Pan et al., 2015) and abstractive summarization (Mihalcea et al., 2015). Work by Garg et al. (2015) is the first one to make use of AMR representation for extracting interactions from biomedical text. They use graph kernel methods to answer the binary question of whether a given AMR subgraph expresses an interaction or not. Our work departs from theirs in that they concentrate only on binary interactions whereas we use AMR to identify complex nested events. Also, our approach additionally makes use of distant supervision to cope with the problem of limited annotated data.

Distant supervision techniques have been successfully used before for relation extraction (Mintz et al., 2009) in general domain. Recent work by (Liu et al., 2014) uses minimal supervision strategy for extracting relations particularly in biomedical texts. Our work departs from theirs in that we introduce a novel AMR path based heuristic to selectively sample the sentences obtained from distant supervision.

7 Conclusion

In this work, we show the effectiveness of using a deep semantic representation based on Abstract Meaning Representations for extracting complex nested events expressed in biomedical text. We hypothesize that an event structure is an AMR subgraph and empirically validate our hypothesis. For learning to extract such event subgraphs from AMR automatically, we develop two Recurrent Neural Network based models: one for identifying the theme, and the other for identifying the cause of the event. To overcome the dearth of manually annotated data in biomedical domain, which explains the low recall of event extraction systems, we train our model on additional training data gathered automatically using a selective distant supervision strategy. Our experiments strongly suggest that AMR parsing improvements, which are expected given the youth of this scientific field of inquiry, and the exploitation of larger, manually curated Biopax-like models and collections of biomolecular texts will be easy to capitalize on catalysts for driving future improvements in this task.

References

Antti Airola, Sampo Pyysalo, Jari Björne, Tapio Pahikkala, Filip Ginter, and Tapio Salakoski. 2008. All-paths graph kernel for protein-protein interaction extraction with evaluation of cross-corpus learning. *BMC bioinformatics* 9(11):1.

Laura Banarescu, Claire Bonial, Shu Cai, Madalina Georgescu, Kira Griffitt, Ulf Hermjakob, Kevin Knight, Philipp Koehn, Martha Palmer, and Nathan Schneider. 2013. Abstract meaning representation for sembanking. In *Association for Computational Linguistics*.

Jari Björne and Tapio Salakoski. 2013. Tees 2.1: Automated annotation scheme learning in the bionlp 2013 shared task. In *Proceedings of the BioNLP Shared Task 2013 Workshop*. pages 16–25.

Quoc-Chinh Bui, David Campos, Erik van Mulligen, and Jan Kors. 2013. A fast rule-based approach for biomedical event extraction. In *Proceedings of the BioNLP Shared Task 2013 Workshop*. Association for Computational Linguistics, pages 104–108.

Razvan Bunescu, Ruifang Ge, Rohit J Kate, Edward M Marcotte, Raymond J Mooney, Arun K Ramani, and Yuk Wah Wong. 2005. Comparative experiments on learning information extractors for proteins and their interactions. *Artificial intelligence in medicine* 33(2):139–155.

K Bretonnel Cohen, Karin Verspoor, Helen L Johnson, Chris Roeder, Philip V Ogren, William A Baumgartner Jr, Elizabeth White, Hannah Tipney, and Lawrence Hunter. 2009. High-precision biological event extraction with a concept recognizer. In *Proceedings of the Workshop on Current Trends in Biomedical Natural Language Processing: Shared Task*. Association for Computational Linguistics, pages 50–58.

Ronan Collobert and Jason Weston. 2008. A unified architecture for natural language processing: Deep neural networks with multitask learning. In *Proceedings of the 25th international conference on Machine learning*. ACM, pages 160–167.

Emek Demir, Michael P Cary, Suzanne Paley, Ken Fukuda, Christian Lemer, Imre Vastrik, Guanming Wu, Peter D'Eustachio, Carl Schaefer, Joanne Luciano, et al. 2010. The biopax community standard for pathway data sharing. *Nature biotechnology* 28(9):935–942.

Jeffrey Flanigan, Sam Thomson, Jaime Carbonell, Chris Dyer, and Noah A Smith. 2014. A discriminative graph-based parser for the abstract meaning representation. In *In ACL*. Citeseer.

Sahil Garg, Aram Galstyan, Ulf Hermjakob, and Daniel Marcu. 2015. Extracting biomolecular interactions using semantic parsing of biomedical text. *arXiv preprint arXiv:1512.01587* .

Kai Hakala, Sofie Van Landeghem, Tapio Salakoski, Yves Van de Peer, and Filip Ginter. 2013. Evex in st?13: Application of a large-scale text mining resource to event extraction and network construction. In *Proceedings of the BioNLP Shared Task 2013 Workshop*. Association for Computational Linguistics, pages 26–34.

Sepp Hochreiter and Jürgen Schmidhuber. 1997. Long short-term memory. *Neural computation* 9(8):1735–1780.

Kaarel Kaljurand, Gerold Schneider, and Fabio Rinaldi. 2009. Uzurich in the bionlp 2009 shared task. In *Proceedings of the Workshop on Current Trends in Biomedical Natural Language Processing: Shared Task*. Association for Computational Linguistics, pages 28–36.

Halil Kilicoglu and Sabine Bergler. 2011. Adapting a general semantic interpretation approach to biological event extraction. In *Proceedings of the BioNLP Shared Task 2011 Workshop*. Association for Computational Linguistics, pages 173–182.

Jin-Dong Kim, Tomoko Ohta, Sampo Pyysalo, Yoshinobu Kano, and Jun'ichi Tsujii. 2009. Overview of bionlp'09 shared task on event extraction. In *Proceedings of the Workshop on Current Trends in Biomedical Natural Language Processing: Shared Task*. Association for Computational Linguistics, pages 1–9.

Jin-Dong Kim, Yue Wang, Nicola Colic, Seung Han Baek, Yong Hwan Kim, and Min Song. 2016. Refactoring the genia event extraction shared task toward a general framework for ie-driven kb development. *ACL 2016* page 23.

Jin-Dong Kim, Yue Wang, Toshihisa Takagi, and Akinori Yonezawa. 2011. Overview of genia event task in bionlp shared task 2011. In *Proceedings of the BioNLP Shared Task 2011 Workshop*. Association for Computational Linguistics, pages 7–15.

Jin-Dong Kim, Yue Wang, and Yamamoto Yasunori. 2013. The genia event extraction shared task, 2013 edition-overview. In *Proceedings of the BioNLP Shared Task 2013 Workshop*. Association for Computational Linguistics, pages 8–15.

Diederik Kingma and Jimmy Ba. 2014. Adam: A method for stochastic optimization. *arXiv preprint arXiv:1412.6980* .

Haibin Liu, Karin Verspoor, Donald C Comeau, Andrew MacKinlay, and W John Wilbur. 2013. Generalizing an approximate subgraph matching-based system to extract events in molecular biology and cancer genetics. In *Proceedings of the BioNLP Shared Task 2013 Workshop*. pages 76–85.

Mengwen Liu, Yuan Ling, Yuan An, and Xiaohua Hu. 2014. Relation extraction from biomedical literature with minimal supervision and grouping strategy. In *Bioinformatics and Biomedicine (BIBM), 2014*

IEEE International Conference on. IEEE, pages 444–449.

Andrew MacKinlay, David Martinez, Antonio Jimeno Yepes, Haibin Liu, W John Wilbur, and Karin Verspoor. 2013. Extracting biomedical events and modifications using subgraph matching with noisy training data. In *Proceedings of the BioNLP Shared Task 2013 Workshop. Association for Computational Linguistics, Sofia, Bulgaria*. pages 35–44.

Jonathan May. 2016. Semeval-2016 task 8: Meaning representation parsing. *Proceedings of SemEval* pages 1063–1073.

Rada Mihalcea, Joyce Yue Chai, and Anoop Sarkar, editors. 2015. *NAACL HLT 2015, The 2015 Conference of the North American Chapter of the Association for Computational Linguistics: Human Language Technologies, Denver, Colorado, USA, May 31 - June 5, 2015*. The Association for Computational Linguistics.

Tomas Mikolov, Ilya Sutskever, Kai Chen, Greg S Corrado, and Jeff Dean. 2013. Distributed representations of words and phrases and their compositionality. In *Advances in neural information processing systems*. pages 3111–3119.

Mike Mintz, Steven Bills, Rion Snow, and Dan Jurafsky. 2009. Distant supervision for relation extraction without labeled data. In *Proceedings of the Joint Conference of the 47th Annual Meeting of the ACL and the 4th International Joint Conference on Natural Language Processing of the AFNLP: Volume 2-Volume 2*. Association for Computational Linguistics, pages 1003–1011.

Makoto Miwa, Rune Sætre, Jin-Dong Kim, and Jun'ichi Tsujii. 2010. Event extraction with complex event classification using rich features. *Journal of bioinformatics and computational biology* 8(01):131–146.

Makoto Miwa, Paul Thompson, and Sophia Ananiadou. 2012. Boosting automatic event extraction from the literature using domain adaptation and coreference resolution. *Bioinformatics* 28(13):1759–1765.

Xiaoman Pan, Taylor Cassidy, Ulf Hermjakob, Heng Ji, and Kevin Knight. 2015. Unsupervised entity linking with abstract meaning representation. In (Mihalcea et al., 2015), pages 1130–1139. http://aclweb.org/anthology/N/N15/N15-1119.pdf.

Michael Pust, Ulf Hermjakob, Kevin Knight, Daniel Marcu, and Jonathan May. 2015. Using syntax-based machine translation to parse english into abstract meaning representation. In *EMNLP*.

Sebastian Riedel and Andrew McCallum. 2011. Robust biomedical event extraction with dual decomposition and minimal domain adaptation. In *Proceedings of the BioNLP Shared Task 2011 Workshop*. Association for Computational Linguistics, pages 46–50.

Benjamin Roth, Tassilo Barth, Michael Wiegand, and Dietrich Klakow. 2013. A survey of noise reduction methods for distant supervision. In *Proceedings of the 2013 workshop on Automated knowledge base construction*. ACM, pages 73–78.

Richard Socher, Eric H Huang, Jeffrey Pennin, Christopher D Manning, and Andrew Y Ng. 2011. Dynamic pooling and unfolding recursive autoencoders for paraphrase detection. In *Advances in Neural Information Processing Systems*. pages 801–809.

Shingo Takamatsu, Issei Sato, and Hiroshi Nakagawa. 2012. Reducing wrong labels in distant supervision for relation extraction. In *Proceedings of the 50th Annual Meeting of the Association for Computational Linguistics: Long Papers-Volume 1*. Association for Computational Linguistics, pages 721–729.

Chuan Wang, Nianwen Xue, Sameer Pradhan, and Sameer Pradhan. 2015. A transition-based algorithm for amr parsing. In *HLT-NAACL*. pages 366–375.

Detecting Personal Medication Intake in Twitter:
An Annotated Corpus and Baseline Classification System

Ari Z. Klein, Abeed Sarker, Masoud Rouhizadeh, Karen O'Connor, Graciela Gonzalez
Department of Biostatistics, Epidemiology, and Informatics
Perelman School of Medicine
University of Pennsylvania
{ariklein,abeed,mrou,karoc,gragon}@upenn.edu

Abstract

Social media sites (e.g., Twitter) have been used for surveillance of drug safety at the population level, but studies that focus on the effects of medications on specific sets of individuals have had to rely on other sources of data. Mining social media data for this information would require the ability to distinguish indications of personal medication intake in this media. Towards that end, this paper presents an annotated corpus that can be used to train machine learning systems to determine whether a tweet that mentions a medication indicates that the individual posting has taken that medication (at a specific time). To demonstrate the utility of the corpus as a training set, we present baseline results of supervised classification.

1   Introduction

Social media allows researchers and public health professionals to obtain relevant information in large amounts directly from populations and/or specific cohorts of interest, and it has evolved into a useful resource for performing public health monitoring and surveillance. According to a Pew report (Greenwood et al., 2016), nearly half of adults worldwide and two-thirds of all American adults (65%) use social media, including over 90% of 18-29 year olds. Recent studies have attempted to utilize social media data for tasks such as pharmacovigilance (Leaman et al., 2010), identifying user behavioral patterns (Struik and Baskerville, 2014), analyzing social circles with common behaviors (Hanson et al., 2013b), and tracking infectious disease spread (Broniatowski et al., 2015).

A large subset of the public health-related research using social media data, including our prior work in the domain, focuses on mining information (e.g., adverse drug reactions, medication abuse, and user sentiment) from posts mentioning medications (Korkontzelos et al., 2016; Hanson et al., 2013b; Nikfarjam et al., 2015). Typically, these and similar studies focus on information at the population level, but processing and deriving information from individual user posts poses significant challenges from the natural language processing (NLP) perspective. Researchers attempt to overcome the noise and inaccuracies in the data by relying on large amounts of data. For example, Hanson et al. (2013b; 2013a) attempted to estimate the abuse of Adderall® using Twitter by detecting the total number of mentions of the medication. The authors did not attempt to assess if a mention represented personal intake or not.

While such a strategy may suffice for deriving estimates "by proxy" at the population level (e.g., higher volume of chatter means higher rates of use), it has at least two limitations: (i) the actual number of tweets representing personal intake within a given sample of tweets is unknown, and (ii) it is not possible to assess the effects of medication intake on subsets of users of interest who take the medication. Studies focusing on specific subsets of individuals rely on other sources of data, such as electronic health records and published literature from clinical trials, where information about the individuals' medication intake is explicit (e.g., Akbarov et al., 2015; Zhou et al., 2016; Romagnoli et al., 2017). Harnessing social media for studying the effects of medications on specific cohorts would require developing systems that can automatically distinguish posts that express personal intake from those that do not.

Due to the very recent incorporation of social media data in healthcare systems, published research on our target task of creating a corpus for automatic detection of personal medication intake

Proceedings of the BioNLP 2017 workshop, pages 136–142,
Vancouver, Canada, August 4, 2017. ©2017 Association for Computational Linguistics

information is scarce. The study by Alvaro et al. (2015) is perhaps the most closely related work to ours. The authors annotated 1,548 tweets for whether they contain "first-hand experiences" of adverse drug reactions (ADRs) to prescription medications, and they used this annotated data in a supervised classification framework aimed at automatically identifying tweets that report personal usage. As far as we are aware, however, they have not made their annotated data public; nonetheless, we do not believe that it would have been exactly the right training set for our classification task. Because our focus is to help set the groundwork for using social media data in medication-related cohort studies, we included a subtle but key factor in our criteria for identifying personal intake: *when* the medication was taken. We will discuss this factor in more detail in the next section. In this paper, we present (i) an analysis of medication-mentioning chatter on Twitter, (ii) a publicly available, annotated corpus of tweets that can be used to advance automatic systems, and (iii) baseline supervised classification results to validate the utility of the annotated data.

2 Method

We chose Twitter as the data source for this study because of its growing popularity in public health research, and its easy-to-use public APIs. We discuss the three primary tasks—data collection, annotation, and classification—in the following subsections.

2.1 Data Collection

To build the corpus, we queried 73,800 Twitter user timelines (that we collected for related work) for 55 medication names, including both prescription and over-the-counter medications, brand and generic names, and types of medications (e.g., *steroid*). Using a tool that was developed by Pimpalkhute et al. (2014), we generated frequent misspellings of the medications in order to expand the query. We then tokenized all of the tweets, using the ARK Twokenizer (O'Connor et al., 2010; Owoputi et al., 2013), and identified 35,075 tweets containing a target medication. To account for the linguistic idiosyncrasies of how Twitter users might express their medication intake, we randomly selected one medication tweet from the 18,033 timelines that included such a tweet, and we prepared them for annotation. For this paper,

10,260 tweets were annotated, with overlapping annotations for 1,026 (10%).

2.2 Annotation

In order to control for studying the effects of medication intake on subsets of individuals in a social media setting, we decided that tweets of interest should not only represent the author's *personal usage* of the target medication in the tweet; they should also indicate the *specific instance* in which the user took the mentioned medication, since researchers using social media data cannot physically observe and record when medications were taken. Only if the tweets provide this additional information about the time of intake can we potentially use Twitter data to assess causal associations between users' health information (also mined from social media data) and the usage of particular medications. As we mentioned earlier, the way that time factors into our definition of "intake" marks an important distinction between our annotated data and Alvaro et al.'s (2015).

We found that, under minimal guidance, intuitively agreeing on what constituted a personal intake of medication, given the above criteria, was very difficult. We attribute this difficulty to the wide range of linguistic patterns in which we found medication mentions occurring. In an effort to obtain high inter-annotator agreement and address the human disagreement that Alvaro et al. seek to overcome, we analyzed linguistic patterns in samples of the data and used this analysis to inform the development of annotation guidelines;[1] in addition, we limited the number of annotation classes to the three high-level classes that we thought were most directly relevant to the classification task at hand: *intake, possible intake*, and *no intake*.

We will summarize our analysis of the three classes of tweets here. *Intake* tweets indicate that (i) the medication was actually taken, (ii) the author of the tweet personally took the medication, and (iii) the medication was taken at a specific time. To illustrate (i), consider the following tweets:

(a) Migraine from hell… **Took** 6 Motrin and nothing's touching it

(b) I've **been off** adderall about a month now and I'm so much happier, but COMPLETELY useless. I'm like a child again.

---

[1] The annotation guidelines and a sample of the annotated data are available at:
https://healthlanguageprocessing.org/twitter-med-intake/

(c) A lot of people hate on prednisone but **I feel better already**. #stuffworksforme

(d) this ibuprofen still ain't **kicked in** my head poundin

While only (a) uses a verb phrase that explicitly indicates intake (*took…*), we can infer from features of the other tweets that the medication was taken: (b) *being off* the medication, (c) experiencing the effects of the medication, and (d) waiting for the medication to *kick in* all entail that the medication was taken.

Moreover, *intake* tweets should indicate that the *author* of the tweet took the medication:

(e) Sorry for this rant thingy, **I** took my Vyvanse today lol

(f) Sick and only had a Tylenol PM at work so now **i** feel better but **i** am fighting sleep☺

(g) Just threw back these Xanax

(h) In soooo much pain tonight and Tylenol just isn't cutting it. Literally hurting all over

Through the use of the first-person reference *I*, (e) explicitly states that the author took the medication, and (f) explicitly attributes the experiential effects of the medication (*feel better, but fighting sleep*) to the author. While (g) and (h) do not explicitly reveal that the *author* took the medication (*threw back*) or is (not) experiencing the effect of the medication (*isn't cutting it*), respectively, the high degree of self-presentation in social media (e.g., Kaplan and Haenlein, 2010; Papacharissi, 2012; Seidman, 2013) allows us to infer that the authors are writing about their own intake and experiences.

Finally, *intake* tweets also specify *when* the medication was taken:

(i) I've been sick **for the last 3 days** taking Ibuprofen just feel better and to fight Infection "swelling"

(j) Tylenol is my bestfriend **at the moment**

(k) maybe i'm tired as had 2 tramadol my bk is sore sore sore... #scoliosis

(l) Prednisone headache! Ahhhh

Tweet (i) uses a temporal marker that explicitly specifies an instance of intake, and, similarly, (j) explicitly indicates when the effect of the intake occurred. Although (k) and (l) do not explicitly specify instances of intake, Twitter's real-time na-

ture (Sakaki et al., 2010) gives us reason to believe that the author of (k) recently *had* the medication and that the effect in (l) is being currently experienced, which represents an intake in the recent past (i.e., a specific instance).

Unlike *intake* tweets, some tweets do not specify that the author actually took the medication or when the medication was taken, but, unlike *no intake* tweets, are generally about the author's intake. Consider the following tweets:

(m) I want to cry it's that painful 😖**gonna** take codeine this morning for sure

(n) 800 mg of Advil **cause this headache is real**

(o) **I need** a Xanax like right now

(p) Codeine is **one hell of a drug**. 😵😵😵

(q) 😵😵😵 I never understood why I get so angryyyy omg **I was so mellow** on Xanax ⚫

(r) I pretty much eat Advil **like it's candy.**✏️📷

We consider a tweet to be a *possible intake* if it expresses the intake as a future event (m); it contains merely a purpose for intake (n); it expresses a present-tense need for the medication (o); it abstractly praises (or criticizes) the medication without describing a concrete effect (p); it indicates that the author has used the medication in the past, but does not specify when (q); or, similarly, it indicates that the author uses the medication frequently, but does not specify an instance of intake (r). We decided to distinguish *possible intake* tweets because they can direct us to a user's timeline for manual probing, where we may find, for example, that a series of tweets aggregate to form a sort of composite *intake* tweet.

In contrast to *intake* and *possible intake* tweets, *no intake* tweets are not about the author's intake of the medication. While some *no intake* tweets are not about intake at all, some may be about the intake by others, not the author:

(s) @[Username redacted] Mine hurt for days last year!! **Take** some paracetamol **hun** ☺

(t) **Gave James** 2 ibuprofen pm and I'm being repaid by the sound of him snoring penetrating through my earplugs

The act of suggesting a medication (s) or giving someone a medication (t) might be interpreted as implying that the author has taken the medication in the past (i.e., a *possible intake*), but, because the tweets are not primarily about the author's intake, we consider this inferential leap to be too large to warrant the same classification as other *possible intake* tweets.

While (s) and (t) are explicitly not about the author's intake, other tweets may not be as obvious, such as tweets that contain merely the name of a medication:

(u) @[Username redacted] @[Username redacted] @[Username redacted] @[Username redacted] **methadone** !

Although (u) also might be interpreted as indicating the author's use of the medication, the textual evidence does not seem to favor this interpretation over other possible ones, such as mere question-answering. We classify tweets that contain merely the name of a medication as *no intake* because, unlike *intake* and *possible intake* tweets, they do not contain enough information for us to conclude that they are about the author's intake.

The "addressivity" (Bakhtin, 1986) markers "@" in (u) reflect the "dialogic" (Bakhtin, 1981) space of social media, wherein the linguistic data that we are mining is not only textual, but "intertextual" (Kristeva, 1980)—that is, oriented to what has already been said by others. Tweets also mark this social orientation to others through features of "reported speech" (Voloshinov, 1973). Consider the following tweets:

(v) @[Username redacted] "I don't either cause these Tylenol aren't doing crap!" Lol

(w) I just wanna give a shoutout to adderall for helping me get through the semester - **Florida State**

While (v) and (w) would otherwise be classified as *intake* tweets, the quotation marks in (v) and the hyphen in (w) mark that the authors are directly reporting the words of others—in (w), a student at Florida State—not their own medication intake.

Other cases of reported speech involve tweets that make cultural references about taking medications—for example, song lyrics or lines from movies. As our analysis of the three classes suggests, identifying indications of personal medication intake in social media required grappling with a number of annotation issues, which forecast the challenges of using this data to train classifiers.

2.3 Classification

We performed supervised classification experiments using several algorithms. The goal for these experiments was not to identify the best performing classification strategy, but to (i) verify that automatic classifiers could be trained using this data, and (ii) generate baseline performance estimates.

We used stratified 80-20 (training/test) split of the annotated set for the experiments. As features, we used only word n-grams (n = 1, 2, and 3) following standard preprocessing (e.g., stemming using the Porter stemmer (Porter, 1980) and lowercasing). We experimented with four classifiers—naïve bayes (NB), support vector machines (SVM), random forest (RF), logistic regression (LR), and a majority-voting based ensemble of the last three. Pairwise classification (i.e., 1-vs-1) is used to adapt the SVMs to the multiclass problem. Parameter optimization for the individual classifiers was performed via 10-fold cross validation over the training set, with an objective function that maximizes the F-score for the *intake* class.

Following the classification experiments, we performed brief error and feature analyses to identify common misclassification patterns and possible future approaches for improving classification performance. To identify informative n-grams for the intake class, we applied the *Information Gain* feature evaluation technique, which computes the importance of an attribute with respect to a class according to the following equation:

$$IG(Class, Attribute) = H(Class) - H(Class|Attribute)$$

$H()$ represents the information entropy for a given state (Yang and Pedersen, 1997). We used the Weka 3 tool[2] for all machine learning and feature analysis experiments. We present the results for these experiments in the next section.

3 Results and Discussion

In this section, we present and discuss the results of annotation and the baseline classification experiments, including a brief error analysis of misclassified *intake* tweets and a feature analysis to identify informative n-grams.

3.1 Annotation

For the corpus that we present in this paper, two expert annotators have annotated 10,260 tweets,

---

[2] Available at: http://www.cs.waikato.ac.nz/ml/weka/. Accessed: 5/25/2017.

with overlapping annotations for 1,026 (10%). Their inter-annotator agreement was κ = 0.88 (Cohen's Kappa). They disagreed on 81 tweets, which the first author of this paper resolved through independent annotation. In total, 1,952 tweets (19%) were annotated as *intake*; 3,219 (31%) were annotated as *possible intake*; and 5,089 (50%) were annotated as *no intake*. These frequencies suggest that a minority of tweets that mention medications represent personal intake, which substantiates the need for this classification when mining large amounts of social media data for drug safety surveillance.

3.2   Classification

Table 1 presents the performances of the different classifiers. The overall accuracy (Acc) over the three classes and the F-scores (F) for each of the three classes are shown. The *no intake* (NI) class has the best F-score due to the larger number of training instances. SVMs, RF and LR classifiers have comparable accuracies, and they outperform the NB baseline. SVMs have the highest F-score for the *intake* (I) class, suggesting that it might be the most suitable classifier for this task.

The voting-based ensemble of the three classifiers does not improve performance over the SVMs. Post-classification analyses revealed that this is because the individual classifiers in the ensemble, particularly the LR and SVMs classifiers, make almost identical predictions given the feature set of n-grams. The confusion matrices for the classifiers' predictions are also alike, with strong inter-classifier agreements in terms of false and true positives and negatives. The results and the analyses suggest that incorporating/generating features that are more informative is more likely to improve performance on this task, rather than combining multiple classifiers on the same feature vectors.

	I (F)	PI (F)	NI (F)	Acc (%)	95% CI
NB	0.59	0.58	0.73	64.4	62.4-66.3
SVM	0.67	0.69	0.80	73.4	71.5-75.1
RF	0.60	0.68	0.80	72.2	70.4-74.0
LR	0.65	0.68	0.79	72.5	70.7-74.3
Ensemble	0.67	0.69	0.80	73.3	71.4-75.1

Table 1: Class-specific F-scores and accuracies for four classifiers and ensemble

The promising results obtained from automatic classification verify that our annotated dataset may indeed be used for training automated classification systems. Including more informative features is likely to further improve performance, particularly for the smallest (*intake*) class.

3.3   Error and Feature Analyses

An analysis of the false negative results of the *intake* class from the SVM classifier suggests that the majority of the errors (62%) could be attributed to the *implicit* indication that (i) the medication was taken, (ii) the author of the tweet personally took the medication, or (iii) the medication was taken at a specific time. In 69% of these cases, the *intake* tweet did not explicitly state (i), that the medication was taken. The next largest set of misclassified *intake* tweets comprised instances where the *intake* tweets contain lexical features that seem to frequently occur in the other classes (e.g., negation). Incorporating semantic features into the SVM classifier is likely to improve classification of the *intake* tweets.

Table 2 presents the 15 most informative n-grams for distinguishing the *intake* class from the others, as identified by the information gain measure. The table suggests that certain personal pronouns and explicit markers of personal consumption (e.g., *I took*), information about effectiveness (e.g., *not working*), and expressions indicating the need for a medication (e.g., *need a*) are useful n-grams for the classification task.

i	*not helping*	*i ve taken*
took	*i need*	*not working*
i took	*ve been taking*	*still in*
took some	*took two*	*need a*
to kick in	*i ve taken*	*just took*

Table 2: Most informative n-grams that distinguish the *intake* class from the others

4   Conclusion

In this paper, we presented a brief analysis of what we consider to be linguistic representations of personal medication intake on Twitter. This linguistic analysis informed our manual annotation of 10,260 tweets. We presented baseline supervised classification results that suggest that this annotated corpus can be used for training automated classification systems to detect personal medication intake in large amounts of social media data, and we will seek to improve the performance of our classifiers in future work. We believe that this classification is an important step towards broadening the use of social media for surveillance of drug safety.

References

Artur Akbarov, Evangelos Kontopantelis, Matthew Sperrin, Susan J. Stocks, Richard Williams, Sarah Rodgers, Anthony Avery, Iain Buchan, and Darren M Ashcroft. 2015. Primary care medication safety surveillance with integrated primary and secondary care electronic health records: A cross-sectional study. *Drug Safety*, 38(7):671–682, July.

Nestor Alvaro, Mike Conway, Son Doan, Christoph Lofi, John Overington, and Nigel Collier. 2015. Crowdsourcing Twitter annotations to identify first-hand experiences of prescription drug use. *Journal of Biomedical Informatics*, 58: 280-287, December.

Mikhail M. Bakhtin. 1981. *The Dialogic Imagination.* University of Texas Press, Austin, TX.

Mikhail M. Bakhtin. 1986. Speech Genres & Other Essays. University of Texas Press, Austin, TX.

David Andre Broniatowski, Mark Dredze, Michael J. Paul, and Andrea Dugas. 2015. Using social media to perform local influenza surveillance in an inner-city hospital: A retrospective observational study. *JMIR Public Health Surveill 2015; 1(1):e5 https://publichealth.jmir.org/2015/1/e5/*, 1(1):e5.

Shannon Greenwood, Andrew Perrin, and Maeve Duggan. 2016. PEW Research Center Social Media Update 2016.

Carl L. Hanson, Scott H. Burton, Christophe Giraud-Carrier, Josh H. West, Michael D. Barnes, and Bret Hansen. 2013a. Tweaking and tweeting: exploring Twitter for nonmedical use of a psychostimulant drug (Adderall) among college students. *Journal of Medical Internet Research*, 15(4):e62, April.

Carl Lee Hanson, Ben Cannon, Scott Burton, and Christophe Giraud-Carrier. 2013b. An exploration of social circles and prescription drug abuse through Twitter. *Journal of Medical Internet Research*, 15(9):e189, January.

Andreas M. Kaplan and Michael Haenlein. 2010. Users of the world, unite! The challenges and opportunities of social media. *Business Horizons*, 53(1):59-68, January-February.

Ioannis Korkontzelos, Azadeh Nikfarjam, Matthew Shardlow, Abeed Sarker, Sophia Ananiadou, and Graciela H. Gonzalez. 2016. Analysis of the effect of sentiment analysis on extracting adverse drug reactions from tweets and forum posts. *Journal of Biomedical Informatics*, 62:148–158, August.

Julia Kristeva. 1980. *Desire in Language: A Semiotic Approach to Literature and Art.* Columbia University Press, New York.

Robert Leaman, Laura Wojtulewicz, Ryan Sullivan, Annie Skariah, Jian Yang, and Graciela Gonzalez. 2010. Towards Internet-Age Pharmacovigilance: Extracting Adverse Drug Reactions from User Posts to Health-Related Social Networks. In *Workshop on Biomedical Natural Language Processing*, pages 117–125, Uppsala, Sweden.

Azadeh Nikfarjam, Abeed Sarker, Karen O'Connor, Rachel Ginn, and Graciela Gonzalez. 2015. Pharmacovigilance from social media: Mining adverse drug reaction mentions using sequence labeling with word embedding cluster features. *Journal of the American Medical Informatics Association*, 22(3):671–81, March.

Brendan O'Connor, Michael Krieger, and David Ahn. 2010. TweetMotif: Exploratory search and topic summarization for Twitter. In Proceedings of the *Fourth International AAAI Conference on Weblogs and Social Media*, pages 384-385, Washington, DC.

Olutobi Owoputi, Brendan O'Connor, Chris Dyer, Kevin Gimpel, Nathan Schneider, and Noah A. Smith. 2013. Improved part-of-speech tagging for online conversational text with word clusters. In Proceedings of the *Conference of the North American Chapter of the Association for Computational Linguistics: Human Language Technologies*, pages 380-391, Atlanta, GA.

Zizi Papacharissi. 2012. Without you, I'm nothing: Performances of the self on Twitter. *International Journal of Communication*, 6:1989-2006,

Pranoti Pimpalkhute, Apurv Patki, Azadeh Nikfarjam, and Graciela Gonzalez. 2014. Phonetic spelling filter for keyword selection in drug mention mining from social media. In *AMIA Joint Summits on Translational Science proceedings AMIA Summit on Translational Science*, pages 90–5.

M. F. Porter. 1980. An algorithm for suffix stripping. *Program*, 14(3):130–137.

Katrina M. Romagnoli, Scott D. Nelson, Lisa Hines, Philip Empey, Richard D Boyce, and Harry Hochheiser. 2017. Information needs for making clinical recommendations about potential drug-drug interactions: a synthesis of literature review and interviews. *BMC Medical Informatics and Decision Making*, 17(21).

Takeshi Sakaki, Makoto Okazaki, and Yutaka Matsuo. 2010. Earthquake shakes Twitter users: Real-time event detection by social sensors. In Proceedings of the *19th International World Wide Web Conference*, pages 851-860, Raleigh, NC.

Gwendolyn Seidman. 2013. Self-presentation and belonging on Facebook: How personality influences social media use and motivations. *Personality and Differences*, 54(3):402-407, February.

Laura Louise Struik and Neill Bruce Baskerville. 2014. The role of Facebook in Crush the Crave, a mobile- and social media-based smoking cessation intervention: Qualitative Framework Analysis of posts. *Journal of Medical Internet Research*, 16(7):e170, July.

Valentin N. Voloshinov. 1973. *Marxism and the Philosophy of Language*. Seminar Press, New York.

Yiming Yang, and Jan O. Pedersen. A Comparative Study on Feature Selection in Text Categorization. In Proceedings of the Fourteenth International Conference on Machine Learning, pages 412–420.

Li Zhou, Neil Dhopeshwarkar, Kimberly G Blumenthal, Foster R. Goss, Maxim Topaz, Sarah P. Slight, and David W. Bates. 2016. Drug allergies documented in electronic health records of a large healthcare system. *Allergy*, 71(9):1305–1313, September.

Unsupervised Context-Sensitive Spelling Correction of Clinical Free-Text with Word and Character N-Gram Embeddings

Pieter Fivez[*]   **Simon Šuster**[*†]   **Walter Daelemans**[*]

[*]CLiPS, University of Antwerp, Antwerp, Belgium
[†]Antwerp University Hospital, Edegem, Belgium
`firstname.lastname@uantwerpen.be`

Abstract

We present an unsupervised context-sensitive spelling correction method for clinical free-text that uses word and character n-gram embeddings. Our method generates misspelling replacement candidates and ranks them according to their semantic fit, by calculating a weighted cosine similarity between the vectorized representation of a candidate and the misspelling context. We greatly outperform two baseline off-the-shelf spelling correction tools on a manually annotated MIMIC-III test set, and counter the frequency bias of an optimized noisy channel model, showing that neural embeddings can be successfully exploited to include context-awareness in a spelling correction model. Our source code, including a script to extract the annotated test data, can be found at `https://github.com/pieterfivez/bionlp2017`.

1 Introduction

The genre of clinical free-text is notoriously noisy. Corpora contain observed spelling error rates which range from 0.1% (Liu et al., 2012) and 0.4% (Lai et al., 2015) to 4% and 7% (Tolentino et al., 2007), and even 10% (Ruch et al., 2003). This high spelling error rate, combined with the variable lexical characteristics of clinical text, can render traditional spell checkers ineffective (Patrick et al., 2010).

Recently, Lai et al. (2015) have achieved nearly 80% correction accuracy on a test set of clinical notes with their noisy channel model. However, their model does not leverage any contextual information, while the context of a misspelling can provide important clues for the spelling correction process, for instance to counter the frequency bias of a context-insensitive corpus frequency-based system. Flor (2012) also pointed out that ignoring contextual clues harms performance where a specific misspelling maps to different corrections in different contexts, e.g. _iron deficiency due to ~~enemia~~_ → _anemia_ vs. _fluid injected with ~~enemia~~_ → _enema_. A noisy channel model like the one by Lai et al. will choose the same item for both corrections.

Our proposed method exploits contextual clues by using neural embeddings to rank misspelling replacement candidates according to their semantic fit in the misspelling context. Neural embeddings have recently proven useful for a variety of related tasks, such as unsupervised normalization (Sridhar, 2015) and reducing the candidate search space for spelling correction (Pande, 2017).

We hypothesize that, by using neural embeddings, our method can counter the frequency bias of a noisy channel model. We test our system on manually annotated misspellings from the MIMIC-III (Johnson et al., 2016) clinical notes. In this paper, we focus on already detected non-word misspellings, i.e. where the misspellings are not real words, following Lai et al.

2 Approach

2.1 Candidate Generation

We generate replacement candidates in 2 phases. First, we extract all items within a Damerau-Levenshtein edit distance of 2 from a reference lexicon. Secondly, to allow for candidates beyond that edit distance, we also apply the Double Metaphone matching popularized by the open source spell checker Aspell.[1] This algorithm converts lexical forms to an approximate phonetic consonant skeleton, and matches all Dou-

---

[1] `http://aspell.net/metaphone/`

Proceedings of the BioNLP 2017 workshop, pages 143–148,
Vancouver, Canada, August 4, 2017. ©2017 Association for Computational Linguistics

ble Metaphone representations within a Damerau-Levenshtein edit distance of 1. As reference lexicon, we use a union of the UMLS® SPECIALIST lexicon[2] and the general dictionary from Jazzy[3], a Java open source spell checker.

2.2 Candidate Ranking

Our setup computes the cosine similarity between the vector representation of a candidate and the composed vector representations of the misspelling context, weights this score with other parameters, and uses it as the ranking criterium. This setup is similar to the contextual similarity score by Kilicoglu et al. (2015), which proved unsuccessful in their experiments. However, their experiments were preliminary. They used a limited context window of 2 tokens, could not account for candidates which are not observed in the training data, and did not investigate whether a bigger training corpus leads to vector representations which scale better to the complexity of the task.

We attempt a more thorough examination of the applicability of neural embeddings to the spelling correction task. To tune the parameters of our unsupervised context-sensitive spelling correction model, we generate tuning corpora with self-induced spelling errors for three different scenarios following the procedures described in section 3.2. These three corpora present increasingly difficult scenarios for the spelling correction task. **Setup 1** is generated from the same corpus which is used to train the neural embeddings, and exclusively contains corrections which are present in the vocabulary of these neural embeddings. **Setup 2** is generated from a corpus in a different clinical subdomain, and also exclusively contains in-vector-vocabulary corrections. **Setup 3** presents the most difficult scenario, where we use the same corpus as for Setup 2, but only include corrections which are not present in the embedding vocabulary (OOV). In other words, here our model has to deal with both domain change and data sparsity.

Correcting OOV tokens in Setup 3 is made possible by using a combination of word and character n-gram embeddings. We train these embeddings with the fastText model (Bojanowski et al., 2016), an extension of the popular Word2Vec model (Mikolov et al., 2013), which creates vec-

[2]https://lexsrv3.nlm.nih.gov/
LexSysGroup/Projects/lexicon/current/
web/index.html
[3]http://jazzy.sourceforge.net

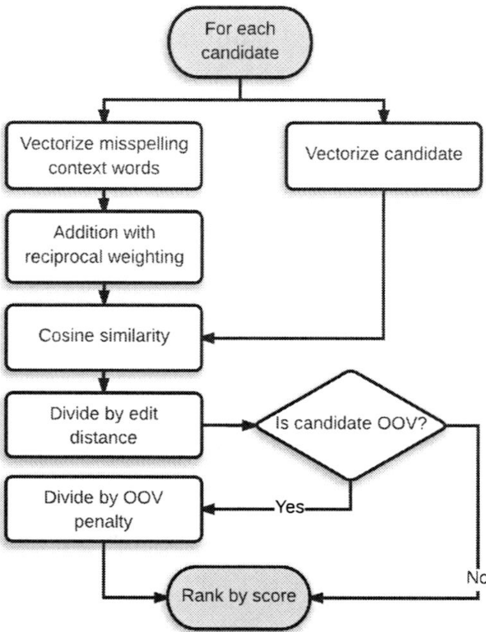

Figure 1: The final architecture of our model. It vectorizes every context word on each side within a specified scope if it is present in the vector vocabulary, applies reciprocal weighting, and sums the representations. It then calculates the cosine similarity with each candidate vector, and divides this score by the Damerau-Levenshtein edit distance between the candidate and misspelling. If the candidate is OOV, the score is divided by an OOV penalty.

tor representations for character n-grams and sums these with word unigram vectors to create the final word vectors. FastText allows for creating vector representations for misspelling replacement candidates absent from the trained embedding space, by only summing the vectors of the character n-grams.

We report our tuning experiments with the different setups in 4.1. The final architecture of our model is described in Figure 1. We evaluate this model on our test data in section 4.2.

3 Materials

We tokenize all data with the Pattern tokenizer (De Smedt and Daelemans, 2012). All text is lowercased, and we remove all tokens that include anything different from alphabetic characters or hyphens.

3.1 Neural embeddings

We train a fastText skipgram model on 425M words from the MIMIC-III corpus, which contains medical records from critical care units. We use the default parameters, except for the dimensionality, which we raise to 300.

3.2 Tuning corpora

In order to tune our model parameters in an unsupervised way, we automatically create self-induced error corpora. We generate these tuning corpora by randomly sampling lines from a reference corpus, randomly sampling a single word per line if the word is present in our reference lexicon, transforming these words with either 1 (80%) or 2 (20%) random Damerau-Levenshtein operations to a non-word, and then extracting these misspelling instances with a context window of up to 10 tokens on each side, crossing sentence boundaries. For **Setup 1**, we perform this procedure for MIMIC-III, the same corpus which we use to train our neural embeddings, and exclusively sample words present in our vector vocabulary, resulting in 5,000 instances. For **Setup 2**, we perform our procedure for the THYME (Styler IV et al., 2014) corpus, which contains 1,254 clinical notes on the topics of brain and colon cancer. We once again exclusively sample in-vector-vocabulary words, resulting in 5,000 instances. For **Setup 3**, we again perform our procedure for the THYME corpus, but this time we exclusively sample OOV words, resulting in 1,500 instances.

3.3 Test corpus

No benchmark test sets are publicly available for clinical spelling correction. A straightforward annotation task is costly and can lead to small corpora, such as the one by Lai et al., which contains just 78 misspelling instances. Therefore, we adopt a more cost-effective approach. We spot misspellings in MIMIC-III by looking at items with a frequency of 5 or lower which are absent from our lexicon. We then extract and annotate instances of these misspellings along with their context, resulting in 873 contextually different instances of 357 unique error types. We do not control for the different genres covered in the MIMIC-III database (e.g. physician-generated progress notes vs. nursing notes). However, in all cases we make sure to annotate actual spelling mistakes and typos as opposed to abbre-viations and shorthand, resulting in instances such as *phebilitis* → *phlebitis* and *sympots* → *symptoms*. We provide a script to extract this test set from MIMIC-III at `https://github.com/pieterfivez/bionlp2017`.

4 Results

For all experiments, we use accuracy as the metric to evaluate the performance of models. Accuracy is simply defined as the percentage of correct misspelling replacements found by a model.

4.1 Parameter tuning

To tune our model, we investigate a variety of parameters:

Vector composition functions

(a) addition
(b) multiplication
(c) max embedding by Wu et al. (2015)

Context metrics

(a) Window size (1 to 10)
(b) Reciprocal weighting
(c) Removing stop words using the English stop word list from scikit-learn (Pedregosa et al., 2011)
(d) Including a vectorized representation of the misspelling

Edit distance penalty

(a) Damerau-Levenshtein
(b) Double Metaphone
(c) Damerau-Levenshtein + Double Metaphone
(d) Spell score by Lai et al.

We perform a grid search for Setup 1 and Setup 2 to discover which parameter combination leads to the highest accuracy averaged over both corpora. In this setting, we only allow for candidates which are present in the vector vocabulary. We then introduce OOV candidates for Setup 1, 2 and 3, and experiment with penalizing them, since their representations are less reliable. As these representations are only composed out of character n-gram vectors, with no word unigram vector, they are susceptible to noise caused by the particular nature of the n-grams. As a result, sometimes the semantic similarity of OOV vectors to other vectors can be inflated in cases of strong orthographic overlap.

Table 1: Correction accuracies for our 3 tuning setups.

	Setup 1	Setup 2	Setup 3
Context	90.24	88.20	57.00
Noisy Channel	85.02	85.86	39.73

Since OOV replacement candidates are more often redundant than necessary, as the majority of correct misspelling replacements will be present in the trained vector space, we try to penalize OOV representations to the extent that they do not cause noise in cases where they are redundant, but still rank first in cases where they are the correct replacement. We tune this OOV penalty by maximizing the accuracy for Setup 3 while minimizing the performance drop for Setup 1 and 2, using a weighted average of their correction accuracies.

All parameters used in our final model architecture are described in Figure 1. The optimal context window size is 9, whereas the optimal OOV penalty is 1.5.

To compare our method against a reference noisy channel model in the most direct and fair way, we implement the ranking component of Lai et al.'s model in our pipeline (**Noisy Channel**), and optimize it with the same MIMIC-III materials that we use to train our embeddings. We perform the optimization by extracting corpus frequencies, which are used to estimate the prior probabilities in the ranking model, from this large data containing 425M words. In comparison, Lai et al.'s own implementation uses corpus frequencies extracted from data containing only 107K words, which is a rather small amount to estimate reliable prior probabilities for a noisy channel model. In the optimized setting, our context-sensitive model (**Context**) outperforms the noisy channel for each corpus in our tuning phase, as shown in Table 1.

4.2 Test

Table 2 shows the correction accuracies for **Context** and **Noisy Channel**, as compared to two baseline off-the-shelf tools. The first tool is Hun-Spell, a popular open source spell checker used by Google Chrome and Firefox. The second tool is the original implementation of Lai et al.'s model, which they shared with us. The salient difference in performance with our own implementation of their noisy channel model highlights the influence

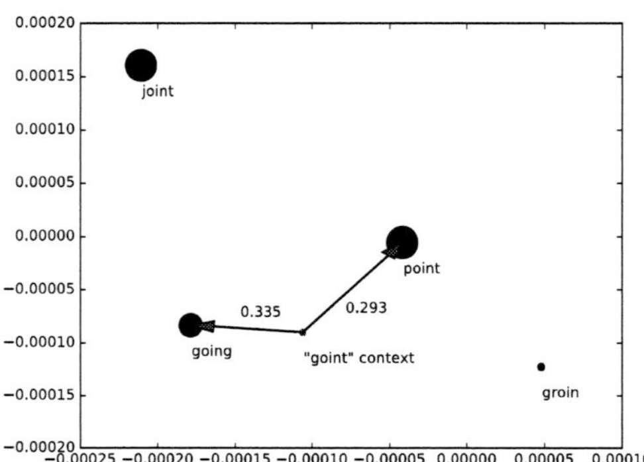

Figure 2: 2-dimensional t-SNE projection of the context of the test misspelling "goint" and 4 replacement candidates in the trained vector space. Dot size denotes corpus frequency, numbers denote cosine similarity. The misspelling context is "new central line lower extremity bypass with sob now [goint] to [be] intubated". While the noisy channel chooses the most frequent "point", our model correctly chooses the most semantically fitting "going".

of training resources and tuning decisions on the general applicability of spelling correction models.

The performance of our model on the test set is slightly held back by the incomplete coverage of our reference lexicon. Missing corrections are mostly highly specialized medical terms, or inflections of more common terminology. Table 2 shows the scenario where these corrections are added to the reference lexicon, leading to a score which is actually higher than those for the tuning corpora.

To analyze whether our context-sensitive model successfully counters the frequency bias of our optimized noisy channel model, we divide the instances of the test set into three scenarios according to the relative frequency of the correct replacement compared to the other replacement candidates. In cases where the correct replacement is the most or second most frequent candidate, the noisy channel scores slightly better. In all other cases, however, our method is more stable. Figure 2 visualizes an example.

Nevertheless, some issues have to be raised. First of all, for the cases with low relative frequency of the correct replacement, the small sample size should be kept in mind: the difference between both models concerns 6 correct instances on

Table 2: The correction accuracies for our test set, evaluated for two different scenarios. *Off-the-shelf*: gives the accuracies of all off-the-shelf tools. *With completed lexicon*: gives the accuracies of our implemented models for the scenario where correct replacements missing from the lexicon are included in the lexicon before the experiment.

Evaluation	HunSpell	Lai et al.	Context	Noisy Channel
OFF-THE-SHELF	52.69	61.97	88.21	87.85
WITH COMPLETED LEXICON			93.02	92.66

a total of 243. While the difference is very pronounced in the much larger tuning corpora, the artificial nature of those corpora does not lead to strong evidence. Moreover, considering the similarity of the general performance of both models on the test set, more test data is needed to make a strong empirical claim about this specific aspect of our model.

While we have optimized the prior probabilities of Lai et al.'s ranking model, the posterior probabilities are still estimated with Lai et al.'s rudimentary spell score, which is a weighted combination of Damerau-Levenshtein and Double Metaphone edit distance. While this error model leads to a noisy channel model which is robust in performance, as shown by our test results, an empirical error model derived from a large confusion matrix can for example help correct the instance described in Figure 2, by capturing that the word-final transformation $t \rightarrow g$ is more probable than the word-initial transformation $g \rightarrow p$. As of now, however, such a resource is not available for the clinical domain.

The errors that our model makes concern, predictably, misspellings for which the contextual clues are too unspecific or misguiding. These cases remain challenging for the concept of our method. While our tuning experiments have explicitly tried to maximize the scope and efficiency of our model, there is still room for improvement, especially for OOV corrections, even as we handle them more effectively than context-insensitive frequency-based methods.

5 Conclusion and future research

In this article, we have proposed an unsupervised context-sensitive model for clinical spelling correction which uses word and character n-gram embeddings. This simple ranking model, which can be tuned to a specific domain by generating self-induced error corpora, counters the frequency bias of a noisy channel model by exploiting contextual clues. As an implemented spelling correction tool, our method greatly outperforms two baseline off-the-shelf spelling correction tools, both a broadly used and a domain-specific one, for empirically observed misspellings in MIMIC-III.

Future research can investigate whether our method transfers well to other genres and domains. Secondly, handling corrections which are not observed in any training data still proves to be a tough task, which might benefit from new conceptual insights. Lastly, it is worthwhile to investigate how our model interacts with the misspelling detection task compared to existing models.

6 Acknowledgements

This research was carried out in the framework of the Accumulate VLAIO SBO project, funded by the government agency Flanders Innovation & Entrepreneurship (VLAIO). We would also like to thank Stéphan Tulkens for his logistic support with coding.

References

Piotr Bojanowski, Edouard Grave, Armand Joulin, and Tomas Mikolov. 2016. Enriching word vectors with subword information. *arXiv preprint arXiv:1607.04606* .

Tom De Smedt and Walter Daelemans. 2012. Pattern for Python. *Journal of Machine Learning Research* 13:2031–2035.

Michael Flor. 2012. Four types of context for automatic spelling correction. *TAL* 53(3):61–99.

Alistair E.W. Johnson, Tom J. Pollard, Lu Shen, Li wei H. Lehman, Mengling Feng, Mohammad Ghassemi, Benjamin Moody, Peter Szolovits, Leo Anthony Celi, and Roger G. Mark. 2016. MIMIC-III, a freely accessible critical care database. *Scientific Data* 3.

Halil Kilicoglu, Marcelo Fiszman, Kirk Roberts, and Dina Demner-Fushman. 2015. An ensemble method for spelling correction in consumer health questions.

AMIA Annual Symposium Proceedings pages 727–73.

Kenneth H. Lai, Maxim Topaz, Foster R. Goss, and Li Zhou. 2015. Automated misspelling detection and correction in clinical free-text records. *Journal of Biomedical Informatics* 55:188–195.

Hongfang Liu, Stephen T. Wu, Dingcheng Li, Siddharta Jonnalagadda, Sunghwan Sohn, Kavishwar Wagholikar, Peter J. Haug, Stanley M. Huff, and Christopher G Chute. 2012. Towards a semantic lexicon for clinical natural language processing. *AMIA Annual Symposium Proceedings* .

Tomas Mikolov, Kai Chen, Greg Corrado, and Jeffrey Dean. 2013. Efficient estimation of word representations in vector space. *Proceedings of Workshop at International Conference on Learning Representations* .

Harshit Pande. 2017. Effective search space reduction for spell correction using character neural embeddings. *Proceedings of the 15th Conference of the European Chapter of the Association for Computational Linguistics: Volume 2, Short Papers* pages 170–174.

J. Patrick, M. Sabbagh, S. Jain, and H. Zheng. 2010. Spelling correction in clinical notes with emphasis on first suggestion accuracy. *2nd Workshop on Building and Evaluating Resources for Biomedical Text Mining* pages 2–8.

Fabrian Pedregosa, Gaël Varoquaux, Alexandre Gramfort, Vincent Michel, Bertrand Thirion, Olivier Grisel, Mathieu Blondel, Peter Prettenhofer, Ron Weiss, Vincent Dubourg, and et al. 2011. Scikit-learn: Machine learning in Python. *The Journal of Machine Learning Research* 12:2825–2830.

Patrick Ruch, Robert Baud, and Antoine Geissbühler. 2003. Using lexical disambiguation and named-entity recognition to improve spelling correction in the electronic patient record. *Artificial Intelligence in Medicine* 29:169–184.

Vivek Kumar Rangarajan Sridhar. 2015. Unsupervised text normalization using distributed representations of words and phrases. *Proceedings of NAACL-HLT 2015* pages 8–16.

William F. Styler IV, Steven Bethard, Sean Finan, Martha Palmer, Sameer Pradhan, Piet C de Groen, Brad Erickson, Timothy Miller, Chen Lin, Guergana Savova, and James Pustejovsky. 2014. Temporal annotation in the clinical domain. *Transactions of the Association for Computational Linguistics* 2:143–154.

Herman D. Tolentino, Michael D. Matters, Wikke Walop, Barbara Law, Wesley Tong, Fang Liu, Paul Fontelo, Katrin Kohl, and Daniel C. Payne. 2007. A UMLS-based spell checker for natural language processing in vaccine safety. *BMC Medical Informatics and Decision Making* 7(3).

Yonghui Wu, Jun Xu, Yaoyun Zhang, and Hua Xu. 2015. Clinical abbreviation disambiguation using neural word embeddings. *Proceedings of the 2015 Workshop on Biomedical Natural Language Processing (BioNLP)* pages 171–176.

Characterization of Divergence in Impaired Speech of ALS Patients

Archna Bhatia[1], Bonnie J. Dorr[1], Kristy Hollingshead[1], Samuel L. Phillips[2], and Barbara McKenzie[2]

[1]Florida Institute for Human and Machine Cognition, 15 SE Osceola Ave, Ocala, FL 34471
[2]James A. Haley VA Hospital, 13000 Bruce B. Downs, Tampa, FL 33612
[1]{abhatia,bdorr,kseitz}@ihmc.us
[2]{samuel.phillips,barbara.mckenzie}@va.gov

Abstract

Approximately 80% to 95% of patients with Amyotrophic Lateral Sclerosis (ALS) eventually develop speech impairments (Beukelman et al., 2011), such as defective articulation, slow laborious speech and hypernasality (Duffy, 2013). The relationship between impaired speech and asymptomatic speech may be seen as a *divergence* from a baseline. This relationship can be characterized in terms of measurable combinations of phonological characteristics that are indicative of the degree to which the two diverge. We demonstrate that divergence measurements based on phonological characteristics of speech correlate with physiological assessments of ALS. Speech-based assessments offer benefits over commonly-used physiological assessments in that they are inexpensive, non-intrusive, and do not require trained clinical personnel for administering and interpreting the results.

1 Introduction

Amyotrophic lateral sclerosis (ALS) or Lou Gehrig's Disease, the most common form of motor neuron disease, is a rapidly progressive, neurodegenerative condition. It is characterized by muscle atrophy, muscle weakness, muscle spasticity, hyperreflexia, difficulty speaking (dysarthria), difficulty swallowing (dysphagia), and difficulty breathing (dyspnea). Mean survival time for ALS patients is three to five years from the time it is diagnosed; however, death may occur within months, or survival may last decades.

Most physiological assessments used to determine the functional status of patients with ALS are invasive, involving the use of expensive equipment and requiring trained clinical personnel to administer the tests and interpret the results. This is the case for a number of standardized objective assessments of bulbar function in ALS patients (Green et al., 2013), for example: breathing patterns, articulatory patterns, and voice loudness. These are generally measured by technologies that record chest wall movements, oral pressures and flows, oral movement and strength, and speech acoustics.

This paper lays the foundation for the development of less invasive phonologically-inspired measures that correlate strongly with (more invasive) physiological measures of ALS. Speech impairments eventually affect 80% to 95% of patients with ALS (Beukelman et al., 2011). In fact, Yorkston et al. (1993) noted that speech impairments may be present up to 33 months prior to diagnosis of ALS. Several previous studies (Yunusova et al., 2016) have shown that speech impairments correlate with physiological changes associated with ALS. Thus, we focus on correlating measures based on phonological features with standard physiological measures, thus enabling new, non-invasive measures for assessing the functionality of an ALS patient without significant overhead for personnel training and administration.

To bring this about, we determine the degree of *divergence* of symptomatic speech from asymptomatic speech taken as a baseline.[1] This determination is based on phonological features in speech, most of which have been previously identified in the literature as being associated with ALS, e.g., *monoloudness*, *hypernasality* and *distorted vowels*, see (Duffy, 2013). These are annotated, for the current study, by specialists, i.e., a phonolo-

---

[1]As part of a longitudinal study, we are exploring individual baselines for each ALS speaker: speech from each speaker's first visit is taken as an individual baseline for the speaker.

Proceedings of the BioNLP 2017 workshop, pages 149–158,
Vancouver, Canada, August 4, 2017. ©2017 Association for Computational Linguistics

gist and a speech therapist experienced in working with ALS patients. The degree of divergence is correlated with physiological assessments of ALS, namely %FVC (Forced Vital Capacity) in sitting (%FVC-SIT) as well as supine (%FVC-SUP) positions.[2]

The rest of the paper is organized as follows: In Section 2 we discuss related work that motivates and informs our research. Section 3 describes data used for our experiments. A discussion of speech divergence is presented in Section 4. Section 5 presents an assessment of the degree to which divergent characteristics in the speech match the level of progress of the ALS condition. This is followed by a discussion of future work and conclusions in Section 6.

2   Related Work

A number of past studies have investigated the utility of measuring the "voice signal" in order to answer questions about a speaker's state from their speech (Schuller et al., 2015, 2011). One such study attempts to distinguish classes of individuals with various speech impairments, such as stuttering (Nöth et al., 2000), aphasia (Fraser et al., 2014), and developmental language disorders (Gorman et al., 2016). The recognition of impaired speech has been employed to detect Alzheimer's (Rudzicz et al., 2014). Various speech-related features have been employed to detect whether the speech is affected by Parkinson's Disease (Bocklet et al., 2011). Relatedly, variations in speech properties under intoxicated and sober conditions have also been conducted (Biadsy et al., 2011).

Our work differs from prior approaches in that we explore perceivable phonological characteristics through the analysis of language divergences. One of the motivations for using phonological features exclusively rather than also using other features employed in prior studies was that phonological features did not require expensive equipment to collect data from speakers as e.g., a feature like *maximum subglottal pressure* would require. Since the goal of this work is to develop a measure that is completely based on speech features that can be identified with a simple click on a

device such as a phone, we focused on phonological features on which a machine can be trained to analyze automatically. Our focus on correlations with phonological features—tied to the notion of divergence from a baseline—is a significant contribution beyond what has been investigated previously.

The notion of *divergence* itself is not a new one in natural language processing. The characterization of divergence classes (Dorr, 1994) has been at the heart of solutions to many different problems ranging from word alignment (Dorr et al., 2002) to machine translation (Habash and Dorr, 2002) to acquisition of semantic lexicons (Olsen et al., 1998). Finding the minimal primitive units—and determining their possible combinations—was the foundation for this earlier work. However, in these earlier studies, primitives consisted of properties that were syntactic, lexical, or semantic in nature, whereas the primitives for the current work consist of properties that are phonological in nature.

Beukelman et al. (2011), Duffy (2013), Green et al. (2013), and Orimaye et al. (2014) have established that pronunciation varies systematically within categories of speech impairment. (Silbergleit et al., 1997; Carrow et al., 1974) have shown that ALS speech shows deviant characteristics. For example, (Ball et al., 2001) observe that ALS speakers manifest altered voice quality. A number of speaker-level characteristics associated with impaired speech studied in prior work have been leveraged for our speech-related divergence detection. For example, Duffy (2013) specifically has enumerated speaker-level characteristics, such as *monopitch* and *monoloudness*.

Rong et al. (2015; 2016), and Yunusova et al. (2016) have previously attempted to identify measures of speech motor function for ALS speech. While certain components of speech such as speaking rate, breathing patterns, and voice loudness have proven too variable to provide a reliable marker (Green et al., 2013), we demonstrate that divergence measurements based on phonological characteristics of speech correlate with physiological assessments of ALS.

In addition to speaker-level characteristics and associated properties, our work defines divergence in terms of speech/span-level characteristics, as described in Section 3. Smaller vowel space areas have been found in ALS speech (Turner et al., 1995; Weismer et al., 2001) which suggests that

---

[2]%FVC-SUP refers to the percent value of the Forced Vital Capacity while the person is in supine position, and %FVC-SIT refers to the percent value of the Forced Vital Capacity while the person is in sitting position. See (Brinkmann et al., 1997; Czaplinski et al., 2006) for additional information about use of FVC in ALS assessments.

vowels may be distorted in ALS speech. Similarly, Kent et al (Kent et al., 1990) found place and manner of articulation for some consonants, and regulation of tongue height for vowels to diverge from asymptomatic speech; these were expected to result in imprecise consonants and distorted vowels. Caruso and Burton (1987) observed that ALS speakers and asymptomatic speakers exhibited significant differences in stop-gap durations as well as in vowel durations.

Yunusova et al. (2016) have also previously shown a correlation between speaking rate and physiological measures of ALS, specifically ALS Functional Rating Scale (ALSFRS).[3] Our own work differs from this prior work in that we define *divergence* in terms of a wider range of speech characteristics—beyond speaking rate—and then demonstrate that divergence measures correlate with physiological measures of ALS.

3 Data: Transcriptions and Phonological Annotations

The data for our experiments consist of recorded speech of 16 recruited subjects with ALS in a clinical setting, collected quarterly for each subject. The subjects range between 35-74 years of age. Their age distribution is as follows: one subject in the 30s, two subjects in their 40s, one subject in their 50s, five subjects in their 60s, and seven subjects in their 70s. Out of the 16 subjects, only one of them is female, the other 15 subjects are male. In terms of race of the subjects, we have the following distribution: White (12), Asian (1), African-American (1), Not reported (2).

The criteria for the recruitment of a particular subject are that the subject: (1) has been diagnosed with ALS; (2) is a native monolingual speaker of American English; (3) has bulbar involvement identified during initial ALS inpatient evaluation; (4) has a forced vital capacity (FVC) of greater than 50% of the expected value for age; and (5) has an ALSFRS-R score[4] of 40 or greater. Excluded from the study are those who have received a diagnosis of dementia, FVC of less than 50%, inability to speak, or inability to follow directions.

As part of their standard clinical care, each ALS-diagnosed subject reports for a quarterly

evaluation during which the following measures are collected: Forced Vital Capacity (FVC; (Brinkmann et al., 1997; Czaplinski et al., 2006)), Penetration-Aspiration Scale (a paper-pencil screen; (Rosenbek et al., 1996)), ALSFRS-R (Cedarbaum and Stambler, 1997; Cedarbaum et al., 1999), and Speech Intelligibility Test (SIT (Beukelman et al., 2011; Yorkston et al., 2007)).

1. She held your dark suit in greasy wash water all year.
2. Don't ask me to hold an oily rag like that.
3. The big dog loved to chew on the old rag doll.
4. Chocolate and roses never fail as a romantic gift.

Table 1: Four TIMIT Sentences

Speech recordings of the same four sentences, that have been preselected, are made during each (quarterly) visit of each of the patients.[5] The four sentences, presented in Table 1, are selected from the Texas-Instrument/MIT (TIMIT) corpus (Garofolo et al., 1993) and were designed to be phonetically rich, thus providing solid coverage of the phonetic space from each subject.[6]

The data also include recordings of four control (asymptomatic) subjects, two of whom are female and two are male, reading the same four TIMIT sentences in the same setting as the symptomatic subjects. These are used as the baseline speech against which divergence scores (defined in the next section) are calculated for the ALS symptomatic speech.

Our hypothesis is that a higher divergence is indicative of the progression of the ALS condition. This study focuses on divergence with respect to asymptomatic speech—taken as a baseline—to determine whether the divergence is speaker dependent or whether it is more generally indicative of ALS progression. If the latter, this would help diagnose patients for which no previous/longitudinal data is available.[7]

ALS speech data for the 16 subjects was transcribed and annotated via speech-analysis software called Praat (Boersma and van Heuven, 2001) for the 14 phonological characteristics enumerated in Table 2. These characteristics

---

[3]ALSFRS is a standard questionnaire-based scale to measure functionality of a person in carrying out daily activities, see (Cedarbaum and Stambler, 1997; Brooks, 1997).

[4]ALSFRS-R is a revised ALSFRS that incorporates assessments of respiratory function (Cedarbaum et al., 1999).

[5]All uses of these data as reported in this paper have been approved by the relevant Institutional Review Board (IRB).

[6]Note the TIMIT sentences 1 and 2 are slightly different from the original TIMIT sentences; the original TIMIT sentences are as follows: (1) She **had** your dark suit in greasy wash water all year; (2) Don't ask me to **carry** an oily rag like that.

[7]However, as longitudinal data becomes available to us in our future work, we will also look at divergence based on speaker dependent baselines.

Speaker level characteristics	
monopitch	Voice lacks inflectional changes; pitch does not change much.
monoloudness	Voice for which the volume/loudness does not change, hence lacking normal variations in loudness.
Speech/span related characteristics	
harshness	Voice seems harsh, rough and raspy—sometimes referred to as pressed voice—similar to what happens when a person talks while lifting a heavy load.
imprecise consonants	Consonant sounds lack precision. There may be slurring, inadequate sharpness, distortions, lack of crispness, and clumsiness in transitioning from one consonant to another. For example, a "w" may be produced instead of a "b".
distorted vowels	Vowel sounds distorted throughout their total duration. For example, a "a" may be produced instead of "i".
prolonged phonemes	A phoneme (i.e., a consonant or a vowel) is prolonged, i.e., its sound (when it is produced) continues over an unusual period of time.
inappropriate silences	Pauses that are produced not at syntactic or prosodic boundaries.
hypernasality	Vowels that are supposed to be non-nasalized are instead nasalized in speech.
strained or strangled quality	Tenseness in voice (as with overall muscular tension). Perceived as increased effort, may seem tense or harsh as if talking and lifting at the same time or as if talking with breath held.
breathiness	Voice seems breathy, weak and thin. May seem like a sighing sound. There may be non-modulated turbulence noise in the produced sound, i.e., audible air escape in voice or bursts of breathiness.
audible inspiration/stridor	Noisy breathing and wheezing may accompany inhaling. There may be a harsh, crowing, or vibratory sound of variable pitch resulting from turbulent air flow caused by partial obstruction of the respiratory passages.
unusual stress	Speech sounds where most important parts of a sentence may be de-stressed or all parts of a sentence are stressed as if all are important or speech sounds may be perceived as robotic, with the same stress—where there is no variation in stress throughout sentence/phrase/word/syllable.
hoarseness	Abnormal voice changes, where voice may sound breathy, raspy, strained, or there may be changes in volume (loudness) or pitch (how high or low the voice is).
vocal fry	Popping or rattling sound of a very low frequency—also known as a creaky voice.

Table 2: Phonological characteristics annotated in symptomatic speech

were pre-identified mostly based on the clinical research literature on ALS speech, e.g., see Duffy (2013): p248. The phonological annotations were made by two specialists: one of whom was a phonologist and the other was a speech therapist with experience working with ALS speakers.

Two classes of phonological characteristics served as the basis of annotations, each with a set of primitive phonological features: *speaker level characteristics* and *speech/span related characteristics*. *Speaker level characteristics* refer to features in speech that are more characteristic of a specific speaker's voice—independent of individual sounds/spans, e.g., *monopitch* which indicates the lack of inflectional changes in voice. These were annotated only once for each speaker.

Speech/span related characteristics, on the other hand, refer to features in speech that are characteristic of a specific sound or are observed for a portion of speech—as opposed to features that are characteristic of the voice itself. For example, the feature *imprecise consonants* refers to the portion of speech where a specific consonant is produced imprecisely, it may involve slurring or inadequate sharpness, e.g., producing a "w" instead of a "b". For these annotations, spans in speech were marked over which these features

were observed.

For each of these characteristics, the annotators also assigned a 1-10 Likert scale (Likert, 1932) rating to indicate the severity of the characteristic when it is observed, where 10 indicates "very severe" and 1 indicates "negligible."

4   Divergence in Speech

Understanding the relationship between impaired speech and asymptomatic speech is facilitated by measuring the degree to which symptomatic speech diverges from a baseline. For the current study, asymptomatic speech—which serves as a baseline—was created from a combination of recordings from asymptomatic speakers as described in Section 3. Simplistically, the degree of divergence is defined as the sum of the changes in a speech utterance from its asymptomatic equivalent. For a correlation to be supported, a large number of changes in speech (i.e., a strong divergence from asymptomatic speech) would correspond to advanced progression of the disease. The relationship between impaired speech and asymptomatic speech is characterized in terms of measurable combinations of phonological characteristics that are indicative of the degree to which the two diverge. The degree of divergence can be used as a diagnostic tool at regular intervals for checking the severity of physiological changes.

Multiple methods have been applied in order to calculate divergence scores:

1. **Feature count based divergence score:** *Feature count* refers to the number of characteristics observed in the speech samples.[8] The *Feature count based divergence score* for each ALS speaker is the difference between the *feature count* for the ALS speech and the *feature count* for the control asymptomatic speech.
   Four variations of this score are used based on how the *feature count* for the control asymptomatic speech is obtained:

   (a) **Average feature count for controls:** It is assumed that asymptomatic speakers may display characteristics identified in Table 2 but to a much smaller extent. Thus, taking a simple average of the *feature count* for control speakers is taken to be most representative of all asymptomatic speakers. The *feature count* for the control asymptomatic speech is the average of the *feature count* for all the control speakers.

   (b) **Minimum feature count for controls:** The control speaker with the minimum number of characteristics in his/her speech is assumed to be the most asymptomatic. Hence, the *feature count* for the control asymptomatic speech is the minimum of the *feature counts* for all control speakers.

   (c) **Gender dependent average feature count for controls:** The presence (or absence) of characteristics may be dependent on the gender of the speaker. To calculate divergence scores, it is best if speakers of the same gender are compared. Hence, the *feature count* for the control asymptomatic speech is the average of the *feature count* for all the control speakers of the same gender as the ALS speaker.

   (d) **Gender dependent minimum feature count for controls:** It is assumed the control speaker with the minimum number of characteristics in his/her speech is the most asymptomatic, but the presence (or absence) of characteristics may be gender dependent. To calculate divergence scores, it is best if speakers of the same gender are compared. Hence, the *feature count* for the control asymptomatic speech is the minimum of the *feature count* for all the control speakers of the same gender as the ALS speaker.

2. **Total frequency based divergence score:** For each speaker, an *observed frequency score* is computed as an aggregate of the frequencies of all observed characteristics in the speech of the speaker.[9] The average of the *observed frequency score* of both the annotators for a given speaker is taken as the *frequency score* for the speaker. The *divergence score* for an ALS speaker is the difference between the *frequency score* for the ALS speech and the *frequency score* for the control asymptomatic speech.
   The same four variations of this score are examined as described in the case of *Feature count based divergence score*, depending on how the *frequency score* for the control asymptomatic speech is obtained.

3. **Likert Scale rating based divergence score:** It is assumed that each of the characteristics may be as indicative of the condition as other characteristics in various ALS speakers. It is also assumed that the severity of the characteristics indicates progression of ALS. An *observed Likert score of the speech samples from a speaker* is taken to be an aggregate of the multiples of Likert Scale rating assigned by an annotator for each occurrence of a characteristic with the weight of the characteristic (which is uniformly taken to be 1 for all the characteristics in the current analysis). A *Likert score for a speaker* is calculated as an average of the two annotators' *observed Likert scores of the speech samples from the speaker*. A *Likert Scale rating based divergence score* for each ALS speaker is then taken to be the difference between the *Likert score* for the ALS speech and the *Likert score* for the control asymptomatic speech. The same four variations of this score are examined as described in the case of *Feature count based divergence score*, depending on how the *Likert score* for the control asymptomatic speech is obtained.

For each of the three divergence measures defined above, a higher score indicates that the patients speech diverges from an asymptomatic

---

[8] The counts from the two annotators were combined together in five different ways described in Section 5.1.

[9] For example, if a characteristic, say *distorted vowels*, is observed 6 times, the frequency for *distorted vowels* is 6.

Spkr	%FVC-SUP	%FVC-SIT	Feature count	Total frequency	Likert scale
18	79	88	-0.75	-5	52.5
9	77	78	4.25	7	53.25
1	77	77	0.25	-2.5	37.75
6	75	79	0.5	0.5	29.5
14	66	41	4.5	10.25	83.5
10	53	64	7	13.75	98.75
5	52	56	7.5	22	127.25
4	51	44	8.75	29.5	189.75
8	50	56	5.75	13.75	53.25
17	50	52	4.25	16	150.25
2	38	78	9.25	87	593
11	32	26	6.25	7	82.75
13	29	46	9.25	36	263.46
19	29	40	3.5	1.25	52.5
7	29	29	6.75	36.5	297.25
12	25	26	4.75	12	123.25

Table 3: Physiological Scores (%FVC) and Divergence Scores (D.S.) variant (d) for average of all four utterances for ALS speakers. The feature count was based on a union of features across the two annotators. Total frequency and Likert scale values were computed as a maximum across the two annotators.

speech baseline more than would be indicated by a lower score. Divergence scores are expected to correlate with physiological measures of changes associated with ALS. Increasing divergence scores would thus serve as an indicator of the disease progression, analogous to decreasing physiological outputs (lower scores) associated with ALS— thus, the two measures are expected to be negatively correlated.

5 Results and Discussion

Table 3 presents two physiological assessment scores (%FVC-SUP and %FVC-SIT) and three divergence scores (defined above) for the 16 ALS speakers.[10] The scores are sorted by %FVC-SUP. The table indicates that the %FVC scores tend to drop as the divergence scores go up. As expected, a decrease in %FVC scores indicates disease progression, and similarly, a higher divergence score indicates disease progression.

5.1 Dealing with differences in Annotations to Calculate Divergence Scores

Since the nature of the annotated phonological characteristics was such that multiple characteristics might share various aspects of speech, annotators were asked to mark all characteristics that

---

[10]Only the variant (d) for each of the divergence scores computed using the three methods is presented in the table to maintain clarity. Note variant (d) refers to the divergence scores calculated with **Gender dependent minimum feature count for controls** setting, as described in Section 4 above.

Divergence Score Type	Correlation	p-value
Feature ct D.S.(d)-Union	0.65	0.007
Feature ct D.S.(d)-Avg	0.58	0.017
Feature ct D.S.(d)-Max	0.58	0.018
Feature ct D.S.(d)-Min	0.58	0.019
Feature ct D.S.(d)-Intersection	0.51	0.045
Likert D.S.(a)-Max	0.49	0.055

Table 4: Correlations between the Physiological Scores (%FVC) and Divergence Scores (D.S.) for all four variants

seemed relevant to them. The general descriptions provided in Table 2 were used as heuristics by the annotators, providing additional help in identifying the characteristics.[11] In order to resolve differences across annotations, we used five different methods to combine the two sets of annotations. Table 3 shows a representative combination of the first case below for the feature count measure and the third case below for both total frequency measure and Likert scale measure:

1. *Union:* The characteristics identified by both the annotators were considered only once.
2. *Intersection:* Only the features annotated by both the annotators were considered.
3. *Max:* The maximum of the two annotators' feature counts was used.
4. *Min:* Minimum of the two annotators' feature counts was used.
5. *Avg:* An average of the two annotators' feature counts was used.

5.2 Association between Divergence Scores and Physiological Scores

To determine whether there was an association between any or all of the divergence scores and the physiological measures of ALS, we correlated the divergence scores with the physiological assessment scores, %FVC-SUP and %FVC-SIT, using Pearson's correlation coefficient. The results are presented in Table 4. For simplicity, we report the correlations in the table as $-1 * \langle correlation \rangle$. Refer Section A for correlations with all the divergence scores.

We observe that while divergence scores do not seem to correlate with the %FVC-SIT score, they do show a moderate correlation with the %FVC-

---

[11]Although these descriptions were somewhat coarse-grained, the idea was to start at this level and to learn more precise features associated with acoustic inputs corresponding to these characteristics. These precise features are expected to be critical for automatic classification of speech samples with respect to ALS progression and, correspondingly, predictive of the physiological scores for patients.

SUP score ($0.49 < r < 0.66$) with moderate p-values ($p < 0.05$). The stronger correlation effect we observe with %FVC-SUP than with %FVC-SIT may be due to higher difficulty in breathing that a patient may experience when (s)he is in supine position than in sitting position.

Consistent with the point above, patients with other pulmonary conditions have also been reported to experience higher difficulty in breathing when in supine position than in sitting or standing positions. Since the patients need to exert higher effort to achieve the same result in supine position than in sitting position, they may not be physiologically able to perform the same in the two positions, i.e., %FVC-SUP may be more sensitive than %FVC-SIT to the condition's progression. Since speech symptoms have also been found to be more readily apparent than other physiological symptoms (Yorkston et al., 1993), this results in a stronger correlation of the speech divergence scores with %FVC-SUP than with %FVC-SIT.

The table also indicates that divergence scores based on a simple measure—counts of features observed in ALS speech—correlate even better with %FVC-SUP scores than divergence scores that are based on slightly more complicated measures such as features' frequency or the Likert Scale ratings.

6 Conclusion and Future Work

This paper has presented a case for viewing the relationship between impaired speech and asymptomatic speech as a *divergence* from a baseline. Novel divergence measures have been developed for distinguishing asymptomatic speech from symptomatic speech, and these have been tested for correlations with physiological measures of ALS progression.

These speech divergence measures are a first step toward developing automated speech-based assessments of progression of the ALS condition that are both less expensive and less intrusive than their physiological counterparts. The current approach has enabled the identification speech-based measures that correlate well with other physiological measures currently used to monitor the progression of the ALS condition. The next step is to test if these measures can be used to predict the values for the currently used physiological measures including %FVC.

Also, the current study is based on manual annotations provided by human specialist annotators. Future research will involve exploration of approaches that can be trained to produce such annotations automatically. These could, in turn, be used to calculate divergence scores and eventually to predict values for other physiological measures.

The theoretical groundwork for developing speech-based measures defines speech divergence in terms of clinically-informed phonological speech characteristics associated with ALS symptomatic speech. We presented three methods, with four variants apiece, to compute speech divergence scores for symptomatic speech. We also showed that speech divergence scores are indeed correlated with physiological assessment scores for the progression of the disease.

Future research will investigate other methods to compute divergence between the symptomatic and asymptomatic speech that yield even stronger correlations with the physiological assessments measures. For example, it would be useful to explore whether the proportion of speech that is affected by the characteristics listed in Table 2 has any relation to the progression of the disease. Divergence scores that incorporate characteristics related to a proportion of the span are expected to be strongly correlated with the progression of ALS.

Two possible variants of how one may compute divergence scores based on such proportion-related information are as follows:

(1) Take *proportion* to be the proportion of speech that is affected by any of the characteristics.[12] One may calculate a divergence score for each ALS speaker as the difference between the *proportion* of speech of the ALS speaker affected by these characteristics and the *proportion* of controls' speech affected by these characteristics. An average of the *proportion* in the annotators' annotations may be used for the calculation of the divergence score.

(2) As a simple analytic, one may also consider proportion-based divergence scores corresponding to each of the characteristics for each ALS speaker. This analytic may be useful for providing a direct relation between a specific characteristic and the progression of the condition. However, it may also be useful to explore divergence classes based on groupings of characteristics that are similarly affected due to the progression of the condition, if any.[13]

---

[12]Note there may be overlapping spans for more than one characteristics.

[13]For the calculation of this variant, an average across the portions of speech for which a characteristic is annotated may

Some characteristics may be grouped to further explore divergences. Green et al. (2013) grouped features according to the speech subsystem involved (e.g., respiratory, phonatory, resonatory and articulatory). A reviewer also mentioned that gender-specific degree of severity of certain features would be interesting to explore. For example, there seems to be evidence that voicing control is more vulnerable in male patients (Kent et al., 1994). Such findings suggest that characteristics such as gender and possibly age may also need to be considered while developing speech divergence-based measures.

In addition, for the current study, each of the characteristics was treated uniformly with respect to ALS. Future work will explore the hypothesis that certain characteristics are more indicative than others with respect to the progression of ALS.

Finally, while prior studies indicate that prosodic recognition is not affected in ALS speakers (Zimmerman et al., 2007), articulatory or phonatory deficits might alter the correct production of interrogative, imperative, or declarative sentences (Congia et al., 1987). These may be found to be useful in the development of speech-based measures of ALS. Thus, future work will investigate the extent to which these variables would be more or less difficult to analyze automatically.

Acknowledgments

This work is supported in part by the VA Office of Rural Health Program "Increasing access to pulmonary function testing for rural veterans with ALS through at home testing" (N08-FY15Q1-S1-P01346) and Award Number 1I21RX001902 titled "DESIPHER: Speech Degradation as an Indicator of Physiological Degeneration in ALS." Additionally, the work is supported by resources and the use of facilities at the James A. Haley Veterans' Hospital. The contents of this paper represents solely the views of the authors and not the views of the Department of Veterans Affairs or the United States Government. The authors are also grateful for constructive feedback from anonymous reviewers. An additional thank you is extended to the two specialists for their time and expertise in providing us with annotations of the phonological features in ALS speech that enabled us to conduct the analyses reported herein.

---

be used to determine the *proportion* of speech for a specific speaker that is affected by the characteristics.

References

L.J. Ball, A. Willis, D.R. Beukelman, and G.L. Pattee. 2001. A protocol for identification of early bulbar signs in als. *Journal of the Neurological Sciences* 191:43–53.

D. Beukelman, S. Fager, and A. Nordness. 2011. Communication support for people with ALS. *Neurology Research International* .

F. Biadsy, W.Y. Wang, and A. Rosenberg. 2011. Intoxication detection using phonetic, phonotactic and prosodic cues. In *Proceedings of INTERSPEECH*.

T. Bocklet, E. Nöth, G. Stemmer, H. Ruzickova, and J. Rusz. 2011. Detection of persons with Parkinson's disease by acoustic, vocal, and prosodic analysis. In *Proceedings of ASRU*.

P. Boersma and V. van Heuven. 2001. Praat, a system for doing phonetics by computer. *Glot International* 5(9/10):341–345.

J.R. Brinkmann, P. Andres, M. Mendoza, and M. Sanjak. 1997. Guidelines for the use and performance of quantitative outcome measures in als clinical trials. *Journal of the Neurological Sciences* 147(1):97–111.

B.R. Brooks. 1997. Amyotrophic lateral sclerosis clinimetric scales guidelines for administration and scoring. In Herndon R., editor, *Handbook of clinical neurologic scales*, Demos Vermande, pages 27–80.

E. Carrow, V. Rivera, M. Mauldin, and Shamblin L. 1974. Deviant speech characteristic in motor neuron disease. *Archives of Otolaryngology - Head and Neck Surgery Journal* 100:212–218.

A.J. Caruso and E.K. Burton. 1987. Temporal acoustic measures of dysarthria associated with amyotrophic lateral sclerosis. *Journal of Speech, Language, and Hearing Research* 30(1):80–87.

J.M. Cedarbaum and N. Stambler. 1997. Performance of the amyotrophic lateral sclerosis functional rating scale (ALSFRS) in multicenter clinical trials. *Journal of Neurological Sciences* 152(Suppl 1):S1–S9.

J.M. Cedarbaum, N. Stambler, E. Malta, C. Fuller, D. Hilt, B. Thurmond, and A. Nakanishi. 1999. The ALSFRS-R: a revised ALS functional rating scale that incorporates assessments of respiratory function. *Journal of the Neurological Sciences* 169(1-2):13–21.

S. Congia, S. Di Ninni, F. Zonza, and M. Melis. 1987. Prosody alterations in amyotrophic lateral sclerosis. In *Amyotrophic Lateral Sclerosis*, Springer US, pages 121–123.

A. Czaplinski, A.A. Yen, and S.H. Appel. 2006. Forced vital capacity (fvc) as an indicator of survival and disease progression in an als clinic population. *Journal of Neurology, Neurosurgery, and Psychiatry* 77(3):390–392.

B.J. Dorr. 1994. Machine translation divergences: A formal description and proposed solution. *Computational Linguistics* 20(4):597–633.

B.J. Dorr, L. Pearl, R. Hwa, and N. Habash. 2002. DUSTer: A method for unraveling cross-language divergences for statistical word-level alignment. In *Proceedings of the Fifth Conference of the Association for Machine Translation in the Americas (AMTA)*. pages 31–43.

J. Duffy. 2013. *Motor Speech Disorders: Substrates, Differential Diagnosis, and Management*. Elsevier Health Sciences, 3rd edition.

K.C. Fraser, G. Hirst, N.L. Graham, J.A. Meltzer, S.E. Black, and E. Rochon. 2014. Comparison of different feature sets for identification of variants in progressive aphasia. In *Proceedings of the ACL CLPsych Workshop*. pages 17–26.

J. Garofolo, L. Lamel, W. Fisher, J. Fiscus, D. Pallett, N. Dahlgren, and V. Zue. 1993. TIMIT acoustic-phonetic continuous speech corpus. `http://web.mit.edu/course/6/6.863/share/nltk_lite/timit/allsenlist.txt`.

K. Gorman, L. Olson, A.P. Hill, R. Lunsford, P.A. Heeman, and J.P.H. van Santen. 2016. Uh and um in children with autism spectrum disorders or language impairment. *Autism Research* 9(8):854–865.

J.R. Green, Y. Yunusova, M.S. Kuruvilla, J. Wang, G.L. Pattee, L. Synhorst, L. Zinman, and J.D. Berry. 2013. Bulbar and speech motor assessment in ALS: Challenges and future directions. *Amyotrophic Lateral Sclerosis and Frontotemporal Degeneration* 14:494–500.

N. Habash and B.J. Dorr. 2002. Handling translation divergences: Combining statistical and symbolic techniques in generation-heavy machine translation. In *Proceedings of the Fifth Conference of the Association for Machine Translation in the Americas, AMTA-2002*. pages 84–93.

R.D. Kent, J.F. Kent, G. Weismer, R.L. Sufit, J.C. Rosenbek, R.E. Martin, and B.R. Brooks. 1990. Impairment of speech intelligibility in men with amyotrophic lateral sclerosis. *Journal of Speech and Hearing Disorders* 55:721–728.

R.D. Kent, H. Kim, G. Weismer, J.F. Kent, J.C. Rosenbek, B.R. Brooks, and M. Workinger. 1994. Laryngeal dysfunction in neurological disease: Amyotrophic lateral sclerosis, parkinson disease, and stroke. *Journal of Medical Speech-Language Pathology* 2(3).

R. Likert. 1932. A technique for the measurement of attitudes. *Archives of Psychology* 140:1–55.

E. Nöth, H. Niemann, T. Haderlein, M. Decher, U. Eysholgt, F. Rosanowski, and T. Wittenberg. 2000. Automatic stuttering recognition using hidden Markov models. In *Proceedings of ICSLP*.

M.B. Olsen, B.J. Dorr, and S. Thomas. 1998. Enhancing automatic acquisition of thematic structure in a large-scale lexicon for mandarin chinese. In *Proceedings of the Fifth Conference of the Association for Machine Translation in the Americas, AMTA-2002*. Langhorne, PA, pages 41–50.

S.O. Orimaye, J. Wong, and K.J. Golden. 2014. Learning predictive linguistic features for Alzheimer's disease and related dementias using verbal utterances. In *Proceedings of the ACL Workshop on Computational Linguistics and Clinical Psychology: From Linguistic Signal to Clinical Reality*. pages 78–87.

P. Rong, Y. Yunusova, J. Wang, and J.R. Green. 2015. Predicting early bulbar decline in amyotrophic lateral sclerosis: A speech subsystem approach. *Behavioral Neurology* .

P. Rong, Y. Yunusova, J. Wang, L. Zinman, G.L. Pattee, J.D. Berry, B. Perry, and J.R. Green. 2016. Predicting speech intelligibility decline in amyotrophic lateral sclerosis based on the deterioration of individual speech subsystems. *PLoS ONE* 11(5).

J.C. Rosenbek, J.A. Robbins, E.B. Roecker, J.L. Coyle, and J.L. Wood. 1996. A penetration-aspiration scale. *Dysphagia* 11(2):93–98.

F. Rudzicz, L. Chan Currie, A. Danks, T. Mehta, and S. Zhao. 2014. Automatically identifying trouble-indicating speech behaviors in Alzheimer's disease. In *Proceedings of ASSETS*. ACM, pages 241–242.

B. Schuller, S. Steidl, A. Batliner, S. Hankte, F. Hönig, J.R. Orozco-Arroyave, E. Nöth, Y. Zhang, and F. Weninger. 2015. The interspeech 2015 computational paralinguistics challenge: nativeness, parkinson's & eating condition.

B. Schuller, S. Steidl, A. Batliner, F. Schiel, and J. Krajewski. 2011. The INTERSPEECH 2011 speaker state challenge. In *Proceedings of INTERSPEECH*. pages 3201–3204.

A.K. Silbergleit, A.F. Johnson, and B.H. Jacobson. 1997. Acousticanalysis of voice in individuals with amyotrophic lateral sclerosis and perceptually normal vocal quality. *Journal of Voice* 11:222–231.

G.S. Turner, K. Tjaden, and G. Weismer. 1995. The influence of speaking rate on vowel space and speech intelligibility for individuals with amyotrophic lateral sclerosis. *Journal of Speech, Language, and Hearing Research* 38:1001–1013.

G. Weismer, J.-Y. Jeng, J.S. Laures, R.D. Kent, and J.F. Kent. 2001. Acoustic and intelligibility characteristics of sentence production in neurogenic speech disorders. *Folia Phoniatr Logop* 53:1–18.

K. Yorkston, D. Beukelman, M. Hakel, and M. Dorsey. 2007. Sentence intelligibility test, speech intelligibility test.

K.M. Yorkston, E. Strand, R. Miller, A. Hillel, and K. Smith. 1993. Speech deterioration in amyotrophic lateral sclerosis: implications for the timing of intervention. *Journal of medical speech-language pathology* 1:35–46.

Y. Yunusova, N.L. Graham, S. Shellikeri, K. Phuong, M. Kulkarni, E. Rochon, D.F. Tang-Wai, T.W. Chow, S.E. Black, L.H. Zinman, and J.R. Green. 2016. Profiling speech and pausing in amyotrophic lateral sclerosis (ALS) and frontotemporal dementia (FTD). *PLoS ONE* 11(1).

E.K. Zimmerman, P.J. Eslinger, Z. Simmons, and A.M. Barrett. 2007. Emotional perception deficits in amyotrophic lateral sclerosis. *Cognitive and Behavioral Neurology* 20(2):70–82.

A Supplemental Material

The correlation results between each of the three types of speech divergence scores with all of their variants and the %FVC-SUP are presented in Tables 5, 6, and 7. As mentioned before, for simplicity, we report the correlations in the tables as $-1 * \langle correlation \rangle$.

Divergence Score Type	Correlation	p-value
Feature ct D.S.(a)-Max	0.58	0.018
Feature ct D.S.(a)-Min	0.58	0.019
Feature ct D.S.(a)-Ave	0.58	0.017
Feature ct D.S.(a)-Union	0.65	0.007
Feature ct D.S.(a)-Intersection	0.51	0.045
Feature ct D.S.(b)-Max	0.58	0.018
Feature ct D.S.(b)-Min	0.58	0.019
Feature ct D.S.(b)-Ave	0.58	0.017
Feature ct D.S.(b)-Union	0.65	0.007
Feature ct D.S.(b)-Intersection	0.51	0.045
Feature ct D.S.(c)-Max	0.58	0.018
Feature ct D.S.(c)-Min	0.58	0.019
Feature ct D.S.(c)-Ave	0.58	0.017
Feature ct D.S.(c)-Union	0.64	0.007
Feature ct D.S.(c)-Intersection	0.50	0.045
Feature ct D.S.(d)-Max	0.58	0.018
Feature ct D.S.(d)-Min	0.58	0.019
Feature ct D.S.(d)-Avg	0.58	0.017
Feature ct D.S.(d)-Union	0.65	0.007
Feature ct D.S.(d)-Intersection	0.51	0.045

Table 5: Correlations between the Physiological Scores (%FVC) and Feature Count Based Divergence Scores (D.S.) for all four variants

Divergence Score Type	Correlation	p-value
Feature freq D.S.(a)-Max	0.45	0.077
Feature freq D.S.(a)-Min	0.42	0.103
Feature freq D.S.(a)-Ave	0.44	0.085
Feature freq D.S.(b)-Max	0.45	0.077
Feature freq D.S.(b)-Min	0.42	0.103
Feature freq D.S.(b)-Ave	0.44	0.085
Feature freq D.S.(c)-Max	0.45	0.083
Feature freq D.S.(c)-Min	0.42	0.106
Feature freq D.S.(c)-Ave	0.44	0.09
Feature freq D.S.(d)-Max	0.46	0.075
Feature freq D.S.(d)-Min	0.42	0.103
Feature freq D.S.(d)-Avg	0.45	0.083

Table 6: Correlations between the Physiological Scores (%FVC) and Feature Frequency Based Divergence Scores (D.S.) for all four variants

Divergence Score Type	Correlation	p-value
Likert Scale D.S.(a)-Max	0.49	0.055
Likert Scale D.S.(a)-Min	0.44	0.089
Likert Scale D.S.(a)-Ave	0.47	0.068
Likert Scale D.S.(b)-Max	0.48	0.061
Likert Scale D.S.(b)-Min	0.43	0.095
Likert Scale D.S.(b)-Ave	0.46	0.074
Likert Scale D.S.(c)-Max	0.48	0.059
Likert Scale D.S.(c)-Min	0.43	0.094
Likert Scale D.S.(c)-Ave	0.46	0.071
Likert Scale D.S.(d)-Max	0.47	0.067
Likert Scale D.S.(d)-Min	0.43	0.094
Likert Scale D.S.(d)-Avg	0.46	0.076

Table 7: Correlations between the Physiological Scores (%FVC) and Likert Scale Based Divergence Scores (D.S.) for all four variants

Deep Learning for Punctuation Restoration in Medical Reports

Wael Salloum, Greg Finley, Erik Edwards, Mark Miller, and David Suendermann-Oeft
EMR.AI Inc
90 New Montgomery St #400
San Francisco, CA 94105, USA
`david@emr.ai`

Abstract

In clinical dictation, speakers try to be as concise as possible to save time, often resulting in utterances without explicit punctuation commands. Since the end product of a dictated report, e.g. an out-patient letter, does require correct orthography, including exact punctuation, the latter need to be restored, preferably by automated means. This paper describes a method for punctuation restoration based on a state-of-the-art stack of NLP and machine learning techniques including B-RNNs with an attention mechanism and late fusion, as well as a feature extraction technique tailored to the processing of medical terminology using a novel vocabulary reduction model. To the best of our knowledge, the resulting performance is superior to that reported in prior art on similar tasks.

1 Introduction

Medical dictation has been a major instrument in clinical settings to minimize the administrative burden on physicians (Johnson et al., 2014; Hammana et al., 2015; Hodgson and Coiera, 2016). Rather than having to fill forms in electronic medical record systems (EMRs) or typing out-patient letters, such labor is often outsourced to medical transcription providers, many of which make use of automated speech recognition (ASR), coupled with a manual correction step, to increase effectiveness and speed of transcription (Salloum et al., 2017). Despite the fact that medical dictation reduces time physicians spend on clinical documentation substantially, an average dictation still takes about three minutes (Edwards et al., 2017). In an attempt to dictate as efficiently as possible, often physicians (a) speak extremely fast, (b) use pre-

dictated paragraphs (so-called *physician normals*), (c) make massive use of abbreviations, and (d) include very limited (if any) instructions regarding formatting and punctuation.

While the ASR system is in charge of turning spoken words into their textual representation, a sophisticated NLP unit, the post-processor, takes care of formatting and structuring the output to produce a draft resembling the out-patient letter as well as possible. Among other responsibilities (such as formatting numerical expressions, dates, section headers, etc.), the post-processor is also charged with restoring punctuation in the letter's narrative. This paper focuses on the automated punctuation restoration in clinical reports, drawing on the latest advances in the NLP sector.

To achieve best possible results in this study, we paid particular attention to the specific challenges faced in medical texts. Foremost among these is a large domain-specific vocabulary, which makes it difficult if not impossible to apply tools developed for general-domain text. When building a system from scratch, however, several factors conspire to make it hard to obtain enough training data: the large medical vocabulary increases problems related to data sparsity and the handling of out-of-vocabulary (OOV) terms; the data often contain sensitive information and have restricted access or availability; and modern methods, such as neural networks as used here, typically require large amounts of data.

We overcame these issues by developing a text pre-processing strategy to reduce vocabulary size, collapsing particular roots and exploiting the fact that many medical terms are built from relatively few morphemes. Our method, which we call the *vocabulary reduction model*, effectively allows the punctuation restoration neural network to focus on morphosyntactic features of words rather than their full semantic representation, as usually cap-

Proceedings of the BioNLP 2017 workshop, pages 159–164,
Vancouver, Canada, August 4, 2017. ©2017 Association for Computational Linguistics

Set	Normalized Text			Reduced Text				
	types	OOVs	tokens	types	tokens	PERIOD	COLON	COMMA
Training	57,046	n/a	15,886,158	11,766	15,933,901	1,803,626	631,452	760,444
Dev	28,509	1,561	2,243,187	10,321	2,248,305	268,374	89,647	111,571
Blind Test	31,806	3,108	2,944,787	10,767	2,952,873	325,549	103,693	127,895

Table 1: Corpus statistics after normalization and vocabulary reduction. No OOVs are reported on the reduced text since the vocabulary reduction algorithm will map OOVs to classes. The last three columns show the counts of each punctuation tag per set.

tured by word embeddings, being less important to the placement of punctuation.

After reviewing the prior art in the field of punctuation restoration in Section 2, we describe the corpus used in this study in Section 3. The system's general architecture based on bidirectional recurrent neural networks with attention mechanism and late fusion is discussed in Section 4, followed by Section 5 providing details on the vocabulary reduction model. Evaluation results are covered in Section 6, and conclusion and future outlook in Section 7.

2 Related Work

Early efforts in this field used hidden-event n-gram language modeling to predict where punctuation should be inserted (Stolcke et al., 1998; Beeferman et al., 1998). Numerous other strategies have also been devised: combining n-grams with constituency parse information (Shieber and Tao, 2003); maximum entropy using n-gram and part-of-speech features (Huang and Zweig, 2002); conditional random fields (CRFs) (Ueffing et al., 2013); feed-forward neural networks and CRFs on n-gram and lexical features (Cho et al., 2015); even reframing the problem as monolingual machine translation (Peitz et al., 2011).

Most recently, it has been demonstrated that recurrent neural networks can restore punctuation very effectively (Tilk and Alumäe, 2015, 2016). Such methods are promising because they should be able to handle long-distance dependencies that are missed by other methods.

There has been little work on punctuation restoration in the medical domain specifically. While using pauses showed to help in punctuation restoration for rehearsed speech such as TED Talks (Tilk and Alumäe, 2016), Deoras and Fritsch (2008) note that medical dictations pose a particular challenge because the speech is often delivered rapidly and without typical prosodic cues, such as

pauses where one would write commas or other punctuation. Thus, although acoustic information has been successfully incorporated for other domains (Huang and Zweig, 2002; Christensen et al., 2001), the same may not be feasible for medical text, so it is especially desirable to have a reliable text-only method.

3 Corpus

The corpus we are using in this study is composed of 32,275 medical reports (i.e., out-patient letters), which we converted into a sequence of tokens with punctuation as tags (since they are the most relevant to medical dictations, we focused on three punctuation marks: colon, comma, and period, represented in the tag set {COLON, COMMA, PERIOD}). We randomly split our corpus into training set, development set, and blind test set. Detailed corpus statistics are given in Table 1.

To reduce the size of the vocabulary, we performed two layers of text preprocessing. First, we performed several text normalization steps such as converting all digits to "D", normalizing numbers, dates, and times into familiar formats (e.g., "D.D", "DD/DDDD", "DD/DD", "DD/D-D/DDDD", "DD:DD"), as well as other tokens of the medical domain into normalized formats (e.g., "DDD/DD" for blood pressure, "lD-lD" for lumbar spinal discs, and "q.D+h" meaning "every D+ hours"). Normalization also included lowercasing, unifying abbreviations (e.g., "p.r.n" and "p.r.n."), and performing simple segmentation (e.g., splitting "'s" from a word). Second, we ran a vocabulary reduction algorithm, as detailed in Section 5, that maps infrequent and OOV words to word classes. The combination of these two layers dramatically reduced the vocabulary size, as shown in Table 1.

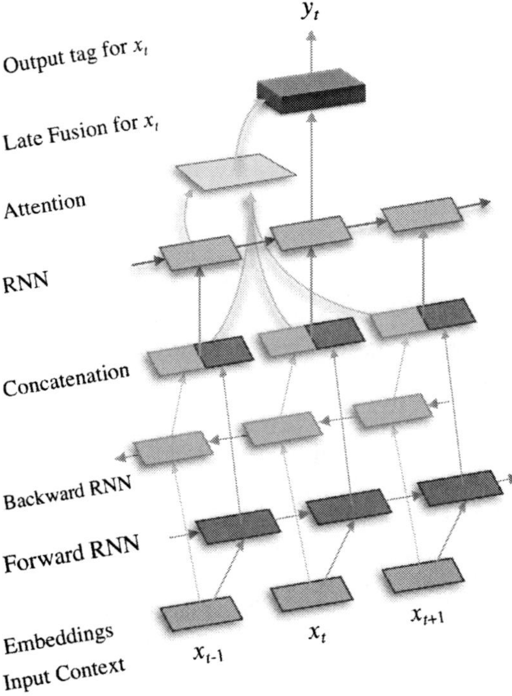

Figure 1: Neural network design for punctuation restoration. The diagram shows an input context for the word x_t and the stack of layers that result in the tag y_t representing the punctuation decision for x_t.

4 The Neural Network Model

We define punctuation restoration as a tagging problem. We try to tag every word in the input sequence with one of four tags: {NONE, COLON, COMMA, PERIOD}. Tagging a word by a punctuation means that the punctuation should be inserted after this word, while tagging with NONE means that the word does not have a punctuation after it. Our neural network approach is based on the work of Tilk and Alumäe (2016). Inspired by Bahdanau et al. (2016), our deep neural network model uses a bidirectional recurrent neural network (B-RNN) (Schuster and Paliwal, 1997) with gated recurrent units (Cho et al., 2014). B-RNNs help in learning long range dependencies on the left and right of the current input word. The B-RNN is composed of a forward RNN and a backward RNN that are preceded by the same word embedding layer. A sliding window of 256 words are passed to the shared embedding layer as one-hot vectors.

On top of the B-RNN, we stack a unidirec-

tional RNN with an attention mechanism (Bahdanau et al., 2016) to assist in capturing relevant contexts that support punctuation restoration decisions. Finally, we use late fusion (Wang and Cho, 2015) to combine the output of the attention mechanism with the current position in the B-RNN without interfering with its memory.

5 The Vocabulary Reduction Model

To improve the modeling of rare words and to deal with OOV words in the test and development sets, we implemented a step that maps rare words to common word classes, reducing the overall size of the vocabulary. This vocabulary reduction allows us to reduce the number of model parameters, which is crucial for fast decoding in a live recognizer.

Table 2 shows examples of prefixes and suffixes that capture the semantic and morpho-syntactic information of infrequent words in our training data such as medical terminology and proper names. For every input word consisting of alphabetical characters only, our vocabulary reduction algorithm goes through the prefix and suffix lists starting from the longer affixes to the shorter ones and tries to match them to the beginning or end of the word, while ensuring that the stem is at least four letters long. If the word starts with a prefix $p+$ of the prefix list we replace it with "pAAAA" (where "AAAA" represents an alphabetical stem). If it starts with a suffix $+q$, we replace it with "AAAAq". Finally, if the word matches a prefix $p+$ and a suffix $+q$, we split it into two tokens "pAA+" and "+AAq", respectively, to ensure that the information in them gets modeled separately. Every rare word consisting of alphabetical characters only that does not match any suffix or prefix is replaced with a token that represents its length range. The length range is computed with a step of five characters resulting in tokens like AAAA_5 for words shorter than five characters, AAAA_10 for words shorter than ten characters, etc. For example, "angiotensinconvertingenzyme" is replaced with AAAA_30. All other rare words (e.g., "t1cn0m0") are replaced with the token "RARE". These handcrafted rare classes allow us to increase the threshold for considering a word rare. This technique not only significantly reduces the size of the vocabulary, but also allows us to better model rare classes with a higher number of tokens.

Size	Prefix	Suffix
4	inte+, anti+, post+, tran+, over+, intr+, peri+, hype+, para+, neur+, hypo+, micr+, rein+, mult+, card+, comp+, retr+, reco+, self+, gran+, extr+, medi+, hemi+, well+, semi+, endo+, radi+, hemo+, fibr+, oste+, elec+	+tion, +ions, +type, +ness, +ized, +date, +able, +gery, +tive, +sult, +tomy, +ated, +tory, +sion, +ates, +ular, +ical, +osis, +ment, +nary, +rate, +ings, +arge, +onal, +itis, +ents, +like, +lity, +ance, +berg
3	non+, pre+, per+, pro+, mar+, sub+, sch+, str+, tri+, ben+	+ing, +ion, +ted, +ate, +lly, +ive, +tic, +ers, +ble, +ies, +ity, +cal, +man, +sis, +son, +ial, +ous, +ell, +ary, +lar, +tes, +ton, +dez
2	re+, de+, mc+, un+, le+, la+, vi+	+ed, +er, +es, +al, +ry, +te, +ic, +ly, +le

Table 2: Examples of affixes of medical terminology and proper names that capture the semantic and/or morpho-syntactic information of infrequent words in our training data.

Punctuation	Precision	Recall	F-Score
COLON	98.6%	98.6%	98.6%
COMMA	84.0%	82.2%	83.1%
PERIOD	96.1%	96.4%	96.3%
Overall	94.2%	94.0%	94.1%

Table 3: Evaluation of punctuation restoration performance on the blind test set.

We replace a word with its rare class whenever we find it 20 or fewer times in the training data, and we perform the affix-based replacement described above whenever the word occurred less than 100 times. These thresholds were tuned on a held-out development set. Running this algorithm on top of the normalized text results in lowering the vocabulary size in our training data to 11,766 types, meaning that four out of five types are replaced with a class.

6   Evaluation

For the present study, we used Keras with Tensor-Flow backend (Chollet, 2015; Abadi et al., 2016; Chollet, 2017). We evaluated on the blind test set by passing the whole set to our system as a sequence of about three million tokens without any indication of beginning or end of sentence, paragraph, or report. All words were lowercased, as described earlier, to avoid giving out any hint of sentence or section header start or end. We report the results in Table 3.

We achieve 96.3% F-Score on periods, which we consider the most important as they define sentence boundaries. The latter are crucial for virtually any subsequent NLP process, such as automatic coding of medical reports (Suendermann-Oeft et al., 2016).

The second most important punctuation type in medical reports is colons, as they define section headers and, thus, help format the report structure. We achieve 98.6% F-Score on colons.

Finally, we get 83.1% F-Score on commas, the hardest tag to predict due to human inconsistency in using them. This inconsistency affects the accuracy of the training data as well as the fairness in the evaluation against the test set. The overall performance of the system on all tags is 94.1% in terms of F-Score. Refer to Table 4 for examples of our system's output.

7   Conclusion and Future Work

Although prior work on punctuation restoration has used different corpora from the work presented in this paper, our result (F-Score 94.1%) compares very favorably with previous publications. For example, Cho et al. (2015) achieve an F-Score of 61.8% on a meeting and lecture corpus, Tilk and Alumäe (2016) produce 64.4% on TED talk transcripts, and Ueffing et al. (2013) report an F-Score of 66.8% on one of Nuance's in-house dictation corpora.

While we have tested the performance of the presented punctuation restoration algorithm on naturalistic medical dictations, we have not yet measured the impact the speech recognizer's word error rate has on the F-Score, a task we plan to address in the near future. We are also interested to learn whether analyzing the speech waveform and characteristic pauses and prosodic patterns in medical dictations can be exploited in a hybrid speech/text punctuation restoration system to enhance accuracy even further. We also plan to replace the vocabulary reduction model by fusing a morphology-aware neural network such as a

Input	... review of systems general positive for fatigue excessive perspiration feeling sick ...
Gold	... review of systems: general: positive for fatigue, excessive perspiration, feeling sick. ...
Punctuated	... review of systems COLON general COLON positive for fatigue COMMA excessive perAA+ +AAtion COMMA feeling sick PERIOD ...
Input	... chronic pruritus dermatology felt that this was neurodermatosis and neurotic excoriations ...
Gold	... chronic pruritus. dermatology felt that this was neurodermatosis and neurotic excoriations. ...
Punctuated	... chronic pruritus PERIOD deAAAA felt that this was neurAA+ +AAosis and neurAA+ +AAtic AAAAions PERIOD ...
Input	... it is available review of systems positive for still some ongoing lower extremity weakness tremulousness and unsteadiness otherwise review of ...
Gold	... it is available. review of systems: positive for still some ongoing lower extremity weakness, tremulousness and unsteadiness. otherwise, review of ...
Punctuated	... it is available PERIOD review of systems COLON positive for still some ongoing lower extremity weakness COMMA AAAAness and unAA+ +AAness PERIOD otherwise COMMA review of ...
Input	... severe clinical depression including hopelessness helplessness worthlessness difficulty focusing concentration and a lot of thoughts of death and dying ...
Gold	... severe clinical depression including hopelessness, helplessness, worthlessness, difficulty focusing, concentration, and a lot of thoughts of death and dying. ...
Punctuated	... severe clinical depression including AAAAness COMMA AAAAness COMMA AAAAness COMMA difficulty AAAAing COMMA concentration COMMA and a lot of thoughts of death and dying PERIOD ...
Input	... is reasonable we will optimize his medications by adding low dose angiotensinconvertingenzyme inhibitors which he currently is not on if the ...
Gold	... is reasonable. we will optimize his medications by adding low dose angiotensinconvertingenzyme inhibitors, which he currently is not on. if the ...
Punctuated	... is reasonable PERIOD we will optimize his medications by adding low dose AAAA_30 inhibitors COMMA which he currently is not on PERIOD if the ...

Table 4: Examples of the output of our system on word sequences of the input. The first example shows the correct handling of consecutive colons indicating a section header and a subsection header. The second example shows the preprocessing of infrequent medical terminology like "neurodermatosis", "neurotic", and "excoriations" by capturing their semantic and part-of-speech information. The third and fourth examples emphasize the case of parallelism captured by mapping "tremulousness and unsteadiness" to "AAAAness and unAA+ +AAness" and "hopelessness helplessness worthlessness" to "AAAAness AAAAness AAAAness", thus predicting commas when needed since the meaning is irrelevant to the punctuation task. The fourth example also shows the correct prediction of coordinated lists, separating them with commas. The final example presents the mapping of a very long word, "angiotensinconvertingenzyme", into "AAAA_30", which reduces the confusion of the network and results in the correct prediction.

character-based convolutional network.

References

M Abadi, A Agarwal, P Barham, E Brevdo, Z Chen, C Citro, GS Corrado, A Davis, J Dean, M Devin, S Ghemawat, I Goodfellow, A Harp, G Irving, M Isard, Y Jia, R Jozefowicz, L Kaiser, M Kudlur, J Levenberg, D Mané, R Monga, S Moore, D Murray, C Olah, M Schuster, J Shlens, B Steiner, I Sutskever, K Talwar, P Tucker, V Vanhoucke, V Vasudevan, F Viégas, O Vinyals, P Warden, M Wattenberg, M Wicke, Y Yu, and X Zheng. 2016. Tensorflow: large-scale machine learning on heterogeneous distributed systems. *arXiv* 1603(04467):1–19.

D Bahdanau, K Cho, and Y Bengio. 2016. Neural machine translation by jointly learning to align and translate [conference paper at iclr 2015]. *arXiv* 1409(0473):1–15.

D Beeferman, A Berger, and JD Lafferty. 1998. Cyberpunc: a lightweight punctuation annotation system for speech. In *Proc ICASSP*. IEEE, volume 2, pages 689–692.

E Cho, K Kilgour, J Niehues, and A Waibel. 2015. Combination of nn and crf models for joint detection of punctuation and disfluencies. In *Proc Interspeech*. ISCA, pages 3650–3654.

K Cho, B van Merriënboer, D Bahdanau, and Y Bengio. 2014. On the properties of neural machine translation: encoder-decoder approaches. *arXiv* 1409(1259):1–9.

F Chollet. 2015. Keras: deep learning library for theano and tensorflow. https://keras.io/.

F Chollet. 2017. *Deep learning with Python*. Manning, Shelter Island, NY.

H Christensen, Y Gotoh, and S Renals. 2001. Punctuation annotation using statistical prosody models. In *Proc ITRW on Prosody in Speech Recognition and Understanding*. ISCA, paper 6, pages 1–6.

A Deoras and J Fritsch. 2008. Decoding-time prediction of non-verbalized punctuation. In *Proc Interspeech*. ISCA, pages 1449–1452.

E Edwards, W Salloum, GP Finley, J Fone, G Cardiff, M Miller, and D Suendermann-Oeft. 2017. Medical speech recognition: reaching parity with humans. In *Proc SPECOM*. Springer, pages 1–10.

I Hammana, L Lepanto, T Poder, C Bellemare, and M-S Ly. 2015. Speech recognition in the radiology department: a systematic review. *HIM J* 44(2):4–10.

T Hodgson and EW Coiera. 2016. Risks and benefits of speech recognition for clinical documentation: a systematic review. *J Am Med Inform Assoc* 23(e1):169–179.

J Huang and G Zweig. 2002. Maximum entropy model for punctuation annotation from speech. In *Proc ICSLP*. ISCA, pages 917–920.

M Johnson, S Lapkin, V Long, P Sanchez, H Suominen, J Basilakis, and L Dawson. 2014. A systematic review of speech recognition technology in health care. *BMC Med Inform Decis Mak* 14(94):1–14.

S Peitz, M Freitag, A Mauser, and H Ney. 2011. Modeling punctuation prediction as machine translation. In *Proc IWSLT*. pages 238–245.

W Salloum, E Edwards, S Ghaffarzadegan, D Suendermann-Oeft, and M Miller. 2017. Crowdsourced continuous improvement of medical speech recognition. In *Proc AAAI Wrkshp Crowdsourcing*. AAAI, San Francisco, CA.

M Schuster and KK Paliwal. 1997. Bidirectional recurrent neural networks. *IEEE Trans Signal Process* 45(11):2673–2681.

SM Shieber and X Tao. 2003. Comma restoration using constituency information. In *Proc HLT-NAACL*. ACL, volume 1, pages 142–148.

A Stolcke, E Shriberg, R Bates, M Ostendorf, D Hakkani, M Plauche, G Tür, and Y Lu. 1998. Automatic detection of sentence boundaries and disfluencies based on recognized words. In *Proc ICSLP*. ISCA, paper 0059, pages 1–4.

D Suendermann-Oeft, S Ghaffarzadegan, E Edwards, W Salloum, and M Miller. 2016. A system for automated extraction of clinical standard codes in spoken medical reports. In *Proc Wrkshp SLT*. IEEE, San Diego, CA.

O Tilk and T Alumäe. 2015. Lstm for punctuation restoration in speech transcripts. In *Proc Interspeech*. ISCA, pages 683–687.

O Tilk and T Alumäe. 2016. Bidirectional recurrent neural network with attention mechanism for punctuation restoration. In *Proc Interspeech*. ISCA, pages 3047–3051.

N Ueffing, M Bisani, and P Vozila. 2013. Improved models for automatic punctuation prediction for spoken and written text. In *Proc Interspeech*. ISCA, pages 3097–3101.

T Wang and K Cho. 2015. Larger-context language modelling. *arXiv* 1511(03729):1–14.

Unsupervised Domain Adaptation for Clinical Negation Detection

Timothy A. Miller[1], Steven Bethard[2], Hadi Amiri[1], Guergana Savova[1]
[1] Boston Children's Hospital Informatics Program, Harvard Medical School
`{firstname.lastname}@childrens.harvard.edu`
[2] School of Information, University of Arizona
`bethard@email.arizona.edu`

Abstract

Detecting negated concepts in clinical texts is an important part of NLP information extraction systems. However, generalizability of negation systems is lacking, as cross-domain experiments suffer dramatic performance losses. We examine the performance of multiple unsupervised domain adaptation algorithms on clinical negation detection, finding only modest gains that fall well short of in-domain performance.

1 Introduction

Natural language processing applied to health-related texts, including clinical reports, can be valuable for extracting information that does not exist in any other form. One important NLP task for clinical texts is concept extraction and normalization, where text spans representing medical concepts are found (e.g., *colon cancer*) and mapped to controlled vocabularies such as the Unified Medical Language System (UMLS) (Bodenreider and Mc-Cray, 2003). However, clinical texts often refer to concepts that are explicitly not present in the patient, for example, to document the process of ruling out a diagnosis. These *negated* concepts, if not correctly recognized and extracted, can cause problems in downstream use cases. For example, in phenotyping, a concept for a disease (e.g., *asthma*) is a strong feature for a classifier finding patients with asthma. But if the text *ruled out asthma* occurs and the negation is not detected, this text will give the exact opposite signal that its inclusion intended.

There exist many systems for negation detection in the clinical domain (Chapman et al., 2001, 2007; Harkema et al., 2009; Sohn et al., 2012; Wu et al., 2014; Mehrabi et al., 2015), and there are also a variety of datasets available (Uzuner et al., 2011; Albright et al., 2013). However generalizability of negation systems is still lacking, as cross-domain experiments suffer dramatic performance losses, even while obtaining F1 scores over 90% in the domain of the training data (Wu et al., 2014).

Prior work has shown that there is a problem of generalizability in negation detection, but has done little to address it. In this work, we describe preliminary experiments to assess the difficulty of the problem, and evaluate the efficacy of existing domain adaptation algorithms on the problem. We implement three unsupervised domain adaptation methods from the machine learning literature, and find that multiple methods obtain similarly modest performance gains, falling well short of in-domain performance. Our research has broader implications, as the general problem of generalizabiliy applies to all clinical NLP problems. Research in unsupervised domain adaptation can have a huge impact on the adoption of machine learning-based NLP methods for clinical applications.

2 Background

Domain adaptation is the task of using labeled data from one domain (the *source* domain) to train a classifier that will be applied to a new domain (the *target* domain). When there is some labeled data available in the target domain, this is referred to as *supervised domain adaptation*, and when there is no labeled data in the target domain, the task is called *unsupervised domain adaptation* (UDA). As the unsupervised version of the problem more closely aligns to real-world clinical use cases, we focus on that setting.

One common UDA method in natural language processing is *structural correspondence learning* (SCL; Blitzer et al. (2006)). SCL hypothesizes that some features act consistently across domains (so-called *pivot features*) while others are still informative but are domain-dependent. The SCL method

165

Proceedings of the BioNLP 2017 workshop, pages 165–170,
Vancouver, Canada, August 4, 2017. ©2017 Association for Computational Linguistics

combines source and target extracted feature sets, and trains classifiers to predict the value of pivot *features*, uses singular value decomposition to reduce the dimensionality of the pivot feature space, and uses this reduced dimensionality space as an additional set of features. This method has been successful for part of speech tagging (Blitzer et al., 2006), sentiment analysis (Blitzer et al., 2007), and authorship attribution (Sapkota et al., 2015), among others, but to our knowledge has not been applied to negation detection (or any other biomedical NLP tasks). One difficulty of SCL is in selecting the pivot features, for which most existing approaches use heuristics about what features are likely to be domain independent.

Another approach to UDA, known as *bootstrapping* or *self-training*, uses a classifier trained in the source domain to label target instances, and adds confidently predicted target instances to the training data with the predicted label. This method has been successfully applied to POS tagging, spam email classification, named entity classification, and syntactic parsing (Jiang and Zhai, 2007; McClosky et al., 2006).

Clinical negation detection has a long history because of its importance to clinical information extraction. Rule-based systems such as Negex (Chapman et al., 2001) and its successor, ConText (Harkema et al., 2009) contain manually curated lists of negation cue words and apply rules about their scopes based on word distance and intervening cues. While these methods do not learn, the word distance parameter can be tuned by experts to apply to their own datasets. The DepNeg system (Sohn et al., 2012) used manually curated dependency path features in a rule-based system to abstract away from surface features. The Deepen algorithm (Mehrabi et al., 2015) algorithm also uses dependency parses in a rule-based system, but applies the rules as a post-process to Negex, and only to the concepts marked as negated.

Machine learning approaches typically use supervised classifiers such as logistic regression or support vector machines to label individual concepts based on features extracted from surrounding context. These features may include manually curated lists, such as those from Negex and ConText, as well as features intended to emulate the rules of those systems, as well as more exhaustive contextual features common to NLP classification problems. The 2010 i2b2/VA Challenge (Uzuner

et al., 2011) had an "assertion classification" task, where concepts had mutually exclusive *present, absent (negated), possible, conditional, hypothetical, and non-patient* attributes, and this task had a variety of approaches submitted that used some kind of machine learning. The top-performing system (de Bruijn et al., 2011) used a multi-level ensemble classifier, classifying assertion status of each word with three different machine learning systems, then feeding those outputs into a concept-level multi-class support vector machine classifier for the final prediction. In addition to standard bag of words features for representing context, this system used Brown clusters to abstract away from surface feature representations. The MITRE system (Clark et al., 2011) used conditional random fields to tag cues and their scopes, then incorporated cue information, section features, semantic and syntactic class features, and lexical surface features into a maximum entropy classifier. Finally, Wu et al. (2014) incorporated many of the dependency features from rule-based DepNeg system (Sohn et al., 2012) and the best features from the i2b2 Challenge into a machine learning system.

3 Methods

In this work, we apply unsupervised domain adaptation algorithms to machine learning systems for clinical negation detection, evaluating the extent to which performance can be improved when systems are trained on one domain and applied to a new domain. We make use of the (Wu et al., 2014) system in these experiments, as it is freely available as part of the Apache cTAKES (Savova et al., 2010)[1] clinical NLP software, and can be easily retrained.

Unsupervised domain adaptation (UDA) takes place in the setting where there is a *source* dataset $D_s = \{\mathbf{X}, \vec{y}\}$, and a *target* dataset $D_t = \{\mathbf{X}\}$, where feature representations $\mathbf{X} \in \mathbb{R}^{N \times D}$ for N instances and D feature dimensions and labels $\vec{y} \in \mathbb{R}^N$. Our goal is to build classifiers that will perform well on instances from D_s as well as D_t, despite having no gold labels from D_t to use at training time. Here we describe a variety of approaches that we have implemented.

The baseline cTAKES system that we use is a support vector machine-based system with L1 and L2 regularization. Regularization is a penalty term added to the classifier's cost function during training that penalizes "more complex" hypotheses, and

---

[1]http://ctakes.apache.org

is intended to reduce overfitting to the training data. L2 regularization adds the L2 norm to the classifier cost function as a penalty and tends to favor smaller feature weights. L1 regularization adds the L1 norm as a penalty and favors sparse feature weights (i.e., setting many weights to zero).

Before attempting any explicit UDA methods, we evaluate the simple method of increasing regularization. While regularization is already intended to reduce overfitting, it may still overfit on a target domain since its hyper-parameter is tuned on the source domain. In a real unsupervised domain adaptation scenario it is not possible to tune this parameter on the target domain, so for this work we use heuristic methods to set the adapted regularization parameter. We first find the optimal regularization hyperparameter C using cross-validation on the source data, then increase it by an order of magnitude and retrain before testing on target data. For example, if we find that the best F1 score occurs when $C = 1$ for a 5-fold cross-validation experiment on the source data, we retrain the classifier at $C = 0.1$ before applying to target test data.[2] Changing this parameter by one order of magnitude is purely a heuristic approach, chosen because that is how we (the authors) typically would vary this parameter during tuning. Future work may explore whether this parameter on target data without supervision, perhaps by using some information about the data distribution in the target domain.

The first UDA algorithm we implement is structural correspondence learning (SCL) (Blitzer et al., 2006). Following Blitzer et al. we select as pivot features those features that occur more than 50 times in both the source and target data. Then, for each data instance i in $\mathbf{X}_c = \{\mathbf{X}_s \cup \mathbf{X}_t\}$, and each pivot feature p, we extract the non-pivot features of i (non-pivot features are simply all features not selected as pivot features), $\vec{x}_i = \mathbf{X}_c[i, \textit{non-pivots}]$, and a classification target, $y_i[p] = [\![\mathbf{X}_c[i, p] > 0.5]\!]$.[3] For each pivot feature p, we train a linear classifier on the $(\vec{x}_i, y_i[p])$ classification instances, take the resulting feature weights, w_p, and concatenate them into a matrix W. We decompose W using singular value decomposition: $W = U\Sigma V^T$, and construct θ as the first d dimensions of U. This matrix θ represents a projection from non-pivot features to a reduced dimensionality version of the

---

[2] Note that since C is the cost of misclassifying training instances, increasing regularization means lowering C.

[3] We use $[\![expr]\!]$ to denote the indicator function, which returns 1 if $expr$ is true and 0 otherwise.

| | Test corpus | | | |
Train corpus	Seed	Stratified	Mipacq	i2b2
Seed	**0.88**	0.76	0.65	0.79
Stratified	0.66	**0.83**	0.67	0.79
Mipacq	0.73	0.78	**0.75**	0.85
i2b2	0.65	0.59	0.64	**0.93**

Table 1: Results (F1 scores) of baseline cross-domain experiments. Bold diagonals indicate in-domain results, which were obtained with 5-fold cross-validation. Off-diagonal elements were trained on source data and tested on target data.

pivot-feature space. At training and test time, features are extracted normally, and non-pivot feature values are multiplied by θ to create *correspondence features* in the reduced-dimensionality pivot space. Following Sapkota et al. (2016), we experiment with two methods of combining correspondence features with the original features: *All+New*, which combines all the original features with the correspondence features, and *Pivot+New* which combines only the pivot features from the original space with the correspondence features.

The next UDA algorithm we implement is bootstrapping. Jiang and Zhai (2007) introduced a variety of methods for UDA, under the broad heading of *instance weighting*, but the method they call *bootstrapping* was the only one which does not rely on any target domain labeled data. This method creates pseudo-labels for a portion of the target data by running a classifier trained only on source data on the target data, and adding confidently classified target instances to the training data, labeled with whatever the classifier decided. Jiang and Zhai experiment with the weights of these instances, either giving higher weights to target instances or weighting them the same as source instances. We implemented a simpler version of bootstrapping that does not modify instance weights, and adds instances based on the initial classifier score (rather than iteratively re-training and adding additional instances). We allow up to 1% of the target instances to be added.

In addition to adding the highest-scoring instances, we also experiment with adding only high-scoring instances from the minority class. In many NLP tasks, including negation detection, the label of interest has low prevalence, and there is a danger that the classifier will be most confident on the majority class and only add target instances with that

Source	Target	None	10xReg	SCL A+N	SCL P+N	BS-All	BS-Minority	ISF
Seed (L1)	Strat	0.76	**0.8**	**0.8**	0.69	0.79	0.79	**0.8**
	Mipacq	0.65	0.66	0.69	0.6	0.69	**0.7**	0.69
	i2b2	0.79	**0.83**	**0.83**	0.71	**0.83**	**0.83**	**0.83**
Strat (L1)	Seed	0.66	0.66	0.66	0.58	0.66	**0.67**	0.66
	Mipacq	0.67	**0.68**	**0.68**	0.65	**0.68**	0.66	**0.68**
	i2b2	0.79	0.79	0.79	0.71	0.79	**0.8**	0.79
Mipacq (L2)	Seed	**0.73**	0.59	**0.73**	0.71	**0.73**	0.71	**0.73**
	Strat	0.78	0.76	0.78	0.71	0.78	**0.79**	0.78
	i2b2	**0.85**	0.77	**0.85**	0.84	0.84	**0.85**	**0.85**
i2b2 (L1)	Seed	0.65	**0.72**	**0.72**	0.67	**0.72**	**0.72**	**0.72**
	Strat	0.59	0.68	**0.69**	0.62	0.68	0.68	0.68
	Mipacq	0.64	**0.69**	**0.69**	0.68	**0.69**	**0.69**	**0.69**
Average		0.71	0.72	**0.74**	0.68	**0.74**	**0.74**	**0.74**

Table 2: Results of unsupervised domain adaptation algorithms (F1 scores). None=No adaptation, 10xReg=Regularization with 10x penalty, SCL A+N is structural correspondence learning with all features in addition to projected (new) features, SCL P+N is SCL with pivot features and projected features, BS-All=Bootstrapping with instances of all classes added to source, BS-Minority=Bootstrapping with only instances of minority class added to source, ISF=Instance similarity features.

label. We therefore experiment with only adding minority class instances, enriching the training data to have a more even class distribution.

The final UDA algorithm we experiment with uses instance similarity features (ISF) (Yu and Jiang, 2015). This method extends the feature space in the source domain with a set of similarity features computed by comparison to features extracted from target domain instances. Formally, the method selects a random subset of K *exemplar instances* from D_t and normalizes them as $\hat{\bar{e}} = \frac{\bar{e}}{\|\bar{e}\|}$. Similarity feature k for instance i in the source data set is computed as the dot product $\mathbf{X}_t[i] \cdot \hat{\bar{e}}[k]$. Following Yu and Jiang, we set $K = 50$ and concatenate the similarity features to the full set of extracted features for each source instance at training. These exemplar instances must be kept around past training time, so that at test time similarity features can be similarly created for test instances.

4 Evaluation

Our evaluation makes use of four corpora of clinical notes with negation annotations – i2b2 (Uzuner et al., 2011), Mipacq (Albright et al., 2013), SHARP (Seed), and SHARP (Stratified). We first perform cross-domain experiments in the no adaptation setting to replicate Wu et al.'s experiments.[4] One difference to Wu et al. is that we evaluate on

---

[4]See that paper for an discussion of corpus differences.

the training split of the target domain – we made this choice because the development and test sets for some of the corpora are quite small and the training data gives us a more stable estimate of performance. We tune two hyperparameters, L1 vs. L2 regularization and the values of regularization parameter C, with five-fold cross validation on the source corpus. We record results for training on all four corpora, testing on all three target domains, as well as a cross-validation experiment to measure in-domain performance. Table 1 shows these results, which replicate Wu et al. in finding dramatic performance declines across corpora.

In our domain adaptation experiments, we also use all four corpora as source domains, and for each source domain we perform experiments where the other three corpora are target domains. This result is reported in Table 2.

5 Discussion and Conclusion

These results show that unsupervised domain adaptation can provide, at best, a small improvement to clinical negation detection systems.

Strong regularization, while not obtaining the highest average performance, provides nominal improvements over no adaptation in all settings except when the source corpus is Mipacq, in which case performance suffers severely. Mipacq has two unique aspects that might be relevant; first, it is the largest training set, and second, it pulls documents from a very diverse set of sources (clinical notes, clinical questions, and medical encyclopedias), while the other corpora only contain clinical notes. Perhaps because the within-corpus variation is already quite high, the regularization parameter that performs best during tuning is already sufficient to prevent overfitting on any target corpus with less variation, and increasing it leads to underfitting and thus poor target domain performance. Future work may explore this hypothesis, which must include some attempt to relate the within- and between-corpus variation.

Four different systems all obtain the highest average performance, with BS-All (standard bootstrapping), BS-Minority (bootstrapping with minority class enrichment), structural correspondence learning (SCL A+N), and instance similarity features (ISF) all showing 3% gain in performance (71% to 74%). While the presence of some improvement is encouraging, the improvements within any given technique are not consistent, so that without labeled data from the target domain it would not be possible to know which UDA technique to use. We set aside the question of "statistical significance," as that is probably too low of a bar – whether or not these results reach that threshold, they are still disappointingly low and likely to cause issues if applied to new data.

In summary, selecting a method is difficult, and many of these methods have hyper-parameters (e.g., pivot selection for SCL, number of bootstrapping instances, number of similarity features) that could potentially be tuned, yet in the unsupervised setting there are no clear metrics to use for tuning performance. Future work will explore the use of unsupervised performance metrics that can serve as proxies to test set performance for optimizing hyperparameters and selecting UDA techniques for a given problem.

Acknowledgments

This work was supported by National Institutes of Health grants R01GM114355 from the National Institute of General Medical Sciences (NIGMS) and U24CA184407 from the National Cancer Institute (NCI). The content is solely the responsibility of the authors and does not necessarily represent the official views of the National Institutes of Health.

References

Daniel Albright, Arrick Lanfranchi, Anwen Fredriksen, Will F Styler, Colin Warner, Jena D Hwang, Jinho D Choi, Dmitriy Dligach, Rodney D Nielsen, James H Martin, Wayne Ward, Martha Palmer, and Guergana K Savova. 2013. Towards comprehensive syntactic and semantic annotations of the clinical narrative. *Journal of the American Medical Informatics Association: JAMIA* 20(5):922–930. https://doi.org/10.1136/amiajnl-2012-001317.

J Blitzer, M Dredze, and F Pereira. 2007. Biographies, Bollywood, Boom-boxes and Blenders: Domain Adaptation for Sentiment Classification. In *ACL 2007*. page 440. http://www.cs.brandeis.edu/ marc/misc/proceedings/acl-2007/ACLMain/pdf/ACLMain56.pdf.

J Blitzer, R McDonald, and F Pereira. 2006. Domain adaptation with structural correspondence learning. *Proceedings of the 2006 Conference on Empirical Methods in Natural Language Processing (EMNLP 2006)* http://dl.acm.org/citation.cfm?id=1610094.

Olivier Bodenreider and Alexa T. McCray. 2003. Exploring semantic groups through visual approaches. *Journal of Biomedical Informatics* 36(6):414–432. https://doi.org/10.1016/j.jbi.2003.11.002.

W W Chapman, W Bridewell, P Hanbury, G F Cooper, and B G Buchanan. 2001. A simple algorithm for identifying negated findings and diseases in discharge summaries. *Journal of biomedical informatics* 34(5):301–310. https://doi.org/10.1006/jbin.2001.1029.

Wendy W. Chapman, David Chu, and John N. Dowling. 2007. ConText : An Algorithm for Identifying Contextual Features from Clinical Text. *Proceedings of the Workshop on BioNLP 2007: Biological, Translational, and Clinical Language Processing* (June):81–88.

Cheryl Clark, John Aberdeen, Matt Coarr, David Tresner-Kirsch, Ben Wellner, Alexander Yeh, and Lynette Hirschman. 2011. MITRE system for clinical assertion status classification. *Journal of the American Medical Informatics Association : JAMIA* 18(5):563–567. https://doi.org/10.1136/amiajnl-2011-000164.

Berry de Bruijn, Colin Cherry, Svetlana Kiritchenko, Joel Martin, and Xiaodan Zhu. 2011. Machine-learned solutions for three stages of clinical information extraction: the state of the art at i2b2 2010. *Journal of the American Medical Informatics Association* 18(5):557–562. https://doi.org/10.1136/amiajnl-2011-000150.

Henk Harkema, John N. Dowling, Tyler Thornblade, and Wendy W. Chapman. 2009. ConText: an algorithm for determining negation, experiencer, and temporal status from clinical reports. *Journal of biomedical* 42(5):839–851.

J Jiang and CX Zhai. 2007. Instance weighting for domain adaptation in NLP. *ACL* .

David McClosky, Eugene Charniak, and Mark Johnson. 2006. Effective Self-Training for Parsing. In *Proceedings of {NAACL HLT} 2006*. New York City, USA, pages 152–159.

Saeed Mehrabi, Anand Krishnan, Sunghwan Sohn, Alexandra M. Roch, Heidi Schmidt, Joe Kesterson, Chris Beesley, Paul Dexter, C. Max Schmidt, Hongfang Liu, and Mathew Palakal. 2015. DEEPEN: A negation detection system for clinical text incorporating dependency relation into NegEx. *Journal of Biomedical Informatics* 54:213–219. https://doi.org/10.1016/j.jbi.2015.02.010.

U Sapkota, T Solorio, M Montes-y Gómez, and S Bethard. 2016. Domain Adaptation for Authorship Attribution: Improved Structural Correspondence Learning. In *Proceedings of the 54th Annual Meeting of the Association for Computational Linguistics*. pages 2226–2235. https://www.aclweb.org/anthology/P/P16/P16-1210.pdf.

Upendra Sapkota, Steven Bethard, and Manuel Montes-y g. 2015. Not All Character N -grams Are Created Equal : A Study in Authorship Attribution. In *Proceedings of NAACL 2015*. pages 93–102.

GK Savova, JJ Masanz, and PV Ogren. 2010. Mayo clinical Text Analysis and Knowledge Extraction System (cTAKES): architecture, component evaluation and applications. *Journal of the American Medical Informatics Association* .

Sunghwan Sohn, Stephen Wu, and Christopher G Chute. 2012. Dependency Parser-based Negation Detection in Clinical Narratives. *AMIA Joint Summits on Translational Science proceedings AMIA Summit on Translational Science* 2012:1–8.

Özlem Uzuner, Brett R South, Shuying Shen, and Scott L DuVall. 2011. 2010 i2b2/VA challenge on concepts, assertions, and relations in clinical text. *Journal of the American* http://jamia.oxfordjournals.org/content/18/5/552.short.

S Wu, T Miller, J Masanz, M Coarr, and S Halgrim. 2014. Negation's not solved: generalizability versus optimizability in clinical natural language processing. *PloS one* .

J Yu and J Jiang. 2015. A Hassle-Free Unsupervised Domain Adaptation Method Using Instance Similarity Features. *ACL (2)* .

BioCreative VI Precision Medicine Track: creating a training corpus for mining protein-protein interactions affected by mutations

Rezarta Islamaj Doğan[1], Andrew Chatr-aryamontri[2], Sun Kim[1], Chih-Hsuan Wei[1],
Yifan Peng[1], Donald C. Comeau[1] and Zhiyong Lu[1]

[1]National Center for Biotechnology Information, National Library of Medicine, National
Institutes of Health, Bethesda, MD 20894, USA

[2]Institute for Research in Immunology and Cancer, Université de Montréal, Montréal, QC
H3C 3J7, Canada

Abstract

The Precision Medicine Track in BioCreative VI aims to bring together the BioNLP community for a novel challenge focused on mining the biomedical literature in search of mutations and protein-protein interactions (PPI). In order to support this track with an effective training dataset with limited curator time, the track organizers carefully reviewed PubMed articles from two different sources: curated public PPI databases, and the results of state-of-the-art public text mining tools. We detail here the data collection, manual review and annotation process and describe this training corpus characteristics. We also describe a corpus performance baseline. This analysis will provide useful information to developers and researchers for comparing and developing innovative text mining approaches for the BioCreative VI challenge and other Precision Medicine related applications.

1 Introduction

Genomic technologies now make possible the routine sequencing of individual genomes and such data makes possible to understand how genetic variations are distributed in healthy and sick populations. On the other hand, proteomics and metabolomics approaches are charting the metabolic and interactions maps of the cell. Such data deluge has generated great expectations in the cure of human diseases. Nonetheless, it is still difficult to predict the phenotypic outcome of a specific genome and designing the most appropriate treatment or establishing preventive programs. Linking allelic varia-

tion and genomic mutations to protein-protein interactions (PPI) is crucial to understand how cellular networks rewire and to support personalized medicine approaches.

To date, no tool is available to facilitate the specific retrieval of such information that remains buried in the unstructured text within the biomedical literature. Our goal is to foster the development of text mining algorithms that specialize in scanning the published biomedical literature and to extract the reported discoveries of protein interactions changing in nature due to the presence of a genomic variations or artificial mutations.

The Precision Medicine Track in BioCreative VI is a community challenge that addresses this problem in the form of two tasks:

- Document Triage: Identification of relevant PubMed citations describing mutations affecting protein-protein interactions
- Relation Extraction: Extraction of experimentally verified PPI pairs affected by the presence of a genetic mutation

Traditionally biological database curators have contributed to the various BioCreative challenges (Hirschman, Yeh et al. 2005, Chatr-aryamontri, Kerrien et al. 2008, Krallinger, Morgan et al. 2008, Lu and Hirschman 2012) supporting the identification of stages in the curation workflow suitable for text mining applications and manually annotating the training and test corpora. Because the manual curation of the current exponentially growing body of biomedical literature is an impossible task, the insertion of robust text mining tools in the curation pipeline represent a feasible and sustainable solution to this problem (Hirschman, Burns et al. 2012).

Proceedings of the BioNLP 2017 workshop, pages 171–175,
Vancouver, Canada, August 4, 2017. ©2017 Association for Computational Linguistics

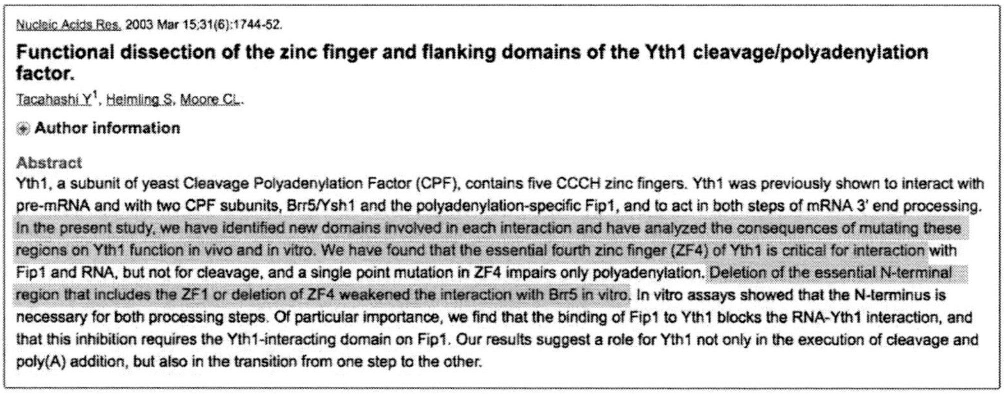

Nucleic Acids Res. 2003 Mar 15;31(6):1744-52.

Functional dissection of the zinc finger and flanking domains of the Yth1 cleavage/polyadenylation factor.

Tacahashi Y[1], Helmling S, Moore CL.

⊕ Author information

Abstract

Yth1, a subunit of yeast Cleavage Polyadenylation Factor (CPF), contains five CCCH zinc fingers. Yth1 was previously shown to interact with pre-mRNA and with two CPF subunits, Brr5/Ysh1 and the polyadenylation-specific Fip1, and to act in both steps of mRNA 3' end processing. In the present study, we have identified new domains involved in each interaction and have analyzed the consequences of mutating these regions on Yth1 function in vivo and in vitro. We have found that the essential fourth zinc finger (ZF4) of Yth1 is critical for interaction with Fip1 and RNA, but not for cleavage, and a single point mutation in ZF4 impairs only polyadenylation. Deletion of the essential N-terminal region that includes the ZF1 or deletion of ZF4 weakened the interaction with Brr5 in vitro. In vitro assays showed that the N-terminus is necessary for both processing steps. Of particular importance, we find that the binding of Fip1 to Yth1 blocks the RNA-Yth1 interaction, and that this inhibition requires the Yth1-interacting domain on Fip1. Our results suggest a role for Yth1 not only in the execution of cleavage and poly(A) addition, but also in the transition from one step to the other.

Figure 1 A PubMed article describing a protein-protein interaction affected by mutation

As we prepared to create our corpus we faced the common situation of limited reviewer time. We took two steps to maximize this limited, valuable resource: First, we reviewed annotations readily available from manually curated PPI databases (Orchard, Ammari et al. 2014) and marked the relevant publications that could be used for the purposes of this challenge; next, we expanded the training set using a set of publically available text mining tools (Kim, Kwon et al. 2012, Wei, Harris et al. 2013) specifically for the retrieval of literature reporting protein interaction and mutation data.

Both of these sets were manually reviewed and categorized as: 1) Articles describing PPI and mutations affecting those molecular interactions, 2) Articles describing mutations and molecular interactions, with no affect or no relation between the two events, 3) Articles describing PPI, 4) Articles describing mutations or genetic variation, and 5) Articles not relevant for either molecular interaction or mutation information. In addition, the database extracted interactions were carefully reviewed and validated in two important aspects: 1) the annotated PPI were described in the PubMed abstract of the corresponding article, as opposed to the full text, and 2) the extracted interactions were affected by a mutation, and this was stated in the abstract.

All manually selected, categorized and carefully reviewed articles make up a set of 4,082 PubMed abstracts. All of these articles can be used for building machine learning methods and other innovative applications for the Precision Medicine Track in BioCreative VI. Of these, 598 PubMed articles are annotated with specific interactions. This smaller set can be used to develop algorithms for the Relation Extraction task and other similar biomedical text mining problems.

We provide here a detailed description of the assembly of this dataset and report the on-going efforts of building the test corpus.

2 Training Corpus

The Precision Medicine track training corpus was generated as a result of two data selection and validation methods:

- Data repurposing
- Text mining triage and manual validation

These approaches are different and as noticed in the article composition resulting from each of them, they are both important contributors to this dataset. Here we describe the procedure followed in each of these approaches, starting with our annotation guidelines and a detailed view of the corpus characteristics. Figure 1 shows an example article in our dataset.

2.1 Annotation guidelines

All selected articles were manually annotated to answer these questions:

- Does this article describe experimentally verified protein-protein interactions?

- Does this article describe a disease known mutation or a mutational analysis experiment?

- Are the database curated PPI pairs for this article mentioned in the abstract?

Table 1 Training Set annotation and distribution amongst different categories

Annotation Category	Curated database selected articles (PPI set)	Text mining tools selected articles (TM set)	Complete Training Set	
True positives	1079	651	1,730	42%
True Negatives	55	322	377	9%
Negative, Yes PPI, No Mutation	1538	82	1,620	40%
Negative, No PPI, Yes Mutation	136	87	99	2.4%
Negative, No PPI, No Mutation	12	120	256	6.3%
Total	2820	1262	4082	100%

- Is the PPI affected by the mutation?

Then, based on the above annotations, articles are carefully categorized as 1) True Positives, for articles specifically describing PPI influenced by genetic mutations, 2) True Negatives, for articles describing both PPIs and genetic variation analysis with no inference of relation between them, 3) articles containing PPI but no mutations, 4) articles containing mutations but no PPI, and 5) articles mentioning neither.

2.2 Curated Database article selection

The IntAct Molecular Interaction Database (Orchard, Ammari et al. 2014) is a freely available, open source database system and analysis tool for molecular interaction data. It currently lists 14,584 manually annotated PubMed full-text articles with 720,711 molecular interactions for 98,289 different interactors. The curation of these molecular interactions is captured at a required level of detail and frequent updates include mapping to binding regions, point mutations and post-translational modifications to a specified sequence with a reference protein sequence database.

A set of 2,852 articles, containing in-the-abstract information about binding interfaces and mutations influencing the interactions, was retrieved from IntAct and these articles went through a careful review and validation round by an experienced curator. Each one of these articles was carefully considered for their suitability for the precision medicine task.

A second manual validation round was then performed on all positively annotated articles of the first round. As a result, 598 articles were identified as relevant for the Relation Extraction task, with experimentally verified interactions influenced by mutations and with explicit interactors in the abstract. All of these interactors were expressed with both their UniProt ID and Gene Entrez ID. The non-relevant articles were further categorized into the more specific categories as described above.

2.3 Text Mining based article selection

The Text Mining approach used two well-known publically available text mining tools: PIE the search (Kim, Kwon et al. 2012) and tmVar (Wei, Harris et al. 2013). PIE[1] the search is a web service that provides an alternate way of querying PubMed for biologists and database curators. The returned articles are ranked based on their probability of describing protein-protein interactions, using a very competitive algorithm and the winner of BioCreative III ACT competition (Krallinger, Vazquez et al. 2011). tmVar[2] is another text mining tool that is the current gold-standard for recognizing sequence variants in PubMed literature. An article marked by tmVar signals the presence of a sequence variant of a mutation in the title and abstract.

These tools were used as follows:

- Step 1: PIE the search was used to select the top scoring (for PPI) PubMed articles published in the last 10 years. This method selected over 13,000 articles.

- Step 2. tmVar was used on the resulting set of Step 1 to select all articles which had a sequence variant in the title or abstract. This method selected around 1,200 articles.

---

[1] https://www.ncbi.nlm.nih.gov/CBBresearch/Wilbur/IRET/PIE/

[2] https://www.ncbi.nlm.nih.gov/CBBresearch/Lu/Demo/tmTools/#tmVar

Table 2 Document Triage Task results

Methods	Avg. Prec.	Precision	Recall	F1	Positive	Negative	Ratio
10-fold CV (PPI set)	0.7577	0.7184	0.6321	0.6725	1079	1741	38%
Validation (TM set)	0.6551	0.6210	0.6897	0.6536	651	611	52%
10-fold CV (all data)	0.7225	0.6891	0.6260	0.6561	1730	2352	42%

- Step 3. All articles in Step 2 were manually annotated as described in the annotation guidelines.

3 Results and Discussion

3.1 Precision Medicine Task Training Corpus Characteristics

The Precision Medicine Task training corpus contains 4,082 selected PubMed abstracts that come from two different sources: curated databases and text mining tool selection. It is important to see the dataset as a whole and to notice the different composition of classified articles coming from both sources as detailed in Table 1.

In addition, we looked at the PIE score distribution of all articles in the dataset. We noticed that the PubMed articles selected via text mining tools had a higher PIE score average than the articles retrieved from curated databases. In particular, while the PIE scores of the articles selected from the curated databases form a normal distribution, the scores of the text mining selected articles are skewed towards high scores.

On a different experiment, we ran the tmVar tool on all curated database selected articles. Interestingly, only 311 out of 1079 positives articles were marked by tmVar.

Thus, if novel algorithm developers only gave more importance to articles selected via text mining tools, or only the text mining tools used in our experiment, they risk biasing curators to only a particular set of articles. Innovative text mining tools should make use of both sets of articles in order to ensure a better coverage of curatable articles.

3.2 Benchmark results and corpus use

A baseline SVM method was designed using unigram and bigram features from titles and abstracts of the training corpus, as shown in the results in Table 2. A first experiment used articles from the curated database for training in a 10-fold cross validation (CV) setting, and tested on the text mining selected articles. And a second experiment mixed all articles in a 10-fold cross validation setting. Results are detailed in Table 2.

The test dataset for BioCreative VI Precision Medicine Track will be a set selected by database curators. First articles will be retrieved via text mining tools and then each article will be manually evaluated by four experienced curators.

4 Conclusions and Public Availability

A vast amount of precision medicine related information can be found in published literature and extracted by skilled domain expert curators. The BioCreative VI Precision Medicine Track corpus characteristics provide important insights on 1) understanding the structure of biological information and why it is relevant for precision medicine purposes, and 2) the best practices for designing computational automatic methods capable of extracting such information from unstructured text.

By releasing this data we aim to facilitate the curation of precision medicine related information available in published literature. This corpus fosters development of innovative text mining algorithms that may help database curators in identifying molecular interactions that differ based on the presence of a specific genetic variant, information which could be translated to clinical practice.

This data comes from two realistic, important data sources: 1) articles retrieved from expert curated PPI databases, re-evaluated and found useful for precision medicine purposes, and 2) articles retrieved via state-of-the-art text mining tools trained to identify articles describing PPI and containing identifiable sequence variants. Both sets of data have slightly different, but useful characteristics and as such, novel text mining tools need to use both sources of information for best application in this new domain.

The BioCreative VI Precision Medicine training corpus will be available to task participants from the BioCreative website and later to the whole scientific community.

5 References

Chatr-aryamontri, A., S. Kerrien, J. Khadake, S. Orchard, A. Ceol, L. Licata, L. Castagnoli, S. Costa, C. Derow, R. Huntley, B. Aranda, C. Leroy, D. Thorneycroft, R. Apweiler, G. Cesareni and H. Hermjakob (2008). "MINT and IntAct contribute to the Second BioCreative challenge: serving the text-mining community with high quality molecular interaction data." Genome Biol **9 Suppl 2**: S5.

Hirschman, L., G. A. Burns, M. Krallinger, C. Arighi, K. B. Cohen, A. Valencia, C. H. Wu, A. Chatr-Aryamontri, K. G. Dowell, E. Huala, A. Lourenco, R. Nash, A. L. Veuthey, T. Wiegers and A. G. Winter (2012). "Text mining for the biocuration workflow." Database (Oxford) **2012**: bas020.

Hirschman, L., A. Yeh, C. Blaschke and A. Valencia (2005). "Overview of BioCreAtIvE: critical assessment of information extraction for biology." BMC bioinformatics **6 Suppl 1**: S1.

Kim, S., D. Kwon, S. Y. Shin and W. J. Wilbur (2012). "PIE the search: searching PubMed literature for protein interaction information." Bioinformatics **28**(4): 597-598.

Krallinger, M., A. Morgan, L. Smith, F. Leitner, L. Tanabe, J. Wilbur, L. Hirschman and A. Valencia (2008). "Evaluation of text-mining systems for biology: overview of the Second BioCreative community challenge." Genome Biol **9 Suppl 2**: S1.

Krallinger, M., M. Vazquez, F. Leitner, D. Salgado, A. Chatr-aryamontri, A. Winter, L. Perfetto, L. Briganti, L. Licata, M. Iannuccelli, L. Castagnoli, G. Cesareni, M. Tyers, G. Schneider, F. Rinaldi, R. Leaman, G. Gonzalez, S. Matos, S. Kim, W. Wilbur, L. Rocha, H. Shatkay, A. Tendulkar, S. Agarwal, F. Liu, X. Wang, R. Rak, K. Noto, C. Elkan and Z. Lu (2011). "The Protein-Protein Interaction tasks of BioCreative III: classification/ranking of articles and linking bio-ontology concepts to full text." BMC bioinformatics **12**(Suppl 8): S3.

Lu, Z. and L. Hirschman (2012). "Biocuration workflows and text mining: overview of the BioCreative 2012 Workshop Track II." Database : the journal of biological databases and curation: bas043.

Orchard, S., M. Ammari, B. Aranda, L. Breuza, L. Briganti, F. Broackes-Carter, N. H. Campbell, G. Chavali, C. Chen, N. del-Toro, M. Duesbury, M. Dumousseau, E. Galeota, U. Hinz, M. Iannuccelli, S. Jagannathan, R. Jimenez, J. Khadake, A. Lagreid, L. Licata, R. C. Lovering, B. Meldal, A. N. Melidoni, M. Milagros, D. Peluso, L. Perfetto, P. Porras, A. Raghunath, S. Ricard-Blum, B. Roechert, A. Stutz, M. Tognolli, K. van Roey, G. Cesareni and H. Hermjakob (2014). "The MIntAct project--IntAct as a common curation platform for 11 molecular interaction databases." Nucleic Acids Res **42**(Database issue): D358-363.

Wei, C. H., B. R. Harris, H. Y. Kao and Z. Lu (2013). "tmVar: a text mining approach for extracting sequence variants in biomedical literature." Bioinformatics **29**(11): 1433-1439.

Painless Relation Extraction with Kindred

Jake Lever and Steven JM Jones
Canada's Michael Smith Genome Sciences Centre
570 W 7th Ave, Vancouver
BC, V5Z 4S6, Canada
{jlever,sjones}@bcgsc.ca

Abstract

Relation extraction methods are essential for creating robust text mining tools to help researchers find useful knowledge in the vast published literature. Easy-to-use and generalizable methods are needed to encourage an ecosystem in which researchers can easily use shared resources and build upon each others' methods. We present the Kindred Python package[1] for relation extraction. It builds upon methods from the most successful tools in the recent BioNLP Shared Task to predict high-quality predictions with low computational cost. It also integrates with PubAnnotation, PubTator, and BioNLP Shared Task data in order to allow easy development and application of relation extraction models.

1 Introduction

Modern biomedical research is beginning to rely on text mining tools to help search and curate the ever-growing published literature and to interpret large numbers of electronic health records. Many text mining tools employ information extraction (IE) methods to translate knowledge discussed in free text into a form that can be easily searched, analyzed and used to build valuable biomedical databases. Examples of applications of IE methods include building protein-protein interaction networks (Donaldson et al., 2003) and automatically retrieving information about proteins (Rebholz-Schuhmann et al., 2007).

Information extraction relies on several key technologies including relation extraction. Relation extraction focuses on understanding the relation between two of more biomedical terms in a

stretch of text. This may be understanding how one protein interacts with another protein, whether a drug treats or causes a particular symptom and many other uses. Most methods assume that entities (e.g. gene and drug names) in the sentence have already been identified, either through a named entity recognition tools (e.g. BANNER (Leaman et al., 2008)) or basic dictionary matching against a word list. The method must then use linguistic cues within the sentence to predict whether or not a relation exists between each pair or group of entities and exactly which type of relation it is.

The BioNLP Shared Task has catalyzed research in relation extraction tools by providing an environment for friendly competition between different relation extraction approaches. The organizers of the relation extraction subtasks provide text from published literature with entities and relations annotated. The participating researchers build relation extraction models and predicted relations on a test set. The participants' predictions are then analyzed by the organizers and the results presented to all. The BioNLP Shared Task has been held in 2009, 2011, 2013 and recently in 2016. The recent 2016 relation extraction problems focused on two areas: bacteria biotopes (BB3 subtask) and seed development (SeeDev subtask). The BB3 subtask required participants to predict relations between bacteria and their habitats. The SeeDev subtask involved prediction of over twenty different relation types related to seed development.

Two main approaches to relation extraction have been taken, a rule-based method and a vector-based method. A rule-based approach identifies common patterns that capture a relation. For instance, two gene names with the word "regulates" between them generally implies a regulation relation between the two entities. The BioSem method

---

[1]http://www.github.com/jakelever/kindred

Proceedings of the BioNLP 2017 workshop, pages 176–183,
Vancouver, Canada, August 4, 2017. ©2017 Association for Computational Linguistics

(Bui et al., 2013) identifies common patterns of words and parts-of-speech between biomedical terms and performed well in the BioNLP Shared Task in 2013.

The vector-based approach transforms a span of text and candidate relation into a numerical vector that can be used in a traditional machine learning classification approach. Support vector machines (SVM) have commonly been used. The TEES (Björne and Salakoski, 2013) and VERSE (Lever and Jones, 2016) methods, which were successful in many of the shared tasks, use this approach with different approaches for creating the vectors and selecting the parameters for classification.

Deep learning, already very popular in natural language processing (LeCun et al., 2015), has begun to be used in the biomedical text mining field with one entry in the BioNLP Shared Task using a recurrent neural network approach (Mehryary et al., 2016). The paper examined the use of long short-term memory (LSTM) networks for relation extraction, especially in situations with small training dataset sizes. Given such a complicated model, the problem of overfitting becomes very large. They proposed approaches to reduce overfitting and the entry performed very well, coming second in the competition.

The VERSE method came first in the BB3 event subtask and third in the SeeDev binary subtask in the BioNLP Shared Task 2016. An analysis of the two systems that outperformed VERSE in the SeeDev subtask points to interesting directions for further development. The SeeDev subtask differs greatly from the BB3 subtask as there are 24 relation types compared to only 1 in BB3 and the training set size for each relation is drastically smaller. The LitWay approach, which came first, uses a hybrid approach of rule-based and vector-based (Li et al., 2016). For "simpler" relations, defined using a custom list, a rule-based approach is used using a predefined set of patterns. The UniMelb approach created individual classifiers for each relation type and was able to predict multiple relations for a candidate relation (Panyam et al., 2016). This approach of treating relation types differently suggests that there may be large differences in how a relation should be treated in terms of the linguistic cues used to identify it and the best algorithm approach to identify it.

There are several shortcomings in the approaches to the BioNLP Shared Tasks, the greatest of all is the poor number of participants that provide code. It is also clear that the advantages of some of the most successful tools are tailored specifically to these datasets and may not be able to generalize easily to other relation extraction tasks. Some tools that do share code such as TEES and VERSE have a large number of dependencies, though TEES ameliorates this problem with an excellent installing tool that manages dependencies. These tools can also be computationally costly, with both TEES and VERSE taking a parameter optimization strategy that requires a cluster for reasonable performance.

The biomedical text mining community is endeavoring to improve consistency and ease-of-use for text mining tools. In 2012, the Biocreative BioC Interoperability Initiative (Comeau et al., 2014) encouraged researchers to develop biomedical text mining tools around the BioC file format (Comeau et al., 2013). More recently, one of the Biocreative BeCalm tasks focuses on "technical interoperability and performance of annotation servers" for a named entity recognition systems. This initiative encourages an ecosystem of tools and datasets that will make text mining a more common tool in biology research. PubAnnotation (Kim and Wang, 2012), which is part of this approach, is a public resource for sharing annotated biomedical texts. The hope of this resource is to provide data to improve biomedical text mining tools and as a launching point for future shared tasks. The PubTator tool (Wei et al., 2013b) provides PubMed abstracts with various biomedical entities annotated using several named entity recognition tools including tmVar (Wei et al., 2013a) and DNorm (Leaman et al., 2013).

In order to overcome some of the challenges in the relation extraction community in terms of ease-of-use and integration, we present Kindred. Kindred is an easy-to-install Python package for relation extraction using a vector-based approach. It abstracts away much of the underlying algorithms in order to allow a user to easily start extracting biomedical knowledge from sentences. However, the user can easily use individual components of Kindred in conjunction with other parsers or machine learning algorithms. It integrates seamlessly with PubAnnotation and PubTator to allow easy access to training data and text to be applied to. Furthermore, we show that it per-

Example Relation:

$$\text{disease} \longleftarrow \text{causes} \longrightarrow \text{gene}$$

The colorectal cancer was caused by mutations in APC

Standoff Format:

```
example.txt:
The colorectal cancer was caused by
mutations in APC
example.a1:
T1  disease 4 21 colorectal cancer
T2  gene 49 52    APC
example.a2:
E1  causes subj:T2 obj:T1
```

JSON Format:

```
{
  "text": "The colorectal cancer was
caused by mutations in APC",
  "denotations":
    [{"id":"T1", "obj":"disease",
      "span":{"begin":4,"end":21}},
     {"id":"T2", "obj":"gene",
      "span":{"begin":49,"end":52}}],
  "relations":
    [{"id":"R1","pred":"causes",
      "subj":"T2", "obj":"T1"}]
}
```

BioC Format:

```
<?xml version='1.0' encoding='UTF-8'?><!DOCTYPE
collection SYSTEM 'BioC.dtd'>
<collection>
  <document>
    <offset>0</offset>
    <text>The colorectal cancer was caused by
mutations in APC</text>
    <annotation id="T1">
      <infon key="type">disease</infon>
      <location offset="4" length="17"/>
      <text>colorectal cancer</text>
    </annotation>
    <annotation id="T2">
      <infon key="type">gene</infon>
      <location offset="49" length="3"/>
      <text>APC</text>
    </annotation>
    <relation id="R1">
      <infon key="type">causes</infon>
      <node refid="T2" role="subj"/>
      <node refid="T1" role="obj"/>
    </relation>
  </passage>
  </document>
</collection>
```

Simple Tag Format:

```
The <disease id="T1">colorectal cancer</disease> was caused by mutations in <gene
id="T2">APC</gene><relation type="causes" subj="T2" obj="T1" />
```

Figure 1: An example of a relation between two entities in the same sentence and the representations of the relation in four input/output formats that Kindred supports.

forms very well on the BioNLP Shared Task 2016 relation subtasks.

2 Methods

Kindred is a Python package that builds upon the Stanford CoreNLP framework (Manning et al., 2014) and the scikit-learn machine learning library (Pedregosa et al., 2011). The decision to build a package was based on the understanding that each text mining problem is different. It seemed more valuable to make the individual features of the relation extraction system available to the community than a bespoke tool that was designed to solve a fixed type of biomedical text mining problem. Python was selected due to the excellent support for machine learning and the easy distribution of Python packages.

The ethos of the design is based on the scikit-learn API that allows complex operations to occur in very few lines of code, but also gives detailed control of the individual components. Individual computational units are encapsulated in separate classes to improve modularity and allow easier testing. Nevertheless, the main goal was to allow the user to download annotated data and build a relation extraction classifier in as few lines of code as possible.

2.1 Package development

The package has been developed for ease-of-use and reliability. The code for the package is hosted on Github. It was also developed using the continuous integration system Travis CI in order to improve the robustness of the tool. This allows regular tests to be run whenever code is committed to the repository. This will enable further development of Kindred and ensure that it continues to work with both Python 2 and Python 3. Coveralls and the Python coverage tool are used to evaluate code coverage and assist in test evaluation.

These approaches were in line with the recent recommendations on improving research software (Taschuk and Wilson, 2017). We hope these techniques will allow for and encourage others to make use of and contribute to the Kindred package.

2.2 Data Formats

As illustrated in Figure 1, Kindred accepts data in four different formats: the standoff format used by BioNLP Shared Tasks, the JSON format used by PubAnnotation, the BioC format (Comeau et al., 2013) and a simple tag format. The standoff format uses three files, a TXT file that contains the raw text, an A1 file that contains information on the tagged entities and an A2 file that contains information on the relations between the entities. The JSON, BioC and simple tag formats integrate this information into single files. The input text in each of these formats must have already been annotated for entities.

The simple tag format was implemented primarily for simple illustrations of Kindred and for easier testing purposes. It is parsed using an XML parser to identify all tags. A relation tag should contain a "type" attribute that denotes the relation type (e.g. causes). All other attributes are assumed to be arguments for the relation and their values should be IDs for entities in the same text. A non-relation tag is assumed to be describing an entity and should have an ID attribute that is used for associating relations.

2.3 Parsing and Candidate Building

The text data is loaded, and where possible, the annotations are checked for validity. In order to prepare the data for classification, the first step is sentence splitting and tokenization. We use the Stanford CoreNLP toolkit for this which is also used for dependency parsing for each sentence.

Once parsing has completed, the associated entity information must then be matched with the corresponding sentences. An entity can contain non-contiguous tokens as was the case for the BB3 event dataset in the BioNLP 2016 Shared Task. Therefore each token that overlaps with an annotation for an entity is linked to that entity.

Any relations that occur entirely within a sentence are associated with that sentence. The decision to focus on relations contained within sentence boundaries is based on the poor performance of relation extraction systems in the past. The VERSE tool explored predicting relations that spanned sentence boundaries in the BioNLP Shared Task and found that the false positive rate was too high. The sentence is also parsed to generate a dependency graph which is stored as a set of triples $(token_i, token_j, dependency_{ij})$

where $dependency_{ij}$ is the type of edge in the dependency graph between tokens i and j. The edge types use the Universal Dependencies format (Nivre et al., 2016).

Relation candidates are then created by finding every possible pair of entities within each sentence. The candidates that are annotated relations are stored with a class number for use in the multiclass classifier. The class zero denotes no relation. All other classes denote relations of specific types. The types of relations and therefore how many classes are required for the multiclass classifier are based on the training data provided to Kindred.

2.4 Vectorization

Each candidate is then vectorized in order to transform the tokenized sentence and set of entity information into a numerical vector that can be processed using the scikit-learn classifiers. In order to keep Kindred simple and improve performance, it only generates a small set of features as outlined below.

- Entity types in the candidate relation

- Unigrams between entities

- Bigrams for the full sentence

- Edges in dependency path

- Edges in dependency path that are next to each entity.

For the entity type and edge relations, they are stored in a one-hot format. For the entity specific relations, features are created for each entity. For instance, if there are three relation types for relations between two arguments, then six binary features would be required to capture the entity types.

The unigrams and bigrams use a bag-of-words approach. Term-frequency inverse-document frequency (TF-IDF) is used for all bag-of-words based features. The dependency path, using the same method as VERSE, is calculated as the minimum spanning tree between the nodes in the dependency graph that are associated with the entities in the candidate relation.

2.5 Classification

Kindred has in-built support for the support vector machine (SVM) and logistic regression classifiers implemented in scikit-learn. By default, the

SVM classifier is used with the vectorized candidate relations. The linear kernel has shown to give good performance and is substantially faster to train than alternative SVM kernels such as radial basis function or exponential.

The success of the LitWay and UniMelb entries to the SeeDev shared task suggested that individual classifiers for unique relation types may give improved performance. This may be due to the significant differences in complexity between different relation types. For instance, one relation type may require information from across the sentence for good classification, whereas another relation type may require only the neighboring word.

Using one classifier per relation type, instead of a single multiclass classifier, means that a relation candidate may be predicted to be multiple relation types. Depending on the dataset, this may be the appropriate decision as relations may overlap. Kindred offers this functionality of one classifier per relation type. However, for the SeeDev dataset, we found that the best performance was actually through a single multiclass classifier.

2.6 Filtering

The predicted set of relations is then filtered using the associated relation type and types of the entities in the relation. Kindred uses the set of relations in the training data to infer the possible argument types for each relation.

2.7 Precision-recall tradeoff

The importance of precision and recall depends on the specific text mining problem. The BioNLP Shared Task has favored the F1-score, giving an equal weighting to precision and recall. Other text mining projects may prefer higher precision in order to avoid biocurators having to manually filter out spurious results. Alternatively, projects may require higher recall in order to not miss any possibly important results. Kindred gives the user the control of a threshold for making predictions. In this case, the logistic regression classifier is used as it allows for easier thresholding. This is because the underlying predicted values can be interpreted as probabilities. We found that logistic regression achieved performance very close to the SVM classifier. By selecting a higher threshold, the classifier will become more conservative, decrease the number of false positives and therefore improve precision at the cost of recall. By using cross-validation, the user can get an idea of

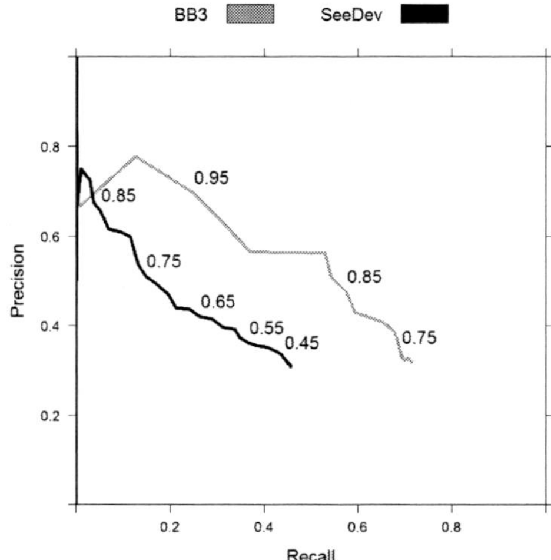

Figure 2: The precision-recall tradeoff when trained on the training set for the BB3 and SeeDev results and evaluating on the development set using different thresholds. The numbers shown on the plot are the thresholds.

the precision-recall tradeoff. The tradeoffs for the BB3 and SeeDev tasks are shown in 2. This allows the user to select the appropriate threshold for their task.

2.8 Parameter optimization

TEES took a grid-search approach to parameter optimization and focused on the parameters of the SVM classifier. VERSE had a significantly larger selection of parameters and grid search was not computationally feasible so a stochastic approach was used. Both approaches are computationally expensive and generally need a computer cluster.

Kindred takes a much simpler approach to parameter optimization and can work out of the box with default values. To improve performance, the user can choose to do minor parameter optimization. The only parameter optimized by Kindred is the exact set of features used for classification. This decision was made with the hypothesis that some relations potentially require words from across the sentence and other need only the information from the dependency parse.

The feature choice optimization uses a greedy algorithm. It calculates the F1-score using cross validation for each feature type. It then selects the best one and tries adding the remaining feature types to it. It continues growing the feature

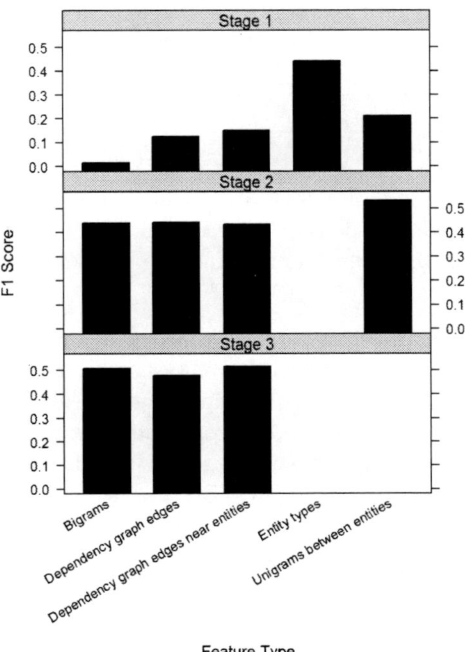

Figure 3: An illustration of the greedy approach to selecting feature types for the BB3 dataset.

set until the cross-validated F1 score does not improve.

Figure 3 illustrates the process for the BB3 subtask using the training set and evaluating on the development set. At the first stage, the entity types feature is selected. This is understandable as the types of entity are highly predictive of whether a candidate relation is reasonable for a particular candidate type, e.g. two gene entities are unlikely to be associated in a 'IS_TREATMENT_FOR' relation. At the next stage, the unigrams between entities feature is selected. And on the third stage, no improvement is made. Hence for this dataset, two features are selected. We use this approach for the BB3 dataset but found that the default feature set performed best for the SeeDev dataset.

2.9 Dependencies

The main dependencies of Kindred are the scikit-learn machine learning library and the Stanford CoreNLP toolkit. Kindred will check for a locally running CoreNLP server and connect if possible. If none is found, then the CoreNLP archive file will be downloaded. After checking the SHA256 checksum to confirm the file integrity, it is ex-

tracted. It will then launch CoreNLP as a background process and wait until the toolkit is ready before proceeding to send parse requests to it. It also makes sure to kill the CoreNLP process when the Kindred package exits. Kindred also depends on the wget package for easy downloading of files, the IntervalTree python package for identifying entity spans in text and NetworkX for generating the dependency path (Schult and Swart, 2008).

2.10 PubAnnotation integration

In order to make use of existing resources in the biomedical text mining community, Kindred integrates with PubAnnotation. This allows annotated text to be downloaded from PubAnnotation and used to train classifiers.

The PubAnnotation platform provides a RESTful API that allows easy download of annotations from a given project. Kindred will initially download the listing of all available text sources with annotation for a given project. The listing is provided as a JSON data file. It will then download the complete set of texts with annotations.

2.11 PubTator integration

Kindred can also download a set of annotated PubMed abstracts that have already been annotated with named entities through the PubTator framework using the RESTful API. This requires the user to provide a set of PubMed IDs which are then requested from the PubTator server using the JSON data format. The same loader used for PubAnnotation data is then used for the PubTator data.

2.12 BioNLP Shared Task integration

Kindred gives easy access to the data from the most recent BioNLP Shared Task. By providing the name of the test and specific data set (e.g. training, development or testing), Kindred manages the download of the appropriate archive, unzipping and loading of the data. As with the CoreNLP dependency, the SHA256 checksum of the downloaded archive is checked before unzipping occurs.

2.13 API

One of the main goals of Kindred is to open up the internal functionality of a relation extraction system to other developers. The authors are keenly aware that their specific interest in relation extraction, in order to build knowledge bases related to cancer, differs from other researchers. With this

	Precision	Recall	F1 Score
Fold 1	0.319	0.715	0.441
Fold 2	0.460	0.684	0.550
Test Set	0.579	0.443	0.502
VERSE	0.510	0.615	0.558

Table 1: Cross-validated results (Fold1/Fold2) and final test set results for Kindred predictions in Bacteria Biotope (BB3) event subtask with test set results for the top performing tool VERSE.

	Precision	Recall	F1 Score
Fold 1	0.333	0.411	0.368
Fold 2	0.255	0.393	0.309
Test Set	0.344	0.479	0.400
LitWay	0.417	0.448	0.432

Table 2: Cross-validated results (Fold1/Fold2) and final test set results for Kindred predictions in Seed Development (SeeDev) binary subtask with test set results for the top performing tool LitWay.

in mind, the API is designed to give easy access to the different modules of Kindred that may be used independently. For instance, the candidate builder or vectorizer could easily be integrated with functionality from other Python packages, which would allow for other machine learning algorithms or deep learning techniques to be tested. Other parsers could easily be integrated and tested with the other parts of the Kindred in order to understand how the parser performance affects the overall performance of the system. We hope that this ease-of-use will encourage others to use Kindred as a baseline method for comparison in future research.

3 Results and Discussion

In order to show the efficacy of Kindred, we evaluate the performance on the BioNLP 2016 Shared Task data for the BB3 event extraction subtask and the SeeDev binary relation subtask. Parameter optimization was used for BB3 subtask but not for the SeeDev subtask which used the default set of feature types. Both tasks used a single multiclass classifier. Tables 1 and 2 shows both the cross-validated results using the provided training/development split as well as the final results for the test set.

The results are in line with the best performing tools in the shared task. It is to be expected that it does not achieve the best score in either task. VERSE, which achieved the best score in the BB3 subtask, utilized a computational cluster to test out different parameter settings for vectorization as well as classification. LitWay, the winner of the SeeDev subtask, used hand-crafted rules for a number of the relation types. Given the computational speed and simplicity of the system, Kindred is a valuable contribution to the community.

These results suggest several possible extensions of Kindred. Firstly, a hybrid system that mixes a vector-based classifier with some hand-crafted rules may improve system performance. This would need to be implemented to allow customization in order to support different biomedical tasks. Kindred is also geared towards PubMed abstract text, especially given the integration with PubTator. Using PubTator's API to annotate other text would allow Kindred to easily integrate other text sources, including full-text articles where possible. Given the open nature of the API, we hope that these improvements, if desired by the community, could be easily developed and tested.

Kindred has several weaknesses that we hope to improve. It does not properly handle entities that lie within tokens. For example, a token "HER2+", with "HER" annotated as a gene name, denotes a breast cancer subtype that is positive for the HER2 receptor. Kindred will currently associate the full token as a gene entity and will not properly deal the "+". This is not a concern for the BioNLP Shared Task problem but may become important in other text mining tasks.

4 Conclusion

We have presented the Kindred relation extraction package. It is designed for ease-of-use to encourage more researchers to test out relation extraction in their research. By integrating a selection of file formats and connecting to a set of existing resources including PubAnnotation and PubTator, Kindred will make the first steps for a researcher must less cumbersome. We also hope that the codebase will allow researchers to build upon the methods to make further improvements in relation extraction research.

Acknowledgments

This research was supported by a Vanier Canada Graduate Scholarship. The authors would like to thank Compute Canada for computational resources used in this project.

References

Jari Björne and Tapio Salakoski. 2013. TEES 2.1: Automated annotation scheme learning in the BioNLP 2013 Shared Task. In *Proceedings of the BioNLP Shared Task 2013 Workshop*. pages 16–25.

Quoc-Chinh Bui, David Campos, Erik van Mulligen, and Jan Kors. 2013. A fast rule-based approach for biomedical event extraction. In *Proceedings of the BioNLP Shared Task 2013 Workshop*. Association for Computational Linguistics, pages 104–108.

Donald C Comeau, Riza Theresa Batista-Navarro, Hong-Jie Dai, Rezarta Islamaj Doğan, Antonio Jimeno Yepes, Ritu Khare, Zhiyong Lu, Hernani Marques, Carolyn J Mattingly, Mariana Neves, et al. 2014. Bioc interoperability track overview. *Database* 2014.

Donald C Comeau, Rezarta Islamaj Doğan, Paolo Ciccarese, Kevin Bretonnel Cohen, Martin Krallinger, Florian Leitner, Zhiyong Lu, Yifan Peng, Fabio Rinaldi, Manabu Torii, et al. 2013. Bioc: a minimalist approach to interoperability for biomedical text processing. *Database* 2013.

Ian Donaldson, Joel Martin, Berry De Bruijn, Cheryl Wolting, Vicki Lay, Brigitte Tuekam, Shudong Zhang, Berivan Baskin, Gary D Bader, Katerina Michalickova, et al. 2003. PreBIND and Textomy–mining the biomedical literature for protein-protein interactions using a support vector machine. *BMC Bioinformatics* 4(1):11.

Jin-Dong Kim and Yue Wang. 2012. PubAnnotation: a persistent and sharable corpus and annotation repository. In *Proceedings of the 2012 Workshop on Biomedical Natural Language Processing*. Association for Computational Linguistics, pages 202–205.

Robert Leaman, Rezarta Islamaj Doğan, and Zhiyong Lu. 2013. DNorm: disease name normalization with pairwise learning to rank. *Bioinformatics* .

Robert Leaman, Graciela Gonzalez, et al. 2008. BANNER: an executable survey of advances in biomedical named entity recognition. In *Pacific Symposium on Biocomputing*. volume 13, pages 652–663.

Yann LeCun, Yoshua Bengio, and Geoffrey Hinton. 2015. Deep learning. *Nature* 521(7553):436–444.

Jake Lever and Steven JM Jones. 2016. VERSE: Event and Relation Extraction in the BioNLP 2016 Shared Task. *Proceedings of the 4th BioNLP Shared Task Workshop* page 42.

Chen Li, Zhiqiang Rao, and Xiangrong Zhang. 2016. LitWay, Discriminative Extraction for Different BioEvents. *Proceedings of the 4th BioNLP Shared Task Workshop* page 32.

Christopher D Manning, Mihai Surdeanu, John Bauer, Jenny Rose Finkel, Steven Bethard, and David McClosky. 2014. The Stanford CoreNLP Natural Language Processing Toolkit. In *ACL (System Demonstrations)*. pages 55–60.

Farrokh Mehryary, Jari Björne, Sampo Pyysalo, Tapio Salakoski, and Filip Ginter. 2016. Deep Learning with Minimal Training Data: TurkuNLP Entry in the BioNLP Shared Task 2016. *Proceedings of the 4th BioNLP Shared Task Workshop* page 73.

Joakim Nivre, Marie-Catherine de Marneffe, Filip Ginter, Yoav Goldberg, Jan Hajic, Christopher D Manning, Ryan McDonald, Slav Petrov, Sampo Pyysalo, Natalia Silveira, et al. 2016. Universal Dependencies v1: A multilingual treebank collection. In *Proceedings of the 10th International Conference on Language Resources and Evaluation (LREC 2016)*. pages 1659–1666.

Nagesh C Panyam, Gitansh Khirbat, Karin Verspoor, Trevor Cohn, and Kotagiri Ramamohanarao. 2016. SeeDev Binary Event Extraction using SVMs and a Rich Feature Set. *Proceedings of the 4th BioNLP Shared Task Workshop* page 82.

Fabian Pedregosa, Gaël Varoquaux, Alexandre Gramfort, Vincent Michel, Bertrand Thirion, Olivier Grisel, Mathieu Blondel, Peter Prettenhofer, Ron Weiss, Vincent Dubourg, et al. 2011. Scikit-learn: Machine learning in Python. *Journal of Machine Learning Research* 12(Oct):2825–2830.

Dietrich Rebholz-Schuhmann, Harald Kirsch, Miguel Arregui, Sylvain Gaudan, Mark Riethoven, and Peter Stoehr. 2007. EBIMedtext crunching to gather facts for proteins from Medline. *Bioinformatics* 23(2):e237–e244.

Daniel A Schult and P Swart. 2008. Exploring network structure, dynamics, and function using NetworkX. In *Proceedings of the 7th Python in Science Conferences (SciPy 2008)*. volume 2008, pages 11–16.

Morgan Taschuk and Greg Wilson. 2017. Ten Simple Rules for Making Research Software More Robust. *PLOS Computational Biology* 13(4).

Chih-Hsuan Wei, Bethany R Harris, Hung-Yu Kao, and Zhiyong Lu. 2013a. tmVar: a text mining approach for extracting sequence variants in biomedical literature. *Bioinformatics* .

Chih-Hsuan Wei, Hung-Yu Kao, and Zhiyong Lu. 2013b. PubTator: a web-based text mining tool for assisting biocuration. *Nucleic acids research* .

Noise Reduction Methods for
Distantly Supervised Biomedical Relation Extraction

Gang Li[1]   **Cathy H. Wu**[1,2]   **K. Vijay-Shanker**[1]

[1]Department of Computer Information & Sciences
[2]Center for Bioinformatics and Computational Biology
University of Delaware
Newark, DE 19716, USA
{ligang,wuc,vijay}@udel.edu

Abstract

Distant supervision has been applied to automatically generate labeled data for biomedical relation extraction. Noise exists in both positively and negatively-labeled data and affects the performance of supervised machine learning methods. In this paper, we propose three novel heuristics based on the notion of proximity, trigger word and confidence of patterns to leverage lexical and syntactic information to reduce the level of noise in the distantly labeled data. Experiments on three different tasks, extraction of protein-protein-interaction, miRNA-gene regulation relation and protein-localization event, show that the proposed methods can improve the F-score over the baseline by 6, 10 and 14 points for the three tasks, respectively. We also show that when the models are configured to output high-confidence results, high precisions can be obtained using the proposed methods, making them promising for facilitating manual curation for databases.

1   Introduction

Biomedical relation extraction is a widely studied field that is concerned with the detection of different kinds of relations between bio-entities mentioned in text. With the rapid growth of biomedical literature, it has attracted much research interest as it makes possible to automatically extract structured information from large amounts of text. Biomedical relation extraction has helped facilitate manual curation of many biomedical databases as well as biological hypothesis generation.

Various tasks have been studied for biomedical relation extraction, e.g., extraction of protein-protein interaction (Airola et al., 2008), drug-drug interaction (Segura-Bedmar et al., 2013) and mutation-disease association (Singhal et al., 2016). In recent years, community-organized events, such as BioNLP (Kim et al., 2012, 2013) and BioCreative (Arighi et al., 2014; Wei et al., 2015b), provide comprehensive evaluation for extraction systems of a wide range of biomedical relations and events. In these tasks, supervised learning methods are commonly used and achieve state-of-the-art results.

When applying supervised learning methods, a training corpus is required to train the extraction model. The creation of a training corpus usually requires curators with domain knowledge, and is a time-consuming and labor-intensive process. Thus, it is one of the main obstacles in the use of supervised learning methods for relation extraction. To address this issue, recently researchers have been using distant supervision to construct training data automatically.

In distant supervision, a heuristic labeling process is used to label a text corpus using known related entity pairs from a database. Text containing these entity mentions or their different name variations are labeled as positive instances. To illustrate the labeling process, we show two example sentences labeled using interacting protein pairs from the database IntAct (Orchard et al., 2014).

- ⟨NgR, p75⟩: **NgR** interacted with **p75** in lipid rafts

- ⟨Mdm2, p53⟩: As a consequence, N-terminally truncated **Mdm2** binds **p53** and promotes its stability.

The above sentences are labeled as positive instances and express protein-protein interaction re-

Proceedings of the BioNLP 2017 workshop, pages 184–193,
Vancouver, Canada, August 4, 2017. ©2017 Association for Computational Linguistics

lation between the protein mention pair. When a protein pair mentioned in a sentence is not recorded by IntAct, the sentence is then labeled as a negative instance. The positively and negatively-labeled data generated by this process can potentially be used by supervised learning algorithms to train a model. Various existing biological databases and the large amount of Medline abstracts and PMC full-length articles can support applying distant supervision for many biomedical relation extraction tasks. However, the main drawback of distant supervision is that the created data can be very noisy, due to the guideless heuristic labeling process. Wrongly labeled instances exist in both positively and negatively-labeled data. For example, consider the two labeled sentences below for protein-protein interaction.

- $\langle$Mdm2, p53$\rangle$: Ribosomal protein S3: A multi-functional protein that interacts with both **p53** and **MDM2** through its KH domain.

- $\langle$LRAP35a, MYO18A$\rangle$: **LRAP35a** binds independently to **MYO18A** and MRCK.

In the first sentence, although the protein pair $\langle$Mdm2, p53$\rangle$ are interacting with each other according to IntAct, no explicit description in the sentence expresses such an interaction relation. It is labeled as a positive instance by the heuristic labeling process, which is a wrong annotation. On the other hand, if a related entity pair has not been recorded in the database, all the sentences containing their mentions will be labeled as negative instances, which may also contain wrong annotations. As an example, the protein pair $\langle$LRAP35a, MYO18A$\rangle$ in the second sentence is not recorded by IntAct. The sentence is labeled as negative, while it expresses an interaction relation between the two proteins. Thus, it is a wrong annotation in the negatively-labeled data.

In this paper, we propose three novel heuristics that attempt to reduce the noise in the positively-labeled data set P as well as the negatively-labeled data set N. First, noise can be removed from P using lexical and syntactic information of the entity mention pairs. Next, high-confidence patterns can be discovered using the purified P, which can then be used to remove noise from N. Experiments on three tasks, extraction of protein-protein interaction, miRNA-gene regulation relation and protein-localization event, show that our methods can improve the F-score by 6, 10 and 14 points over the

baseline for the three tasks, respectively. Furthermore, we show that our methods obtain 0.71, 0.95 and 0.77 precision at recall level 0.30 for the three tasks, respectively, making them promising for facilitating database curation.

In the rest of the paper, we first discuss the related work in Section 2. Section 3 describes the three tasks for experiments, as well as the databases and text corpora used in our experiments for applying distant supervision. In Section 4, we describe the details of the proposed methods. Experiments results will be reported in Section 5. We conclude with future work in Section 6.

2 Related Work

Distant supervision for relation extraction was first proposed by Craven and Kumlien (1999) to extract protein-localization relation. Mintz et al. (2009) used Freebase relations to annotate articles in Wikipedia and trained a logistic regression model to extract 102 different types of relations. Riedel et al. (2010) proposed to use multi-instance learning to tolerate noise in the positively-labeled data. They relaxed the original assumption in distant supervision that all the positively-labeled sentences of an entity pair express the relation of interest and instead, they assume that at least one of the sentences does. Hoffmann et al. (2011) and Surdeanu et al. (2012) continued to augment the multi-instance model with a multi-label classifier for each entity pair, to exploit correlations and conflicts among different relations to improve performance. In these approaches, researchers focus on developing models that can tolerate noise and improve extraction performance on entity pair level. However, it is important to note that the noise is not explicitly removed from the labeled data, and extraction on sentence level is not optimized directly.

Focusing on explicitly reducing noise from the distantly-labeled training data, Intxaurrondo et al. (2013) proposed three simple heuristics to remove noise from the positively-labeled data. They tried to filter out positively-labeled instances that appear too frequently or have a large distance from their cluster centroid, or positive entity pairs that have a low partial mutual information. Takamatsu et al. (2012) proposed a statistical model to estimate $P(relation|pattern)$, and removed positively-labeled instances that match a low-probability pattern. Xu et al. (2013)

used pseudo-relevance feedback to discover high-confidence related entity pairs which do not exist in the database, and removed negatively-labeled instances of these entity pairs. Roller et al. (2015) tried to reduce noise in the negatively-labeled data by inferring new relations of a knowledge graph using a random-walk algorithm. Roth et al. (2013) gave a nice review of some of the above methods.

Distant supervision has also been applied to extract biomedical relation. Zheng and Blake (2015) used a heuristic based on dependency path frequency to reduce noise in the positively-labeled data for extraction of protein-localization relations. Thomas et al. (2011) used a list of words which are frequently employed to indicate protein interaction to filter out noise for protein-protein interaction extraction. Roller and Stevenson (2015) tried to combine existing hand-labeled data with distantly labeled data to improve the performance for drug-condition relations. Multi-instance learning was used by Roller et al. (2015) to extract two subsets of relations in UMLS database with reduced noise by a path ranking algorithm, and by Lamurias et al. (2017) to extract miRNA-gene relations.

3 Resources

3.1 Task Definition

In this paper, we use three tasks, extraction of protein-protein interaction (PPI), miRNA-gene regulation relation (MIRGENE) and protein-localization event (PLOC), to evaluate our methods. Extraction of PPIs is a well-studied task (Miwa et al., 2009; Peng et al., 2016). We aim to extract interacting protein pairs from text using distant supervision, and evaluate it on one of the public corpora used by previous work. Extraction of miRNA-gene regulation relations have attracted much interest recently because of the rapid growth of miRNA-related literature (Bagewadi et al., 2014; Li et al., 2015). In a MIRGENE relation, a miRNA regulates gene expression via direct binding to the gene's 3' UTR or indirect pathway effect. Extraction of protein-localization event has been a subtask in BioNLP shared task from 2009 to 2013 in the Genia track (Kim et al., 2013). It describes the event that a protein is localized to a subcellular location. We only consider extraction of such events when the sentence mentions the protein and the location, same with Zheng and Blake (2015). We list an example sen-tence for each task below.

- PPI: Interaction of **Shc** with **Grb2** regulates association of Grb2 with mSOS.

- MIRGENE: **MicroRNA-223** regulates **FOXO1** expression and cell proliferation.

- PLOC: The **cyclin G1** protein was localized in **nucleus**.

3.2 Training Data Construction

To construct the training set, we need a database containing related entity pairs and a large amount of text for the heuristic labeling. Table 1 lists the databases, text corpora and numbers of positively/negatively-labeled instances produced by the heuristic labeling process for the three tasks.

Task	Database	Abstracts	Positive / Negative
PPI	IntAct	14,769	67,099 / 108,016
MIRGENE	Tarbase, miRTarBase	30,000	75,632 / 97,118
PLOC	UniProt	30,000	28,985 / 82,132

Table 1: Databases, text corpora and distantly labeled data for the three tasks.

From all the Medline abstracts, we randomly sampled 30,000 abstracts with sentences mentioning a pair of miRNA and gene for miRNA-gene regulation relation, and 30,000 abstracts with sentences mentioning a pair of protein and subcellular location for protein-localization event. We tried sampling more abstracts but the experiment results were not significantly different. For protein-protein interaction, using Medline abstracts leads to a skewed labeled data set (1:7.4 positive/negative ratio), we turned to use all the abstracts that are curated by IntAct database as the text corpus. Although this may result in less noise, we will show that our proposed methods are still able to improve performance over the baseline in the experiments.

In the heuristic labeling process, we need to recognize entity mentions in text and map them to their database entry. For gene/protein, we use the output from GenNorm++ (Wei et al., 2015a). We use simple regular expressions to recognize miRNA mentions, and map them to a miRNA entry in TarBase (Vlachos et al., 2014) or miRTar-Base (Hsu et al., 2014) using the number in the miRNA name. For subcellular location, similar to Zheng and Blake (2015), we use a dictionary from

UniProt (UniProt Consortium, 2014) and perform string matching to find subcellular location mentions. The entry "secreted" is removed as it is not a specific subcellular location. The dictionary contains name variants for each location, and we normalize a matched variant in text to its standard name.

3.3 Test Data

We evaluate the baselines and proposed methods on a test set directly for the three tasks. Note that in the context of distant supervision, we should expect little or no hand-labeled data. Hence, we can not assume the availability of a development set for the purpose of parameter tuning. Thus, when a method has multiple possible choices for a parameter, we will report the results using different parameter values.

For the test set, we use the AIMed corpus (Bunescu et al., 2005) for PPI extraction, same with Bobic et al. (2012). We extend the corpus in our work (Li et al., 2015) to include relation mention annotations, and use the development set to evaluate MIRGENE extraction. For PLOC extraction we use BioNLP 2011 Genia training and development set, same with Zheng and Blake (2015). Gold entity annotations in these corpora are used except for subcellular location, we use the dictionary from UniProt to recognize them, as BioNLP Genia corpus only annotates subcellular locations that participate in an event. The characteristics of the three test corpora are listed in Table 2. We ensure that the test sets do not overlap with the training sets. Specifically, all the abstracts used by the test sets are removed from the document pools from where the training sets are sampled.

Task	Documents	Annotations (P/N)
PPI	225	1,000 / 4,611
MIRGENE	200	464 / 775
PLOC	1,167	125 / 1,783

Table 2: Test sets for the three tasks.

4 Methods

4.1 Model and Feature Set

Logistic regression (LR) model is used for all our proposed methods in the experiments. An example sentence with relevant dependency relations and its extracted features are shown in Fig. 1 and Table 3. E-walk and v-walk features are ⟨edge, stem, edge⟩ and ⟨stem, edge, stem⟩ triples including the direction extracted from the shortest dependency path. They preserve partial structure information and are more generalizable than the full dependency path.

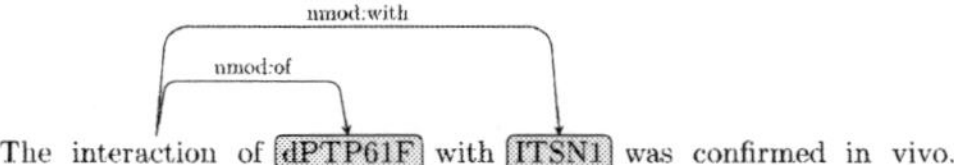

Figure 1: Example sentence for feature extraction.

No.	Feature
1	P1←nmod:of←→nmod:with→P2
2	nmod:of←interact→nmod:with
3	P1←nmod:of→interact interact→nmod:with→P2
4	P1_with_P2 of_P2_with_P2_be interact_of_P1_with_P2_be_confirm
5	2
6	1

Table 3: Features extracted from the example sentence. P1 and P2 represent the two protein mentions. 1: unlexicalized shortest dependency path; 2: e-walk features; 3: v-walks features; 4: three stem sequences, 5: number of edges on the shortest dependency path; 6: number of stems on the first stem sequence.

For all the lexical terms, we use their stems produced by Porter's stemmer (Porter, 1980). Charniak parser (Charniak, 2000; Charniak and Johnson, 2005) with the biomedical model (Mcclosky, 2010) is used to produce constituency parse for each sentence, which is converted to collapsed dependency parse using Stanford CoreNLP converter (Manning et al., 2014) with *CCprocessed* setting. We remove features that only appear once in the whole training set.

4.2 Baselines

The baseline is a LR model trained on the distantly labeled set without any filtering of noise. We also implement two previous methods for comparison. First, we train a LR model on the distantly labeled set filtered by a heuristic (DPFreq) proposed by Zheng and Blake (2015), which removes positively-labeled instances with a shortest dependency path that appear less than k times in the positive set. They hypothesize that rare dependency path is unlikely to express a relation. As we tried different values of k and obtained similar F-scores for the three tasks, we only report the results for $k = 5$ to save space. Note that since different features, text corpus and named entity recognition tool are used, we are not trying to reproduce the exact results reported in Zheng and Blake (2015). In addition, we implement a widely-used multi-instance model described in Surdeanu et al. (2012) and train it on unfiltered distantly labeled data.

4.3 Proposed Heuristics

We propose three novel filtering methods to remove noise from both positively and negatively-labeled data. These methods are applied in a sequential manner so that each step removes more noise based on the filtered data from the previous step.

The first heuristic is concerned with multiple mentions of an entity in a sentence. If the entity is related to another entity mentioned in the sentence, all the binary combinations of their mentions will be labeled as positive by the default labeling process. This usually introduces noise, since not all combinations are likely to be in the relation. For example, consider the sentence below.

Overexpression of **miR-193b** inhibited the expression of **CCND1**, and knock-down of **CCND1** inhibited the proliferation of GC cells, suggesting that **miR-193b** exerted its anti-tumorigenic role in GC cells through targeting **CCND1** gene.

miR-193b regulates CCND1 according to the database TarBase. The six binary combinations between miR-193b and CCND1 in the sentence will be labeled as positive instances. However, the sentence only expresses miRNA-gene regulation relation for the first and the last combination. The other four are wrongly labeled and hence constitute noise in the positively-labeled data.

To remove such noise, we hypothesize that only the closest pair of the entity mentions express the relation. The closest pair is defined as following: for a positively-labeled entity mention pair $\langle e_1, e_2 \rangle$, if their shortest dependency path has the smallest length among all the positively-labeled instances that involve either e_1 or e_2, the pair $\langle e_1, e_2 \rangle$ is considered as a closest pair. When computing the dependency path length, we skip the *appos* relation. The heuristic is described as below.

Heuristic of closest pairs (CP): remove positively-labeled instances that are not closest pair, when multiple mentions of one or both entities are present in the sentence.

For the three tasks and many other biomedical text-mining tasks, the relation or event is often indicated by a small set of trigger words (e.g., *interact/bind* for PPI, *regulate/target* for MIRGENE, and *localize/translocate* for PLOC). Following the usage in the BioNLP Genia corpus, we can term these words as trigger words. With knowledge of a comprehensive set of trigger words, we can hypothesize that sentences without a trigger word are less likely to express the target relation or event. We propose to automatically mine such trigger words from the large distantly-labeled corpus, and use them to remove noise from the positively-labeled data.

Trigger words are usually verbs, or in their nominal or adjectival form. Our target is then to identify stems of verb triggers, which can also be used to match nominal or adjectival form of the verb. A simple procedure is used: first, count all the verb stems on the shortest dependency paths of the positively-labeled instances generated by the heuristic labeling process. As we want to choose triggers that are strongly associated with the relation, we only use dependency paths that contain one token, excluding the two entity mentions. These verb stems are then sorted by frequency and the high-frequency stems are chosen for the trigger list. We list the top 10 verb stems for the three tasks in Table 4.

For each positively-labeled instance, we search for trigger stems in the tokens on its shortest dependency path or in the maximum dominating noun phrase. A maximum dominating noun phrase is defined as the maximally-spanning noun phrase that encloses the two entity mentions, with only noun or prepositional phrases as descendants. For example, in the text fragment "interaction between **FAK** and **PP1** regulates a process", the maximum dominating noun phrase is "interaction between **FAK** with **PP1**" for this protein mention pair. As sentences without a trigger word are less likely to express the target relation or event, we use the heuristic described below to remove noise.

Heuristic of trigger word (TW): remove positively-labeled instances if a trigger stem is not found on the shortest dependency path or in the maximum dominating noun phrase of the entity mention pair.

By using heuristic CP and TW, we can already filter out a substantial part of the positively-labeled

Task	Verb stems	Pattern and example sentence
PPI	interact, bind, associ, phosphoryl, recruit, activ, coloc, coimmunoprecipit, co-immunoprecipit, regul	PROTEIN1←nsubj←interact→nmod:with→PROTEIN2 **mGrb10** <u>interacts</u> with **Nedd4**.
MIRGENE	target, regul, inhibit, downregul, suppress, repress, down-regul, correl, induc, promot	GENE←dobj←target←advcl←root→nsubj→MIRNA **MiR-429** play its role in PDAC by <u>targeting</u> **TBK1**.
PLOC	local, transloc, express, associ, interact, detect, coloc, find, co-loc, target	PROTEIN←nmod:of←transloc→amod→LOCATION Importin beta mediates **nuclear** <u>translocation</u> of **Smad 3**.

Table 4: The top 10 verb stems and top pattern and example sentence for the three tasks.

data. Using heuristic CP+TW with 50 trigger stems, 65% of the positively-labeled data can be removed for PPI. For MIRGENE and PLOC, the removal ratio is 38% and 59%, respectively. We hypothesize that the remaining set will still contain a large amount of data for training and more importantly, it will be of high quality, and thus it would be possible to discover high-confidence patterns from it using pattern occurrence frequency.

Finally, we turn to the last heuristic that we introduce. Recall noisy instances in negatively-labeled data should be labeled as positive but are negatively labeled because of incompleteness of the database used for distant supervision. We try to mine some high-confidence patterns from the purified positively-labeled set after the application of heuristic CP and TW. We define a pattern as a shortest dependency path lexicalized by a trigger stem between the entity mention pair. The pattern frequencies in the positively-labeled data filtered by heuristic CP and TW are counted. The most frequent pattern and an example sentence for each task are shown in Table 4.

Our hypothesis is that any entity mention pair connected by a high-confidence pattern is likely to be related and hence probably constitute noise in the negatively-labeled data. Therefore, we consider the next heuristic described below.

Heuristic of high-confidence patterns (HP): remove negatively-labeled instances which match a high-confidence pattern mined from positively-labeled data.

Note that heuristic DPFreq, CP and TW remove instances from the positively-labeled data, whereas HP is the only heuristic that removes instances from the negatively-labeled data. Heuristic TW depends on the number of trigger stems, while heuristic HP depends on both the number of trigger stems and high-confidence patterns, as it needs the trigger stems to lexicalize the shortest dependency path to form a pattern.

5 Results and Discussions

We use precision, recall and F-score to evaluate the baselines and proposed methods. The top 50 trigger stems were used in heuristic TW, while the top 50 trigger stems and the top 100 patterns were used in heuristic HP. The results are presented in Table 5. Specificity is also presented. We will discuss how different numbers of trigger stems and patterns may affect the results later.

Table 5 shows that the multi-instance model and the use of heuristic DPFreq or CP increased precision compared to the baseline for all the three tasks, indicating that they can effectively remove noise from the positively-labeled data. Using heuristic CP+TW further improved precisions over heuristic CP for the three tasks. However, using heuristic DPFreq, CP or CP+TW did not improve the F-score over the baseline for PPI and MIRGENE, due the decreased recall. By removing noise from the negatively-labeled data using heuristic HP in addition to CP and TW, the recalls can be improved with minor or no decrease in precision, resulting in higher F-scores than the baseline, the MI model and other heuristics for all the three tasks. This suggests that the proposed heuristics can effectively remove noise from both positively and negatively-labeled data, and to obtain better F-scores, it is important to filter both positive and negative set to improve precision and recall simultaneously. Although PLOC extraction did not obtain a good precision in all the experiments, we will show that high precision can be achieved for high-confidence PLOC extraction later in this section.

By applying heuristic CP+TW+HP, the F-score can be improved by 10 points for PPI extraction compared to Bobic et al. (2012), and 11 points for PLOC extraction compared to Zheng and Blake (2015).

Different numbers of trigger stems: as different numbers of trigger stems can be used in heuristic TW and HP, we investigated how they affect

	PPI				MIRGENE				PLOC			
Method	P	R	F	S	P	R	F	S	P	R	F	S
Bobic et al. (2012)	0.26	**0.78**	0.39	-	-	-	-	-	-	-	-	-
Zheng and Blake (2015)	-	-	-	-	-	-	-	-	**0.43**	0.25	0.31	-
Baseline	0.37	0.52	0.43	0.86	0.56	0.58	0.57	0.74	0.18	**0.57**	0.28	0.94
Multi-instance (MI)	0.57	0.35	0.43	0.91	0.64	0.56	0.59	0.78	0.22	0.38	0.29	0.94
DPFreq	0.42	0.41	0.41	0.87	0.63	0.50	0.56	0.78	0.21	0.39	0.29	0.94
CP	0.55	0.34	0.42	0.95	0.68	0.50	0.57	0.81	0.26	0.51	0.35	0.95
CP+TW	**0.69**	0.28	0.40	0.93	0.72	0.44	0.55	0.83	0.34	0.42	0.37	0.95
CP+TW+HP	0.65	0.39	**0.49**	0.93	**0.73**	**0.61**	**0.67**	0.84	0.35	0.53	**0.42**	0.95

Table 5: Precision, recall, F-score and specificity of all the methods for three extraction tasks.

the performance for the three tasks. In Fig. 2 (a)-(c), precisions, recalls and F-scores are shown for applying heuristic CP+TW and CP+TW+HP (using top 100 patterns) with different numbers of trigger stems. PPI and MIRGENE extraction maintained a stable precision with increasing recall when the number of trigger stem increased. For PLOC extraction precision decreased with increased recall when more trigger stems were used, indicating that the quality of the trigger stems can be improved. Using 100 patterns to remove noise resulted in much better recalls and F-scores for all the three tasks across different numbers of trigger stems, further confirming that heuristic HP is an effective method to remove noise from the negatively-labeled data.

Different numbers of patterns: we investigated how different numbers of patterns used by heuristic HP affect the results. In Fig. 2 (d)-(f), precisions, recalls and F-scores are shown for applying CP+TW+HP (using top 50 trigger stems) with different number of patterns. The performances using heuristic CP+TW with 50 trigger stems are included for comparison. We can see that recalls can be consistently improved when more patterns were used, with minor or no decrease in precision. Compared to the results only using heuristic CP+TW, even using small number of patterns can achieve better performance.

A major use case of biomedical relation extraction is to help identify high-confidence entity pairs to facilitate manual curation for databases. Thus, a desired property of a relation extractor is to achieve high precision for such high-confidence extractions. Logistic regression model outputs a probability for each test instance, and high probability indicates high confidence to be positive.

To investigate the performance of the proposed methods for the high-confidence extractions, we draw precision-recall curves using the probability produced by the logistic regression model. By definition, logistic regression model predicts an instance as positive if the probability is greater than 0.5. By varying the threshold, we can calculate precisions at different recall levels. For example, when the threshold is set to 0.9, the model only predicts an instance with probability greater than 0.9 as positive. Ideally the model should achieve better precision when the threshold is high.

For each task, six curves are drawn in Fig. 3. We can see that using heuristic CP+TW+HP obtained higher precisions than the baselines and other heuristics on the left side of the figures, which correspond to the performance for high-confidence extractions. The multi-instance model also obtained better precisions compared to the baseline at lower recall levels. Specifically, by using heuristic CP+TW+HP, PPI, MIRGENE and PLOC extraction can achieve the highest precisions among the six curves, which are 0.71, 0.95 and 0.77, respectively, at recall level 0.30.

6 Conclusion

In this paper, we proposed three novel heuristics that use lexical and syntactic information to remove noise from labeled data generated by distant supervision. Experiments showed that the proposed methods achieved significantly higher F-scores than the baseline and previous works for the three tasks, and high precision can be obtained for high-confidence results. For future work, we plan to improve the trigger stem list by asking curators to remove non-informative stems. Aggregating evidences from all the sentences for entity pair level extraction or incorporating direct supervision (Wallace et al., 2016) are two interesting directions.

The code and data used in the experiments

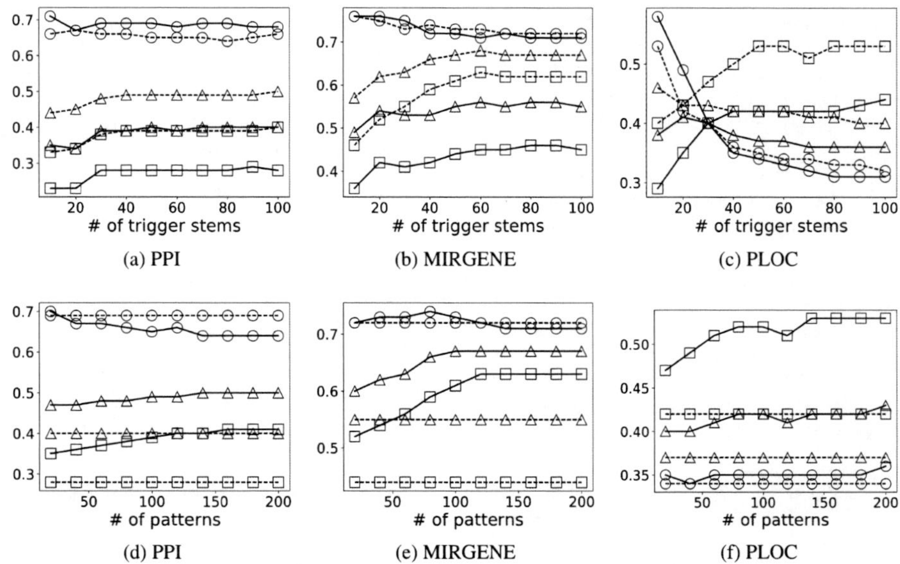

Figure 2: Results of using different numbers of trigger stems (a)-(c) and patterns (d)-(f). Markers: precision (circle), recall (square), F-score (triangle). (a)-(c): CP+TW (solid) and CP+TW+HP (dashed). (d)-(f): CP+TW (dashed) and CP+TW+HP (solid).

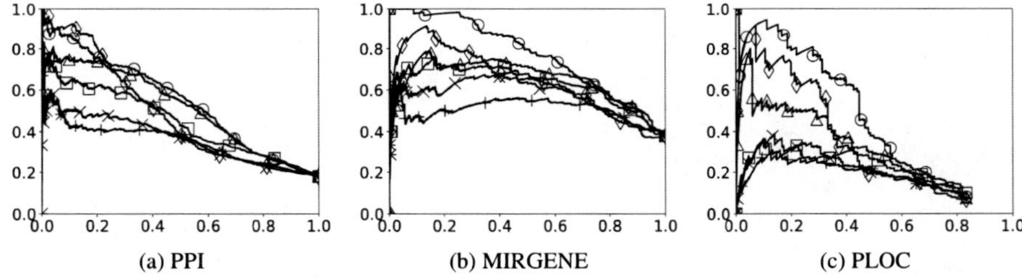

Figure 3: Precision-recall curves for the three tasks. Y-axis represents precision and X-axis represents recall. Markers: baseline (+), multi-instance (diamond), DPFreq (x), CP (square), CP+TW using 50 trigger stems (triangle), CP+TW+HP using 50 trigger stems and 100 patterns (circle).

of this paper are available at `http://biotm.cis.udel.edu/biotm/projects/ds`.

Acknowledgments

Research reported in this manuscript was supported by the National Institutes of Health under the Grant No. U01GM120953. The funders had no role in study design, data collection and analysis, decision to publish, or preparation of the manuscript.

References

Antti Airola, Sampo Pyysalo, Jari Björne, Tapio Pahikkala, Filip Ginter, and Tapio Salakoski. 2008. All-paths graph kernel for protein-protein interaction extraction with evaluation of cross-corpus learning. *BMC Bioinformatics* 9 Suppl 11:S2. https://doi.org/10.1186/1471-2105-9-S11-S2.

Cecilia N Arighi, Cathy H Wu, Kevin B Cohen, Lynette Hirschman, Martin Krallinger, Alfonso Valencia, Zhiyong Lu, John W Wilbur, and Thomas C Wiegers. 2014. BioCreative-IV virtual issue. *Database* 2014. https://doi.org/10.1093/database/bau039.

Shweta Bagewadi, Tamara Bobić, Martin Hofmann-Apitius, Juliane Fluck, and Roman Klinger. 2014. Detecting miRNA mentions and relations in biomedical literature. *F1000Res.* 3. https://doi.org/10.12688/f1000research.4591.3.

Tamara Bobic, Roman Klinger, Philippe Thomas, and Martin Hofmann-Apitius. 2012. *Proceedings of the Joint Workshop on Unsupervised* and *Semi-Supervised Learning in NLP*, Association for Computational Linguistics, chapter Improving Distantly Supervised Extraction of Drug-Drug and Protein-Protein Interactions, pages 35–43. http://aclweb.org/anthology/W12-0705.

Razvan Bunescu, Ruifang Ge, Rohit J Kate, Edward M Marcotte, Raymond J Mooney, Arun K Ramani, and Yuk Wah Wong. 2005. Comparative experiments on learning information extractors for proteins and their interactions. *Artif. Intell. Med.* 33(2):139–155. https://doi.org/10.1016/j.artmed.2004.07.016.

Eugene Charniak. 2000. A maximum-entropy-inspired parser. In *1st Meeting of the North American Chapter of the Association for Computational Linguistics*. http://aclweb.org/anthology/A00-2018.

Eugene Charniak and Mark Johnson. 2005. Coarse-to-fine n-best parsing and MaxEnt discriminative reranking. In *Proceedings of the 43rd Annual Meeting on Association for Computational Linguistics*. Association for Computational Linguistics, Stroudsburg, PA, USA, volume 1 of *ACL '05*, pages 173–180. http://aclweb.org/anthology/P05-1022.

M Craven and J Kumlien. 1999. Constructing biological knowledge bases by extracting information from text sources. *Proc. Int. Conf. Intell. Syst. Mol. Biol.* pages 77–86.

Raphael Hoffmann, Congle Zhang, Xiao Ling, Luke Zettlemoyer, and S. Daniel Weld. 2011. Knowledge-based weak supervision for information extraction of overlapping relations. In *Proceedings of the 49th Annual Meeting of the Association for Computational Linguistics: Human Language Technologies*. Association for Computational Linguistics, pages 541–550. http://aclweb.org/anthology/P11-1055.

Sheng-Da Hsu, Yu-Ting Tseng, Sirjana Shrestha, Yu-Ling Lin, Anas Khaleel, Chih-Hung Chou, Chao-Fang Chu, Hsi-Yuan Huang, Ching-Min Lin, Shu-Yi Ho, Ting-Yan Jian, Feng-Mao Lin, Tzu-Hao Chang, Shun-Long Weng, Kuang-Wen Liao, I-En Liao, Chun-Chi Liu, and Hsien-Da Huang. 2014. miRTarBase update 2014: an information resource for experimentally validated miRNA-target interactions. *Nucleic Acids Res.* 42(Database issue):D78–85. https://doi.org/10.1093/nar/gkt1266.

Ander Intxaurrondo, Mihai Surdeanu, Oier Lopez de Lacalle, and Eneko Agirre. 2013. Removing noisy mentions for distant supervision. *Procesamiento del Lenguaje Natural* 51(0):41–48.

Jin-Dong Kim, Ngan Nguyen, Yue Wang, Jun'ichi Tsujii, Toshihisa Takagi, and Akinori Yonezawa. 2012. The genia event and protein coreference tasks of the BioNLP shared task 2011. *BMC Bioinformatics* 13 Suppl 1(Suppl 11):S1. https://doi.org/10.1186/1471-2105-13-S11-S1.

Jin-Dong Kim, Yue Wang, and Yamamoto Yasunori. 2013. *Proceedings of the BioNLP Shared Task 2013 Workshop*, Association for Computational Linguistics, chapter The Genia Event Extraction Shared Task, 2013 Edition - Overview, pages 8–15. http://aclweb.org/anthology/W13-2002.

Andre Lamurias, Luka A Clarke, and Francisco M Couto. 2017. Extracting microRNA-gene relations from biomedical literature using distant supervision. *PLoS One* 12(3):e0171929. https://doi.org/10.1371/journal.pone.0171929.

Gang Li, Karen E Ross, Cecilia N Arighi, Yifan Peng, Cathy H Wu, and K Vijay-Shanker. 2015. miR-Tex: A text mining system for miRNA-Gene relation extraction. *PLoS Comput. Biol.* 11(9):e1004391. https://doi.org/10.1371/journal.pcbi.1004391.

Christopher Manning, Mihai Surdeanu, John Bauer, Jenny Finkel, Steven Bethard, and David Mc-Closky. 2014. The stanford corenlp natural language processing toolkit. In *Proceedings of 52nd Annual Meeting of the Association for Computational Linguistics: System Demonstrations*. Association for Computational Linguistics, pages 55–60. https://doi.org/10.3115/v1/P14-5010.

David Mcclosky. 2010. *Any Domain Parsing: Automatic Domain Adaptation for Natural Language Parsing*. Ph.D. thesis, Brown University, Providence, RI, USA.

Mike Mintz, Steven Bills, Rion Snow, and Daniel Jurafsky. 2009. Distant supervision for relation extraction without labeled data. In *Proceedings of the Joint Conference of the 47th Annual Meeting of the ACL and the 4th International Joint Conference on Natural Language Processing of the AFNLP*. Association for Computational Linguistics, pages 1003–1011. http://aclweb.org/anthology/P09-1113.

Makoto Miwa, Rune Saetre, Yusuke Miyao, and Jun'ichi Tsujii. 2009. Protein-protein interaction extraction by leveraging multiple kernels and parsers. *Int. J. Med. Inform.* 78(12):e39–46. https://doi.org/10.1016/j.ijmedinf.2009.04.010.

Sandra Orchard, Mais Ammari, Bruno Aranda, Lionel Breuza, Leonardo Briganti, Fiona Broackes-Carter, Nancy H Campbell, Gayatri Chavali, Carol Chen, Noemi del Toro, Margaret Duesbury, Marine Dumousseau, Eugenia Galeota, Ursula Hinz, Marta Iannuccelli, Sruthi Jagannathan, Rafael Jimenez, Jyoti Khadake, Astrid Lagreid, Luana Licata, Ruth C Lovering, Birgit Meldal, Anna N Melidoni, Mila Milagros, Daniele Peluso, Livia Perfetto, Pablo Porras, Arathi Raghunath, Sylvie Ricard-Blum, Bernd Roechert, Andre Stutz, Michael Tognolli, Kim van Roey, Gianni Cesareni, and Henning Hermjakob. 2014. The MIntAct project–IntAct as a common curation platform for 11 molecular interaction databases. *Nucleic Acids Res.* 42(Database issue):D358–63. https://doi.org/10.1093/nar/gkt1115.

Yifan Peng, Cecilia Arighi, Cathy H Wu, and K Vijay-Shanker. 2016. BioC-compatible full-text passage detection for protein-protein interactions using extended dependency graph. *Database* 2016. https://doi.org/10.1093/database/baw072.

M F Porter. 1980. An algorithm for suffix stripping. *Programmirovanie* 14(3):130–137.

Sebastian Riedel, Limin Yao, and Andrew McCallum. 2010. *Modeling Relations and Their Mentions without Labeled Text*, Springer Berlin Heidelberg, volume 6323 of *Lecture Notes in Computer Science*, page 148–163. https://doi.org/10.1007/978-3-642-15939-8_10.

Roland Roller, Eneko Agirre, Aitor Soroa, and Mark Stevenson. 2015. Improving distant supervision using inference learning. In *Proceedings of the 53rd Annual Meeting of the Association for Computational Linguistics and the 7th International Joint Conference on Natural Language Processing (Volume 2: Short Papers)*. Association for Computational Linguistics, pages 273–278. https://doi.org/10.3115/v1/P15-2045.

Roland Roller and Mark Stevenson. 2015. *Proceedings of BioNLP 15*, Association for Computational Linguistics, chapter Making the most of limited training data using distant supervision, pages 12–20. https://doi.org/10.18653/v1/W15-3802.

Benjamin Roth, Tassilo Barth, Michael Wiegand, and Dietrich Klakow. 2013. A survey of noise reduction methods for distant supervision. In *Proceedings of the 2013 workshop on Automated knowledge base construction*. ACM, pages 73–78. https://doi.org/10.1145/2509558.2509571.

Isabel Segura-Bedmar, Paloma Martınez, and Marıa Herrero-Zazo. 2013. Semeval-2013 task 9: Extraction of drug-drug interactions from biomedical texts (ddiextraction 2013). *Proceedings of Semeval* pages 341–350.

Ayush Singhal, Michael Simmons, and Zhiyong Lu. 2016. Text mining Genotype-Phenotype relationships from biomedical literature for database curation and precision medicine. *PLoS Comput. Biol.* 12(11):e1005017. https://doi.org/10.1371/journal.pcbi.1005017.

Mihai Surdeanu, Julie Tibshirani, Ramesh Nallapati, and D. Christopher Manning. 2012. Multi-instance multi-label learning for relation extraction. In *Proceedings of the 2012 Joint Conference on Empirical Methods in Natural Language Processing and Computational Natural Language Learning*. Association for Computational Linguistics, pages 455–465. http://aclweb.org/anthology/D12-1042.

Shingo Takamatsu, Issei Sato, and Hiroshi Nakagawa. 2012. Reducing wrong labels in distant supervision for relation extraction. In *Proceedings of the*

50th Annual Meeting of the Association for Computational Linguistics (Volume 1: Long Papers)*. Association for Computational Linguistics, pages 721–729. http://aclweb.org/anthology/P12-1076.

Philippe Thomas, Illès Solt, Roman Klinger, and Ulf Leser. 2011. *Proceedings of Workshop on Robust Unsupervised and Semisupervised Methods in Natural Language Processing*, Association for Computational Linguistics, chapter Learning Protein Protein Interaction Extraction using Distant Supervision, pages 25–32. http://aclweb.org/anthology/W11-3904.

UniProt Consortium. 2014. Activities at the universal protein resource (UniProt). *Nucleic Acids Res.* 42(Database issue):D191–8. https://doi.org/10.1093/nar/gkt1140.

Ioannis S Vlachos, Maria D Paraskevopoulou, Dimitra Karagkouni, Georgios Georgakilas, Thanasis Vergoulis, Ilias Kanellos, Ioannis-Laertis Anastasopoulos, Sofia Maniou, Konstantina Karathanou, Despina Kalfakakou, Athanasios Fevgas, Theodore Dalamagas, and Artemis G Hatzigeorgiou. 2014. DIANA-TarBase v7.0: indexing more than half a million experimentally supported miRNA:mRNA interactions. *Nucleic Acids Res.* https://doi.org/10.1093/nar/gku1215.

Byron C Wallace, Joël Kuiper, Aakash Sharma, Mingxi Brian Zhu, and Iain J Marshall. 2016. Extracting PICO sentences from clinical trial reports using supervised distant supervision. *J. Mach. Learn. Res.* 17.

Chih-Hsuan Wei, Hung-Yu Kao, and Zhiyong Lu. 2015a. GNormPlus: An integrative approach for tagging genes, gene families, and protein domains. *Biomed Res. Int.* 2015:918710. https://doi.org/10.1155/2015/918710.

Chih-Hsuan Wei, Yifan Peng, Robert Leaman, Allan Peter Davis, Carolyn J Mattingly, Jiao Li, Thomas C Wiegers, and Zhiyong Lu. 2015b. Overview of the BioCreative V chemical disease relation (CDR) task. In *Proceedings of the fifth BioCreative challenge evaluation workshop*. pages 154–166.

Wei Xu, Raphael Hoffmann, Le Zhao, and Ralph Grishman. 2013. Filling knowledge base gaps for distant supervision of relation extraction. In *Proceedings of the 51st Annual Meeting of the Association for Computational Linguistics (Volume 2: Short Papers)*. Association for Computational Linguistics, pages 665–670. http://aclweb.org/anthology/P13-2117.

Wu Zheng and Catherine Blake. 2015. Using distant supervised learning to identify protein subcellular localizations from full-text scientific articles. *J. Biomed. Inform.* https://doi.org/10.1016/j.jbi.2015.07.013.

Role-Preserving Redaction of Medical Records to Enable Ontology-Driven Processing

Seth Polsley, Atif Tahir, Muppala Raju, Akintayo Akinleye, Duane Steward
Texas A&M Health Science Center
College of Medicine Biomedical Informatics Research group
College Station, Texas
`spolsley,atif.tahir,mnpr84,akinleyeakintayo,dsteward@tamu.edu`

Abstract

Electronic medical records (EMR) have largely replaced hand-written patient files in healthcare. The growing pool of EMR data presents a significant resource in medical research, but the U.S. Health Insurance Portability and Accountability Act (HIPAA) mandates redacting medical records before performing any analysis on the same. This process complicates obtaining medical data and can remove much useful information from the record. As part of a larger project involving ontology-driven medical processing, we employ a method of recognizing protected health information (PHI) that maps to ontological terms. We then use the relationships defined in the ontology to redact medical texts so that roles and semantics of terms are retained without compromising anonymity. The method is evaluated by clinical experts on several hundred medical documents, achieving up to a 98.8% f-score, and has already shown promise for retaining semantic information in later processing.

1 Introduction

Medical health records data has immense potential for research in furthering the field of automated healthcare. Unfortunately, one of the challenges facing medical informatics is the dissemination and sharing of digital records for research and analysis due to strict regulations regarding patient confidentiality. Protecting protected health information (PHI) is a critical responsibility of health care providers, with the U.S. Health Insurance Portability and Accountability Act (HIPAA) outlining a number of principles. Removing PHI can also mean removing critical parts of a record, so building redaction techniques that preserve as much information about the original data as possible while still retaining anonymity is an important pre-processing step.

In this work, we discuss a redaction framework for removing PHI from medical records through de-identification. One of the primary goals of this framework is to preserve valuable information like roles, semantics, and time intervals as much as possible. Because this forms the pre-processing stage of future text processing, we elected to model roles according to a formal ontology; this maintains relationships and enables straightforward detection of ontological terms in later phases.

2 Background

Knowledge buried in medical text is valuable, but due to federal law protecting sensitive data, it must be de-identified for distribution. Most existing methods rely on rule-based systems that match patterns and dictionaries of expressions that frequently contain PHI. Sweeny's Scrub tool uses templates and a context window to replace PHI (Sweeney, 1996). Datafly, also by Sweeny, offers user-specific profiles, including a list of preferred fields to be scrubbed (Sweeney, 1997). Thomas developed a method that uses a lexicon of 1.8 million names to identify people along with "Clinical and Common Usage" words from the Unified Medical Language System (UMLS) (Thomas et al., 2002). Miller developed a de-identification system for cleaning proper names from records of indexed surgical pathology reports at the Johns Hopkins Hospital (Miller et al., 2001). Proper names were identified from available lists of persons, places and institutions, or by their proximity to keywords, such as "Dr." or "hospital." The Perl

Proceedings of the BioNLP 2017 workshop, pages 194–199,
Vancouver, Canada, August 4, 2017. ©2017 Association for Computational Linguistics

tool *Deid* is a recent development which combines several of these rule-based and lexical approaches with some additional capabilities like better handling of time (Neamatullah et al., 2008).

While identifying PHI for removal or anonymization remains an open challenge, simply redacting texts overlooks one of the more fundamental aspects of recent biomedical informatics, which has incorporated a focus on ontology-driven development (Mortensen et al., 2012; Ye et al., 2009; Tao et al., 2013; Sari et al., 2013; Omran et al., 2009; Lumsden et al., 2011; Pathak et al., 2009). In a domain like healthcare – where information is dense, diverse, and specialized – an ontology allows representing knowledge in a usable manner, because it describes a framework for clearly defining known terms and their relationships (Hakimpour and Geppert, 2005; Lee et al., 2006; Pieterse and Kourie, 2014; Strohmaier et al., 2013; Kapoor and Sharma, 2010). Once the data has been formally described via an ontology, new applications become apparent. To provide several examples, simply by formalizing electronic records as an ontology, researchers have shared better ways to represent patient care profiles (Riaño et al., 2012), perform risk assessment (Draghici and Draghici, 2000), evaluate elderly care (Hsieh et al., 2015), and more (Rector et al., 2009; Rajamani et al., 2014). Perhaps the greatest promise lies in ontology-driven computational models, where the structure of an ontology makes the data accessible to programmatic operations, and there have been several applications to the problem of automated diagnosis (Bertaud-Gounot et al., 2012; Haug et al., 2013; Hoogendoorn et al., 2016).

Some of these ontology-driven techniques do consider redaction as it relates to the ontology. Of particular note is the extensive work by South et al. in identifying the exact types of PHI present throughout the medical record according to risk (South et al., 2014). Dernoncourt applied recurrent neural networks to the task of identifying PHI by type to remove the need for large dictionaries on the i2b2 dataset (Dernoncourt et al., 2016). In the future, we hope to share a more direct contrast between our role-labeling and South et al.'s, but our goals remain distinct from either South et al. or Dernoncourt. Because our ontology centers around the medical encounter, we must leverage the EMR's dynamic list of patients, caregivers, and providers to ensure roles are preserved according to their specific encounter. In this way, our work is more similar to Douglass' MIMIC dataset, which uses a patient list to assure role (Douglass et al., 2004).

3 Methods

The core reasoning for our methodology is that knowing the role of a redacted name can be vital, and since we will be processing patient records at the encounter-level, tying specific roles to single encounters is necessary. For instance, was a condition reported by the caregiver or by the clinician and at what time? That is just a single question illustrating the potential for confusion when names are redacted without roles or ordering, yet, there is no need to blindly attempt to extract roles from free text. Nearly every EMR maintains structured data like a patient's name, family contact, and attending physician. By leveraging this knowledge, pseudonyms can be constructed that remove confusion regarding roles in the final text.

To formally support role-preservation, we begin by defining a very simple ontology to relate key roles and terms. Patients are treated by clinicians and observed by caregivers. Treatments (or interventions) are given on the basis of a medical encounter, and, depending on the outcome, may lead to more medical encounters or the end of the record of care. This is a very basic means of modeling roles in medical texts, but it supports cross-domain redaction that preserves much of the semantics and relationships after the anonymization stage.

The redaction pipeline operates on data in two stages to support better identification of roles in the text. First, the structured data is used to extract whatever knowledge is available, typically roles like doctors and patients, to perform knowledge-based redaction. Second, the unstructured text undergoes entity recognition to clean missed terms. While this approach requires some insight about the data beforehand, it is a logical means of ensuring we can remove all PHI without damaging roles and relationships.

3.1 Structured

3.1.1 Patient-Centric Role Preservation

Our system initially builds a dictionary of known individuals in each role. A person can have any number of names of any length but all of them are

Table 1: Sample dictionary of names

Patients	Caregivers	Providers
Original		
Ira Jones	Michael Jones	Daniel Moore
	Barbara Davis	Mary Johnson
Redacted		
$Clark$	$Clark_{CAREGIVER1}$	$Clark_{PROVIDER1}$
	$Clark_{CAREGIVER2}$	$Clark_{PROVIDER2}$

drawn directly from the fields in the EMR. In accordance with the ontology, patients will be identified first as the subject of care, a unique field in most systems. Depending on the domain, there will be a personal doctor, an attending physician, or some other clinician name given in a separate field. Caregivers may be drawn from locations like billing or family contacts. For this part, knowledge of the data structure is necessary, but once the source fields are identified, they will be consistent across the other records.

Once the dictionary of names and roles is built, patients are assigned a pseudonym randomly from a list of non-matching family names to provide anonymity and linked to the pseudonym in the dictionary. Subsequently, all individuals associated with that patient are assigned a derivative pseudonym denoting their role. Consider the example shown in Table 1. For this small dictionary of a single patient, we see more than one caregiver and provider listed. The system first replaces the patient's name, Patricia Jones, with a false name, Clark. This identifier then becomes the basis for all subsequent individuals with a connection to the patient.

After the dictionary has been constructed, the system knows all the original names and their new pseudonyms. The medical texts are scanned for any occurrence of any known name, ignoring case or modifiers like possessive forms. Full names will be on file, but given names and family names may appear separately in the record. Regular expressions are used to match variants of names while enforcing order.

3.1.2 Date Offsets

It is worth emphasizing the importance of dates in medical record data. One can simply remove or replace dates to redact PHI, as with names, but just like names, we wished to preserve more information in support of the ontology. In particular, intervals between encounters or patient ages under 89 are compliant with HIPAA and useful for tasks like association mining. A common solution is to use offsets for dates because the original date will be erased from the document without losing intervals. However, an unconstrained random offset still loses information. For instance, intervals given in the free text will be broken if a day of the week is mentioned and then a date given. Our system ensures intervals are undamaged by constraining date offsets in week-long intervals. Thus, even if the dates are moved by years, there's no loss in day-granular intervals.

The date offset is applied across all records of a single patient uniformly to maintain interval and continuity of encounters. Furthermore, the system is very flexible about handling dates in free text, using as much knowledge as possible to piece together correct, redacted dates. For example, a snippet of a medical note may read: "A surgery was performed in 2005 to correct the issue; on March 4, the patient..." Because the redaction system makes use of the structured fields, it would extract the date of entry for this medical note. Assuming that date is *March 7, 2006*, the system will move forward labeling unspecified years as *2006*, giving a means of differentiating the vague dates *2005* and *March 7*.

3.2 Unstructured

The second pass of de-identification also operates over free text, but it does not make use of known information such as the dictionary of names or the dates of an entry. Instead, general attributes of potential PHI are used to locate and remove sensitive data. Email addresses, phone numbers, mailing addresses, and medical case numbers are located through common regular expressions. ZIP codes are retained because they are not considered PHI and can be useful for location-based operations.

Unknown entities appear frequently in the text due to other names of people or places being written that are not listed in the dictionary of names. To account for these entities, Stanford's *CoreNLP* is used to detect any remaining entities in the text which do not belong to a linked pseudonym (Manning et al., 2014). All entities are redacted according to their determined type, e.g. $NAME1$ for a person or $LOCATION1$ for a place. Even in the unstructured phase, sequential naming schemes ensure unknown people and places do not become confounded with any other entities.

3.3 Complete Pipeline

By the time the pipeline has finished, the text has been run through two rounds of de-identification. First, any useful knowledge is pulled from the data in the EMR to build a dictionary for rule-based redaction that preserves roles. Second, operating without any knowledge, a set of regular expressions and more sophisticated entity recognition methods are employed to clear other sensitive data without adding ambiguity or destroying valuable non-PHI information. The inclusion of *CoreNLP* in the final part supports more advanced language models than simply using rules and regular expressions. This allows the complete pipeline to capture almost any potential PHI while still recognizing known entities, particularly those relevant to the ontology, or types of entities, such as contact numbers of locations.

4 Evaluation

We worked with data sets from two different domains – veterinary and hospice care. Fortunately, due to the cross-domain design of our ontology, there was little difficulty in identifying fields that mapped to elements of the ontology. Upon defining this mapping, huge portions of text from both domains were pushed through the full pipeline. The resulting text included ontological terms and other marked regions, e.g. ZIP codes, while removing as little other information as possible.

Ideally, the final medical texts appear identical to the original files with only the PHI removed. To evaluate this, a team of clinical experts reviewed hundreds of documents, marking missed PHI or text that was unnecessarily redacted in each. From the veterinarian domain, where we studied complete discharge summaries (DS), two medical doctors reviewed 122 cases. From the hospice domain, which operated on shorter clinical notes (CN), the same experts reviewed 500 notes. To provide a simple baseline for comparison, we also tested a single rule-based approach for matching patient names against a data set of 15 documents.

As we see in Table 2, the system performed very well at correctly identifying PHI and non-PHI, especially in contrast with the patient-names baseline. In the discharge summaries, the majority of false negatives were due to previously-unnamed doctors who were neither in the dictionary nor detected during entity recognition. Only one misspelling of a patient name was detected.

Table 2: Word-level metrics for baseline (BL), discharge summaries (DS), and clinical notes (CN)

Count	BL	DS	CN
False Negatives	498	76	4
False Positives	0	5	250
True Positives	63	3391	1655
True Negatives	17191	75694	50460

Table 3: Performance of baseline (BL), discharge summaries (DS), and clinical notes (CN).

Metric	BL	DS	CN
Specificity	100%	99.9%	99.5%
Sensitivity	11.2%	97.8%	99.8%
Precision	100%	99.9%	86.9%
F-Score	20.2%	98.8%	92.9%

In the clinical notes, there were a great deal more false positives. Because the final step incorporates *CoreNLP*, certain texts will include many entities that are not PHI. Table 3 shows that specificity, sensitivity/recall, and precision are high for both, although the precision for clinical notes suffers due to the many false positives. While the baseline achieves high precision by matching only patient names, the lower sensitivity and f-score demonstrate the high number of PHI belonging to other categories that the full system captures.

5 Conclusion and Ongoing Work

Medical records can provide a wealth of information for data scientists but due to their sensitive nature, are often limited in availability. Effective, reliable redaction is the best known solution to the problem, but most techniques will lose exact details like encounter-level roles. In this work, we integrate knowledge and model-based approaches to augment redaction. In future works, we seek to share some of the benefits we have seen using roles to create better semantic clusters and word models than achieved through only pseudonyms. We hope that such de-identification pipelines, highly cognizant of the original data structure, will encourage a future of richer and more capable ontology-driven analysis.

Acknowledgments

We wish to thank Girish Kasiviswanathan, Pratik Mrinal, Amber Schulze, Preston Baker, Ruihong Huang, and Tracy Hammond for their support.

References

Valérie Bertaud-Gounot, Régis Duvauferrier, and Anita Burgun. 2012. Ontology and medical diagnosis. *Informatics for Health and Social Care* 37(2):51–61.

Franck Dernoncourt, Ji Young Lee, Ozlem Uzuner, and Peter Szolovits. 2016. De-identification of patient notes with recurrent neural networks. *Journal of the American Medical Informatics Association* page ocw156.

M Douglass, GD Clifford, Andrew Reisner, GB Moody, and RG Mark. 2004. Computer-assisted de-identification of free text in the mimic ii database. In *Computers in Cardiology, 2004*. IEEE, pages 341–344.

Anca Draghici and George Draghici. 2000. Cross-disciplinary approach for the risk assessment ontology design. *Information Resources Management Journal (IRMJ)* 26(1):37–53.

Farshad Hakimpour and Andreas Geppert. 2005. Resolution of semantic heterogeneity in database schema integration using formal ontologies. *Information Technology and Management* 6(1):97–122.

Peter J Haug, Jeffrey P Ferraro, John Holmen, Xinzi Wu, Kumar Mynam, Matthew Ebert, Nathan Dean, and Jason Jones. 2013. An ontology-driven, diagnostic modeling system. *Journal of the American Medical Informatics Association* 20(e1):e102–e110.

Mark Hoogendoorn, Peter Szolovits, Leon MG Moons, and Mattijs E Numans. 2016. Utilizing uncoded consultation notes from electronic medical records for predictive modeling of colorectal cancer. *Artificial intelligence in medicine* 69:53–61.

Nan-Chen Hsieh, Rui-Dong Chiang, and Wen-Pin Hung. 2015. Ontology based integration of residential care of the elderly system in long-term care institutions. *Journal of Advances in Information Technology* 6(3).

Bhaskar Kapoor and Savita Sharma. 2010. A comparative study ontology building tools for semantic web applications. *International Journal of Web & Semantic Technology (IJWesT)* 1(3):1–13.

Yugyung Lee, Kaustubh Supekar, and James Geller. 2006. Ontology integration: Experience with medical terminologies. *Computers in Biology and Medicine* 36(7):893–919.

Jim Lumsden, Hazel Hall, and Peter Cruickshank. 2011. Ontology definition and construction, and epistemological adequacy for systems interoperability: A practitioner analysis. *Journal of Information Science* 37(3):246–253.

Christopher D Manning, Mihai Surdeanu, John Bauer, Jenny Rose Finkel, Steven Bethard, and David Mc-Closky. 2014. The stanford corenlp natural language processing toolkit. In *ACL (System Demonstrations)*. pages 55–60.

R Miller, JK Boitnott, and GW Moore. 2001. Web-based free-text query system for surgical pathology reports with automatic case deidentification. *Arch Pathol Lab Med* 125:1011.

Jonathan Mortensen, Matthew Horridge, Mark A Musen, and Natalya Fridman Noy. 2012. Applications of ontology design patterns in biomedical ontologies. In *AMIA*.

Ishna Neamatullah, Margaret M Douglass, H Lehman Li-wei, Andrew Reisner, Mauricio Villarroel, William J Long, Peter Szolovits, George B Moody, Roger G Mark, and Gari D Clifford. 2008. Automated de-identification of free-text medical records. *BMC medical informatics and decision making* 8(1):32.

Esraa Omran, Albert Bokma, Shereef Abu Al-Maati, and David Nelson. 2009. Implementation of a chain ontology based approach in the health care sector. *Journal of Digital Information Management* 7(5).

Jyotishman Pathak, Harold R Solbrig, James D Buntrock, Thomas M Johnson, and Christopher G Chute. 2009. Lexgrid: a framework for representing, storing, and querying biomedical terminologies from simple to sublime. *Journal of the American Medical Informatics Association* 16(3):305–315.

Vreda Pieterse and Derrick G Kourie. 2014. Lists, taxonomies, lattices, thesauri and ontologies: Paving a pathway through a terminological jungle. *Knowledge Organization* 41(3).

Sripriya Rajamani, Elizabeth S Chen, Mari E Akre, Yan Wang, and Genevieve B Melton. 2014. Assessing the adequacy of the hl7/loinc document ontology role axis. *Journal of the American Medical Informatics Association* pages amiajnl–2014.

Alan L Rector, Rahil Qamar, and Tom Marley. 2009. Binding ontologies and coding systems to electronic health records and messages. *Applied Ontology* 4(1):51–69.

David Riaño, Francis Real, Joan Albert López-Vallverdú, Fabio Campana, Sara Ercolani, Patrizia Mecocci, Roberta Annicchiarico, and Carlo Caltagirone. 2012. An ontology-based personalization of health-care knowledge to support clinical decisions for chronically ill patients. *Journal of biomedical informatics* 45(3):429–446.

Anny Kartika Sari, Wenny Rahayu, and Mehul Bhatt. 2013. An approach for sub-ontology evolution in a distributed health care enterprise. *Information Systems* 38(5):727–744.

Brett R South, Danielle Mowery, Ying Suo, Jianwei Leng, Óscar Ferrández, Stephane M Meystre, and Wendy W Chapman. 2014. Evaluating the effects of machine pre-annotation and an interactive annotation interface on manual de-identification of clinical text. *Journal of biomedical informatics* 50:162–172.

Markus Strohmaier, Simon Walk, Jan Pöschko, Daniel Lamprecht, Tania Tudorache, Csongor Nyulas, Mark A Musen, and Natalya F Noy. 2013. How ontologies are made: Studying the hidden social dynamics behind collaborative ontology engineering projects. *Web Semantics: Science, Services and Agents on the World Wide Web* 20:18–34.

Latanya Sweeney. 1996. Replacing personally-identifying information in medical records, the scrub system. In *Proceedings of the AMIA annual fall symposium*. American Medical Informatics Association, page 333.

Latanya Sweeney. 1997. Guaranteeing anonymity when sharing medical data, the datafly system. In *Proceedings of the AMIA Annual Fall Symposium*. American Medical Informatics Association, page 51.

Cui Tao, Guoqian Jiang, Thomas A Oniki, Robert R Freimuth, Qian Zhu, Deepak Sharma, Jyotishman Pathak, Stanley M Huff, and Christopher G Chute. 2013. A semantic-web oriented representation of the clinical element model for secondary use of electronic health records data. *Journal of the American Medical Informatics Association* 20(3):554–562.

Sean M Thomas, Burke Mamlin, Gunther Schadow, and Clement McDonald. 2002. A successful technique for removing names in pathology reports using an augmented search and replace method. In *Proceedings of the AMIA Symposium*. American Medical Informatics Association, page 777.

Yan Ye, Zhibin Jiang, Xiaodi Diao, Dong Yang, and Gang Du. 2009. An ontology-based hierarchical semantic modeling approach to clinical pathway workflows. *Computers in biology and medicine* 39(8):722–732.

Annotation of pain and anesthesia events for surgery-related processes and outcomes extraction

Wen-wai Yim
Palo Alto Veterans Affairs
Stanford University
Palo Alto, CA 94305, USA
wwyim@stanford.edu

Dario Tedesco
University of Bologna
Bologna, 40126, Italy
Stanford University
Palo Alto, CA 94305, USA
dariot@stanford.edu

Catherine Curtin
Palo Alto Veterans Affairs
Stanford University
Palo Alto, CA 94305, USA
ccurtin@stanford.edu

Tina Hernandez-Boussard
Stanford University
Palo Alto, CA 94305, USA
boussard@stanford.edu

Abstract

Pain and anesthesia information are crucial elements to identifying surgery-related processes and outcomes. However pain is not consistently recorded in the electronic medical record. Even when recorded, the rich complex granularity of the pain experience may be lost. Similarly, anesthesia information is recorded using local electronic collection systems; though the accuracy and completeness of the information is unknown. We propose an annotation schema to capture pain, pain management, and anesthesia event information.

1 Introduction

Post surgical pain continues to be a challenging problem for the health system. Firstly, continued pain after surgery, or chronic persistent postsurgical pain, is common with about 20% of patients having pain long after the wounds have healed (Neil and Macrae, 2009; Kehlet et al., 2006). Secondly, inadequate acute post operative pain control contributes to adverse events such as impaired pulmonary function and impaired immune function (White and Kehlet, 2010). Finally, post surgical pain can be a gateway to addiction, which has taken on increased urgency with the current opioid crisis (Waljee et al., 2017). To improve these problems, it is crucial to have a clear understanding of the patients' pain and its treatments.

There is some evidence that different interventions such as the use of multi-modal pain management and different anesthesia types, e.g. use of regional anesthesia and nonsteroidal anti-inflammatory drugs, can improve pain management (Baratta et al., 2014). However, different analgesic treatments have different side-effect profiles; moreover, some treatment combinations are not appropriate for certain populations. Furthermore, genetics, age, prior exposure to surgery, and social norms influences the experience of pain. Therefore, there is a clear need to capture anesthesia and pain information and relate them to individual history, social, and genetic factors to improve surgical outcomes.

Even with mandated collection, pain is not always recorded (Lorenz et al., 2009). Even when recorded as structured data, there are a variety of scales that are institution-dependent, e.g. a site-specific 0-10 numeric rating scale or a multidimensional questionnaire such as the Brief Pain Inventory. Additionally, it is difficult to capture the rich complex characteristics of pain in structured ways. Anesthesia type, on the other hand, may be recorded or inferred from procedures, medications, or structured input as part of surgery documentation. However, such recording practices differ by institution and local software.

In this work, we present annotation schemas for pain, pain treatment, and anesthesia events for text extraction, as well as report on inter-annotator agreement and corpus statistics. The ultimate goal is to build a new system or adapt an existing system, using this annotated corpus, to automatically extract such information from clinical free text. The extracted data could then be used to complement missing structured information, facilitating greater opportunities for longitudinal study of patients' pain experience long after initial surgery.

Proceedings of the BioNLP 2017 workshop, pages 200–205,
Vancouver, Canada, August 4, 2017. ©2017 Association for Computational Linguistics

2   Related work

To our knowledge, there is no systematic creation of a pain annotation schema for text extraction, however we reference two extraction systems that identify pain information based on their own targeted needs. (Heintzelman et al., 2013) created a system that extracted pain mentions, severity, start date, end date. Their annotation was based on a created 4-value severity of pain created by the development team. Items were identified using the Unified Medical Language System (UMLS) vocabularies for dictionary look-up (Bodenreider, 2004). Dates and locations were extracted by developed contextual rules. In another work, (Redd et al., 2016) used a series of regular expressions to extract pain score in intensive care unit notes. In contrast to previous works, our work provides a more detailed set of annotations that include different clinical aspects of pain, as well as two other event types (treatment and anesthesia) important for studying outcomes. Similarly, there has not been any work on anesthesia-specific annotation and extraction.

Relating this work to a larger context, our pain, treatment, and anesthesia event annotations can be thought of as more specific reincarnations of the CLEF corpus and i2b2 event annotations (Roberts et al., 2008; Uzuner et al., 2011). For example, under the CLEF annotation schema, pain would fall under the condition entity, with the pain's location aligning to CLEF's locus/sub-location/locality schema. Drug, intervention, and negation for conditions are also elements we capture in our annotation schema. Under the i2b2/VA 2010 concepts, assertions, and relations challenge schema, pain would be considered a medical problem and pain treatments or anesthesia could be identified treatments. Our annotation of status' are related to assertion and relations between pain and treatment function similarly to their medical problem treatment relations. Pain and treatment annotation can also be compared to medication and adverse drug events, where instead the focus of events are on pain symptoms and treatment concepts (Uzuner et al., 2010; Karimi et al., 2015).

3   Corpus creation

We drew data from two sources (1) Stanford University's (SU) Clarity electronic medical record database, a component of the Epic Systems software, and (2) MTSamples.com, a online source of anonymized dictated notes. With approval of an institutional review board, we identified a cohort of surgical patients that underwent 5 procedures associated with high pain: distal radius fracture, hernia replacement, knee replacement, mastectomy, and thoracotomy. We focused on three note types: anesthesia, operative, and outpatient clinic visit notes. Anesthesia and operative notes were sampled from the day of surgery, whereas clinic notes were randomly sampled within 3 months prior and 1 year after the surgery. Because of the variation in clinic notes, we performed stratified random sampling per sub-note type and per surgery category.

From MTsamples, we isolated operative (surgery) and clinic visit notes. Clinic notes were considered those not grouped into specialized categories, e.g. surgery, autopsy, discharge. Frequencies by type are shown in Table 1.

Corpus	Anesthesia	Clinic	Operative
MTsamples	-	90	75
SU	90	90	75
TOTAL	90	180	150

Table 1: Breakdown of note types

4   Guideline Creation

Annotation guidelines were created iteratively with a medical general practitioner as well as a biomedical informatics scientist. The initial pain event schema was derived from existing literature (Fink, 2000) and cues from Stanford Health Care's pain collection practices. Schemas were designed and altered according to feedback from a surgical attendee and an anesthesiologist.

Our annotation focuses on three event types: pain, treatment, and anesthesia events. Below is a description of the entities (in some cases phrasal highlights) for each type of event. Those concepts marked with a * are event heads for which other entities may attach to.

Pain information:
Pain* - indication of pain including signs and symptoms that denote pain or diseases definitionally characterized as pain, e.g. *"myalgia"*, with attributes **_Goal_**:{*binary*} and **_Status_**:{*Current, Past, None, Unknown, Not Patient*}
Description - descriptive characteristics of the indicated pain, e.g. *"burning"*
Frequency - information regarding periodic oc-

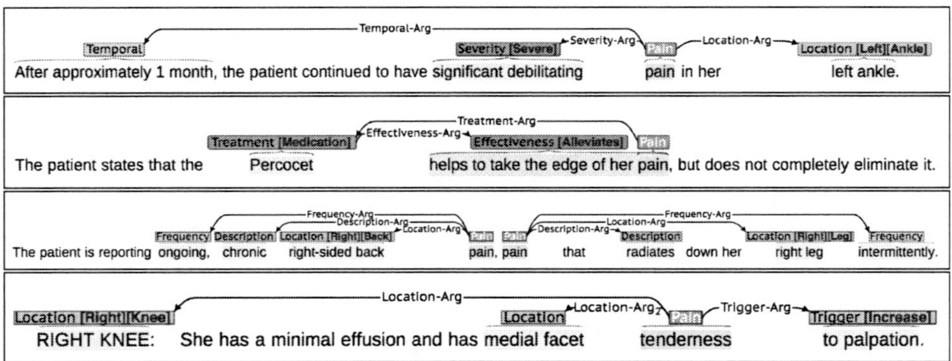

Figure 1: Example pain and treatment events

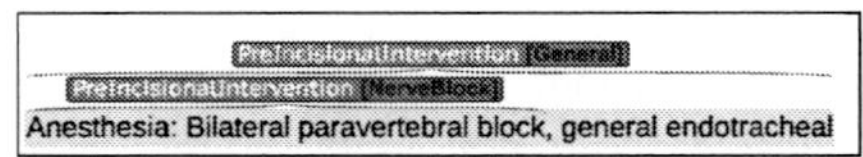

Figure 2: General and nerve block anesthesia text

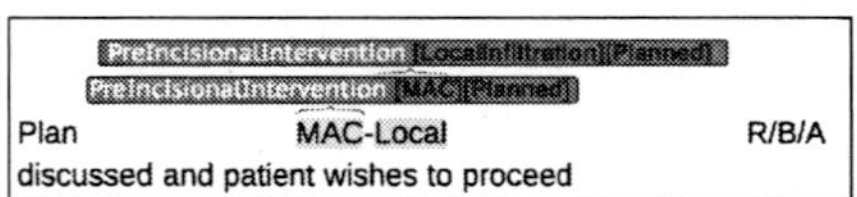

Figure 3: MAC and local anesthesia text

curence of the indicated pain, e.g. *"occasional"*

Location - location of pain, with attributes *Laterality:*{*Bilateral, Left, Right, Unspecified*} and *Type:*{*Abdomen, Ankle, Arm, Back, Back-lower, Back-upper, Breast, Buttocks, ChestArea, Ear, Elbow, Eye, Foot, Generalized, Groin, Hand, Head, Hip, Incisional, Jaw, Knee, Leg, Mouth, Neck, Nose, Pelvis, Shoulder, Throat, Wrist, Other*} (This attribute is useful for matching with structured data that pre-specify locations)

Severity - severity of pain, with attribute *Severityattribute:* {*0,1,..10, mild, moderate, severe* }

Temporal - demarkations of time points at which pain occurs, including time relative to events

Treatment - interventions used on patient (see next section for more information)

Trend - trend of pain with attribute *TrendAttribute:* {*Increasing, Decreasing, No change*}

Trigger - events that cause some change in pain, with attribute *TriggerAttribute:*{*Increase,Decrease*}

Treatment information:

Effectiveness - Effectiveness of treatment with attributes *EffectivenessAttribute:* {*Alleviates, Worsens, No change*}

Treatment* - possible treatments for pain with attributes *Type:*{*Acupuncture, Electrotherapy, Heat/cold therapy, Medication, No further action, Other, Physical Therapy, Steroid injection, Surgical procedure*} and *Status:*{*Current, Past, None, Planned, Requested, Recommended, ConditionalRecommended, NotPatient*}

Temporal - demarkations of time points at which treatment occurs, including time relative to events

Anesthesia information:

Pre-incisional intervention* - anesthetic intervention that occurs prior to incision, with attributes *Status:*{*Current, Past, None, Planned, Requested, Recommended, NotPatient*} and *Type:*{*General, Regional-unspecified, Nerve block, Spinal block, Epidural, MAC (monitored anesthesia care), Local infiltration*}

Event heads, e.g. treatment, were always annotated whereas event arguments, e.g. effectiveness, were only annotated when an event head was present. Only pain medications defined in a curated list (or its synonyms) were annotated as treatment entities to avoid medical knowledge reliance. To avoid annotation fatigue, Status attributes were unmarked if Current.

5 Annotation

After development of an initial schema, a random sample of documents from each SU and MTSamples of anesthesia, operative, and clinical notes were drawn to measure inter-annotator agreement between a general practitioner and a biomedical informatics scientist. Pain and treatment events were annotated for clinical notes, whereas only pre-incisional intervention events were annotated

Field	Set1	Set2	Set1+2	Full
Description	1.00	0.250	0.625	36
Effectiveness	–	0.833	0.769	22
Frequency	0.889	0.909	0.900	36
Location	0.800	0.870	0.832	512
Pain	0.912	0.947	0.929	613
Severity	0.966	0.914	0.921	88
Temporal	0.500	0.698	0.628	200
Treatment	0.686	0.832	0.791	671
Trend	0.770	0.00	0.625	21
Trigger	0.884	0.851	0.839	128
ALL	0.797	0.858	0.831	2327

Table 2: IAA and counts for clinic note entities

Field	Set1	Set2	Set1+2	Full
EffectivenessAttribute	–	0.333	0.308	21
LateralityAttribute	0.758	0.804	0.774	101
LocationAttribute	0.737	0.716	0.700	457
Goal	–	0.920	0.911	16
Pain:StatusAttribute	0.756	0.885	0.822	201
SeverityAttribute	0.966	0.778	0.843	87
Treatment:Type	0.647	0.773	0.744	654
Treatment:StatusAttribute	0.595	0.569	0.597	499
TrendAttribute	0.769	0.00	0.625	21
TriggerAttribute	0.465	0.766	0.602	126
ALL	0.697	0.766	0.749	2183

Table 3: IAA and counts for clinic note attributes

Field	Set1	Set2	Set1+2	Full
Description-Arg	0.667	0.250	0.533	38
Effectiveness-Arg	–	0.909	0.909	23
Frequency-Arg	0.923	0.769	0.846	37
Location-Arg	0.738	0.864	0.795	520
Severity-Arg	0.968	0.889	0.909	91
Temporal-Arg	0.449	0.738	0.620	221
Treatment-Arg	0.800	0.500	0.522	41
Trend-Arg	0.769	0.00	0.625	21
Trigger-Arg	0.883	0.773	0.800	131
ALL	0.744	0.797	0.760	1123

Table 4: IAA and counts for clinic note relations

Field	Set1	Set2	Set1+2	Full
Type	0.906	–	0.906	257
StatusAttribute	0.898	–	0.898	40
ALL	0.902	–	0.902	297

Table 5: IAA and counts for anesthesia note attributes

Field	Set1	Set2	Set1+2	Full
Type	0.935	–	0.935	237
StatusAttribute	0.860	–	0.860	5
ALL	0.897	–	0.897	242

Table 6: IAA counts for operative note attributes

for anesthesia and surgery notes.

An initial set (Set1) included 15 clinic and 15 operative notes from MTSamples; and 30 anesthesia, 15 clinic, and 15 operative notes from SU. Two rounds of revision and agreement were performed on this set. Changes or adjustments to annotation guidelines were made as necessary during annotator agreement cycles. Because clinic notes presented more complexity, we drew another 15 documents from MTSamples and 15 from SU resulting in a new subset (Set2). EffectivenessAttribute and Goal attributes were added from the second set onwards. Two rounds of revisions were performed on this set. Finally, the combined set was revised. The remaining corpus (60 anesthesia, 120 clinic, 120 operative notes) was evenly split and single-annotated by the two annotators. We used brat, a web-based software, for our annotation (Stenetorp et al., 2012).

Inter-annotator agreement (IAA) was evaluated using F1 measure, the harmonic mean of positive predictive value and sensitivity, for entities, relations, and attributes (Hripcsak and Rothschild, 2005). All reported measures are based on partial matches (text spans need only to overlap). For this, relations require that corresponding entity arguments overlap with accurate relation labels.

6 Results

Tables 2-6 show final agreement levels for the separate sets of inter-annotator documents and then for the full inter-annotator corpus for the entities, attributes, and relation levels. We also report the frequencies of each field for the full corpus.

For clinic notes, 125 documents had at least one entity, with 19 ± 19 entities, 10 ± 11 relations per non-empty report. Table 7 shows the top 90% of unique co-occurring relation combinations attached to the same pain entity. Most pain entities appeared either without attached relations or with a Location-Arg. For treatment entities not attached to pain entities as an argument (632 entities), 74% had no attachments, 24% were attached to a Temporal-Arg alone, the rest had either an Effectiveness-Arg relation alone or both. Most relations existed within a close context, however a small number did appear at 2 or more sentences away. This included 10% of Trigger-Arg, 7% of Treatment-Arg, 2% of Severity-Arg, and 2% of Temporal-Arg relations. The remaining relations appeared on the same or one sentence away.

Identification of pain and treatment events for clinical notes was relatively challenging. Ten entities with their related attributes, as well as 8 relation types were involved. Moreover, clinical

Top co-occurring relations for same pain	Count	Fraction	Cum. Fract.
{Location-Arg}	285	0.465	0.465
{}	45	0.073	0.538
{Trigger-Arg}	35	0.057	0.595
{Location-Arg, Trigger-Arg}	28	0.046	0.641
{Location-Arg, Temporal-Argv}	26	0.042	0.684
{Severity-Arg}	22	0.036	0.719
{Location-Arg, Severity-Arg}	18	0.029	0.749
{Description-Arg, Location-Arg}	16	0.026	0.775
{Frequency-Arg, Location-Arg}	16	0.026	0.801
{Severity-Arg, Trigger-Arg}	12	0.020	0.821
{Location-Arg, Treatment-Arg}	9	0.015	0.835
{Temporal-Arg}	9	0.015	0.850
{Treatment-Arg}	8	0.013	0.863
{Trend-Arg}	8	0.013	0.876
{Location-Arg, Severity-Arg, Trigger-Arg}	7	0.011	0.887
{Effectiveness-Arg, Treatment-Arg}	5	0.008	0.896
{Location-Arg, Trend-Arg}	5	0.008	0.904

Table 7: Frequency of relation-combinations connecting to same pain entity

notes tend to contain unpredictable expressions, e.g. *"pain [...] waxing and waning"* or *"worse with hiking"*, and narrative information that spans over several sentences, the conclusion of which could communicate a resolved status. Thirteen out of 613 mentions of pain were attributed as past. Out of 126 marked TriggerAttributes, 114 were aggravating factors (Increase), with only 12 mentions of alleviating factors (Decrease). Interestingly, many severity attributes were qualitative descriptions with 22 for mild, 13 for moderate, and 23 for severe out of 87 total marked. For treatment types, of 654 identified treatment types, 428 were surgical procedures, 116 medication, 82 physical therapy, 12 steroid injection. The remaining had frequencies of 3-5 each.

Ideologically, there were nuances to annotating pain information. While the easiest references to pain were trivial, e.g. *pain*, some required referencing dictionaries, e.g. *myalgia*, or reading context, e.g. *discomfort*. Distinguishing between cause of and timing for pain was not always clear. For example, in *"pain is worse in the morning"* and *"pain [...] when running"*, both underlines could be considered as either Trigger or Temporal. Our final decision was to mark as a Trigger when believed to be causal of the pain rather than delineating chronology. Some pain attributes had multiple connotations. For example, *"chronic pain"*, defined as presence of pain for longer than 3 months, has both a duration and frequency context. We decided to assign *chronic* as a description attribute. Extent of decisions were specified in annotation guidelines. Finally, there are unavoidable limitations in text interpretation. For example, in *"patient is very tender to palpation"*, *very* may be normalized to moderate or severe based on anno-

tator subjectivity. Furthermore, pain may be suggested but not explicitly stated, e.g. *"woman [...] with [...] debilitating abdominal wall hernias"* (most likely painful), and therefore not captured.

Anesthesia and operative note entity agreement was at 0.923 F1 and 0.934 F1. There was a total of 235 and 254 entities for anesthesia and operative notes. For anesthesia reports, 72 had at least one entity, with 4 ± 5 entities each; operative reports, 130 had at least one entity, with 2 ± 1 entities each. 15% of Pre-incisional intervention entities were marked as Planned for anesthesia reports; 1% for operative reports. Agreements for operative and anesthesia entities and attributes were high (Table 5 and 6). This is due to the focused nature of these domains. However, our annotation schema did not include implicit references, e.g. *"skin was anesthetized with 1% lidocaine solution"* where *lidocaine* is often used for local anesthesia.

To improve IAA, further annotation would benefit from pre-annotation of entities trained on this starting set. This would increase consistency and throughput. Additional annotation of a larger corpus would provide larger samples sizes to estimate task challenge for less populated classes.

7 Conclusions and Future Work

In this work, we present a rich annotation schema for pain and pain interventions, as well as an annotation categorization for anesthesia types. Although this work was developed in the surgical setting, the pain annotation schema presented here can be adapted for other settings. Future work includes building our extraction system and applying these data to assess important patient outcomes and health services research.

Annotation guidelines and the MTSamples portion of our corpus is available through our group's website (med.stanford.edu/boussard-lab.html).

Acknowledgments

This project was partially funded from the Veterans Affairs Big Data Scientist Training Enhancement Program. This project was supported by grant number R01HS024096 from the Agency for Healthcare Research and Quality. The content is solely the responsibility of the authors and does not necessarily represent the official views of the Agency for Healthcare Research and Quality or the Department of Veterans Affairs.

References

Jaime L Baratta, Eric S Schwenk, and Eugene R Viscusi. 2014. Clinical consequences of inadequate pain relief: barriers to optimal pain management. *Plastic and reconstructive surgery* 134(4S-2):15S–21S.

Olivier Bodenreider. 2004. The unified medical language system (umls): integrating biomedical terminology. *Nucleic acids research* 32(suppl 1):D267–D270.

Regina Fink. 2000. Pain assessment: the cornerstone to optimal pain management. In *Baylor University Medical Center. Proceedings*. Baylor University Medical Center, volume 13, page 236.

Norris H Heintzelman, Robert J Taylor, Lone Simonsen, Roger Lustig, Doug Anderko, Jennifer A Haythornthwaite, Lois C Childs, and George Steven Bova. 2013. Longitudinal analysis of pain in patients with metastatic prostate cancer using natural language processing of medical record text. *Journal of the American Medical Informatics Association* 20(5):898–905.

George Hripcsak and Adam S Rothschild. 2005. Agreement, the f-measure, and reliability in information retrieval. *Journal of the American Medical Informatics Association* 12(3):296–298.

Sarvnaz Karimi, Alejandro Metke-Jimenez, Madonna Kemp, and Chen Wang. 2015. Cadec: A corpus of adverse drug event annotations. *Journal of biomedical informatics* 55:73–81.

Henrik Kehlet, Troels S Jensen, and Clifford J Woolf. 2006. Persistent postsurgical pain: risk factors and prevention. *The Lancet* 367(9522):1618–1625.

Karl A Lorenz, Cathy D Sherbourne, Lisa R Shugarman, Lisa V Rubenstein, Li Wen, Angela Cohen, Joy R Goebel, Emily Hagenmeier, Barbara Simon, Andy Lanto, et al. 2009. How reliable is pain as the fifth vital sign? *The Journal of the American Board of Family Medicine* 22(3):291–298.

Michael JE Neil and William A Macrae. 2009. Post surgical pain-the transition from acute to chronic pain. *Reviews in pain* 3(2):6–9.

Douglas Redd, Cynthia Brandt, Kathleen Akgun, Jinqiu Kuang, and Qing Zheng-Treitler. 2016. Improving pain assessment in medical intensive care unit through natural language processing. In *American Medical Informatics Association Annual Conference 2016*.

Angus Roberts, Robert Gaizauskas, Mark Hepple, George Demetriou, Yikun Guo, Andrea Setzer, and Ian Roberts. 2008. Semantic annotation of clinical text: The clef corpus. In *Proceedings of the LREC 2008 workshop on building and evaluating resources for biomedical text mining*. pages 19–26.

Pontus Stenetorp, Sampo Pyysalo, Goran Topić, Tomoko Ohta, Sophia Ananiadou, and Jun'ichi Tsujii. 2012. Brat: a web-based tool for nlp-assisted text annotation. In *Proceedings of the Demonstrations at the 13th Conference of the European Chapter of the Association for Computational Linguistics*. Association for Computational Linguistics, pages 102–107.

Özlem Uzuner, Imre Solti, and Eithon Cadag. 2010. Extracting medication information from clinical text. *Journal of the American Medical Informatics Association* 17(5):514–518.

Özlem Uzuner, Brett R South, Shuying Shen, and Scott L DuVall. 2011. 2010 i2b2/va challenge on concepts, assertions, and relations in clinical text. *Journal of the American Medical Informatics Association* 18(5):552–556.

Jennifer F Waljee, Linda Li, Chad M Brummett, and Michael J Englesbe. 2017. Iatrogenic opioid dependence in the united states: Are surgeons the gatekeepers? *Annals of surgery* 265(4):728–730.

Paul F White and Henrik Kehlet. 2010. Improving postoperative pain managementwhat are the unresolved issues? *The Journal of the American Society of Anesthesiologists* 112(1):220–225.

Identifying Comparative Structures in Biomedical Text

Samir Gupta[1], A. S. M. Ashique Mahmood[1], Karen E. Ross[2], Cathy H. Wu[1,3], K. Vijay-Shanker[1]
[1]Department of Computer & Information Sciences
[3]Center for Bioinformatics and Computational Biology
University of Delaware, Newark, DE
{sgupta, ashique, wuc, vijay}@udel.edu
[2]Department of Biochemistry and Molecular & Cellular Biology
Georgetown University Medical Center, Washington, DC
ker25@georgetown.edu

Abstract

Comparison sentences are very commonly used by authors in biomedical literature to report results of experiments. In such comparisons, authors typically make observations under two different scenarios. In this paper, we present a system to automatically identify such comparative sentences and their components i.e. the compared entities, the scale of the comparison and the aspect on which the entities are being compared. Our methodology is based on dependencies obtained by applying a parser to extract a wide range of comparison structures. We evaluated our system for its effectiveness in identifying comparisons and their components. The system achieved a F-score of 0.87 for comparison sentence identification and 0.77-0.81 for identifying its components.

1 Introduction

Biomedical researchers conduct experiments to validate their hypotheses and infer associations between biological concepts and entities, such as mutation and disease or therapy and outcome. It is often not enough to simply report the effects of an intervention; instead, the most common way to validate such observations is to perform comparisons. In such studies, researchers make observations under two different scenarios (e.g., disease sample vs. control sample). When the differences between the groups are statistically significant, association can be inferred.

Comparative studies are prevalent in nearly every field of biomedical/clinical research. For example, in the experimental approach known as "reverse genetics", researchers draw inferences about gene function by comparing the pheno- type of a gene knockdown sample to that of a sample expressing the gene at the normal level. In clinical trial studies, researchers study the effectiveness or side-effects of a drug compared to a placebo. A simple PubMed query "compared[TIAB] OR than[TIAB] OR versus[TIAB]" returned 3,149,702 citations, which provides a rough estimate of the pervasive nature of comparisons in the biomedical literature. Thus, development of automated techniques to identify such statements would be highly useful.

Comparative sentences typically contain two (or more) entities, which are being compared with respect to some common aspect. Consider sentence (1), which compares gene expression level in cancerous vs. non-cancerous tissues:

(1) The expression of GPC5 gene was lower in lung cancer tissues compared with adjacent noncancerous tissues.

Typically, the entities, which we will refer as **compared entities**, are of the same type. In the example, the entities being compared are two tissues: "lung cancer tissues" and "adjacent noncancerous issues", which are separated by the phrase "compared with". "Expression of GPC5 gene", which we call the **compared aspect**, is the aspect on which comparison between the two entities is being made. The word "lower" indicates the scale of the comparison, thereby providing an ordering of the compared entities with respect to the compared aspect. These definitions are similar to those described in (Park and Blake, 2012).

In this paper, we describe a system to automatically identify comparative structures from text. We have developed patterns based on sentence syntactic dependency information to identify comparison sentences and also extract the various components (compared aspect, compared entities

Proceedings of the BioNLP 2017 workshop, pages 206–215,
Vancouver, Canada, August 4, 2017. ©2017 Association for Computational Linguistics

and scale). The developed system identifies explicit comparative structures at the sentence level, where all the components of the comparison are present in the sentence. The main challenge is to capture patterns at a sufficiently high level given the sheer variety of comparative structures. In the rest of the paper we will define the task, describe our approach and comparison patterns and present the results of our evaluation. We achieved a F-score of 0.87 for identifying comparison sentences and 0.78, 0.81, 0.77 for extracting the compared aspect, scale indicator and compared entities, respectively. Thus the major contributions of this work are:

- Development of a general approach for identifying comparison sentences using syntactic dependencies.
- Development of methods to extract all of the components of the comparative structure.

2 Related Works

The sentence constructions used to make comparisons in English are complex and variable. Bresnan (1973) discussed the syntax of comparative clause construction in English and noted its syntactic complexity, 'exhibiting a variety of grammatical processes'. Friedman (1989) reported a general treatment of comparative structures based on basic linguistic principles and noted that automatically identifying them is computationally difficult. They also noted that comparative structures resemble and can be transformed into other syntactic forms such as general coordinate conjunctions, relative clauses, and certain subordinate and adverbial clauses and thus 'syntactically the comparative is extraordinarily diverse'. In (Staab and Hahn, 1997), the authors proposed a model of comparative interpretation that abstracts from textual variations using descriptive logic representation.

The above studies provide an analysis of comparative sentences from a linguistic point of view. Computational systems for identifying comparisons have also been developed. Jindal and Liu (2006a) proposed a machine learning approach to identify comparative sentences from text. The system first categorizes comparative sentences into different types, and then presents a pattern discovery and supervised learning approach to classify each sentence into two classes: comparative and non-comparative. Class sequential rules

based on words and part-of-speech tags automatically generated while learning the model were used as features in this work. The authors evaluated their classifier on product review sentences containing comparison between products and reported a precision of 79% and a recall of 81%. The authors extended their work (Jindal and Liu, 2006b) to extract comparative relations i.e. the compared entities and their features, and comparison keywords from the identified comparison sentences. In (Xu et al., 2011), the authors described a machine learning approach to extract and visualize comparative relations between products from Amazon customer reviews. They describe a comparative relation as a 4-tuple, containing the two compared products, the compared aspect and a comparison direction (better, worse, same). They reported a F-score of 38.81% using multi-class SVM and 56.68% using Conditional Random Fields (CRF). (Jindal and Liu, 2006b; Xu et al., 2011) are the only works that extract the different components of the comparison. In (Ganapathibhotla and Liu, 2008), the authors focused on mining opinions from comparative sentences from product review sentences and extracting the preferred product. Yang and Ko (2009) proposed a machine learning approach to identify comparative sentences from Korean web-based text but did not address the extraction of the comparison arguments. They first constructed a set of comparative keywords manually and extracted candidate comparative sentences and then used Maximum Entropy Model (MEM) and Naive Bayes (NB) to eliminate non-comparative sentences from the candidates.

Relatively few works on identifying comparative sentences and/or its components from biomedical text have been developed. Park and Blake (2012) reported a machine learning approach to identify comparative claims automatically from full-text scientific articles. They introduced a set of semantic and syntactic features for classifications using three different classifiers: Naive Bayes (NB), a Support Vector Machine (SVM) and a Bayesian network (BN). They evaluated their approach on full-text toxicology articles and achieved F1 score of 0.76, 0.65, and 0.74 on a validation set for the NB, SVM and BN, respectively. The focus of this work was on identifying comparison sentences and the extraction of its components was not addressed.

Fiszman et al. (2007) described a technique to identify comparative constructions in MEDLINE citations using under-specified semantic interpretation. The authors used textual patterns combined with semantic predications extracted from the semantic processor SemRep (Rindflesch and Fiszman, 2003; Rindflesch et al., 2005). The predications extracted by SemRep are based on the Unified Medical Language System (UMLS) (Humphreys et al., 1998). Their system extracts the compared entities (limited to drugs) and the scale of the comparison. They reported an average F-score of 0.78 for identifying the compared drug names, scale and scale position. To the best of our knowledge, (Fiszman et al., 2007) is the only reported work that goes beyond identification of comparison sentences to identify the different components of the comparison in biomedical text. But unlike our work, theirs is limited to comparison between drugs, does not extract the comparison aspect and appears to be limited in their coverage of comparison structures.

3 Method

3.1 Task Definition

Basic comparison sentences contain two or more **compared entities (CE)** and a **comparison aspect (CA)** on which compared entities are being compared. Additionally, there are two parts in such sentences indicating the comparison. The first is the presence of a word that indicates the scale of the comparison and the other separates the two compared entities. The former is often comparative adjectives or adverbs (such as "higher", "lower", "better", etc.), while the latter can be expressed with phrases or words (such as "than", "compared with", "versus" etc.). We will refer to the former comparative word indicating the scale as the **Scale Indicator (SI)** and the latter, separating the entities, as the **Entity Separator (ES)**. In example (2) below the key parts of such a comparison structure are highlighted.

(2) [Arteriolar sclerosis]$_{CA}$ was significantly higher$_{SI}$ in addicts$_{CE}$ than$_{ES}$ controls$_{CE}$.

Jindal and Liu (2006b) categorized comparative structures into four classes: (1) Non-Equal Gradable, (2) Equative, (3) Superlative and (4) Non-Gradable. *Non-Equal Gradable* comparison indicate relations of the type greater or less than, providing an ordering of the compared entities. *Equa-*

tive structures indicate equal relation between the two entities with respect to the aspect. Comparisons where one entity is " better" than all other entities are termed as *Superlative*. Sentences in which the compared entities are not explicitly graded are called *Non-Gradable*.

Based on our previous discussion, we will be addressing only the first two types: **Non-Equal Gradable** and **Equative** comparison. First, we consider processing at the sentence-level only. While there are cases of comparisons, where the context provided by a larger body of text might provide the information about all the components, they are not considered in this work. Thus most of the superlative cases will not be considered because all the compared entities are rarely mentioned within a single sentence. It also rules out cases such as in Example (3a), where the second compared entity must be inferred from previous sentences. Second, we consider only those sentences where the authors mention the result or conclusion of an experiment/study. Thus, we will not consider sentences such as in Example (3b), since it only mentions the intention to perform a comparison but does not indicate the result of the experiment. While such sentences can still be captured with minor changes to our existing patterns, our goal here is to only consider sentences that indicate the results of experiment by means of comparison. The patterns developed in this work identify explicit comparative structures at the sentence level and extract all components of the comparison relations, i.e., the compared aspect, entities and the scale indicator.

(3) a. Mean procedure time was significantly shorter for the percutaneous procedure.

    b. We compared lesion growth between placebo and tissue plasminogen activator-treated patients.

3.2 Approach

The different steps of our system are depicted in Figure 1. Given an input text, typically a Medline abstract, we first tokenize and split the text into sentences using the Stanford CoreNLP toolkit (Manning et al., 2014). We then use the Charniak-Johnson parser (Charniak, 2000; Charniak and Johnson, 2005) with David McClosky's adaptation to the biomedical domain (Mcclosky, 2010) to obtain constituency parse trees for each sentence. Next we use the Stanford conversion tool (Manning et al., 2014; De Marneffe et al., 2014) to convert the parse tree to into the syntactic dependency graph (SDG). We use the "CCProcessed" representation, which collapses and propagates dependencies allowing for an appropriate treatment of sentences that involve conjunctions. Note that "CCProccessed" is helpful as dependencies involving preposition, conjuncts, as well as referent of relative clauses are "collapsed" to get direct dependencies between context words. Thus, as seen in Figure 2, which shows the "CCProcessed" SDG, there is a direct edge from "lower" to the cells in the Noun Phrase (NP) "Hep3B cells" rather than a path with two edges where the first reaches the preposition "in" and the second from "in" word to the word "cells". This simplifies the pattern development in relation extraction.

Based on this syntactic dependencies representation, we have developed patterns to identify the different arguments of the comparison relation. Next we use Semgrex, which is a part of the Stanford NLP Toolkit, to specify the translated patterns as regular expressions based on lemmas, part-of-speech tags, and dependency labels, which will automatically match with the sentence dependency parse structure. We have developed a total of 35 and 8 patterns to identify Non-Equal Gradable and Equative comparisons respectively. The developed Semgrex rules as well as the evaluation test set can be found at the link below[1]. Each Semgrex rule/pattern identifies all components of the comparison, specifically the head of the comparison aspect, entities and scale. Since the components are typically Noun Phrases (NPs), we look at the outgoing edges from the head nouns to obtain the NPs corresponding to the comparison components. In the next subsection, we will discuss the development of different comparison patterns.

3.3 Comparative Patterns

As discussed earlier in subsection 3.1, the two key parts in a basic comparison sentence are a Scale Indicator (SI), indicating the scale of the comparison and a Entity Separator (ES), separating the compared entities. We will use dependencies from these SI and ES words to extract the compared aspect and the compared entities. We have developed rules based on syntactic dependencies for various combinations of the two keys parts. We broadly categorize our comparison patterns based

---

[1] http://biotm.cis.udel.edu/biotm/projects/comparison/

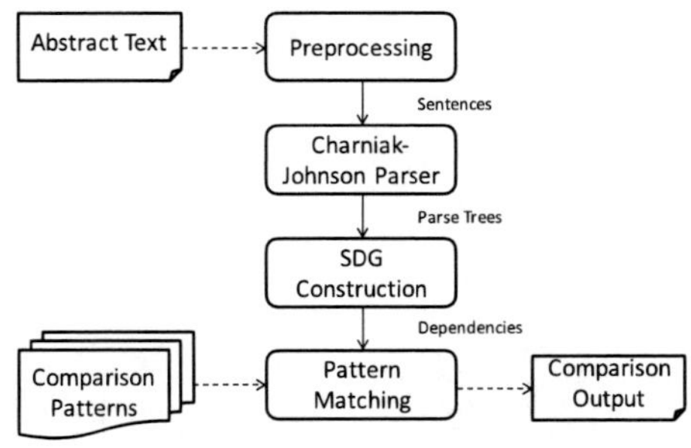

Figure 1: Comparison Pipeline.

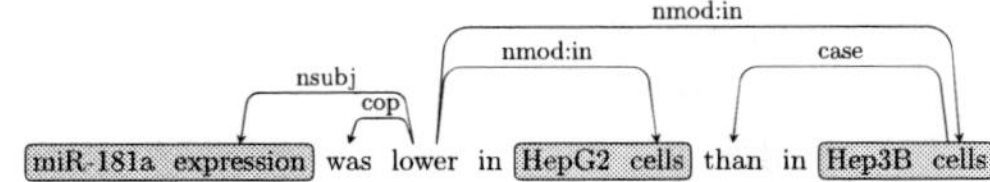

Figure 2: Example SDG

on the Scale Indicator word indicating either Non-Equal Gradable or Equative Comparison.

3.3.1 Non-Equal Gradable

Non-Equal Gradable comparison indicates a difference between the compared entities. Based on three part-of-speech tags (POS) of the Scale Indicator, different syntactic structures are possible, as described below. Note that in all the figures depicting the dependency graph the compared aspect is highlighted in blue and the compared entities in yellow.

Comparative Adjective: Starting with the most frequent case for Scale Indicator, which is a *comparative adjective(JJR)* such as "better", "higher", "lower" etc., there are two broad categories of syntactic structures which we consider. The **first** category involves copular structures, where the JJR serves as the predicate of the comparison relation. The compared aspect is typically the subject of the JJR as shown in Figure 3a. Thus we follow the *nsubj* edge from the JJR to get the head of compared aspect. We use the *nmod:than* from JJR to extract one of the compared entities. The second entity will also have an edge from the JJR, which can be prepositional edge (*nmod:in* as in Figure 3a). Thus we use *nmod* edges from the predicate JJR to determine the second compared entity. Note all prepositional edges such as "with", "for", "during" etc. are considered. Additionally, the second compared entity will be separated by an Entity Separator ("than" in this case) from the first com-

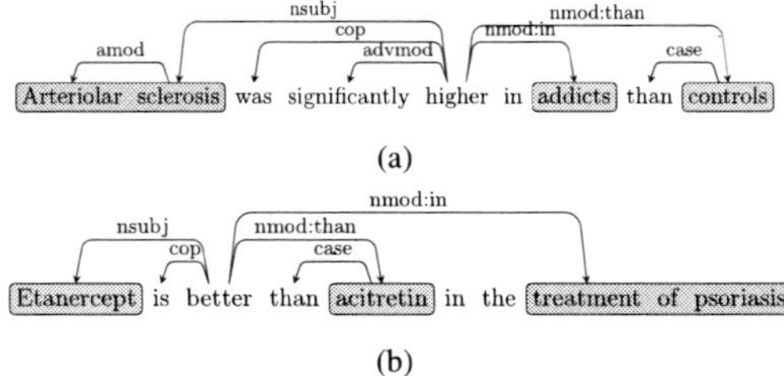

Figure 3: Comparative Adjective copular forms.

pared entity. Thus we further verify that the extracted compared entities are separated by an ES.

The position of the entity separator "than" is critical for determining the second compared entity as well as the first compared entity. As shown in Figure 3b, despite the similar copular structure to the sentence in Figure 3a, the subject of the JJR ("better" in this case) is the compared entity rather than the aspect. This is due to the fact that the JJR is followed by the ES "than". Thus ordering of the words is an important clue when differentiating between these cases.

The **second** category involves sentences, where the comparative adjective modifies a head noun and this modified noun provides the compared aspect, as shown in Figures 4 and 5. Since the compared aspect is modified by the JJR, we used the *amod* edge to detect the aspect. In these cases, the noun phrase containing the Scale Indicator will be connected to a verb and typically serves as the predicate of the comparison relation. The entity separator in the sentence in Figure 4 is "compared to" and we can extract one of the compared entities ("intravenous morphine") by following the *advcl:compared_to* edge from the predicate verb ("offers").

Note that in the first example (Figure 4), the Verb Group ("offer") is in the active form and in the second example (Figure 5), it is in the passive form ("was observed in"). Due to the active/passive form difference, the aspect is in the object position and one of the compared entities in the subject position in the first example, while the reverse is true for the second example. In the dependency representation, the *nsubj* edge and the *nmod:in* edge provide the subjects in active and passive cases and *dobj* and *nsubjpass* provide the possible objects. Note that in certain cases, the author might use an adjective (JJ) instead of the comparative form ("high" instead of "higher"). We

treat such cases in the same way we treat the comparative adjective (JJR) form.

Note that the Semgrex patterns only identifies the head words of the various components, which are typically NPs. We follow outgoing dependency edges from these head words to extract phrases corresponding to each comparison component. For example, in Figure 3a "sclerosis" is identified as the aspect head and we follow the edge *amod* to extract the aspect phrase "Arteriolar sclerosis". In Figure 5, we extract "TP expression" as the aspect phrase and not "Higher TP expression" as "higher" is the trigger of the comparison and identified as the scale.

Comparative Adverb: In these sentences, the comparison scale is indicated through comparative adverbs (RBR) such as "more", "less" etc.. Typically, the RBR modifies an adjective (JJ) as shown in Figure 6, where the adjective is "effective". This adjective serves as the predicate of the comparison and dependency edges from it are used to determine the aspect and entities. The syntactic structure and our rules are very similar to the first category of the Comparative Adjective case. Thus we use the *nsubj* and *advcl:compared_to* edges from "effective" to determine the compared entities. Note that the compared aspect in this example is a clause headed by a VBG ("reducing MCP-1 levels") and thus in addition to *nmod* edges, we need to consider the adverbial clause modifier (*advcl*) edge to determine the aspect.

Verbs: Certain verbs such as "increased", "decreased" as well as "improved" indicate differences and can be used as a SI. This verb serves as the predicate of the comparison relation and outgoing dependencies can be used to determine the arguments of the comparison. We have observed two categories based on the voice (**passive** vs. **active**) of the Verb Group containing this verb. The passive case is depicted in Figure 7a ("was increased in"). In this case, we follow the *nsubjpass* edge to determine the compared aspect. In Figure 7b, since the scale indicator "improved" is in active voice, the direct object of the verb will instead provide the aspect. Extraction and verification of the compared entities is similar to the cases described previously (e.g. *nmod:in* in Figure 7a; *dobj* and *advcl:compared_with* in Figure 7b).

Note that a verb in past participle tense (VBN) can be used as an adjective and modify a noun (e.g., *Increased* TP expression was found in . . .).

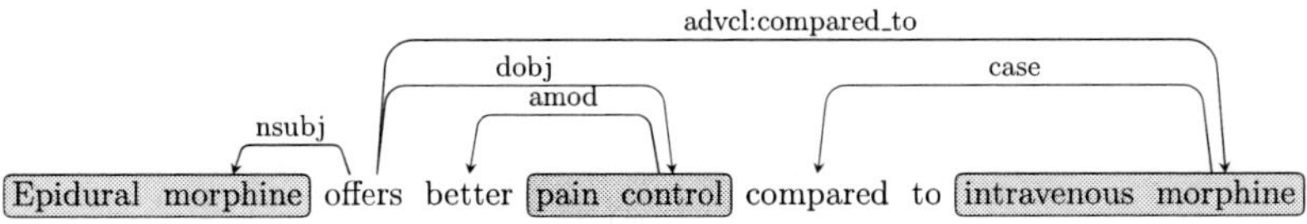

Figure 4: Comparative Adjective modifier form 1.

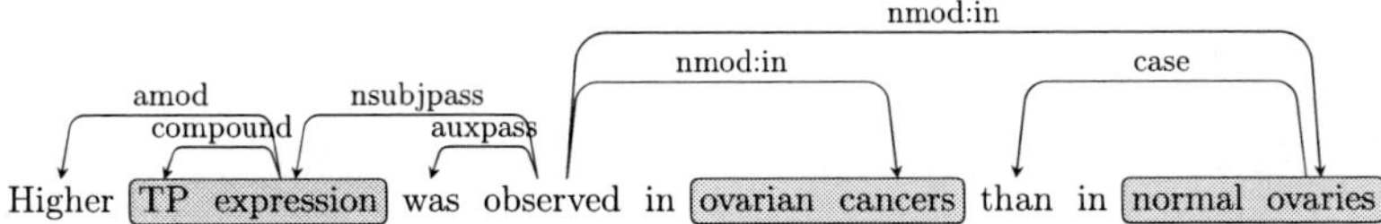

Figure 5: Comparative Adjective modifier form 2.

We treat cases when the scale indicator verb is used as a modifier of a NP like the second category of Comparative Adjectives.

3.3.2 Equative

A sentence with Equative comparison corresponds to cases, where the result of comparison indicates no difference between the compared entities (as in Figure 8). In these cases, it is very rare to find the usual Entity Separator (ES) and instead words such as conjuctions ("and", "or"), " between" and "among" play the role of the ES. We have observed three frequently occurring types of such Equative comparative structures.

The **first** category involves the structure **"X as JJ as Y"**, where JJ is an adjective. In these cases, the adjective serves as the predicate of the comparison. Figure 8 depicts such a case, where the adjective is "effective". Here one of the compared entity "botox" is the subject of the JJ "effective". The second compared entity "oral medication" is preceded by the ES "as" and a *nmod:as* edge from the JJ to the entity is present. The compared aspect is typically attached to the second compared entity through a nmod edge (*nmod:for* in this case). Note that the ES "as" need not appear immediately after the JJ (e.g. "Botox is as effective for overactive bladder as oral medication"). Due to the "CCProcessed" representation of collapsing edges we can still consider the *nmod:as* from "effective" to determine the second compared entity. The only difference in this case is that the *nmod:for* edge used to determine the aspect is from the predicate "effective".

The **second** case involves the Scale Indicator phrase **"similar to"** as shown in Figure 9. Here the subject of the adjective "similar" is the compared aspect. The *nmod* edges (*nmod:in* in this example) from "similar" are used to determine the compared entities. The entities in these cases are separated through conjunctions. Note that the SI "similar" can also modify the compared aspect (e.g. "Similar CA was observed in CE1 and CE2"). This case closely resembles the second category of comparative adjectives and similar rules are used.

The **third** category involves Scale Indicator phrases such **"no differences", "no changes"** etc. Similar to the case of the second category comparative adjectives, here the SI "difference" is part of a NP and hence is connected to a verb, which serves as the predicate. Typically these verbs can be "linking" verbs ('is', "was" etc.) in active form or certain verbs indicating presence ("found in", "noted in", "observed in") in the passive form. In active voice case, as shown in Figure 10, the SI typically follows an existential such as "there". In these cases, the *nmod:between* from the predicate verb ("was" in this case) is used to determine the compared entities. Other nmod edges we consider are *nmod:among* and *nmod:in*. The compared aspect is attached to the second compared entity though *nmod* edges (*nmod:for* in this example). A large proportion of Equative structures do not mention the compared entities explicitly, and as per the definition of our task, we do not extract the comparison components in these cases.

4   Evaluation

We evaluated our system for its effectiveness in identifying comparative sentences and its components on a test set of 189 comparisons from 125 abstracts annotated by a co-author, who was not involved in the design and development of the sys-

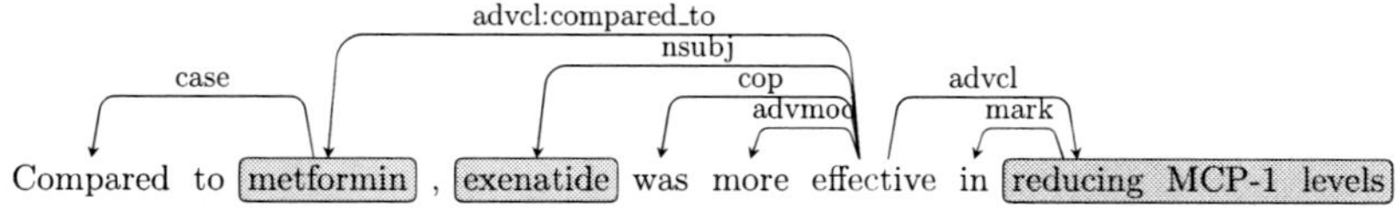

Figure 6: Comparative Adverb copular form.

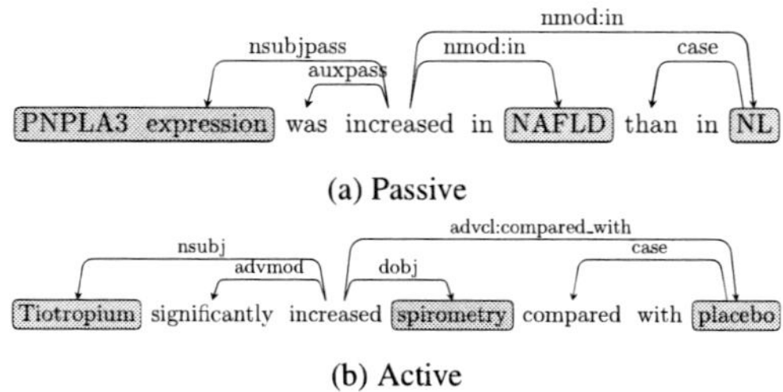

(a) Passive

(b) Active

Figure 7: Comparative Verb forms.

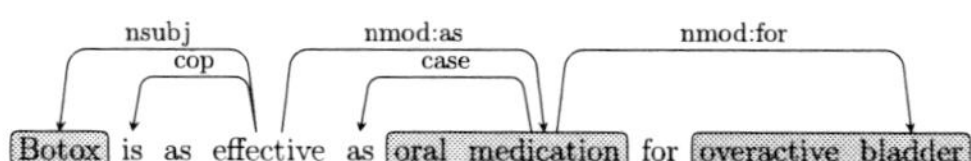

Figure 8: Equative Form 1.

tem. Note that the annotator also annotated an additional 50 abstracts, which was used in the development of the comparison patterns. Although the work by Fiszman et al. (2007) attempts to tackle the similar task of identifying comparison sentences and its components, we do not directly compare with their results. This is due to the fact that their implementation is limited to "direct comparisons of the pharmacological actions of two drugs". We ran their system on our annotated test data and only 8 out of the 189 comparisons were identified by their system as their implementation only detects comparison if the two compared entities (CEs) are drugs. We also ran their system on some artificially created sentences obtained by replacing CEs with drugs and observed that their system seemed limited in the coverage of comparison structures. In the subsequent sections, we will describe the evaluation methodology, present the results and provide an analysis of errors.

4.1 Experimental Setup

To evaluate our system's performance, we have created a test set of 125 abstracts. We selected abstracts that usually draw conclusions by means of comparing between two contrasting situations. Randomized controlled trials (RCT), which compare the outcome between two randomly selected groups, fit this definition very well. For this reason, we searched for RCTs in PubMed with the query "(Randomized Controlled Trial[Publication Type]). This query yielded 431,226 abstracts. However, we noticed that this set lacked abstracts concerning gene expression studies. Thus, we appended to our initial dataset with abstracts related to the effect of differential expression of genes on diseases. As we target to identify comparison sentences, we chose abstracts tagged as "comparative study" in PubMed because they tend to contain comparisons. We used the PubMed query: "(Comparative Study [Publication Type]) AND expression[TIAB] AND (cancer[TI] OR carcinoma[TI])", restricting the comparative studies to gene expressions and cancer related studies. This query yielded 8,479 abstracts.

From this initial set of abstracts, we randomly selected 125 abstracts for annotation by a biomedical researcher expert who did not take part in the development of the system. 150 sentences from the 125 abstracts were annotated as comparison sentences and included 189 comparisons. Our guidelines required the annotation of the four components for each comparison: the compared aspect (CA), the two compared entities (CE1 and CE2) and a word or phrase that indicates the scale of comparison (SI). Additionally, they (the guidelines) required annotation at a sentence level for sentences which had a explicit conclusion i.e. indicated the scale of comparison and is not a mention of a planned investigation.

4.2 Results and Discussion

Annotations of the test set of 125 abstracts yielded 189 comparisons, each containing a compared aspect, a scale indicator and two compared entities. We ran our system on the test set and evaluated its performance on correctly identifying the (1) comparison sentences, (2) compared aspect, (3) scale indicator and (4) compared entities. When computing true positives, we compared the head word of the annotated components with the head words extracted by our system. A mismatch resulted

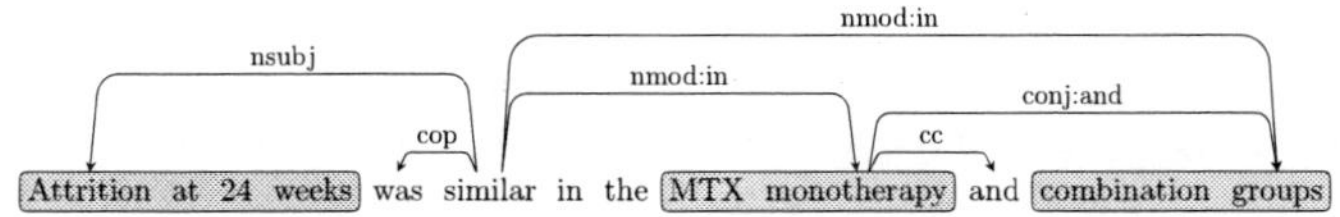

Figure 9: Equative Form 2.

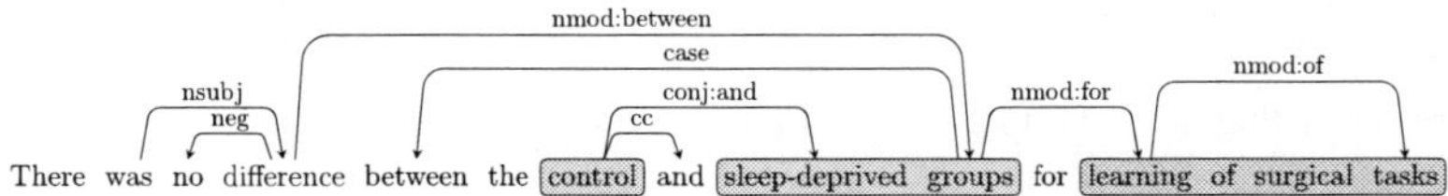

Figure 10: Equative Form 3.

Table 1: Evaluation Results.

Type	Precison	Recall	F-Score
Sentence	0.91	0.83	0.87
Comparison Aspect	0.85	0.72	0.78
Scale Indicator	0.87	0.75	0.81
Compared Entities	0.84	0.72	0.77

in both a false negative and false positive. We computed Precision (P), Recall (R), and F-score (F) measures for each evaluation type, results of which are shown in Table 1.

We analyzed the errors made by our system and majority of the errors (more than 80%) encountered were due to incorrect parsing of complicated sentences. For example, in sentence (4), the clause modifier edge *acl* to "compared" was from "feed" instead of the aspect "palatable". If the clause "with significantly less consumption of treated feed" is removed, thereby simplifying the sentence, the parse is correct and we correctly extract the comparison.

(4) Pro-Dynam was significantly less palatable, with significantly less consumption of treated feed compared with either Equipalazone Powder or Danilon Equidos

A second but rarer category of error involves cases, where we did not consider certain Scale Indicators (SI) such as "superior", "non-inferior", "extra" as in sentence (5). In such examples, the parser tagged the SI as adjective (JJ) and not a comparative adjective (JJR) even though these words indicate a comparison. Since our treatment of such patterns was limited to JJR scale indicators, we missed these cases. It is important to note that our system will identify such structures if we replace such JJ scale indicators by a JJR.

(5) Moxifloxacin was non-inferior to ceftriaxone/metronidazole in terms of clinical response at test-of-cure in the PP population

The third category involved cases missed due to missing patterns such as seen in sentences (6). In sentence (6a), two set of patients are being compared with respect to improvement extent, while sentence (6b) compares the concentration of "plasma F2-isoprostane" before and after drug administration. These cases where a comparison sentence was not detected due to missing patterns were very few.

(6) a. Both paroxetine and placebo-treated patients improved to a similar extent on selfrated pain measures

   b. Maximal plasma F2-isoprostane concentrations after IS + C (iron sucrose + vitamin C) were significantly elevated from baseline

More than 90% of the false positive cases, where we detected a component of a comparison incorrectly was due to parsing error. For example, in sentence (7), the compared aspect is incorrectly identified as "Sixty minutes" as the parser detects it as the subject of "higher" rather than "FEV(1)% increase". If the phrase "Sixty minutes after" is removed, the parse is correct and we correctly identify the aspect. We would like to emphasis that most of the errors, either FN or FP, were due to incorrect parsing of complicated sentences rather than the incompleteness of our developed patterns.

(7) Sixty minutes after the bronchodilator inhalation, the FEV(1)% increase was higher in OXI groups than in the IB group.

5 Conclusion

We have presented a system to identify comparison sentences and extract their components from literature using syntactic dependencies. The significance of developing a system to identify comparisons arises from the prevalent nature of comparative structures in the biomedical literature. We have observed that in a sample of abstracts describing randomized controlled trials or comparative studies, almost every abstract contained at least one comparison. Moreover, other text-mining applications might rely on extracting the arguments of a comparison. For example, this approach could be applied to mining reports of differential expression experiments, which are inherently comparisons. In (Yang et al., 2010), the authors defined seven comparative classes of differential expression analyses relevant to the processes of neoplastic transformation and progression, including cancer vs. normal tissue, high grade vs. low grade samples, and metastasis vs. primary cancer. Because comparative statements are often used to summarize the results of a study, these sentences are often of high interest to the reader. To the best of our knowledge, ours is the only work that attempts to cover a wide range of comparisons, capture all comparison components, and does not impose any restrictions on the type of compared entities. Our system achieved F-scores of 0.87, 0.78, 0.81 and 0.77 for identifying comparison sentences, aspects, scale and entities respectively. We plan to extend this work to consider situations, where one of the entities is implied and needs to be extracted from context.

Acknowledgments

Research reported in this manuscript is supported by the National Institutes of Health under Grants No. U01GM120953 and No. 1 U01 HG008390-01. The funders had no role in study design, data collection and analysis, decision to publish, or preparation of the manuscript.

References

Joan W Bresnan. 1973. Syntax of the comparative clause construction in english. *Linguist. Inq.* 4(3):275–343.

Eugene Charniak. 2000. A maximum-entropy-inspired parser. In *Proceedings of the 1st North American Chapter of the Association for Computational Linguistics Conference*. Association for Computational Linguistics, Stroudsburg, PA, USA, NAACL 2000, pages 132–139.

Eugene Charniak and Mark Johnson. 2005. Coarse-to-fine n-best parsing and MaxEnt discriminative reranking. In *Proceedings of the 43rd Annual Meeting on Association for Computational Linguistics*. Association for Computational Linguistics, Stroudsburg, PA, USA, ACL '05, pages 173–180.

Marie-Catherine De Marneffe, Timothy Dozat, Natalia Silveira, Katri Haverinen, Filip Ginter, Joakim Nivre, and Christopher D Manning. 2014. Universal stanford dependencies: A cross-linguistic typology. In *LREC*. volume 14, pages 4585–4592.

Marcelo Fiszman, Dina Demner-Fushman, Francois M Lang, Philip Goetz, and Thomas C Rindflesch. 2007. Interpreting comparative constructions in biomedical text. In *Proceedings of the Workshop on BioNLP 2007: Biological, Translational, and Clinical Language Processing*. Association for Computational Linguistics, Stroudsburg, PA, USA, BioNLP '07, pages 137–144.

Carol Friedman. 1989. A general computational treatment of the comparative. In *Proceedings of the 27th Annual Meeting on Association for Computational Linguistics*. Association for Computational Linguistics, Stroudsburg, PA, USA, ACL '89, pages 161–168.

Murthy Ganapathibhotla and Bing Liu. 2008. Mining opinions in comparative sentences. In *Proceedings of the 22Nd International Conference on Computational Linguistics - Volume 1*. Association for Computational Linguistics, Stroudsburg, PA, USA, COLING '08, pages 241–248.

Betsy L Humphreys, Donald A B Lindberg, Harold M Schoolman, and G Octo Barnett. 1998. The unified medical language system. *J. Am. Med. Inform. Assoc.* 5(1):1–11.

Nitin Jindal and Bing Liu. 2006a. Identifying comparative sentences in text documents. In *Proceedings of the 29th Annual International ACM SIGIR Conference on Research and Development in Information Retrieval*. ACM, New York, NY, USA, SIGIR '06, pages 244–251.

Nitin Jindal and Bing Liu. 2006b. Mining comparative sentences and relations. *AAAI* .

Christopher D Manning, Mihai Surdeanu, John Bauer, Jenny Rose Finkel, Steven Bethard, and David Mc-Closky. 2014. The stanford corenlp natural language processing toolkit. In *ACL (System Demonstrations)*. pages 55–60.

David Mcclosky. 2010. *Any Domain Parsing: Automatic Domain Adaptation for Natural Language Parsing*. Ph.D. thesis, Providence, RI, USA.

Dae Hoon Park and Catherine Blake. 2012. Identifying comparative claim sentences in full-text scientific articles. In *Proceedings of the Workshop on Detecting Structure in Scholarly Discourse*. Association for Computational Linguistics, Stroudsburg, PA, USA, ACL '12, pages 1–9.

Thomas C Rindflesch and Marcelo Fiszman. 2003. The interaction of domain knowledge and linguistic structure in natural language processing: interpreting hypernymic propositions in biomedical text. *J. Biomed. Inform.* 36(6):462–477.

Thomas C Rindflesch, Marcelo Fiszman, and Bisharah Libbus. 2005. Semantic interpretation for the biomedical research literature. In Hsinchun Chen, Sherrilynne S Fuller, Carol Friedman, and William Hersh, editors, *Medical Informatics*, Springer US, Integrated Series in Information Systems, pages 399–422.

Steffen Staab and Udo Hahn. 1997. Comparatives in context. In *Proceedings of the Fourteenth National Conference on Artificial Intelligence and Ninth Conference on Innovative Applications of Artificial Intelligence*. AAAI Press, Providence, Rhode Island, AAAI'97/IAAI'97, pages 616–621.

Kaiquan Xu, Stephen Shaoyi Liao, Jiexun Li, and Yuxia Song. 2011. Mining comparative opinions from customer reviews for competitive intelligence. *Decis. Support Syst.* 50(4):743–754.

Seon Yang and Youngjoong Ko. 2009. Extracting comparative sentences from korean text documents using comparative lexical patterns and machine learning techniques. In *Proceedings of the ACL-IJCNLP 2009 Conference Short Papers*. Association for Computational Linguistics, Stroudsburg, PA, USA, ACLShort '09, pages 153–156.

Zhen Yang, Fei Ren, Changning Liu, Shunmin He, Gang Sun, Qian Gao, Lei Yao, Yangde Zhang, Ruoyu Miao, Ying Cao, Yi Zhao, Yang Zhong, and Haitao Zhao. 2010. dbDEMC: a database of differentially expressed miRNAs in human cancers. *BMC Genomics* 11 Suppl 4:S5.

Tagging Funding Agencies and Grants in Scientific Articles using Sequential Learning Models

Subhradeep Kayal **Zubair Afzal** **George Tsatsaronis** **Sophia Katrenko**
Pascal Coupet **Marius Doornenbal** **Michelle Gregory**

Content and Innovation Group
Operations Division
Elsevier B.V.
The Netherlands

{d.kayal, m.afzal.1, g.tsatsaronis, s.katrenko, p.coupet, m.doornenbal, m.gregory}@elsevier.com

Abstract

In this paper we present a solution for tagging funding bodies and grants in scientific articles using a combination of trained sequential learning models, namely conditional random fields (*CRF*), hidden markov models (*HMM*) and maximum entropy models (*MaxEnt*), on a benchmark set created in-house. We apply the trained models to address the *BioASQ* challenge 5c, which is a newly introduced task that aims to solve the problem of funding information extraction from scientific articles. Results in the dry-run data set of *BioASQ* task 5c show that the suggested approach can achieve a micro-recall of more than 85% in tagging both funding bodies and grants.

1 Introduction and Description of the BioASQ Task 5c

The scientific research and development market is a \$136bn industry in the US alone, with a 5-year growth of 2.3%, as recorded in 2017[1]. Within this economy, organizations which fund research need to ensure that they are awarding funds to the right research teams and topics so that they can maximize the impact of the associated available funds. As a result, institutions and researchers are required to report on funded research outcomes, and acknowledge the funding source and grants. In parallel, funding bodies should be in a position to trace back these acknowledgements and justify the impact and results of their research allocated funds to their stakeholders and the taxpayers alike. Researchers should also be able to have access to such information, which can help

them make better educated decisions during their careers, and help them discover appropriate funding opportunities for their scientific interests, experience and profile. This situation creates unique opportunities for the affiliated industry, to coordinate and develop low-cost, or cost-free, solutions that can serve funding agencies and researchers. A fundamental problem that needs to be addressed is, however, the ability to automatically extract the funding information from scientific articles, which can in turn become searchable in bibliographic databases.

In this work we address this problem of automating the extraction of funding information from text, using machine learning techniques. We evaluate and combine several state-of-the-art sequential learning approaches, to accept a scientific article as a raw text input and provide the detected funding agencies and associated grant IDs as output.

In order to test our approach, we have participated in the *BioASQ* challenge 5c[2], which is a part of the larger *BioASQ* challenge. *BioASQ* organizes challenges which include tasks relevant to hierarchical text classification, machine learning, information retrieval, QA from texts and structured data, multi-document summarization and many other areas (Tsatsaronis et al., 2015). In this particular task (challenge 5c), the participants are asked to extract grant and funding agency information from full text documents available in PubMed Central[3]. Annotations from PubMed are used to evaluate the information extraction performance of participating systems, with the evaluation criterion being micro-recall. Furthermore, the agencies to be reported must be in a predetermined list as provided by the National Library of Medicine

---

[1] https://www.ibisworld.com/industry/default.aspx?indid=1430

[2] http://participants-area.bioasq.org/general_information/Task5c/

[3] https://www.ncbi.nlm.nih.gov/pmc/

216

Proceedings of the BioNLP 2017 workshop, pages 216–221,
Vancouver, Canada, August 4, 2017. ©2017 Association for Computational Linguistics

(NLM)[4].

2 Background Literature

2.1 Named Entity Recognition

Named entity recognition (*NER*) locates units of information, such as names of organizations, persons and locations and numeric expressions, from unstructured text. Each such unit of information is then known as a *named entity*. In the context of this paper, the named entities that are identified are either *Funding Agencies* (*FA*) or *Grant IDs* (*GR*). As an example, given a text of the form: *"This work was supported by the Funding Organization with grant No. 1234"*, the *NER* task is to label *"Funding Organization"* in text as *FA* and *"1234"* as *GR*. In principle, effective *NER* systems usually employ rule-based (Farmakiotou et al., 2000; Cucerzan and Yarowsky, 1999; Chiticariu et al., 2010), gazetteer (Ritter et al., 2011; Torisawa, 2007) and machine learning approaches (Chieu, 2002; McCallum and Li, 2003; Florian et al., 2003; Zhou and Su, 2002). In this work we utilize several sequential learning (Dietterich, 2002) machine learning approaches for *NER*, which are discussed next. A detailed survey of *NER* techniques for further reading may be found in the work of Nadeau et al. (2007).

2.1.1 Sequential Learning Approaches

Sequential learning approaches model the relationships between nearby data points and their class labels, and can be classified into *generative* or *discriminative*. In the context of *NER*, *Hidden Markov Models* (*HMMs*) are generative models that learn the joint distribution between words and their labels (Bikel et al., 1999; Zhou and Su, 2002). A *HMM* is a *Markov chain* with hidden states, and in *NER* the observed states are words while the hidden states are their labels. Given labelled sentences as training examples, *NER HMMs* find the maximum likelihood estimate of the parameters of the joint distribution, a problem for which many algorithmic solutions are known (Rabiner, 1990). *Conditional Random Fields* (*CRFs*) are discriminative, in contrast to *HMMs*, and find the most likely sequence of labels or entities given a sequence of words. The relationship between the labels is modelled by a *Markov Random Field*. *Linear chain CRFs* are

well suited to sequence analysis and have been applied succssfully in the past in parts-of-speech tagging (Lafferty et al., 2001), shallow parsing (Sha and Pereira, 2003) and *NER* (McCallum and Li, 2003). Finally, another way of modelling data for *NER* is *Maximum Entropy* (*MaxEnt*) models, which select the probability distribution that maximizes entropy, thereby making as little assumptions about the data as possible. Following the seminal work of Berger et al. (1996), maximum entropy estimation has been successfully applied to *NER* in many works (Chieu, 2002; Bender et al., 2003). Essentially, *CRFs* are also maximum entropy models working over the entire sequence, whereas *MaxEnt* models make decisions for each state independently of the other states.

2.1.2 State-of-the-art Open-source Toolkits

Several open-source toolkits implement one or more of the learning approaches mentioned in the previous section. This section discusses three of them in particular, which have been found to be efficient, scalable and robust in practice, and which are used as base approaches in the current work.

The *Stanford CoreNLP toolkit*[5] is a *JVM*-based text annotation framework whose *NER* implementation is based on enhanced *CRFs* with long-distance features to capture more of the structure in text (Finkel et al., 2005). An important feature of the toolkit is the ability to use distributional similarity measures, which assume that similar words appear in similar contexts (Curran, 2003). The toolkit is released with a well-engineered feature extractor, as well as pre-trained models for recognizing *persons*, *locations* and *organizations*.

LingPipe[6] is another *Java*-based *NLP* toolkit, whose efficient *HMM* implementation includes *n*-gram features. The toolkit has been successfully applied in the past in gene recognition in text (Carpenter, 2007).

Finally, in this work we also use the *Apache OpenNLP*[7] toolkit, which implements *NER* either by using discriminative trained *HMMs* (Collins, 2002), or by training *MaxEnt* models (Ratnaparkhi, 1998).

---

[4]https://www.nlm.nih.gov/bsd/grant_acronym.html

[5]http://stanfordnlp.github.io/CoreNLP/
[6]http://alias-i.com/lingpipe/demos/tutorial/read-me.html
[7]https://opennlp.apache.org/

2.2 Related Work

To the best of our knowledge, this is the first piece of research work that systematically explores the concept of extracting funding information from the full text of scientific articles. The next closest category of related published research works mostly aims at extracting names of organizations from affiliation strings, e.g., the works of Jonnalagadda et al. (2010), and Yu et al. (2007), both of which aim at extracting names of organizations from the metadata of published scientific articles. There are, however, several initiatives that started recently and are aiming at a similar direction to the current work, such as the *ERC* project *"Extracting funding statements from full text research articles in the life sciences"*[8].

3 Methodology

3.1 Overview

The suggested approach receives as input a text chunk, e.g., the raw full text of a scientific article, and annotates the input text with entities corresponding to *Funding Agencies* (*FAs*) and *Grant IDs* (*GRs*), where present. A two-step search strategy for finding *FA* and *GR* entities in text has been implemented. The process starts by splitting the input text into paragraphs, which are in turn given sequentially as input to a binary text classifier that identifies only those paragraphs which may contain any funding information. *NER* is performed next, only on the said filtered text paragraphs, to annotate them with *FA* and *GR* labels. This design enjoys several benefits; primarily it minimizes the execution time of the approach, as the most costly component, which is the *NER* part, is only executed in a small selection of paragraphs in which the binary text classifier has detected evidence of funding information. In parallel, it reduces significantly the false positives of the approach, as there are many text segments in a scientific full text article that contain strings which a *NER* component could potentially annotate falsely as *FA*, e.g., the organisation names in the affiliation information of the authors.

3.2 Training Data Gathering

For this task, we have created a *"Gold"* set for training, i.e., a manually curated and annotated set of scientific articles with *FA* and *GR* labels. Such

[8]http://cordis.europa.eu/result/rcn/186297_en.html

a gold set was created, even though *BioASQ* task 5c provides a training set, as several discrepancies were observed in the said training set, the most important being the absence of entity offsets. The *"gold"* set was created with journal articles from a large number of scientific publishers, and comprises $1,950$ articles annotated by three professional annotators, who were provided with comprehensive guidelines explaining the process and the entities. A harmonization process then merged the annotations of the three experts; when all three agreed, annotations were automatically harmonized, whilst the disagreements between the annotators were resolved manually by a subject matter expert (*SME*). From the $1,950$ articles, $1,682$ contained at least one funding-related annotation. As for the individual entities, a total of $3,428$ *FA* and $2,592$ *GR* annotations exist in the set. Pairwise averaged *Cohen's kappa* (Cohen, 1960) was used to calculate the inter-annotators agreement, which for this set was measured at 0.89, suggesting a high-quality dataset. The *"gold"* set was used for two purposes: (i) to train the binary text classifier that detects the paragraphs of text which contain funding information; the number of positive samples were found to be $1,682$, while the number of negative samples had a much higher value at $47,565$, constituting a highly imbalanced set for the task, and, (ii) to train the *NER* components that detect *FA* and *GR* entities.

3.3 Detecting Text with Funding Information

The first step is to separate the parts of the text which contain funding information from the parts which do not. To address this problem, we have used *Support Vector Machines* (*SVMs*), which are known to perform favourably on text classification problems (Joachims, 1998). More precisely, an *L2 regularized linear SVM* has been used, operating on *TF-IDF* vectors extracted from the segments of each input text, based on a bigram bag-of-words representation. The *SVM* was trained on the examples of positive and negative segments, i.e., paragraphs with and without funding information, which could be found in the "gold" set described in the previous section. The regularization parameter for the *SVM* was found to be $C = 2$ based on cross-validation experiments to maximize the final recall.

3.4 Training and Using Sequential Learning Models

As described in section 2.1.1 and 2.1.2, we have employed a variety of complementary techniques to best extract the described entities from text. All of the individual models, namely, a *CRF* implementation from the *Stanford CoreNLP*, a *LingPipe* based enhanced *HMM*, and an *OpenNLP* implementation of the *MaxEnt* tagger, were trained on the said "gold" set using the default hyperparameter settings, as provided by their respective implementations.

Additionally, word clusters were provided to the *Stanford CoreNLP* toolkit, which has the ability to utilize distributional similarity features. The clustering was performed by first extracting word-embedding vectors from the "gold" set, using the unsupervised *Word2Vec* algorithm by Mikolov et al. (2013), followed by performing k-means clustering to create the clusters, based on the cosine-similarity of the word vectors.

For the specific purpose of *BioASQ* challenge 5c, keeping in mind that it is evaluated on micro-recall, the unique outputs of the various models were pooled in, to create the final list of named entities to be provided as output.

3.5 Task Specific Post-processing Detected Entities

In order to perform well on *BioASQ* 5c, some additional post-processing steps were performed.

Extraction of Funding Agency from Grant ID Usually grant IDs contain an acronym from which the corresponding funding agencies can be inferred. As an example, a fictitious grant of the form "MRC123A" would contain the acronym "MRC", signifying that it has been sanctioned by the "Medical Research Council". For task 5c of *BioASQ*, NLM provides a dictionary of acronyms mapped to the respective agency[9], which has been used to retrieve funding agencies from the detected grant IDs.

Corrections to Grants In some cases the prefix of grant numbers was incorrectly published with a letter 'O' rather than the numeric '0'. For example, RO1/AI45338-04 instead of R01/AI45338-04. As NLM has corrected these in their annotations, so did we in a post-processing step.

<hr>

[9] https://www.nlm.nih.gov/bsd/grant_acronym.html

Method	FA μR	GR μR
HMM	80.4	82.3
MaxEnt	81.1	83.9
CRF-distsim	83.3	86.1
Pooled	**85.2**	**86.2**

Table 1: Percentage Micro-recall results for the identification of *Funding Agencies* (*FA*) and *Grant IDs* (*GR*) from the dry-run dataset of *BioASQ* task 5c.

4 Results

As the aforementioned models are trained on a entirely different manually curated "gold" set, evaluations could be made in one pass on the entire dry-run data set of *BioASQ* task 5c, which consisted of $15,205$ documents from PubMed.

Table 1 presents the micro-recall results of the trained models being evaluated on the dry-run dataset. The models listed as *HMM* and *MaxEnt* are self-explanatory, while *CRF-distsim* is the *Stanford CoreNLP* toolkit based *CRF* model which also utilizes distributional similarities, as described in section 3.4. *Pooling* represents the meta-model created by pooling in all the outputs from the individual models. In each case, the outputs undergo the same post-processing step, as described in the previous section.

The table shows that the *CRF* model performs extremely well and is complemented by the other models, all of which better the micro-recall of the *pooled* meta-model, which performs 1.9 percentage points better than the *CRF* in detecting *FA* entities, while performing comparably for *GR* annotations.

5 Conclusions

In this paper we have tackled the problem of funding information extraction from scientific articles, in the context of the *BioASQ* challenge 5c. We have tested and combined state-of-the-art sequential learning models, along with creating a benchmark dataset for training. The results on the dry-run dataset of the challenge indicate the good performance of *Conditional Random Fields* as well as the complementary performance of the other models, whose combination is evaluated at an overall best micro-recall of 85.2% for *Funding Agencies* and 86.2% for *Grant IDs*.

References

Oliver Bender, Franz Josef Och, and Hermann Ney. 2003. Maximum entropy models for named entity recognition. In *Proceedings of the Seventh Conference on Natural Language Learning at HLT-NAACL 2003*. pages 148–151.

Adam L. Berger, Vincent J. Della Pietra, and Stephen A. Della Pietra. 1996. A maximum entropy approach to natural language processing. *Computational Linguistics* 22(1):39–71.

Daniel M. Bikel, Richard Schwartz, and Ralph M. Weischedel. 1999. An algorithm that learns what's in a name. *Machine Learning* 34(1-3):211–231.

Bob Carpenter. 2007. Lingpipe for 99.99% recall of gene mentions. In *Proceedings of the 2nd BioCreative Workshop*.

Hai Leong Chieu. 2002. Named entity recognition: a maximum entropy approach using global information. In *Proceedings of the 2002 International Conference on Computational Linguistics*. pages 190–196.

Laura Chiticariu, Rajasekar Krishnamurthy, Yunyao Li, Frederick Reiss, and Shivakumar Vaithyanathan. 2010. Domain adaptation of rule-based annotators for named-entity recognition tasks. In *Proceedings of the 2010 Conference on Empirical Methods in Natural Language Processing*. pages 1002–1012.

Jacob Cohen. 1960. A Coefficient of Agreement for Nominal Scales. *Educational and psychological measurement* 20(1):37–46.

Michael Collins. 2002. Discriminative training methods for hidden markov models: Theory and experiments with perceptron algorithms. In *Proceedings of the ACL-02 Conference on Empirical Methods in Natural Language Processing*. pages 1–8.

Silviu Cucerzan and David Yarowsky. 1999. Language independent named entity recognition combining morphological and contextual evidence. In *Proceedings of the Joint SIGDAT Conference on EMNLP and VLC*. pages 90–99.

James R. Curran. 2003. *From distributional to semantic similarity*. Ph.D. thesis, University of Edinburgh.

Thomas G. Dietterich. 2002. Machine learning for sequential data: A review. In *Proceedings of the Joint IAPR International Workshop on Structural, Syntactic, and Statistical Pattern Recognition*. pages 15–30.

Dimitra Farmakiotou, Vangelis Karkaletsis, John Koutsias, George Sigletos, Constantine D. Spyropoulos, and Panagiotis Stamatopoulos. 2000. Rule-based named entity recognition for greek financial texts. In *Proceedings of the Workshop on Computational Lexicography and Multimedia Dictionaries (COMLEX 2000*. pages 75–78.

Jenny Rose Finkel, Trond Grenager, and Christopher Manning. 2005. Incorporating non-local information into information extraction systems by gibbs sampling. In *Proceedings of the 43rd Annual Meeting on Association for Computational Linguistics*. pages 363–370.

Radu Florian, Abe Ittycheriah, Hongyan Jing, and Tong Zhang. 2003. Named entity recognition through classifier combination. In *Proceedings of the Seventh Conference on Natural Language Learning at HLT-NAACL 2003 - Volume 4*. pages 168–171.

Thorsten Joachims. 1998. Text categorization with suport vector machines: Learning with many relevant features. In *Proceedings of the 10th European Conference on Machine Learning*. pages 137–142.

Siddhartha Jonnalagadda and Philip Topham. 2010. Nemo: Extraction and normalization of organization names from pubmed affiliation strings. *Journal of Biomedical Discovery and Collaboration* 5:50–75.

John D. Lafferty, Andrew McCallum, and Fernando C. N. Pereira. 2001. Conditional random fields: Probabilistic models for segmenting and labeling sequence data. In *Proceedings of the Eighteenth International Conference on Machine Learning*. pages 282–289.

Andrew McCallum and Wei Li. 2003. Early results for named entity recognition with conditional random fields, feature induction and web-enhanced lexicons. In *Proceedings of the Seventh Conference on Natural Language Learning at HLT-NAACL 2003 - Volume 4*. pages 188–191.

Tomas Mikolov, Ilya Sutskever, Kai Chen, Greg Corrado, and Jeffrey Dean. 2013. Distributed representations of words and phrases and their compositionality. In *Proceedings of the 26th International Conference on Neural Information Processing Systems*. pages 3111–3119.

David Nadeau and Satoshi Sekine. 2007. A survey of named entity recognition and classification. *Linguisticae Investigationes* 30(1):3–26.

Lawrence R. Rabiner. 1990. In *Readings in Speech Recognition*, chapter A Tutorial on Hidden Markov Models and Selected Applications in Speech Recognition, pages 267–296.

Adwait Ratnaparkhi. 1998. *Maximum Entropy Models for Natural Language Ambiguity Resolution*. Ph.D. thesis, University of Pennsylvania.

Alan Ritter, Sam Clark, Mausam, and Oren Etzioni. 2011. Named entity recognition in tweets: An experimental study. In *Proceedings of the Conference on Empirical Methods in Natural Language Processing*. pages 1524–1534.

Fei Sha and Fernando Pereira. 2003. Shallow parsing with conditional random fields. In *Proceedings of the 2003 Conference of the North American Chapter of the Association for Computational Linguistics on Human Language Technology - Volume 1*. pages 134–141.

Kentaro Torisawa. 2007. Exploiting wikipedia as external knowledge for named entity recognition. In *Proceedings of the Joint Conference on Empirical Methods in Natural Language Processing and Computational Natural Language Learning*. pages 698–707.

George Tsatsaronis, Georgios Balikas, Prodromos Malakasiotis, Ioannis Partalas, Matthias Zschunke, Michael R. Alvers, Dirk Weissenborn, Anastasia Krithara, Sergios Petridis, Dimitris Polychronopoulos, Yannis Almirantis, John Pavlopoulos, Nicolas Baskiotis, Patrick Gallinari, Thierry Artiéres, Axel-Cyrille Ngonga Ngomo, Norman Heino, Eric Gaussier, Liliana Barrio-Alvers, Michael Schroeder, Ion Androutsopoulos, and Georgios Paliouras. 2015. An overview of the bioasq large-scale biomedical semantic indexing and question answering competition. *BMC Bioinformatics* 16(1).

Wei Yu, Ajay Yesupriya, Anja Wulf, Junfeng Qu, Marta Gwinn, and Muin J. Khoury. 2007. An automatic method to generate domain-specific investigator networks using pubmed abstracts. *BMC Medical Informatics and Decision Making* 7(1).

GuoDong Zhou and Jian Su. 2002. Named entity recognition using an HMM-based chunk tagger. In *Proceedings of the 40th Annual Meeting on Association for Computational Linguistics*. pages 473–480.

Deep Learning for Biomedical Information Retrieval: Learning Textual Relevance from Click Logs

Sunil Mohan, Nicolas Fiorini, Sun Kim, Zhiyong Lu
National Center for Biotechnology Information
Bethesda, MD 20894, USA
{sunil.mohan, nicolas.fiorini, sun.kim, zhiyong.lu}@nih.gov

Abstract

We describe a Deep Learning approach to modeling the relevance of a document's text to a query, applied to biomedical literature. Instead of mapping each document and query to a common semantic space, we compute a variable-length difference vector between the query and document which is then passed through a deep convolution stage followed by a deep regression network to produce the estimated probability of the document's relevance to the query. Despite the small amount of training data, this approach produces a more robust predictor than computing similarities between semantic vector representations of the query and document, and also results in significant improvements over traditional IR text factors. In the future, we plan to explore its application in improving PubMed search.

1 Introduction

The goal of this research was to explore Deep Learning models for learning textual relevance of documents to simple keyword-style queries, as applied to biomedical literature. We wanted to address two main research questions: (1) Without using a curated thesaurus of synonyms and related terms, or an industry ontology like Medical Subject Headings (MeSH®) (Lu et al., 2009), can a neural network relevance model go beyond measuring the presence of query words in a document, and capture some of the semantics in the rest of the document text? (2) Can a deep learning model demonstrate robust performance despite training on a relatively small amount of labelled data?

We had access to a month of click logs from PubMed®[1], a biomedical literature search engine serving about 3 million queries a day, 20 results per page (Dogan et al., 2009). Most current users of the system are domain experts looking for the most recent papers by an author or search with complex topical boolean query expressions on document aspects. For a small proportion ($\sim 5\%$) of the searches in PubMed, the retrieved articles are sorte by relevance, instead of the default sort order by date. Usage analysis has shown (ibid.) that topic-based queries are a significant part of the search traffic. Such queries often combine two or more entities (e.g. gene and disease), and while users still use short queries, the users are persistent and will frequently reformulate their queries to narrow the search results. So improving the ranking is important to satisfy the needs of PubMed's expanding user base.

Traditional lexical Information Retrieval (IR) factors measure the prominence of query terms in documents treated as bags of words. While such factors like Okapi BM25 (Robertson et al., 1994) and Query Likelihood (Miller et al., 1999) are quite effective, there are several cases where they fail. Two that we wanted to target were: (i) *under-specified query problem*, where even irrelevant documents have prominent presence of the query terms, and relevance requires analysis of the topics and semantics not directly specified in the query, and (ii) the *term mismatch problem* (Furnas et al., 1987), which requires detection of related alternative terms or phrases in the document when the actual query terms are not in the document.

2 Background

Deep Learning models have been applied to various types of text matching problems. Their common goal is to go beyond the lexical bag-of-words

---

[1] http://ncbi.nlm.nih.gov/pubmed

Proceedings of the BioNLP 2017 workshop, pages 222–231,
Vancouver, Canada, August 4, 2017. ©2017 Association for Computational Linguistics

treatment and model text matching as a complex function in a continuous space. An overview of neural retrieval models can be found in (Zhang et al., 2016; Mitra and Craswell, 2017). We review some of this work that motivated our research.

Most text Deep Learning models start with a numeric vector representation of text's lexical units, most commonly terms or words. Ideally these vectors are trained as part of the model, however when training data is limited, many researchers pre-train these word-vectors in an unsupervised manner on a large text corpus, often using one of the word2vec models (Mikolov et al., 2013a,b). We used the SkipGram Hierarchical Softmax method to pre-train our word-vectors on Titles and Abstracts from all documents in PubMed.

Word Mover's Distance (WMD) (Kusner et al., 2015) is an (untrained) model for determining the semantic similarity between two texts by computing the pairwise distances between the words' vectors. It leverages the similarity of vectors of semantically related words. When applied to ad hoc IR, it often successfully tackles the term mismatch problem. We compare our model's performance against WMD, and show that the added complexity produces further improvements in ranking.

Many deep learning text similarity and IR models first project the query and each document to vectors to a common latent semantic space. A second stage then determines the 'match' between the query and document vectors. In the relevance model described in (Huang et al., 2013) the last stage is the cosine similarity function, and in follow-up work (Shen et al., 2014) the authors use a convolutional layer as part of the semantic mapping network, and a feed-forward classification network is trained to compute the similarity. Instead of training word embeddings, their document presentation is based on representing each word as a bag of letter tri-grams. Their model is trained on about 30 million labelled query-document pairs extracted from the click logs of a web search engine. The convolution layer is used to capture a word's context and word n-grams. A similar approach is taken in (Gao et al., 2014). The ARC-I semantic similarity model of (Hu et al., 2014) uses a stack of interleaving convolution and max-pooling layers to map a sentence to a semantic vector. They argue that stacking convolutions of width 3 or more allows them to capture richer compositional semantics than the recurrent

(Mikolov et al., 2010) or recursive (Socher et al., 2011a,b) approaches. However convolutional architectures do have fixed depths that bound the level of composition. Our use of a vertical stack of convolutional layers without interleaving pooling layers is similar to the successful image recognition models AlexNet (Krizhevsky et al., 2012) and VGGNet (Simonyan and Zisserman, 2015).

Severyn & Moschitti's (2015) model to rank short text pairs is trained on small data ($\sim 50k - 100k$ samples). Word embeddings are pretrained using word2vec, a convolutional network maps documents to a semantic vector, followed by a difference matrix and a 3-layer classification network to compute the similarity between the input texts. This is much closer to our final approach, and we compare the performance of our relevance model against this model, but using word-embeddings of size 300 rather than 50 to try to capture richer semantics in biomedical literature.

Another approach to text matching first develops 'local interactions' by comparing all possible combinations of words and word sequences between the two texts. Examples are described in (Hu et al., 2014; Lu and Li, 2013). A recent IR model based on this approach is described in (Guo et al., 2016). Authors argue that the local interaction based approach is better at capturing detail, especially exact query term matches. Our approach simplifies the local-interactions by pairing each document word with a single query word, followed by deep convolutions to attempt to capture some related compositional semantics.

3   The Data

3.1   The Input

We extracted query-document pairs from one month of PubMed click logs where users selected 'Best Match' (relevance) as the retrieval sort order. For each search resulting in a click, the first page of up to 20 documents was recorded. If the clicked document was not on the first page, it was added to this list. The first click on a PubMed search result takes you to a document summary page. Further clicks to the full text of the document were also recorded. Documents that received clicks were labelled as relevant. This binary notion of relevance was used to train our models, and for model evaluation using precision-based ranking metrics. We also experimented with relevance levels, based on a formula hand-tuned to match human-perceived relevance (see appendix). We report NDCG metrics using these relevance levels.

The queries were restricted to simple text searches, of up to seven words, thus eliminating boolean expressions, author searches and queries mentioning document fields. Log extracts were further restricted to queries with at least 21 documents, and at least 3 clicked documents. These filters reduced the the logs to about 33,500 queries.

These queries were randomly split to 60% training, and 20% each for validation and testing. The number of documents available for each query was quite skewed. Since the metrics we use (described below) give equal weight to each query, we further sub-sampled the training and validation datasets to pick at most 20 of the most relevant documents and an equal number of non-relevant documents. This helped balance out the significance of the queries without reducing the data size too much, and improved the mean per-query metrics of the trained models. The resulting training dataset consisted of 634,790 samples (query-document pairs).

3.2   Pre-processing the Input

We used each document's Title and Abstract to form its text. After some experimentation and evaluation on the validation dataset, we found that limiting this to the first 50 words was optimal. Documents shorter than that were padded with 0's, as were queries shorter than 7 words.

We used a simple tokenizer that split words on space and punctuation, while preserving abbreviations and numeric forms, followed by a conversion to lower-case. All punctuation was dropped, which also resulted in a loss of sentence and some grammatical structure, an area to be explored in the future. Numeric forms were collapsed into 7 classes: Integer, Fraction in (0, 1), Real number, year "19xx", year "20xx", Percentage (number followed by "%"), and dollar amount (number preceded by "$"). Removing stop-words from the query and documents did not improve performance of the models.

We leveraged the large PubMed corpus of about 26 million documents to pre-train the word vectors, using the SkipGram Hierarchical Softmax method of `word2vec` (Mikolov et al., 2013b), with a window size of ± 5, a minimum term-frequency of 101, and a word-vector size of 300. This resulted in a vocabulary of 207,716 words. Rare words were replaced with the generic "UNK"

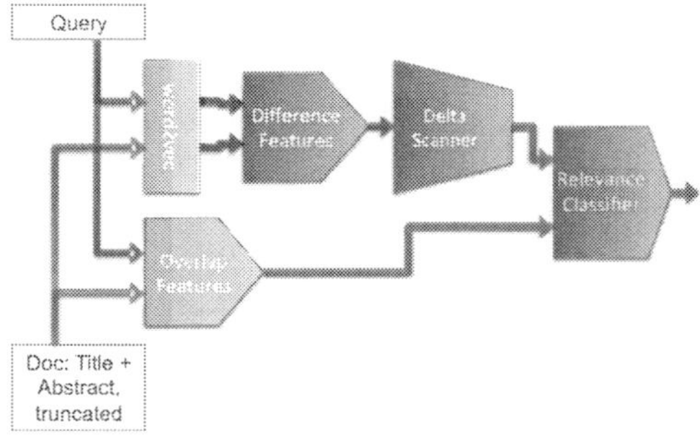

Figure 1: The Delta Relevance Classifier.

token, which was initialized to $\sim U[-0.25, 0.25]$, as in (Severyn and Moschitti, 2015).

4   The Delta Relevance Classifier Model

The components of the Delta Relevance Classifier (figure 1) are described below. Optimal sizes of the various layers were determined by tuning for best accuracy on validation data.

4.1   Note on Convolutional Layers

A convolutional operation (LeCun, 1989) is a series of identical transformations on subsequences of the input obtained by a sliding window on the input. The result is called a feature map, and a convolutional layer will usually involve several feature maps. The width of the input subsequence is called the filter width.

In our application, the input is a sequence of words, each word represented by a real vector of size d. A convolution of filter-width k processes word k-grams. The value of the t-th element of the j-th feature map $\mathbf{c}^j$ is computed as follows:

$$\mathbf{c}^j_{[t]} = \sigma((\mathbf{x} * \mathbf{W}^j)_{[t]} + b_j)$$

$$\mathbf{x} * \mathbf{W}^j = \mathbf{x}_{[t-k+1:t]} \cdot \mathbf{W}^j$$

$$= \sum_{i=1}^{d} \sum_{j=1}^{m} (x_{i,t-k+j} K_{i,j})$$

Here σ is the non-linear activation function, $\mathbf{W}^j \in \mathbb{R}^{d \times k}$, $b_j \in \mathbb{R}$ are the parameters of the j-th feature map, and the input is $\mathbf{x} \in \mathbb{R}^{d \times m}$. In the models in this paper, the feature maps are applied in *full* mode, which effectively pads the input on either side with $k - 1$ d-sized 0-vectors, so $x_{i,j} = 0$ for $j < 1$ or $j > m$, and t ranges from 1 to $m + k - 1$.

Applied to text, a convolutional layer of width 3 will extract features from 3-grams. A second con-

	Full Test Data	Neg20+	OneNewWord	AllNewWords
Nbr. of Queries	6,797	2,600	1,823	1,002
Nbr. of Samples	416,509	208,734	90,353	50,827
Prop. of Samples +ive	45.2%	39.5%	48.9%	48.8%
Prop. of Samples -ive	54.8%	60.5%	51.1%	51.2%
+ives without all Query terms in Title	38.9%	13.9%	34.0%	25.2%
-ives with all Query terms in Title	59.1%	83.6%	65.4%	73.4%

Table 1: Test Data and its subsets

volutional layer of the same width stacked above then extracts features from 5-grams, and so on.

4.2 Query-Document Overlap Features

Following (Severyn and Moschitti, 2015), we compute some overlap features to aid relevance detection when dealing with exact matches, and rare words collapsed to the 'UNK' token. We use the following overlap features, the first two of which are taken directly from that paper: (i) proportion of query and document words in common, (ii) IDF-weighted version of (i), (iii) proportion of query words in the document, and (iv) proportion of query bigrams in the document.

4.3 Difference Features Stage

Instead of developing all pairwise local interactions between query and document terms, we capture interactions between pairs of closest terms. This simplifies the model, and since queries are short, we are unlikely to loose any useful interactions. The difference features are computed in two steps (algorithm 1). First, for each word in the document of a query-document pair, the closest query word in absolute vector distance is identified (skipping all "UNK" words in the query and document). We then output the difference vector, along with its length and the cosine angle between the two vectors. With word-vectors of size d and a document of T words, the output of this stage is a real matrix of size $d \times (T + 2)$. We found $T = 50$ produced the best results for the Delta models.

Algorithm 1 Query-Doc Difference Features

Input: Query text **Q** and Document text **D**.
for each word vector w in **D** s.t. $w \neq$ UNK **do**:
    Find $w_q = \arg\min(u \in \mathbf{Q}, u \neq \text{UNK}) \|w - u\|$
    Output: $(w - w_q)$, $\|w - w_q\|$, $\cos(w, w_q)$
end for

4.4 Delta Scanner Stage

The Delta Scanner stage is a vertical stack of three Convolutional layers of 256 filters each, fol-lowed by a Dropout layer, and then a Global Max-Pooling layer outputting a fixed-width vector. All feature maps use the *ReLU* (Rectified Linear Unit) activation function.

The input to the Delta Scanner stage is the $d \times (T + 2)$ matrix produced by the Difference Features stage. Documents whose text is fewer than T words are right-padded with 0's, and the Delta Scanner supports a mask input that it uses to ignore the padding. The output of this stage is a vector of size 256, representing the semantic difference between the the query and the document in a query-document pair. The remaining hyper-parameters are: Dropout probability, and the L2-regularization coefficient.

4.5 Relevance Classifier Stage

This is a deep fully connected feed-forward logistic regression stage. The input to the Relevance Classifier stage is the combined vectors output from the Overlap Features and Delta Scanner stages, with a total width of $260 = (4+256)$. This data is fed through the following layers:

   i. a Dropout layer,

   ii. two feed-forward layers, each of width 260, using the *ReLU* activation function,

   iii. another Dropout layer, and

   iv. a *sigmoid*-based Classification layer.

The Relevance Classifier's output is an estimate of the probability of the input document's relevance to the query. Documents are ranked in reverse order on this estimated probability.

This stage's hyper-parameters are: Dropout probability (same value used for both Dropout layers), and the L2-regularization coefficient.

4.6 Loss Function and Sample Weighting

The data labels capture a binary sense of relevance, and our models are binary classifiers, so we used the standard binary cross-entropy loss.

In the default mode, the neural network models were trained without any weighting of the training samples. We trained a second set of models with sample weights derived from the non-binary relevance levels (described above). For each relevance level r, a weight of $\max[1, \log(1 + r)]$ was used. This damped the relevance levels, while ensuring that each relevant document received at least the same weight as a non-relevant document.

4.7 Optimization and Implementation Notes

All the neural network models were optimized using Adadelta (Zeiler, 2012), with mini-batches of 256 samples. Mini-batch gradient descent was run for 10 epochs, and the trained values at the end of the epoch producing the best classification accuracy on the Validation dataset were chosen. A greedy search was done in the grid space of the hyper-parameters for the Delta Scanner and Relevance Classifier stages, and the values that produced the best validation accuracy were selected.

5 Experimental Setup

We compare the performance of the relevance models on the following ranking metrics: NDCG at rank 20, Precision at ranks 5, 10 and 20, and Mean Average Precision (MAP). Scoring ties were resolved by sorting on decreasing document-id.

5.1 Methods Compared

We compared the performance of our deep learning model against: BM25; the Unigram Query Likelihood Model (UQLM) with Dirichlet Smoothing (Zhai and Lafferty, 2004); Word Mover's Distance (WMD) that leverages pretrained word-vectors; and a couple of neural network models based on the architecture described in (Severyn and Moschitti, 2015).

We tested BM25 on the document Title, Abstract and Title + Abstract, and found BM25 on Title to give the best ranking performance, with parameters $k_1 = 2.0, b = 0.75$. Similarly, UQLM applied to the document Title and WMD applied to the document Title after removal of stop-words performed better than the other alternatives.

5.1.1 Severyn-Moschitti Model

We tested four variants of the relevance classifier described in (Severyn and Moschitti, 2015). All versions used the same input data and word-vectors as used for the Delta model. In the basic version, which we will refer to as "SevMos-C1", the query and document were fed into a single-layer Convolutional stage as described in

section 4.1, with 256 feature maps and a filter width of 5. This was followed by a Dropout layer and then Global Max-Pooling. The outputs of the query and document convolutions, along with the overlap features described in section 4.2, were fed into a Classifier stage. This stage computed a difference between the query and document features using a difference matrix, and this value along with the other inputs were fed into a deep classification stage identical to that used in the Delta model (section 4.5), sized to match these inputs.

In the "SevMos-C3" variant of this model, we replaced the single-layer convolution stage with a deeper 3-layer stack of convolutions of filter width 3, followed by global max-pooling, just like the Delta model's 'Delta Scanner' stage.

In addition to training the models on unweighted samples, we also trained separate models on relevance-based weighted samples (see section 4.6), which we refer to below as "SevMos-C1 w" and "SevMos-C3 w".

Optimal values for the L2-regularization and Dropout probability hyper-parameters were determined by doing a greedy grid search, as described for the Delta model.

5.2 The Test Data

The test data used to compare performance of the different textual relevance approaches is the held-out 20% split of the data extracted from search logs, as described in section 3.1, without any further sub-sampling. Of the relevant documents ("+ives"), 38.9% did not contain all query terms in the title. Similarly among the non-relevant documents ("-ives"), 59.1% contained all the query terms in the title (see table 1).

In addition to comparing ranking metrics of the different approaches on the test data, we also wanted to explore the main research questions motivating this work: (i) the problems of *under-specified queries* and *term mismatch*, and (ii) *model robustness*. To help answer these questions, we also compare ranking metrics on the following subsets of the test data:

Neg20+: This consists of all queries for which there were at least 20 non-relevant documents that contained all the query words in the title. This helps evaluate performance on under-specified queries.

OneNewWord: The 1,823 test queries which

	NDCG.20	MAP	Prec.5	Prec.10	Prec.20
rev DocID	0.164	0.456	0.344	0.376	0.406
BM25-Title	0.353	0.568	0.592	0.550	0.502
UQLM-Title	0.341	0.561	0.575	0.541	0.500
WMD-Title	0.356	0.579	0.602	0.565	0.516
SevMos-C1	0.345	0.581	0.599	0.569	0.528
SevMos-C3	0.339	0.577	0.597	0.564	0.524
Delta	**0.375**	**0.597**	**0.627**	**0.586**	**0.539**
Delta – WMD	*+5.3%*	*+3.1%*	*+4.2%*	*+3.7%*	*+4.5%*
Delta – SevMos-C1	*+8.7%*	*+2.8%*	*+4.7%*	*+3.0%*	*+2.1%*

Table 2: Ranking metrics on the Full Test Data

contain at least one new word that did not occur in any training or validation queries.

AllNewWords: A smaller subset of queries all of whose words are new: none of the training or validation queries included these words.

The last two subsets will help us evaluate model robustness. The statistics of the test data and its subsets are summarized in table 1.

6 Main Results and Discussion

6.1 Models trained on Un-weighted Samples

Table 2 compares the performance of all the above ranking factors and models on the full test data. The first row shows the metrics obtained by ranking all the documents on reverse order of Document ID. We use this as a score tie-breaker for all the other rankers, so it provides a useful baseline performance of an uninformed ranker.

As also seen in (Shen et al., 2014), BM25 on Title slightly outperforms the Unigram Query Likelihood Model. We have seen other cases where UQLM outperforms BM25. We believe the better performance of BM25 here is partly due to it being a strong factor in the relevance ranking from which these click logs were extracted, thus biasing the click data to some extent.

Word Mover's Distance (WMD-Title) is the first factor in the table that takes non-query words into account, and it does show an improvement over BM25. However WMD relies on the word-vectors obtained by unsupervised training, using a simple Euclidean distance on these vectors as the semantic distance between words. This, and its relatively simple computation, limit how well it performs.

The SevMos-C1 model applies a complex non-linear transformation on the word-vector based text space, in an attempt to better capture comparable semantics of documents. However its NDCG numbers are worse than both WMD and BM25, although its precision numbers, while better than BM25, are about the same as those for WMD. Given that the neural network models in this table were trained on a boolean version of relevance, we expect the main gains to be in the precision-based metrics, which also use a boolean notion of relevance. The lack of improvement in precision metrics over WMD shows that SevMos-C1's non-linear transformations are not doing a better job of capturing query and document semantics.

The SevMos-C3 model learns a more complex non-linear transformation than SevMos-C1, by using a stack of three non-linear convolution layers instead of one in the first part of the model. However its metrics are no better (actually somewhat worse) than SevMos-C1 across the board. So increasing the expressive power of the model did not help. Lack of sufficient training data might be limiting the performance of these models.

The main difference between the Delta model and SevMos-C3 is that the Delta Model starts by computing a difference vector between the Document and Query's word-vector representations. This local interaction vector is inspired by Word Mover's Distance, and in the Delta model we hope to combine the benefits of the WMD and Sev-Mos approaches, while at the same time reducing the complexity of the input space, and thus allowing us to extract more benefit from the small amount of training data. The performance metrics for the Delta model do indeed show sizeable improvements over both WMD and SevMos-C1 (and thus also over BM25 and UQLM). The relative improvements in the metrics are shown in the last two rows of the table[2].

The 'Neg20+' section of the table 3 compares

---

[2]All cited improvements have been verified to be statistically significant to at least a 99% confidence level using a paired t-test.

	NDCG.20	MAP	Prec.5	Prec.10	Prec.20
Subset: Neg20+					
rev DocID	0.098	0.413	0.310	0.335	0.365
BM25-Title	0.252	0.474	0.490	0.461	0.431
UQLM-Title	0.235	0.466	0.473	0.454	0.428
WMD-Title	0.263	0.483	0.501	0.472	0.441
SevMos-C1	0.277	0.499	0.518	0.492	0.462
SevMos-C3	0.272	0.496	0.519	0.490	0.459
Delta	**0.296**	**0.509**	**0.539**	**0.507**	**0.473**
Delta − WMD	*+12.5%*	*+5.4%*	*+7.6%*	*+7.4%*	*+7.3%*
Delta − SevMos-C1	*+6.9%*	*+2.0%*	*+4.1%*	*+3.0%*	*+2.4%*
Subset: OneNewWord					
rev DocID	0.224	0.490	0.366	0.409	0.443
BM25-Title	0.373	0.595	0.606	0.567	0.520
UQLM-Title	0.368	0.595	0.601	0.567	0.526
WMD-Title	0.362	0.600	0.606	0.578	0.531
SevMos-C1	0.363	0.609	0.614	0.588	0.549
SevMos-C3	0.354	0.603	0.613	0.583	0.547
Delta	**0.402**	**0.625**	**0.644**	**0.606**	**0.559**
Delta − WMD	*+11.0%*	*+4.2%*	*+6.3%*	*+4.8%*	*+5.3%*
Delta − SevMos-C1	*+10.7%*	*+2.6%*	*+4.9%*	*+3.1%*	*+1.8%*
Subset: AllNewWords					
rev DocID	0.230	0.509	0.392	0.439	0.466
BM25-Title	0.352	0.585	0.594	0.564	0.519
UQLM-Title	0.348	0.588	0.593	0.566	0.527
WMD-Title	0.333	0.584	0.582	0.567	0.530
SevMos-C1	0.354	0.607	0.608	0.589	0.554
SevMos-C3	0.340	0.600	0.606	0.578	0.550
Delta	**0.386**	**0.622**	**0.642**	**0.609**	**0.565**
Delta − WMD	*+15.9%*	*+6.5%*	*+10.3%*	*+7.4%*	*+6.6%*
Delta − SevMos-C1	*+9.0%*	*+2.5%*	*+5.6%*	*+3.4%*	*+2.0%*

Table 3: Ranking metrics on selected subsets of the Test Data

	NDCG.20	MAP	Prec.5	Prec.10	Prec.20
Full Test Data					
SevMos-C1 w	0.358	0.586	0.609	0.575	0.531
SevMos-C3 w	0.352	0.582	0.602	0.573	0.528
Delta w	**0.383**	**0.597**	**0.628**	**0.588**	**0.538**
Delta w − SevMos-C1 w	*+7.0%*	*+1.9%*	*+3.1%*	*+2.3%*	*+1.3%*
Delta w − Delta	*+2.1%*	*+0.0%*	*+0.2%*	*+0.3%*	*-0.2%*
Neg20+					
SevMos-C1 w	0.404	0.620	0.635	0.608	0.560
SevMos-C3 w	0.396	0.616	0.635	0.605	0.557
Delta w	**0.427**	**0.630**	**0.653**	**0.617**	**0.564**
Delta w − SevMos-C1 w	*+5.7%*	*+1.6%*	*+2.8%*	*+1.5%*	*+0.7%*
Delta w − Delta	*+0.9%*	*-0.2%*	*-0.9%*	*+0.2%*	*-0.4%*
OneNewWord					
SevMos-C1 w	0.378	0.615	0.628	0.595	0.552
SevMos-C3 w	0.364	0.609	0.619	0.587	0.548
Delta w	**0.408**	**0.624**	**0.644**	**0.606**	**0.558**
Delta w − SevMos-C1 w	*+7.9%*	*+1.5%*	*+2.5%*	*+1.8%*	*+1.1%*
Delta w − Delta	*+1.5%*	*-0.2%*	*+0.0%*	*+0.0%*	*-0.2%*
AllNewWords					
SevMos-C1 w	0.368	0.616	0.629	0.597	0.558
SevMos-C3 w	0.353	0.603	0.609	0.582	0.552
Delta w	**0.389**	**0.621**	**0.642**	**0.607**	**0.562**
Delta w − SevMos-C1 w	*+5.7%*	*+0.8%*	*+2.1%*	*+1.7%*	*+0.7%*
Delta w − Delta	*+0.8%*	*-0.2%*	*+0.0%*	*-0.3%*	*-0.5%*

Table 4: Ranking metrics for Relevance-Weighted models

ranking performance on a subset of the test data that should be harder to rank for factors and models that do not give some consideration to the non-query words in the document. In this data a significant number of non-relevant documents contain all the query words in the title. Comparing these numbers with the previous table shows that there is indeed a significant drop in performance for all the factors and models considered here. However while BM25's NDCG metrics drop by 28.6%, the Delta model's NDCG drops by only 21.1%, with the corresponding drops in MAP being 16.5% and 14.7%, respectively. The Delta model still shows the best metrics on this test set, and its degree of improvement over WMD is bigger, as expected from a more complex model.

Model robustness is tested when queries with words not seen during training (i.e. training and validation datasets) are encountered. This is explored in sections '*OneNewWord*' and '*All-NewWords*' of table 3. Both these sub-tables show a consistently better performance by the Delta model over the other approaches compared here. Interestingly, the improvements in the Delta model's NDCG at 20 metrics over the other approaches are quite sizeable, even though for a simple un-weighted relevance classifier, the primary target was precision and not NDCG.

6.2 Relevance Weighted Models

In this section we explore the performance of the Delta model trained on relevance-weighted samples against the corresponding weighted versions of the neural network models SevMos-C1 and SevMos-C3. These metrics are shown in table 4. A quick comparison with previous tables shows that all the models turn in better NDCG numbers than their un-weighted versions. In particular, the "Delta w" model continues to depict statistically significant better metrics than the other weighted neural network models "SevMos-C1 w" and "SevMos-C3 w".

Comparing the Delta weighted model against the unweighted Delta model, we see that there is a statistically significant improvment in the NDCG metrics for all the Test subsets (at the 99% confidence level). However the precision metrics do not show a significant change. So by weighting the samples we have been able to improve the NDCG without hurting the precision.

7 Concluding Remarks

We have demonstrated a Deep Learning approach for learning textual relevance from a fairly small labelled training dataset. We show that this model is robust and it outperforms both traditional IR factors as well as related shallow (WMD) and deep (SevMos) models based on continuous representations of text, with better results on the under-specified query and term mismatch problems.

While the Delta model is comparable to other local-interaction ranking models, we compute fewer and richer interactions. We believe the fewer interactions captured in the difference vector are sufficient for the shorter queries in our data. As a comparison, the model in (Guo et al., 2016) computes a match histogram based on cosine similarity between all document-query word pairs, and also query-term IDF based weighting. We plan to test this model on our data.

The main advantage to the separate semantic vector approach is that document semantic vectors can be pre-computed. Prediction run-time then primarily depends on the complexity of the similarity computation between these semantic vectors. Local-interaction models, including ours, do not allow this pre-computation, significantly increasing the ranker's run-time cost.

We believe the most promising directions for future research include: modeling deeper semantics (see example in appendix), unsupervised training on data auto-generated from the corpus and fine-tuning with supervised training, improving extraction of non-binary relevance levels and using a pair-wise ranking target. Further investigation is also warranted for incorporating these models into PubMed.

Acknowledgments

This research was supported by the Intramural Research Program of the NIH, National Library of Medicine.

References

Rezarta Islamaj Dogan, G Craig Murray, Aurélie Névéol, and Zhiyong Lu. 2009. Understanding PubMed® user search behavior through log analysis. *Database* 2009:bap018.

G. W. Furnas, T. K. Landauer, L. M. Gomez, and S. T. Dumais. 1987. The vocabulary problem in human system communication. *CACM* 30(11):964–971.

Jianfeng Gao, Patrick Pantel, Michael Gamon, Xiaodong He, and Li Deng. 2014. Modeling interestingness with deep neural networks. In *Proceedings of EMNLP 2014*. ACL, Doha, Qatar, pages 2–13.

Jiafeng Guo, Yixing Fan, Qingyao Ai, and W. Bruce Croft. 2016. A deep relevance matching model for ad-hoc retrieval. In *Proceedings of CIKM 2016*. ACM, New York, NY, USA, pages 55–64.

Baotian Hu, Zhengdong Lu, Hang Li, and Qingcai Chen. 2014. Convolutional neural network architectures for matching natural language sentences. In Z. Ghahramani, M. Welling, C. Cortes, N. D. Lawrence, and K. Q. Weinberger, editors, *Advances in NIPS 27*, pages 2042–2050.

Po-Sen Huang, Xiaodong He, Jianfeng Gao, Li Deng, Alex Acero, and Larry Heck. 2013. Learning deep structured semantic models for web search using clickthrough data. In *Proceedings of CIKM 2013*. ACM, New York, NY, USA, pages 2333–2338.

Alex Krizhevsky, Ilya Sutskever, and Geoffrey E Hinton. 2012. Imagenet classification with deep convolutional neural networks. In F. Pereira, C. J. C. Burges, L. Bottou, and K. Q. Weinberger, editors, *Advances in NIPS 25*, pages 1097–1105.

Matt Kusner, Yu Sun, Nicholas Kolkin, and Kilian Weinberger. 2015. From word embeddings to document distances. In *Proceedings of The 32nd International Conference on Machine Learning*. JMLR, ICML 2015, pages 957–966.

Yann LeCun. 1989. Generalization and network design strategies. In R. Pfeifer, Z. Schreter, F. Fogelman, and L. Steels, editors, *Connectionism in Perspective*, Elsevier, Zurich, Switzerland.

Zhengdong Lu and Hang Li. 2013. A deep architecture for matching short texts. In C. Burges, L. Bottou, M. Welling, Z. Ghahramani, and K. Q. Weinberger, editors, *Advances in NIPS 26*, pages 1367–1375.

Zhiyong Lu, Won Kim, and W John Wilbur. 2009. Evaluation of query expansion using MeSH in PubMed. *Information retrieval* 12(1):69–80.

Tomas Mikolov, Kai Chen, Greg Corrado, and Jeffrey Dean. 2013a. Efficient estimation of word representations in vector space. In *Proceedings of the Workshop at ICLR 2013*.

Tomas Mikolov, Martin Karafiát, Lukas Burget, Jan Cernocky, and Sanjeev Khudanpur. 2010. Recurrent neural network based language model. In *Proceedings of INTERSPEECH 2010*.

Tomas Mikolov, Ilya Sutskever, Kai Chen, Greg S Corrado, and Jeff Dean. 2013b. Distributed representations of words and phrases and their compositionality. In C. J. C. Burges, L. Bottou, M. Welling, Z. Ghahramani, and K. Q. Weinberger, editors, *Advances in NIPS 26*, pages 3111–3119.

David R. H. Miller, Tim Leek, and Richard M. Schwartz. 1999. A hidden markov model information retrieval system. In *Proceedings of SIGIR 1999*. ACM, New York, NY, USA, pages 214–221.

Bhaskar Mitra and Nick Craswell. 2017. Neural models for information retrieval. *CoRR* abs/1705.01509. https://arxiv.org/abs/1705.01509.

Stephen E. Robertson, Steve Walker, Susan Jones, Micheline Hancock-Beaulieu, and Mike Gatford. 1994. Okapi at TREC-3. In *Proceedings of TREC 1994*. NIST, Dept. of Commerce, pages 109–126.

Aliaksei Severyn and Alessandro Moschitti. 2015. Learning to rank short text pairs with convolutional deep neural networks. In *Proceedings of SIGIR 2015*. ACM, New York, NY, USA, pages 373–382.

Yelong Shen, Xiaodong He, Jianfeng Gao, Li Deng, and Grégoire Mesnil. 2014. A latent semantic model with convolutional-pooling structure for information retrieval. In *Proceedings of CIKM 2014*. ACM, New York, NY, USA, pages 101–110.

Karen Simonyan and Andrew Zisserman. 2015. Very deep convolutional networks for large-scale image recognition. In *Proceedings of ICLR 2015*.

Richard Socher, Cliff Chiung-Yu Lin, Andrew Ng, and Chris Manning. 2011a. Parsing natural scenes and natural language with recursive neural networks. In Lise Getoor and Tobias Scheffer, editors, *Proceedings of ICML 2011*. ACM, New York, NY, USA, pages 129–136.

Richard Socher, Jeffrey Pennington, Eric H. Huang, Andrew Y. Ng, and Christopher D. Manning. 2011b. Semi-supervised recursive autoencoders for predicting sentiment distributions. In *Proceedings of EMNLP 2011*. ACL, pages 151–161.

Matthew D. Zeiler. 2012. ADADELTA: an adaptive learning rate method. *CoRR* abs/1212.5701.

Chengxiang Zhai and John Lafferty. 2004. A study of smoothing methods for language models applied to information retrieval. *ACM Trans. Inf. Syst.* 22(2):179–214.

Y. Zhang, M. M. Rahman, A. Braylan, B. Dang, H. Chang, H. Kim, Q. McNamara, A. Angert, E. Banner, V. Khetan, T. McDonnell, A. T. Nguyen, D. Xu, B. C. Wallace, and M. Lease. 2016. Neural information retrieval: A literature review. *CoRR* abs/1611.06792. http://arxiv.org/abs/1611.06792.

A Document Relevance Levels

Deriving relevance level of a document to a query from observed clicks is still experimental. We use the following formula:

$$\mu \times \text{AbClicks} + (1 - \mu) \times \text{FTClicks}$$
$$+ \frac{1}{\lambda} \times \text{IsDocWithoutFullText} \times \text{AbClicks}$$

with the parameters $\mu = 0.33, \lambda = 15$, where *AbClicks* is the number of observed clicks to the document summary page in PubMed, *FTClicks* is the number of observed clicks to the document's full text, if available, and the value of *IsDocWithoutFullText* is 1 if the full text for that document is not available, and 0 otherwise. The formula attempts to capture the increased notion of relevance if the user accesses the document's full text, without penalizing documents whose full-text is not available. The parameters were hand-tuned to reflect domain experts' relevance judgments.

B   Rankings on Some Example Queries

Here are some example queries from the test set showing the titles of the top 3 ranked documents for the Delta weighted model, BM25 and WMD. Relevance levels of the documents are indicated inside parentheses before the titles.

B.1   Query: `cryoglobulinemia`

This word did not occur in training or validation queries. Delta w ranks the most relevant document at the top despite its use of an alternative spelling. BM25 and WMD seem to prefer shorter titles with exact matches. Number of documents in the test dataset: relevant = 27, non-relevant = 26. Top three relevance levels: 39.0, 11.0, 4.0.

As ranked by **Delta w**:
i. (39.0) Diagnostics and treatment of cryoglobulinaemia: it takes two to tango.
ii. (0.0) Clinical features of 30 patients with cryoglobulinemia.
iii. (4.0) The diagnosis and classification of the cryoglobulinemic syndrome.

As ranked by **BM25**:
i. (11.0) Cryoglobulinemia Vasculitis.
ii. (3.0) Cryoglobulinemia (review).
iii. (1.0) Role of CXCL10 in cryoglobulinemia.

As ranked by **WMD**:
i. (11.0) Cryoglobulinemia Vasculitis.
ii. (3.0) Cryoglobulinemia (review).
iii. (3.0) Primary cryoglobulinemia with cutaneous features.

B.2   Query: `oesophageal cancer review`

The word *oesophageal* did not occur in training or validation queries. The word *review* does not occur in the title of all relevant documents. Both Delta w and WMD successfully locate alternative spellings of the word. Number of documents in the test dataset: relevant = 22, non-relevant = 28. Top three relevance levels: 7.0, 4.0, 4.0.

As ranked by **Delta w**:
i. (7.0) Esophageal cancer: Recent advances in screening, targeted therapy, and management.
ii. (3.0) Esophageal cancer: A Review of epidemiology, pathogenesis, staging workup and treatment modalities.
iii. (3.0) Esophageal Cancer Staging.

As ranked by **BM25**:
i. (3.0) Imaging of oesophageal cancer with FDG-PET/CT and MRI.
ii. (0.0) Systematic review and network meta-analysis: neoadjuvant chemoradiotherapy for locoregional esophageal cancer.
iii. (0.0) Serum autoantibodies in the early detection of esophageal cancer: a systematic review.

As ranked by **WMD**:
i. (3.0) Esophageal Cancer Staging.
ii. (0.0) Outcomes in the management of esophageal cancer.
iii. (4.0) Endoscopic Management of Early Esophageal Cancer.

B.3   Query: `chronic headache and depression review`

In this example, both WMD and Delta w are able to leverage word vectors to relate headache to migraine. However both miss the most relevant document, whose title is "Psychological Risk Factors in Headache" (relevance level = 6.0). This example demonstrates the need for deeper semantic modeling. Number of documents in the test dataset: relevant = 23, non-relevant = 18. Top three relevance levels: 6.0, 3.0, 3.0.

As ranked by **Delta w**:
i. (3.0) Migraine and depression: common pathogenetic and therapeutic ground?
ii. (3.0) Migraine and depression comorbidity: antidepressant options.
iii. (3.0) Migraine and depression: bidirectional comorbidities?

As ranked by **BM25**:
i. (3.0) Comprehensive management of headache and depression.
ii. (0.0) Chronic daily headache in children and adolescents.
iii. (0.0) Screening for depression and anxiety disorder in children with headache.

As ranked by **WMD**:
i. (3.0) Comprehensive management of headache and depression.
ii. (3.0) Chronic headaches and the neurobiology of somatization.
iii. (3.0) Migraine and depression: biological aspects.

Detecting Dementia through Retrospective Analysis of Routine Blog Posts by Bloggers with Dementia

Vaden Masrani
Dept. of Computer Science
University of British Columbia
vadmas@cs.ubc.ca

Gabriel Murray
Dept. of Computer Information Systems
University of the Fraser Valley
gabriel.murray@ufv.ca

Thalia Field
Dept. of Neurology
University of British Columbia
thalia.field@ubc.ca

Giuseppe Carenini
Dept. of Computer Science
University of British Columbia
carenini@cs.ubc.ca

Abstract

We investigate if writers with dementia can be automatically distinguished from those without by analyzing linguistic markers in written text, in the form of blog posts. We have built a corpus of several thousand blog posts, some by people with dementia and others by people with loved ones with dementia. We use this dataset to train and test several machine learning methods, and achieve prediction performance at a level far above the baseline.

1 Introduction

Dementia is estimated to become a trillion dollar disease worldwide by 2018, and prevalence is expected to double to 74.7 million by 2030 (Prince, 2015). Dementia is a clinical syndrome caused by neurodegenerative illnesses (e.g. Alzheimer's Disease, vascular dementia, Lewy Body dementia). Symptoms can include memory loss, decreased reasoning ability, behavioral changes, and – relevant to our work – speech and language impairment, including fluency, word choice and sentence structure (Klimova and Kuca, 2016).

Recently, there have been attempts to combine clinical information with language analysis using machine learning and NLP techniques to aid in diagnosis of dementia, and to distinguish between types of pathologies (Jarrold et al., 2014; Rentoumi et al., 2014; Orimaye et al., 2014; Fraser et al., 2015; Masrani et al., 2017). This would provide an inexpensive, non-invasive and efficient screening tool to assist in early detection, treatment and institution of supports. Yet, much of the work to date has focused on analyzing spoken language collected during formal assessment, usually with standardized exam tools.

There has been comparatively little work done on analyzing *written* language spontaneously generated by people with dementia. In coming years, there will be an increased number of tech-savvy seniors using the internet, and popular online commentators will continue to age. There will therefore be a growing dataset available in the form of tweets, blog posts, and comments on social media, on which to train a classifier. Provided our writers have a verified clinical diagnosis of dementia, such a dataset would be large, inexpensive to acquire, easy to process, and require no manual transcriptions.

There are downsides to using written language samples as well. Unlike spoken language, written text can be edited or revised by oneself or others. People with dementia may have "good days" and "bad days," and may write only on days when they are feeling lucid, and therefore written samples may be biased towards more intact language. Furthermore, we do not have an accompanying audio file and patients are not constrained to a single topic; people with dementia may have greater facility discussing familiar topics. A non-standardized dataset will also prevent the collection of common test-specific linguistic or acoustic features. However, working with a very large dataset may be able to mitigate the effects of these limitations.

In this work we gather a corpus of blog posts publicly available online, some by people with dementia and others by the loved ones of people with dementia. We extract a variety of linguistic features from the texts, and compare multiple machine learning methods for detecting posts written by people with dementia. All models perform well above the baseline, demonstrating the feasibility of this detection task.

Proceedings of the BioNLP 2017 workshop, pages 232–237,
Vancouver, Canada, August 4, 2017. ©2017 Association for Computational Linguistics

2   Related Work

Early signs of dementia can be detected through analysis of writing samples (Le et al., 2011; Riley et al., 2005; Kemper et al., 2001). In the "Nun Study" researchers analyzed autobiographies written in the US by members of the School Sisters of Notre Dame between 1931-1996. Those nuns who met criteria for dementia had lower grammatical complexity scores and lower "idea density" in their autobiographies.

Le et al. (2011) performed a longitudinal analysis of the writing styles of three novelists: Iris Murdoch who died with Alzheimer's disease (AD), Agatha Christie (suspected AD), and P.D. James (normal brain aging). Measurements of syntactic and lexical complexity were made from 51 novels spanning each of the author careers. Murdoch and Christie exhibited evidence of linguistic decline in later works, such as vocabulary loss, increased repetition, and a deficit of noun tokens (Le et al., 2011).

Despite evidence that linguistic markers predictive of dementia can be found in writing samples, there have been no attempts to train models to classify dementia based on writing alone. Previous work has been successful in training models using transcribed utterances from patients undergoing formal examinations, but this data is difficult to acquire and many models use audio and/or test-specific features which would not be available from online text (Rentoumi et al., 2014; Orimaye et al., 2014; Fraser et al., 2014; Roark et al., 2011). State-of-the-art classification accuracy of 81.92% was achieved by Fraser et al. (2015) with logistic regression using acoustic, textual, and test-specific features on 473 samples from DementiaBank dataset, an American cohort of 204 persons with dementia and 102 controls describing the "Cookie Theft Picture", a component of the Boston Diagnostic Aphasia Examination (Becker et al., 1994; Giles et al., 1996). More recently, these results have been extended via domain adaptation by Masrani et al. (Masrani et al., 2017).

Our methods are similar to Fraser et al. (2015), with the main difference being the dataset used and their inclusion of audio and test-specific features, which are not available in our case. To the best of our knowledge, ours is the first comparison of models trained exclusively on unstructured written samples from persons with dementia.

3   Experimental Design

In this section, we describe the novel blog corpus and experimental setup.

3.1   Corpus

We scraped the text of 2805 posts from 6 public blogs as described in Table 1. Three blogs were written by persons with dementia (First blogger: male, AD, age unknown. Second blogger: female, AD, age 61. Third blogger: Male, Dementia with Lewy Bodies, age 66) and three written by family members of persons with dementia to be used as control (all female, ages unknown). Other demographic information, such as education level, was unavailable. From each of the three dementia blogs, we manually filtered all texts not written by the owner of the blog (e.g. fan letters) or posts containing more images than text. We were left with 1654 samples written by persons with dementia and 1151 from healthy controls. The script to download the corpus is available at `https://github.com/vadmas/blog_corpus/`.

3.2   Classification Features

Following Fraser et al. (2015), we extracted 101 features across six categories from each blog post. These features are described below.

Parts Of Speech (14)   We use the Stanford Tagger (Toutanova et al., 2003) to capture the frequency of various parts of speech tags (nouns, verbs, adjectives, adverbs, pronouns, determiners, etc). Frequency counts are normalized by the number of words in the sentence, and we report the sentence average for a given post. We also count not-in-dictionary words and word-type ratios (noun to verb, pronoun to noun, etc).

Context Free Grammar (45)   Features which count how often a phrase structure rule occurs in a sentence, including NP→VP PP, NP→DT NP, etc. Parse trees come from the Stanford parser (Klein and Manning, 2003).

Syntactic Complexity (28)   Features which measure the complexity of an utterance through metrics such as the depth of the parse tree, mean length of word, sentences, T-Units and clauses and clauses per sentence. We used the L2 Syntactic Complexity Analyzer (Lu, 2010).

Psycholinguistic (5)   Psycholinguistic features are linguistic properties of words that effect word

URL	Posts	Mean words	Start Date	Diagnosis
https://creatingmemories.blogspot.ca/	618	242.22 (s=169.42)	Dec 2003	AD
http://living-with-alzhiemers.blogspot.ca/	344	263.03 (s=140.28)	Sept 2006	AD
http://parkblog-silverfox.blogspot.ca/	692	393.21 (s=181.54)	May 2009	Lewy Body
http://journeywithdementia.blogspot.ca/	201	803.91 (s=548.34)	Mar 2012	Control
http://earlyonset.blogspot.ca/	452	615.11 (s=206.72)	Jan 2008	Control
http://helpparentsagewell.blogspot.ca/	498	227.12 (s=209.17)	Sept 2009	Control

Table 1: Blog Information.

processing and learnability (Salsbury et al., 2011). We used five psycholinguisic features: *Familiarity*, *Concreteness*, *Imageability*, *Age of acquisition*, and the *SUBTL* , which is a measure of the frequency with which a word is used in daily life (Kuperman et al., 2012; Brysbaert and New, 2009a; Salsbury et al., 2011). Psycholinguistic word scores are derived from human ratings[1] while the SUBTL frequency norm[2] is based on 50 million words from television and film subtitles (Brysbaert and New, 2009b).

Vocabulary Richness (4) We calculated four metrics which capture the range of vocabulary in a text: type-token ratio, Brunet's index, a length insensitive version of the type-token ratio, Honore's statistic, and the moving-average type-token ratio (MATTR) (Asp and De Villiers, 2010; Covington and McFall, 2010). These metrics have been shown to be effective in previous AD research (Bucks et al., 2000; Fraser et al., 2015)

Repetitiveness (5) We represent sentences as TF-IDF vectors and compute the cosine similarity between sentences. We then report the proportion of sentence pairs below three similarity thresholds (0, 0.3, 0.5) as well as the min and average cosine distance across all pairs of sentences.

3.3 Training and Testing

We perform a 9-fold cross validation by training each model on all the posts of four blogs and testing on the remaining two, where we assure that each test set contains the posts of one control blog and one dementia blog. Within each fold we perform a feature selection step before training where we select for inclusion into the model the first k features which have the highest absolute correlation with the labels in the training fold.

---

[1] http://websites.psychology.uwa.edu.au/school/MRCDatabase/uwa_mrc.htm
[2] http://subtlexus.lexique.org/

4 Results

For each machine learning model, we calculate the ROC curve and the area under the curve (AUC), comparing with a random performance baseline AUC of 0.5. The AUC results are shown in Figure 1, with all models well above the baseline of 0.5. The best performing models are logistic regression and neural networks, with average AUC scores of 0.815 and 0.848, respectively.

The SUBTL measure of vocabulary richness was the feature most correlated with the outcome variable in eight out of nine folds. Figure 2 shows the SUBTL scores for each blog post in the corpus, arranged by blog and with the bloggers with dementia shown in the top row. A lower score indicates a richer vocabulary. We can see that the bloggers with dementia have a less rich vocabulary. Interestingly, however, the longitudinal trend does not show their vocabularies worsening during the time-period captured in this corpus. The analysis of other features highly informative for the target prediction is ongoing, and additional findings will be discussed at the workshop.

5 Conclusion

We have shown that it is possible to distinguish bloggers with dementia from those without, on a novel corpus of blog data. We extracted linguistic features from the texts and compared a large number of machine learning methods, all of which performed well above the baseline. While feature analysis is ongoing, we have made some interesting observations about the effect of the SUBTL measure of vocabulary richness. Future work will include liaising with patient and caregiver support groups to expand this new dementia corpus, inclusion of a topic clustering preprocessing step to control for variation across content, and further longitudinal analysis.

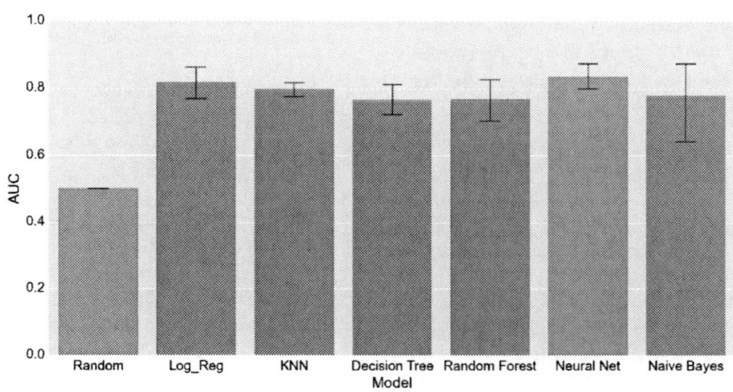

Figure 1: Comparison of models. We show the mean AUC and 90% confidence intervals across a 9-fold CV. All the posts of a blog appear in either the training or test set, but not both.

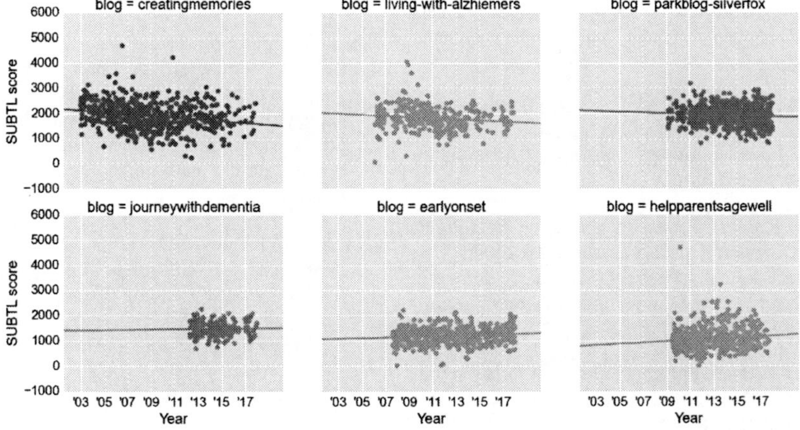

Figure 2: SUBTL word scores for each post in a given blog. Bloggers with dementia (AD or Dementia w/ Lewy Bodies) appear in the top row.

References

Elissa D Asp and Jessica De Villiers. 2010. *When language breaks down: Analysing discourse in clinical contexts*. Cambridge University Press.

James T Becker, François Boiler, Oscar L Lopez, Judith Saxton, and Karen L McGonigle. 1994. The natural history of alzheimer's disease: description of study cohort and accuracy of diagnosis. *Archives of Neurology* 51(6):585–594.

Marc Brysbaert and Boris New. 2009a. Moving beyond Kučera and Francis: A critical evaluation of current word frequency norms and the introduction of a new and improved word frequency measure for american english. *Behavior Research Methods* 41:977–990.

Marc Brysbaert and Boris New. 2009b. Moving beyond kučera and francis: A critical evaluation of current word frequency norms and the introduction of a new and improved word frequency measure for american english. *Behavior research methods* 41(4):977–990.

Romola S Bucks, Sameer Singh, Joanne M Cuerden, and Gordon K Wilcock. 2000. Analysis of spontaneous, conversational speech in dementia of alzheimer type: Evaluation of an objective technique for analysing lexical performance. *Aphasiology* 14(1):71–91.

Michael A Covington and Joe D McFall. 2010. Cutting the gordian knot: The moving-average type–token ratio (mattr). *Journal of Quantitative Linguistics* 17(2):94–100.

Kathleen C Fraser, Graeme Hirst, Naida L Graham, Jed A Meltzer, Sandra E Black, and Elizabeth Rochon. 2014. Comparison of different feature sets for identification of variants in progressive aphasia. *ACL 2014* page 17.

Kathleen C Fraser, Jed A Meltzer, and Frank Rudzicz. 2015. Linguistic features identify alzheimers disease in narrative speech. *Journal of Alzheimer's Disease* 49(2):407–422.

Elaine Giles, Karalyn Patterson, and John R Hodges. 1996. Performance on the boston cookie theft picture description task in patients with early dementia of the alzheimer's type: missing information. *Aphasiology* 10(4):395–408.

William Jarrold, Bart Peintner, David Wilkins, Dimitra Vergryi, Colleen Richey, Maria Luisa Gorno-Tempini, and Jennifer Ogar. 2014. Aided diagnosis of dementia type through computer-based analysis of spontaneous speech. In *Proceedings of the ACL Workshop on Computational Linguistics and Clinical Psychology*. pages 27–36.

Susan Kemper, Lydia H Greiner, Janet G Marquis, Katherine Prenovost, and Tracy L Mitzner. 2001. Language decline across the life span: findings from the nun study. *Psychology and aging* 16(2):227.

Dan Klein and Christopher D Manning. 2003. Accurate unlexicalized parsing. In *Proc. of ACL, Sapporo, Japan*. Association for Computational Linguistics, pages 423–430.

Blanka Klimova and Kamil Kuca. 2016. Speech and language impairments in dementia. *Journal of Applied Biomedicine* 14(2):97–103.

Victor Kuperman, Hans Stadthagen-Gonzalez, and Marc Brysbaert. 2012. Age-of-acquisition ratings for 30,000 english words. *Behavior Research Methods* 44(4):978–990.

Xuan Le, Ian Lancashire, Graeme Hirst, and Regina Jokel. 2011. Longitudinal detection of dementia through lexical and syntactic changes in writing: a case study of three british novelists. *Literary and Linguistic Computing* 26(4):435–461.

Xiaofei Lu. 2010. Automatic analysis of syntactic complexity in second language writing. *International Journal of Corpus Linguistics* 15(4):474–496.

Vaden Masrani, Gabriel Murray, Thalia Field, and Giuseppe Carenini. 2017. Domain adaptation for detecting mild cognitive impairment. In *Proc. of Canadian AI, Edmonton, Canada*.

Sylvester Olubolu Orimaye, Jojo Sze-Meng Wong, and Karen Jennifer Golden. 2014. Learning predictive linguistic features for alzheimer's disease and related dementias using verbal utterances. In *Proc. 1st Workshop. Computational Linguistics and Clinical Psychology (CLPsych)*.

Martin James Prince. 2015. *World Alzheimer Report 2015: the global impact of dementia: an analysis of prevalence, incidence, cost and trends*. London.

Vassiliki Rentoumi, Ladan Raoufian, Samrah Ahmed, Celeste A de Jager, and Peter Garrard. 2014. Features and machine learning classification of connected speech samples from patients with autopsy proven alzheimer's disease with and without additional vascular pathology. *Journal of Alzheimer's Disease* 42(s3).

Kathryn P Riley, David A Snowdon, Mark F Desrosiers, and William R Markesbery. 2005. Early life linguistic ability, late life cognitive function, and neuropathology: findings from the nun study. *Neurobiology of aging* 26(3):341–347.

Brian Roark, Margaret Mitchell, John-Paul Hosom, Kristy Hollingshead, and Jeffrey Kaye. 2011. Spoken language derived measures for detecting mild cognitive impairment. *IEEE Transactions on Audio, Speech, and Language Processing* 19(7):2081–2090.

Tom Salsbury, Scott A Crossley, and Danielle S McNamara. 2011. Psycholinguistic word information in second language oral discourse. *Second Language Research* 27(3):343–360.

Kristina Toutanova, Dan Klein, Christopher D Manning, and Yoram Singer. 2003. Feature-rich part-of-speech tagging with a cyclic dependency network. In *Proc. of NAACL, Edmonton, Canada*. Association for Computational Linguistics, pages 173–180.

Protein Word Detection using Text Segmentation Techniques

G.Devi
Department of CSE
IIT Madras
Chennai-600036, India
gdevi@cse.iitm.ac.in

Ashish V. Tendulkar
Google Inc.,
Hyderabad-500084, India
ashishvt@google.com

Sutanu Chakraborti
Department of CSE
IIT Madras
Chennai-600036, India
sutanuc@cse.iitm.ac.in

Abstract

Literature in Molecular Biology is abundant with linguistic metaphors. There have been works in the past that attempt to draw parallels between linguistics and biology, driven by the fundamental premise that proteins have a language of their own. Since word detection is crucial to the decipherment of any unknown language, we attempt to establish a problem mapping from natural language text to protein sequences at the level of words. Towards this end, we explore the use of an unsupervised text segmentation algorithm to the task of extracting "biological words" from protein sequences. In particular, we demonstrate the effectiveness of using domain knowledge to complement data driven approaches in the text segmentation task, as well as in its biological counterpart. We also propose a novel extrinsic evaluation measure for protein words through protein family classification.

1 Introduction

Research works in the field of Protein Linguistics (Searls, 2002) are largely based on the underlying hypothesis that proteins have a language of their own. However, modeling of protein molecules using linguistic approaches is yet to be explored in depth. This might be due to the structural complexities inherent to protein molecules. Instead of resorting to purely wet lab experiments, we propose to make use of the abundant data available in the form of protein sequences together with knowledge from domain experts to model the protein language. From a linguistic point of view, the first step in deciphering an unknown language will be to identify the independent lexical units or words of the language. This motivates our current attempt to establish a problem mapping from natural language text to protein sequences at the level of words. Towards this end, we explore the use of an unsupervised word segmentation algorithm to the task of extracting "biological words" from protein sequences.

Many unsupervised word segmentation algorithms use compression based techniques ((Chen, 2013), (Hewlett and Cohen, 2011), (Zhikov et al., 2010), (Argamon et al., 2004), (Kityz and Wilksz, 1999)) and are largely centred around the principle of Minimum Description Length (MDL). We use the MDL based segmentation algorithm described in (Kityz and Wilksz, 1999) which makes use of the repeating subsequences present within text corpus to compress it. It is found that the segments generated by this algorithm exhibit close resemblances to words of English language. There are also other non-compression based unsupervised word segmentation and morphology induction algorithms in literature ((Mochihashi et al., 2009), (Hammarström and Borin, 2011), (Soricut and Och, 2015)). However, in this context of protein sequence analysis, we have chosen to use MDL based unsupervised segmentation because it resembles closely the first natural attempt of a linguist in identifying words of an unknown language i.e. looking for repeating subsequences as candidates for words.

As we do not have access to ground-truth knowledge about protein words, we propose to use a novel extrinsic evaluation measure based on protein family classification. SCOPe is an extended database of SCOP hierarchy (Murzin et al., 1995) which classifies protein domains based on the structural and sequence similarities. We have proposed a MDL based classifier for the task of

Proceedings of the BioNLP 2017 workshop, pages 238–246,
Vancouver, Canada, August 4, 2017. ©2017 Association for Computational Linguistics

automatic SCOPe prediction. The performance of this classifier is used as an extrinsic measure of the quality of protein segments.

Finally, the MDL based word segmentation used in (Kityz and Wilksz, 1999) is purely data driven and does not have access to any domain-specific knowledge source. We propose that constraints based on domain knowledge can be profitably used to improve the performance of segmentation algorithms. In English, we use constraints based on pronounceability rules to improve word segmentation. In protein segmentation, we use knowledge of SCOPe Class labels (Fox et al., 2014) to impose constraints. In both cases, constraints based on domain knowledge are seen to improve the segmentation quality.

To summarize, the main contributions of our work are the following :

1. We attempt to establish a mapping from protein sequences to language at the level of words which is a vital step in the linguistic approach to protein language decoding. Towards this end, we explore the use of an unsupervised text segmentation algorithm to the task of extracting "biological words" from protein sequences.

2. We propose a novel extrinsic evaluation measure for protein words via protein family classification.

3. We demonstrate the effectiveness of using domain knowledge to complement data driven approaches in the text segmentation task, as well as in its biological counterpart.

2 Related Work

Protein Linguistics (Searls, 2002) is the study of applying linguistic approaches to understand the structure and function of protein molecules. Research in the field of Protein Linguistics is largely based on the underlying assumption that proteins have a language of their own. David Searls draws many analogies between Linguistics and Molecular Biology to show how a linguistic metaphor can be seen interwoven into many problems of Molecular Biology. The fundamental analogy is that the 20 amino acids of proteins and 4 nucleotides of genes are analogous to the 26 letters in English alphabet.

Literature is abundant with parallels between language and biology (Bralley, 1996; Searls,

2002; Atkinson and Gray, 2005; Gimona, 2006; Tendulkar and Chakraborti, 2013). There are striking similarities between the structure of a protein molecule and a sentence in a Natural Language text some of which have been highlighted in Figure 1.

Gimona (2006) presents an excellent discussion on linguistics-based protein annotation and raises the interesting question of whether compositional semantics could improve our understanding of protein organization and functional plasticity. Tendulkar and Chakraborti (2013) also have drawn many parallels between biology and linguistics.

The wide gap between available primary sequences and their three dimensional structures leads to the thought that the current protein structure prediction methods might struggle due to lack of understanding of the folding code from protein sequence. If biological sequences are analogous to strings generated from a specific but unknown language, then it will be useful to find the rules of the unknown language. And, word identification is fundamental to the task of learning rules of an unknown language.

Motomura et. al ((2012),(2013)) use a frequency based linguistic approach to protein decoding and design. They call the short consequent sequences (SCS) present in protein sequences as words and use availability scores to assess the biological usage bias of SCS. Our approach of using MDL for segmentation is interesting in that it does not require prior fixing of word length as in (Motomura et al., 2012), (Motomura et al., 2013).

3 Word Segmentation

Word is defined as a single distinct conceptual unit of language, comprising inflected and variant forms[1]. In English, though space acts as a good approximation for word delimiter, proper nouns like *New York* or phrases like *once in a blue moon* make sense only when taken as a single unit. Therefore, space is not a good choice for delimiting atomic units of meaning.

Imagine a corpus of English text with spaces and other delimiters removed. Now, word segmentation is the problem of dividing a continuous piece of text into meaningful units. For example, imagine a piece of text in English with delimiters removed such as 'BIRDONTHETREE'. The contin-

---

[1] https://en.oxforddictionaries.com/definition/word

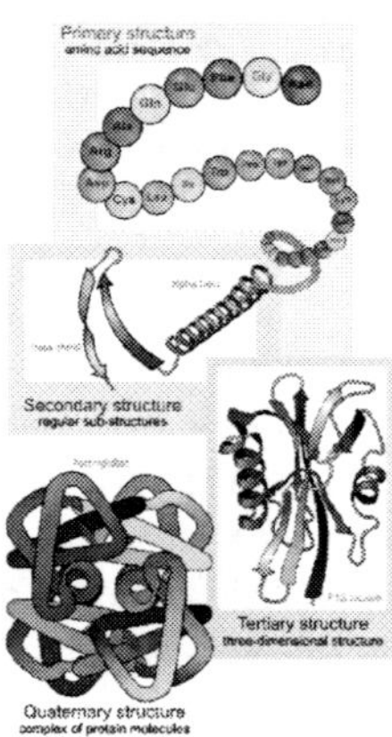

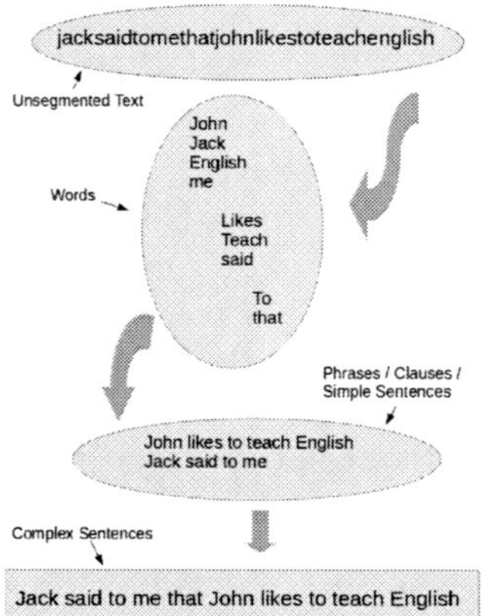

Figure 1: Structural levels in a Protein Molecule [Image source : (Wikipedia, 2017)] vs. Natural Language Sentence

uous text can be segmented into four meaningful units as 'BIRD','ON','THE','TREE'. Analogously, we define protein segmentation as the problem of dividing the amino acid sequence of a protein molecule into biologically meaningful segments. For example, the toy protein sequence 'MATGQKLMRAIRVFEFGG-PEVLKLQSDVVVPVPQSHQ' can consist of three segments 'MATGQKLMRAIR', 'VFEFGGPEV', 'LKLQSDVVVPVPQSHQ'. For our work, we assume that the word segmentation algorithm does not have knowledge about English lexicon. The significance of this assumption can be understood in the context of protein segmentation. Since the ground truth about words in protein language is not known, we consider the problem of protein segmentation to be analogous to unsupervised word segmentation in English.

We begin this section by explaining why MDL can be a good model selection principle for learning words followed by description of the algorithm used and results obtained on Brown corpus.

3.1 MDL for Segmentation

According to the principle of Minimum Description Length (MDL),

$$\text{Data compression} \rightarrow \text{Learning}$$

Any regularity present in data can be used to compress the data which can also be seen as learning of a model underlying the data (Grünwald, 2005). In an unsegmented text corpus, the repetition of words creates statistical regularities. Therefore, the key idea behind using MDL for word segmentation is that we can learn word-like segments by compressing the text corpus.

Description Length (DL) of a corpus X is defined as the number of bits needed to encode it using Shannon Fano coding [(Shannon, 2001),(Kityz and Wilksz, 1999)] and is expressed as given below.

$$DL(X) = -\sum_{x \epsilon V} c(x) \log \frac{c(x)}{|X|} \qquad (1)$$

where, V is the language vocabulary, $c(x)$ is the frequency of word x in the given corpus and $|X|$ is total number of words in X.

As an unsupervised learning algorithm does not have access to language lexicon, the initial DL of the corpus is calculated by using the language alphabet as its vocabulary. When the algorithm learns word-like segments, we can expect the DL of corpus to get reduced. According to MDL, the segmentation model that best minimizes the combined description length of *data + model* (i.e. *corpus+ vocabulary*) is the best approximation of the underlying word segmentation.

An exponential number of candidate segmentations is possible for a piece of unsegmented text. For example, some candidate segmentations for the text 'BIRDONTHETREE' are given below.

'B','IRDONTHETREE'

'BI','RD','ONTHET','R','E','E'

'B','I','R','D','ONTHET','REE'

'BIR','D','ONT','HE','TREE'

'BIRDON','THE','TREE'

'BIRD','ON','THETREE'

(Kityz and Wilksz, 1999) define a goodness measure called *Description Length Gain* (DLG) to quantify the compression effect produced by a candidate segmentation. DLG of a candidate segmentation is equal to the sum of DLGs of individual segments within it. DLG of a segment s is defined as the reduction in description length achieved by retaining this segment as a single lexical unit while aDLG stands for the average description length gain as given below.

$$DLG(s) = DL(X) - DL(X[r \to s] \oplus s)$$
$$aDLG(s) = \frac{DLG(s)}{c(s)}$$

where, $X[r \to s]$ represents the new corpus obtained by replacing all occurrences of the segment s by a single token r, $c(s)$ is the frequency of the segment s in corpus and $\oplus$ represents the concatenation of two strings with a delimiter in between. This is necessary because MDL minimizes the combined DL of corpus and vocabulary. (Kityz and Wilksz, 1999) uses Viterbi algorithm to find the optimal segmentation of a corpus. Time complexity of the algorithm is $O(mn)$ where n is the length of the corpus and m is the maximal word length.

3.2 Imposing Language Constraints

MDL based algorithm as described in (Kityz and Wilksz, 1999) performs uninformed search through the space of word segmentations. We propose to improve the performance of unsupervised algorithm by introducing constraints based on domain knowledge. These constraints help to improve the word-like quality of the MDL segments. For example, in English domain, we have used the following language constraints, mainly inspired by the fact that legal English words are pronounceable.

1. Every legal English word has at least one vowel in it

2. There cannot be three consecutive consonants in the word beginning except when the first consonant is 's'

3. Some word beginnings are impossible. For example, 'db', 'km', 'lp', 'mp', 'ns', 'ms', 'td', 'kd', 'md', 'ld', 'bd', 'cd', 'fd', 'gd', 'hd', 'jd', 'nd', 'pd', 'qd', 'rd', 'sd', 'vd', 'wd', 'xd', 'yd', 'zd'

4. Bigrams having high probability of occurrence at word boundaries are obtained apriori from a knowledge base to facilitate splitting of long segments

3.3 MDL Segmentation of Brown Corpus

The goal of our experiments is twofold. First, we apply an MDL based algorithm to identify word boundaries. Second, we use constraints based on domain knowledge to further constrain the search space and thereby improving the quality of segments.

The following is a sample input text from Brown corpus (Francis and Kucera, 1979) used in our experiment.

implementationofgeorgiasautomobiletitlelaw wasalsorecommendedbytheoutgoingjury iturgedthatthenextlegislatureprovideenab lingfundsandresettheeffectivedatesothata norderlyimplementationofthelawmaybeeffect

The output segmentation obtained after applying MDL algorithm is given below. It can be seen that the segments identified by the MDL algorithm are close to the actual words of English language.

implementationof georgias automobile title l a w wasalso recommend edbythe outgoing jury i tur g edthat thenext legislature provide enabling funds andre s et theeffective d ate sothat anorderly implementationof thelaw maybe effect ed

The segments generated by MDL are improved by applying the language constraints listed in previous section. Sample output is shown below. We can observe the effect of constraints on segments, for example, [l][a][w] is merged into [law] ; [d][ate] is merged into [date].

implementationof georgias automobile title law wasalso recommend edbythe outgoing jury i tur ged that thenext legislature provide enabling funds andre set theeffective date sothat anorderly implementationof thelaw maybe effect ed

Segmentation results are evaluated by averaging the precision and recall over multiple random samples of Brown Corpus. A segment is declared as

Algorithm	Precision	Recall
MDL (Kityz and Wilksz, 1999)	79.24	34.36
MDL + Constraints	82.57	41.06

Table 1: Boundary Detection by MDL Segmentation

Algorithm	Precision	Recall
MDL(Kityz and Wilksz, 1999)	39.81	17.26
MDL + Constraints	52.94	26.36

Table 2: Word Detection by MDL Segmentation

a correct word only if both the starting and ending boundaries are identified correctly by the segmentation algorithm. Word precision and word recall are defined as follows.

$$\text{Word Precision} = \frac{\text{No. of correct segments}}{\text{Total no. of segments}}$$

$$\text{Word Recall} = \frac{\text{No. of correct segments}}{\text{Total no. of words in corpus}}$$

Boundary precision and boundary recall are defined as follows.

$$\text{Boundary Precision} = \frac{\text{\# correct segment boundaries}}{\text{\# segment boundaries}}$$

$$\text{Boundary Recall} = \frac{\text{\# correct segment boundaries}}{\text{\# word boundaries}}$$

The performance of our learning algorithm averaged over 10 samples of size 10,000 characters (from random indices in Brown corpus) is shown in Tables 1 and 2. The reported results are in line with our proposed hypothesis that domain constraints help in improving the performance of unsupervised MDL segmentation.

4 Protein Segmentation

In this section, we discuss our experiments in protein domain. Choice of protein corpus is very critical to the success of MDL based segmentation. If we look at the problem of corpus selection from a language perspective, we know that similar documents will share more words in common than dissimilar documents. Hence, we have chosen our corpus from databases of protein families like SCOPe and PROSITE. We believe that protein sequences performing similar functions will have similar words.

4.1 Qualitative Analysis

The objective of our experiments on PROSITE database (Sigrist et al., 2012) is to qualitatively analyse the protein segments. It can be observed that within a protein family, some regions of the protein sequences have been better conserved than others during evolution. These conserved regions are found to be important for realizing the protein function and/or for the maintenance of its three dimensional structure. As part of our study, we examined if the MDL segments are able to capture the conserved residues represented by PROSITE patterns.

MDL segmentation algorithm was applied to 15 randomly chosen PROSITE families containing varying number of protein sequences. [2] Within a PROSITE family, some sequences get compressed more than others. An interesting observation is that the less compressed sequences are those that have evolved over time and hence have low sequence similarity with other members of the protein family. But, they have the conserved residues intact and MDL segmentation algorithm is able to capture those conserved residues.

For example, consider the PROSITE pattern [3] for Amidase enzyme (PS00571) G-[GAV]-S-[GS](2)-G-x-[GSAE]-[GSAVYCT]-x-[LIVMT]- [GSA]-x(6)-[GSAT]-x- [GA]-x-[DE]-x- [GA]-x-S- [LIVM]-R-x-P-[GSACTL] . The symbol 'x' in a PROSITE pattern is used for a position where any amino acid is accepted. 'x(6)' stands for a chain of five amino acids of any type. For patterns with long chains of x, MDL algorithm captures the conserved regions as a series of adjacent segments. For example, in the protein sequence with UniProtKB id O00519, the conserved residues and MDL segments are shown in Figure 2.

As another example, consider the family PS00319 with pattern G-[VT]-[EK]-[FY]-V-C-C-P . This PROSITE pattern is short and does not contain any 'x'. In such cases, the conserved residues can get captured accurately by MDL segments. The protein sequence with UniProtKB id P14599 has less sequence similarity but its conserved residues GVE-FVCCP are captured exactly in a single MDL segment. We also studied the distribution of segment lengths among the PROSITE families. A single corpus was created combining the sequences from

---

[2]The output segments are available at https://1drv.ms/f/s!AnQHeUjduCq0ae9rWhuoybZoA-U

[3]A PROSITE pattern like [AC]-x-V-x(4)-AV is to be translated as: [Ala or Cys]-any-Val-any-any-any-any-Ala-Val

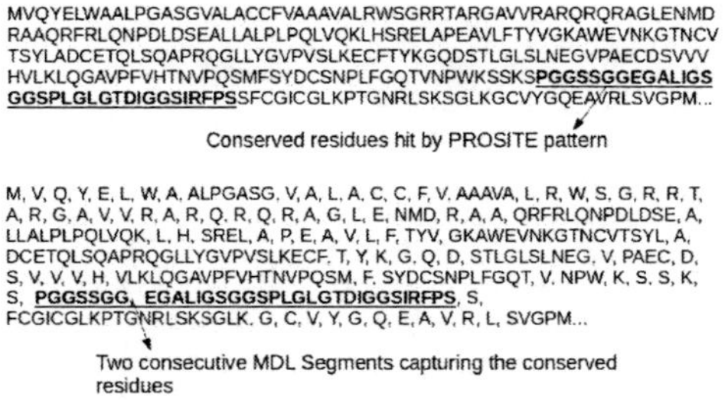

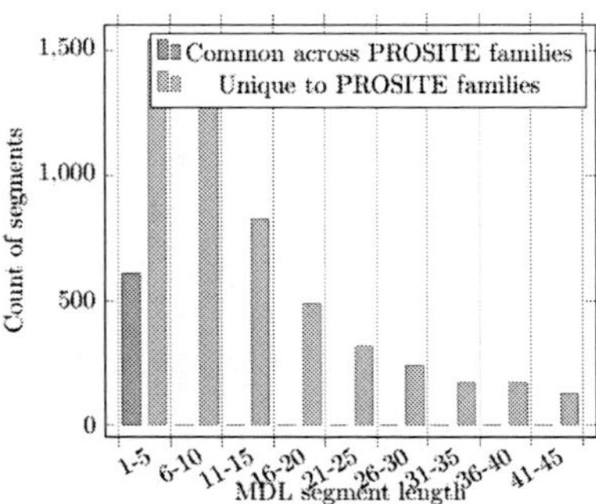

Figure 2: Conserved residues and MDL segments of a protein sequence (UniProtKB id O00519) in PROSITE family PS00571

Figure 3: Distribution of MDL segment lengths among PROSITE families PS00319, PS00460, PS00488, PS00806 and PS00818

5 randomly chosen PROSITE families and the distribution of segment lengths is shown in Figure 3. Protein segments that were common among the families were typically four or five amino acids in length. However, within each individual family there were longer segments unique to that family. Very long segments (length >15) are formed when the corpus contains many sequences with high sequence similarities.

4.2 Quantitative Analysis

Unlike in English language, we do not have access to ground truth about words in proteins. Hence, we propose to use a novel extrinsic evaluation measure based on protein family classification. We describe a compression based classifier that uses the MDL segments (envisaged as words in proteins) for SCOPe predictions.The performance of the MDL based classifier on SCOPe predictions is used as an extrinsic evaluation measure of protein segments.

4.2.1 MDL based Classifier

Suppose we want to classify a protein sequence p into one of k protein families, the MDL based classifier is given by,

$$\text{family (p)} = \underset{\text{family}}{\operatorname{argmax}} DLG(p, \text{family}_{1...k}) \quad (2)$$

where $DLG(p, \text{family}_i)$ is the measure of the compression effect produced by protein sequence p in the protein corpus of family_i. We hypothesize that a protein sequence will be compressed more by the protein family it belongs to, because of the presence of similar words among the same family members.

Experimental Setup The dataset used for protein classification is ASTRAL Compendium (Chandonia et al., 2004). It contains protein domain sequences for domains classified by the SCOPe hierarchy. ASTRAL 95 subset based on SCOPe v2.05 is used as training corpus and the test set is created by accumulating the protein domain sequences that were newly added in SCOPe v2.06. Performance of the MDL classifier is discussed in four SCOPe levels - *Class, Fold, Superfamily and Family*. At all levels, we consider only the protein domains belonging to four SCOPe *classes* A,B,C and D representing *All Alpha, All Beta, Alpha+Beta, Alpha/Beta* respectively. The blind test set contains a total of 4821 protein domain sequences.

SCOPe classification poses the problem of *class imbalance* due to the non-uniform distribution of domains among different classes at all SCOPe levels. Due to this problem, we use macro precision and macro recall (Yang, 1999) as performance measures and are given by the below equations.

$$Precision_{macro} = \frac{1}{q} \sum_{i=1}^{q} \frac{TP_i}{TP_i + FP_i} \quad (3)$$

$$Recall_{macro} = \frac{1}{q} \sum_{i=1}^{q} \frac{TP_i}{TP_i + FN_i} \quad (4)$$

4.2.2 Performance of MDL Classifier

Class Prediction Out of 4821 domain sequences in the test data, the MDL classifier abstains from prediction for 71 sequences due to multiple classes giving the same measure of compression. The MDL Classifier achieves a macro precision of 75.64% and macro recall of 69.63% in *Class* prediction.

SCOPe level	Macro Average Precision	Macro Average Recall
Class	75.64	69.63
Fold	60.59	45.08
Super family	56.65	43.73
Family	43.25	37.7

Table 3: Performance of MDL Classifier in SCOPe Prediction

SCOPe level	Weighted Average Precision	Weighted Average Recall
Class	76.38	69.77
Fold	**81.49**	49.25
Super family	72.80	48.23
Family	45.02	35.85

Table 4: Performance of MDL Classifier in SCOPe Prediction - Weighted Measures

Fold Prediction SCOPe v2.05 contains a total of 1208 *folds* out of which 991 folds belong to *classes* A,B,C and D. The distribution of protein sequences among the *folds* is non-uniform ranging from 1 to 2254 sequences with 250 *folds* containing only one sequence. MDL Classifier achieves a macro precision of 60.59% and macro recall of 45.08% in *fold* classification.

Impact of Corpus Size The number of protein domains per class decreases greatly down the SCOPe hierarchy. The *folds* (or *families, superfamilies*) that have very few sequences should have less contribution in the overall prediction accuracy. We weighted the macro measures based on the number of instances which resulted in the weighted averages reported in Table 4. The MDL classifier achieves a weighted macro precision of 81.49% in SCOPe *fold* prediction which is higher than the precision at any other level. This observation highlights the quality of protein segments generated by MDL algorithm. It is also important to note that *fold* prediction is an important sub task of protein structure prediction just as how word detection is crucial to understanding the meaning of a sentence.

4.3 MDL Classifier as a Filter

The *folds* which are closer to each other in the SCOPe hierarchy tend to compress protein sequences almost equally. Instead of returning a

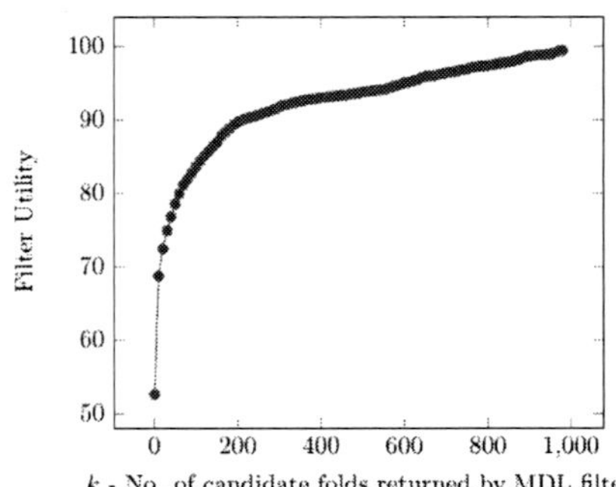

Figure 4: Variation of Filter Utility with Filter Size k

single *fold* giving maximum compression, if the MDL classifier returns the top-k candidates, then we can reduce the search space for manual or high cost inspections. We define utility of the MDL classifier when used as a filter as given below.

$$\text{Utility} = \frac{\text{No. of predictions where correct fold is in top-k list}}{\text{Total no. of predictions}}$$

Figure 4 shows the k versus utility on test data. It can be seen from the graph that at k=400 (which is approximately 33% of the total number of folds), top-k predictions are able to give 93% utility. In other words, in 93% of the test sequences, MDL filter can be used to achieve nearly 67% reduction in the search space of 1208 folds.

4.4 Impact of Constraints based on Domain Knowledge

Similar to experiments in English domain, the MDL algorithm on protein dataset can also be enhanced by including constraints from protein domain knowledge. For example, in a protein molecule, hydrophobic amino acids are likely to be found in the interior, whereas hydrophilic amino acids are more likely to be in contact with the aqueous environment. This information can be used to introduce checks on allowable amino acids at the beginning and end of protein segments. Unlike in English, identifying constraints based on protein domain knowledge is difficult because there are no lexicon or protein language rules readily available. Domain expertise is needed for getting explicit constraints.

As proof of concept, we use the SCOPe *class* labels of protein sequences as domain knowledge and study its impact on the utility of the MDL filter. After introducing *class* knowledge, MDL filter achieves an utility of 93% at k=100, i.e., in 93%

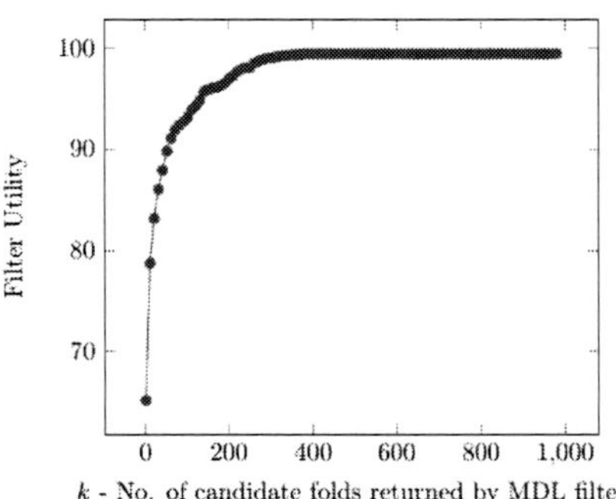

Figure 5: Variation of Filter Utility with Filter Size k after adding constraints based on SCOPe Class labels

of the test sequences, MDL filter can be used to achieve nearly 90% reduction in the search space of 1208 folds. In the absence of *class* knowledge, the same filter utility was obtained at k=400 which is only 67% reduction of search space (Figure 5). Through this experiment, we emphasize that appropriate domain knowledge can help in improving the quality of word segmentation in protein sequences. Such domain knowledge could be imposed in the form of constraints during unsupervised learning of protein words. We would like to emphasize the fact that introducing domain knowledge in the form of class labels as in supervised or semi-supervised learning frameworks may not be appropriate in protein sequences due to our current ignorance of the true protein words.

5 Discussion

In the words of Jones and Pevzner (Jones and Pevzner, 2004), "It stands to reason that if a word occurs considerably more frequently than expected, then it is more likely to be some sort of 'signal' and it is crucially important to figure out the biological meaning of the signal". In this paper, we have proposed protein segments obtained from MDL segmentation as the signals to be decoded.

As part of our future work, we would like to study the performance of SCS words (Motomura et al., 2012), (Motomura et al., 2013) in protein family classification and compare it against MDL words; We would also like to measure the availability scores of MDL segments. It may also be insightful to study the co-occurrence matrix of MDL segments.

6 Conclusion

Given the abundance of unlabelled data, data driven approaches have witnessed significant success over the last decade in several tasks in vision, language and speech. Inspired by the correspondence between biological and linguistic tasks at various levels of abstraction as revealed by the study of Protein Linguistics, it is only natural that there would be a propensity to extend such approaches to several tasks in Computational Biology. A linguist already knows a lot about language however, and a biologist knows lot about biology; so, it does make sense to incorporate what they already know to constrain the hypothesis space of a machine learner, rather than make the learner rediscover what the experts already know. The latter option is not only demanding in terms of data and computational resources, it may need us to solve riddles we just do not have answers to. Classifying a piece of text as humorous or otherwise is hard at the state of the art; there are far too many interactions between variables than we can model, not only do the words interact between them, they also interact with the mental model of the person reading the joke. It stretches our wildest imaginations to think of a purely bottom up Deep Learner that is deprived of common-sense and world knowledge to learn such end-to-end mappings reliably by looking at data alone. The same is true in biological domains where non-linear interactions between a large number of functional units make macro-properties "emerge" out of interactions between individual functional units. We feel that a realistic route is one where top down (knowledge driven) approaches complement bottom up (data driven) approaches effectively. This paper would have served a modest goal if it has aligned itself towards demonstrating such a possibility within the scope of discovering biological words, which is just one small step in the fascinating quest towards deciphering the language in which biological sequences express themselves.

References

Shlomo Argamon, Navot Akiva, Amihood Amir, and Oren Kapah. 2004. Efficient unsupervised recursive word segmentation using minimum description length. In *Proceedings of the 20th international conference on Computational Linguistics*. Association for Computational Linguistics, page 1058.

Quentin D Atkinson and Russell D Gray. 2005. Curious parallels and curious connectionsphylogenetic thinking in biology and historical linguistics. *Systematic biology* 54(4):513–526.

Patricia Bralley. 1996. An introduction to molecular linguistics. *BioScience* 46(2):146–153.

John-Marc Chandonia, Gary Hon, Nigel S Walker, Loredana Lo Conte, Patrice Koehl, Michael Levitt, and Steven E Brenner. 2004. The ASTRAL compendium in 2004. *Nucleic acids research* 32(suppl 1):D189—-D192.

Ruey-Cheng Chen. 2013. An improved mdl-based compression algorithm for unsupervised word segmentation. In *ACL (2)*. pages 166–170.

Naomi K Fox, Steven E Brenner, and John-Marc Chandonia. 2014. Scope: Structural classification of proteinsextended, integrating scop and astral data and classification of new structures. *Nucleic acids research* 42(D1):D304–D309.

W Nelson Francis and Henry Kucera. 1979. The brown corpus: A standard corpus of present-day edited american english. *Providence, RI: Department of Linguistics, Brown University [producer and distributor]* .

Mario Gimona. 2006. Protein linguisticsa grammar for modular protein assembly? *Nature Reviews Molecular Cell Biology* 7(1):68–73.

Peter Grünwald. 2005. A tutorial introduction to the minimum description length principle. *Advances in minimum description length: Theory and applications* pages 23–81.

Harald Hammarström and Lars Borin. 2011. Unsupervised learning of morphology. *Computational Linguistics* 37(2):309–350.

Daniel Hewlett and Paul Cohen. 2011. Fully unsupervised word segmentation with bve and mdl. In *Proceedings of the 49th Annual Meeting of the Association for Computational Linguistics: Human Language Technologies: short papers-Volume 2*. Association for Computational Linguistics, pages 540–545.

Neil C Jones and Pavel Pevzner. 2004. *An introduction to bioinformatics algorithms*. MIT press.

Chunyu Kityz and Yorick Wilksz. 1999. Unsupervised learning of word boundary with description length gain. In *Proceedings of the CoNLL99 ACL Workshop. Bergen, Norway: Association f or Computational Linguistics*. Citeseer, pages 1–6.

Daichi Mochihashi, Takeshi Yamada, and Naonori Ueda. 2009. Bayesian unsupervised word segmentation with nested pitman-yor language modeling. In *Proceedings of the Joint Conference of the 47th Annual Meeting of the ACL and the 4th International Joint Conference on Natural Language Processing of the AFNLP: Volume 1-Volume 1*. Association for Computational Linguistics, pages 100–108.

Kenta Motomura, Tomohiro Fujita, Motosuke Tsutsumi, Satsuki Kikuzato, Morikazu Nakamura, and Joji M Otaki. 2012. Word decoding of protein amino acid sequences with availability analysis: a linguistic approach. *PloS one* 7(11):e50039.

Kenta Motomura, Morikazu Nakamura, and Joji M Otaki. 2013. A frequency-based linguistic approach to protein decoding and design: Simple concepts, diverse applications, and the scs package. *Computational and structural biotechnology journal* 5(6):1–9.

Alexey G Murzin, Steven E Brenner, Tim Hubbard, and Cyrus Chothia. 1995. Scop: a structural classification of proteins database for the investigation of sequences and structures. *Journal of molecular biology* 247(4):536–540.

David B Searls. 2002. The language of genes. *Nature* 420(6912):211–217.

Claude Elwood Shannon. 2001. A mathematical theory of communication. *ACM SIGMOBILE Mobile Computing and Communications Review* 5(1):3–55.

Christian JA Sigrist, Edouard De Castro, Lorenzo Cerutti, Béatrice A Cuche, Nicolas Hulo, Alan Bridge, Lydie Bougueleret, and Ioannis Xenarios. 2012. New and continuing developments at prosite. *Nucleic acids research* page gks1067.

Radu Soricut and Franz Josef Och. 2015. Unsupervised morphology induction using word embeddings. In *HLT-NAACL*. pages 1627–1637.

Ashish Vijay Tendulkar and Sutanu Chakraborti. 2013. Parallels between linguistics and biology. In *Proceedings of the 2013 Workshop on Biomedical Natural Language Processing, Sofia, Bulgaria: Association for Computational Linguistics*. Citeseer, pages 120–123.

Wikipedia. 2017. Protein structure — Wikipedia, the free encyclopedia. http://en.wikipedia.org/w/index.php?title=Protein%20structure&oldid=774730776. [Online; accessed 22-April-2017].

Yiming Yang. 1999. An evaluation of statistical approaches to text categorization. *Information retrieval* 1(1-2):69–90.

Valentin Zhikov, Hiroya Takamura, and Manabu Okumura. 2010. An efficient algorithm for unsupervised word segmentation with branching entropy and mdl. In *Proceedings of the 2010 Conference on Empirical Methods in Natural Language Processing*. Association for Computational Linguistics, pages 832–842.

External Evaluation of Event Extraction Classifiers for Automatic Pathway Curation: An extended study of the mTOR pathway

Wojciech Kusa
AGH University of
Science and Technology
Cracow, Poland
wojciechkusa@gmail.com

Michael Spranger
Sony Computer Science
Laboratories Inc.
Tokyo, Japan
michael.spranger@gmail.com

Abstract

This paper evaluates the impact of various event extraction systems on automatic pathway curation using the popular mTOR pathway. We quantify the impact of training data sets as well as different machine learning classifiers and show that some improve the quality of automatically extracted pathways.

1 Introduction

Biological pathways encode sequences of biological reactions, such as phosphorylation, activation etc, involving various biological species, such as genes, proteins (Aldridge et al., 2006; Kitano, 2002). Studying and analyzing pathways is crucial to understanding biological systems and for the development of effective disease treatments and drugs (Creixell et al., 2015; Khatri et al., 2012). There have been numerous efforts to reconstruct detailed process-based and disease level pathway maps such as Parkinson disease map (Fujita et al., 2014), Alzheimers disease Map (Mizuno et al., 2012), mTOR pathway Map (Caron et al., 2010), and the TLR pathway map (Oda and Kitano, 2006). Traditionally, these maps are constructed and curated by expert pathway curators who manually read numerous biomedical documents, comprehend and assimilate the knowledge in them and construct the pathway.

With increasing number of scientific publications manual pathway curation is becoming more and more impossible. Therefore, Automated Pathway Curation (APC) and semi-automated biological knowledge extraction has been an active research area (Ananiadou et al., 2010; Ohta et al., 2013; Szostak et al., 2015) trying to overcome the limitations of manual curation using various techniques from hand-crafted NLP systems

(Allen et al., 2015) to machine learning techniques (Björne et al., 2011). Machine-learning NLP systems, in particular, show good performance in BioNLP tasks, but they are still performing less good in automated pathway curation, partly because there have been few attempts to measure the performance of NLP systems for APC directly.

Recently, there has been some attempt at remedying the situation and new datasets and evaluation measures have been proposed. For instance, Spranger et al. (2016) use the popular human-generated mTOR pathway map (Caron et al., 2010; Efeyan and Sabatini, 2010; Katiyar et al., 2009) and quantify the performance of a particular APC system and its ability to recreate the complete pathway automatically. Results reported were mixed.

One of the key components in such APC systems is identification of triggers, events and their relationships. These machine learning-based systems are essentially just supervised classification components.

This paper explores whether we can improve results of automated pathway curation for mTOR pathway by using different training datasets and learning algorithms. We show that the choice of event extraction classifiers increases F-score by up to 20% compared to state-of-the-art system. Our results also show that within limits the choice of training data has significantly less impact on results than the choice of classifier. Our results also suggest that additional research is necessary to solve the problem of APC.

2 Automatic Pathway Curation

We constructed an automatic pathway curation system that take as input scientific articles in PDF format and transforms them into SBML encoded, annotated pathway maps. The pipeline has multi-

247

Proceedings of the BioNLP 2017 workshop, pages 247–256,
Vancouver, Canada, August 4, 2017. ©2017 Association for Computational Linguistics

ple steps.

1. PDFs are translated into pure text files using the cermine[1] tool.

2. Preprocessing provides tokenization, POS tagging, dependency and syntax parsing.

3. An *event extraction system* extracts the mentions of entities (genes, proteins etc), reactions (e.g. phosphorylation) and their arguments (theme, cause, product).

4. A converter constructs pathways from the information provided by the event extraction system.

5. An annotation system maps extracted entities and events to Entrez gene identifiers and SBO terms.

The following sections detail steps 3 to 5.

2.1 Event Extraction

We used the TURKU Event Extraction System (TEES) for event extraction (Björne et al., 2010). This system is one of the most successful BioNLP systems. It has not only won 1st place in BioNLP competitions but was also the only one NLP system that participated in all BioNLP-ST 2013 tasks (Björne et al., 2012). The system combines various NLP techniques to extract information from text. TEES workflow consists of four steps:

1. Trigger Detection - detection of named entities and event triggers in a given sentence to construct nodes of the event graph.

2. Edge detection - construction of complex events linking few triggers to create event graph. Output produced during this step is often a directed, typed edge connecting two entity nodes.

3. Unmerging - event nodes from merged event graph are duplicated in order to separate arguments into valid combinations. This step is needed for evaluation of final results in BioNLP Shared Task standard.

4. Modifiers detection - final component that defines additional attributes for events such as speculation and negation modifiers.

By default TEES trains a different instance of multiclass Support Vector Machines (SVM) for each step. Recent versions of TEES (Björne and Salakoski, 2015) allow to easily exchange the SVM classifiers with other supervised classification algorithms. For example, all *scikit-learn* multiclass, supervised learning algorithms that support sparse feature matrices can be applied (Pedregosa et al., 2011). Thanks to this it is possible to test different algorithms for event extraction task and automatic pathway extraction. For this paper, we exchanges classifiers in all steps 1-4s as described in Section 3. The output of TEES is a standoff formatted representation of entities and events.

2.2 Conversion Standoff to SBML pathways

In principle events and entities extracted by TEES correspond to biological species and reactions. We translate the NLP representation into SBML – the standard, XML-based markup language for representing biological models (Hucka et al., 2003). SBML essentially encodes models using biological players called `sbml:species`[2]. `sbml:species` can participate in interactions, called `sbml:reaction`. Species participate in interaction as `sbml:reactant`, `sbml:product` and `sbml:modifier`. The basic idea being that some quantity of reactant is consumed to produce a product. Reactions are influenced by modifiers. The mapping algorithm is adopted from and described in more detail in Spranger et al. (2015).

2.3 SBO/GO, Entrez Gene Annotations

The SBML encoded, automatically extracted pathway is further *annotated* using Systems Biology Ontology (SBO) (Le Novère, 2006) and Gene Ontology (GO) terms. SBO also provides a class hierarchy for reaction types. For instance, the NLP system identify phosphorylation reactions, which are a subclass of conversion reactions. All reactions in the data are automatically annotated with SBO/GO term (coverage 100%) using an annotation scheme detailed in (Spranger et al., 2015).

Species (e.g. proteins, genes) were annotated using the gene/protein named entity recognition and normalization software GNAT (Hakenberg et al., 2011) - a publicly available gene/protein normalization tool. GNAT returns a set of Entrez Gene identifiers (Maglott et al., 2005) for each input string. Species were annotated using all returned Entrez Gene identifiers for a particular

---

[1] http://cermine.ceon.pl/index.html

[2] We refer to SBML vocabulary using the prefix "sbml".

species (organism human). We call the set of Entrez Gene identifiers returned by GNAT for each species *Entrez Gene signature*.

3 Classifiers for Event Extraction

In this paper we evaluate classifiers for event extraction (Section 2.1) and their impact on the overall performance of the automatic pathway extraction system. We compare the following classifiers:

- **Support Vector Machines (SVM)** is the default TEES classifier (Joachims, 1999). It was optimized for linear classification and its performance scales linearly with the number of training examples.

- **Decision Tree (DT)** creates a model that can predict the target value by learning simple decision rules inferred from the training data. Compared to the other techniques they are relatively fast, cost of using tree is logarithmic in the number of examples. We use Gini impurity criterion to evaluate quality of the split.

- **Random Forest (RF)** classifiers fit a number of ensembled decision tree classifiers, each built from a bootstrap sample of a training set. The best split of node is chosen only from a random subset of the features, not all features. Final classifiers are combined by averaging their probabilistic prediction. Single tree have a higher bias but, due to averaging variance of the random forest as a whole decreases.

- **Multinomial Naive Bayes (MNNB)** This is an implementation of the naive Bayes algorithm for multinomial data which is one of the classic variants used in classification of discrete features (e.g. text classification). Additive smoothing parameter was set to 1.

- **Multi-layer Perceptron (MLP)** MLP is a feedforward neural network model. We use hidden layer with 100 neurons and rectified linear unit activation function. We optimize for logarithmic loss using stochastic gradient descent. Learning rate is constant and equal to 0.001.

For DT, RF, MNNB and MLP we use implementations from *scikit-learn* Python library (Pedregosa et al., 2011).

Item	ANN	GE11	PC13
Documents	60	908	260
Words	11960	205729	53811
Entities	1921	11625	7855
Events	1284	10310	5992
Modifiers	71	1382	317
Renaming	101	571	455

Table 1: Corpora statistics

Reaction type	ANN	GE11	PC13
Acetylation	0	0	38
Activation	0	0	359
Binding	211	988	606
Catalysis	87	0	0
Conversion	0	0	124
Deacetylation	0	0	1
Degradation	0	0	49
Demethylation	0	0	4
Dephosphorylation	14	0	22
Deubiquitination	0	0	3
Dissociation	55	0	54
Gene_expression	46	2265	384
Hydroxylation	0	0	1
Inactivation	0	0	76
Localization	27	281	96
Methylation	0	0	7
Negative_regulation	194	1309	801
Pathway	0	0	443
Phosphorylation	252	192	406
Positive_regulation	235	3385	1506
Protein_catabolism	18	110	0
Regulation	132	1113	707
Transcription	8	667	74
Translation	1	0	11
Transport	0	0	189
Ubiquitination	4	0	31

Table 2: Reaction types annotated for training data sets.

4 Datasets

4.1 Training Datasets

In order to quantify the impact of training data, we test the following three training sets.

- **ANN** - consists of 60 abstracts of scientific papers from Pubmed database related to the mTORpathway map. This dataset was human-annotated for NLP system training (Ohta et al., 2011, Corpus annotations (c) GENIA Project) .

- **GE11** consists of 908 abstracts and full texts of scientific papers used in BioNLP ST 2011 GENIA Event Extraction task as training data (Kim et al., 2012).

- **PC13** consists of 260 abstracts of scientific papers used in BioNLP ST 2013 Pathway Curation task as training data (Ohta et al., 2013). The task goal was to evaluate the applicability of event extraction systems to support the automatic curation and evaluation of biomolecular pathway models.

The overall corpora statistics are summarized in Table 1. GE11 and PC13 have the largest number of annotated events. ANN is much smaller in comparison. Also, the distribution of event types differs between data sets (Table 2). GE11 uses more general terms (Binding, Regulation) compared to PC13 where some specific events appear only a few times (Deacetylation, Hydroxylation, Methylation).

We train classifiers on four combinations of the three training datasets: 1) standalone GE11; 2) GE11+ANN - combined GE11 and ANN; 3) combined GE11+PC13+ANN - GE11, PC13 and ANN; 4) PC13+ANN - combined PC13 and ANN. For instance, DT+GE11 refers to a decision tree classifier trained on GE11.

We use GE11-Devel BioNLP ST2011 dataset for hyperparameter optimization of all classifiers.

4.2 Test Data

Performance of classifiers is tested on the mTOR pathway map (Caron et al., 2010). The map was constructed by expert human curators using 522 full text papers from the PubMed database. The experts curated a single large map using CellDesigner (Funahashi et al., 2008) - a software for modeling and executing mechanistic models of pathways. CellDesigner represents information using a heavily customized XML-based SBML format (Hucka et al., 2003).

Target Human expert data We translate the curator map into standard SBML and further enrich the information using SBO/GO and Entrez Gene annotations. For SBO/GO, we use existing annotations provided by curators and extend them by automatic annotations deduced from reactants and products of reactions. For example, if a phosphoryl group is added in a reaction, it is annotated using the SBO term for phosphorylation. Each reaction may be annotated with multiple SBO/GO terms. Also we annotate the curated map with Entrez gene identifiers (similar to the automatic extraction data). We call this pathway *TARGET*.

Testing classifiers The 522 full text papers – used by human curators for the construction of the mTOR pathway – are used for evaluating the different text mining classifiers. For this, we plug in (trained) classifiers into the automatic pathway extraction pipeline which performs preprocessing, event extraction, conversion to SBML and annotation (see also Section 2). The output of this is an annotated SBML file that is subsequently compared to human-curated SBML-encoded pathway data.

5 Evaluation

Evaluation of the classifiers (and the system as a whole) is performed by comparing the automatically extracted pathway with the hand-curated pathway. Spranger et al. (2016) propose a number of graph overlap algorithms for quantifying the difference and similarity of two pathways. Here we employ the same measures. The following summarizes the strategies.

Species In order to decide whether species in two pathways are the same, we use the name of the identifiers and their Entrez gene signatures.

nmeq: Two species are equal if their names are exactly equal. We remove certain prefixes from the names (e.g. phosphorylated).

appeq: Two species are equal if their names are approximately equal. Two names are approximately equal iff their Levenshtein-based string distance is above 90 (Levenshtein, 1966)

enteq: Two species are equal if their entrez gene identifiers are exactly equal. This basically translates to the two species bqbiol:is identifier sets being exactly the same (order does not matter).

entov: Two species are equal if their entrez gene identifiers sets overlap. This basically translates to the two species bqbiol:is identifier sets overlapping.

wc: Human curated data contains complex species that contain other species as constituents (species that consist of various proteins etc). *wc* allows species to match with constituents of complexes.

Reaction match based on their SBO/GO annotations

sboeq: Two reactions are equal iff their signatures are exactly the same. That is, the whole set of SBO/GO terms of one reaction is the same as of the other reaction.

sboov: Two reactions are equal, iff their signatures overlap. That is, the intersection of the set of SBO/GO terms of one reaction is with the set of SBO/GO terms of the other reaction is not empty.

sobisa: Two reactions are equal, iff there is at least one SBO/GO term in each signature that relate in a is_a relationship in the SBO reaction type hierarchy. For instance, if there is a phosphorylation reaction and a conversion reaction, then *sboisa* will match because phosphorylation is a subclass of conversion according to the SBO type hierarchy.

Edges only match if their labels are strictly equal. So if an edge is a reactant, then it has to be a reactant in the other pathway. Same holds for products and modifiers.

Subgraph matching strategies are combinations of matching strategies for species, reactions (and for edges which is always the same). For instance, the matching strategy *nmeq, sboeq* is the most strict and requires that species names are exactly equal and that SBO/GO signatures of reactions are exactly equal. The matching strategy *appeq/enteq/wc, sboisa* is the most loose strategy. In this strategy, two species match if their names are approximately equal or if their Entrez gene identifiers overlap or if any of this applies to one of the constituents of the two species. Two reactions match if any of their SBO/GO terms are in a is_a relationship. We compare a total of 24 matching strategies.

Subgraph overlap is computed as follows. For each subgraph in the extracted pathway we search for subgraphs in the human curated data that match according to some subgraph matching strategy. We use *micro-averaged F-score*, precision and recall (Sokolova and Lapalme, 2009) for quantifying the retrieval results. F-score is used to quantify the overlap of species, reactions and edges. We then macro-average these results to get a *total F-score* quantifying performance of the extraction system as a whole.

6 Results

Some classifiers take long to train, so we only have partial results for MLP. However, all other classi-fiers (DT, MNNB, RF, SVM) finished training on all selected combinations of training data sets.

Since we tested 24 subgraph overlap measures with 18 classifiers, we receive a lot of data that cannot be discussed in detail in this paper. Here, we concentrate on general trends in the data. Code and datasets are published as appropriate[3].

6.1 Extraction Results: Species, Reactions, Subgraphs

Generally speaking the extracted pathways contain two order of magnitudes more species reactions, and edges than the *TARGET* pathway (see Table 3 for all results). This is normal since the extracted pathways consist of all combinations of entity and event mentions in text. The same entities may occur more often in the text then they are referenced in the actual pathway.

Our results show that extraction classifiers perform inconsistent with respect to the identification of compartments. While some classifiers retrieve a lot of compartment information (via localization events), others (especially MNNB trained on ANN and PC13 datasets) do not extract any compartments. MNNB with our parameter choice might not be able to learn many different event types so it skips least frequent reaction types (one of which is localization event).

Measuring how many subgraphs there are per pathway, we can see that more than half of all species extracted by classifiers are isolated and not connected to any reactions. Similarly we see many (small) subgraphs being extracted by the classifiers, whereas *TARGET* consists of essentially one large connected graph (with a few modeling mistakes).

6.2 General Trends Subgraphs overlap

Let us first concentrate on overall performance especially with respect to previous results. For this we compute the best classifiers and their score for different matching strategies. For each matching strategy, we evaluate all classifiers and then choose the best performing one and compare it with the results reported in Spranger et al. (2016)/Spr16. Table 4 shows that the best classifiers outperform Spr16 in all cases and for some subgraph overlap measures by 10 points.

If we analyze the classifiers from this paper in more detail, results (Figure 1, Table 5) show that

---

[3] https://github.com/sbnlp/
2017BioNLPEvaluation/

name	# species	# reactions	# compartments	# edges	# reactant edges	# edges product	# modifiers	# isolated species	# isolated subgraphs
DT+GE11	282361	92899	201	195531	89001	91895	14635	118162	187871
DT+GE11+ANN	284187	95096	188	212490	100529	93886	18075	115427	184542
DT+GE11+PC13+ANN	289504	94496	208	207447	94044	93559	19844	118281	188013
DT+PC13+ANN	279647	82977	20	188325	86802	82469	19054	123309	184698
MLP+GE11+ANN	278510	88502	230	193150	88655	87636	16859	114541	182456
MNNB+GE11	264413	69744	202	137828	61448	69250	7130	139402	198972
MNNB+GE11+ANN	245680	45690	0	86771	40102	45676	993	166712	206606
MNNB+GE11+PC13+ANN	269008	68926	0	142712	70292	68894	3526	151495	203903
MNNB+PC13+ANN	287314	76932	0	183029	94693	76925	11411	154210	199844
RF–GE11	227613	29573	9	50444	20786	29133	525	178233	206874
RF–GE11+ANN	261414	67974	347	130556	57195	67271	6090	136180	199157
RF–GE11+PC13+ANN	203314	32075	1	58083	25312	31704	1067	146342	177371
RF–PC13+ANN	236220	37018	0	68559	30493	36909	1157	168927	204771
SVM+GE11	288421	98938	451	200595	89769	97791	13035	109060	191175
SVM+GE11+ANN	262327	81207	388	169841	73033	80203	16605	109862	177023
SVM+GE11+PC13+ANN	275303	85435	312	179661	77587	84549	17525	114941	184481
SVM+PC13+ANN	275256	82119	59	177651	79239	81512	16900	120729	186122
TARGET	2242	777	7	2457	1044	892	521	15	4

Table 3: General statistics of all datasets. Number of extracted species, reactions and compartments. Total number of edges and of product, reactant and modifier edges. The table also shows the number of isolated species and the number of unconnected subgraphs for each pathway. The human curated mTOR pathway *TARGET* numbers are shown in the last row.

	this	Spr16
	f-score	f-score
nmeq, sboeq	**11.7**	7.6
nmeq, sboov	**15.3**	11.4
nmeq, sboisa	**18.1**	13.6
appeq, sboeq	**12.5**	8.1
appeq, sboov	**16.3**	12.0
appeq, sboisa	**19.4**	14.5
appeq/enteq, sboeq	**16.9**	11.9
appeq/enteq, sboov	**21.7**	17.1
appeq/enteq, sboisa	**26.0**	20.4
appeq/entov, sboeq	**36.2**	26.9
appeq/entov, sboov	**41.9**	34.7
appeq/entov, sboisa	**48.6**	39.5
nmeq/wc, sboeq	**23.3**	15.0
nmeq/wc, sboov	**26.0**	19.6
nmeq/wc, sboisa	**29.1**	22.0
appeq/wc, sboeq	**24.6**	15.7
appeq/wc, sboov	**27.4**	20.4
appeq/wc, sboisa	**30.9**	23.1
appeq/enteq/wc, sboeq	**39.7**	29.1
appeq/enteq/wc, sboov	**45.3**	37.2
appeq/enteq/wc, sboisa	**52.0**	42.2
appeq/entov/wc, sboeq	**39.7**	29.1
appeq/entov/wc, sboov	**45.3**	37.2
appeq/entov/wc, sboisa	**52.0**	42.2

Table 4: This table compares macro F-score performance of the classifiers discussed in this paper with results reported in Spranger et al. (2016)

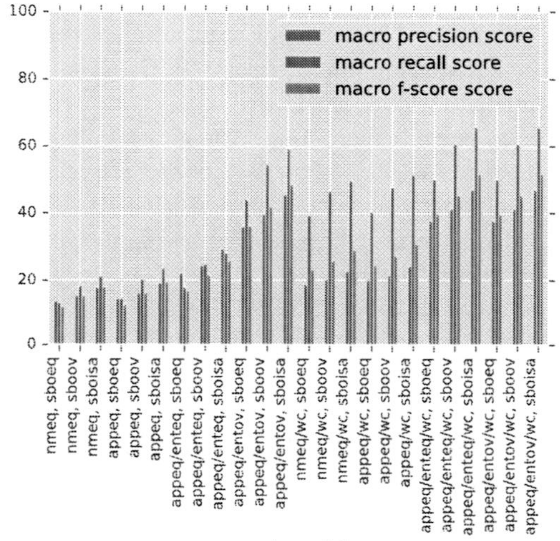

Figure 1: Best performing classifier f-score, precision and recall for each subgraph overlap functions. The x-axis are the different subgraph overlap function. The y-axis shows precision, recall, f-score of the highest classifier for each subgraph overlap function. Notice that these can be different classifiers for each subgraph overlap function (see Table 5 for all results).

for the strictest matching strategy (*nmeq, sboeq*) the best classifiers reach a macro F-score of 12 (with 14 precision, 13 recall scores). For the loosest strategy (*appeq/entov/wc, sboisa*) this goes up to F-score 52 (47 precision, 66 recall). These results show that when it comes to *exact* extraction the classifiers fail badly, whereas with more looser overlap strategies, performance becomes reasonable and there is some overlap between the extracted and the human-curated data. Of course, this also entails that the automatically extracted pathway does not completely capture what humans are constructing from the text.

Generally speaking overlap strategies that are loose with respect to constituents of complex species (*wc*) outperform their non *wc* counterparts. For instance, *nmeq/wc, sboeq* performs much better than *nmeq, sboeq*. This shows that complex species are important for the mTOR pathway but their extraction is not very detailed - which is why the overlap matching strategy has to be lenient with respect to complex species constituents. The increase in F-score for *wc* matching strategies is primarily driven by an increase in recall score. For instance, the difference between *nmeq, sboeq* and *nmeq/wc, sboeq* is more than 20 points, whereas precision does not improve that much. The reasons for that is that the same subgraphs in the extracted pathway overlap with more subgraphs in *TARGET*. So it is not the case that other subgraphs in the extracted pathway overlap with *TARGET*.

Results also show that recall is in general much higher than precision for looser strategies. For instance, *wc* strategies (right hand side of Figure 1) double the recall score w.r.t to their precision scores. This also shows that in principle loosening matching strategies impacts mostly recall as the same subgraphs in the extracted data overlap with the human curated data.

6.3   Classifier Performance in Detail

The bottom figure in Figure 2 shows the best classifiers in terms of precision, recall and F-score. We measured how often a classifier is the best classifier (for each of the 24 subgraph overlap strategies). It is clear that overall Random Forest classifier (RF) performance is the best. For all 24 matching strategies it is a Random Forest classifier that is better than any other competitor with RF trained on PC13 and ANN being the most frequent best classifier overall. Second place is Random Forest trained simply on GE11 (the largest dataset in terms of entities and events). No other classifiers (SVM, MLP, MNNB, DT) outperform RF. Training on all datasets (RF+GE11+PC13+ANN) does not seem to increase success significantly. Performance across different RF classifiers is on par and good (see Table 5)

Results in the top figure of Figure 2 show that RF has the best precision performance.

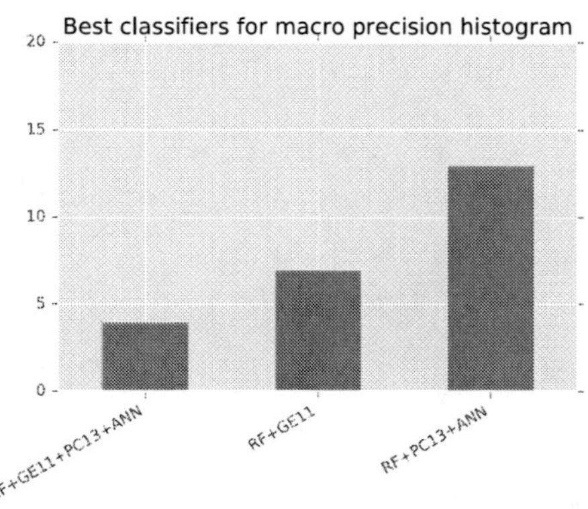

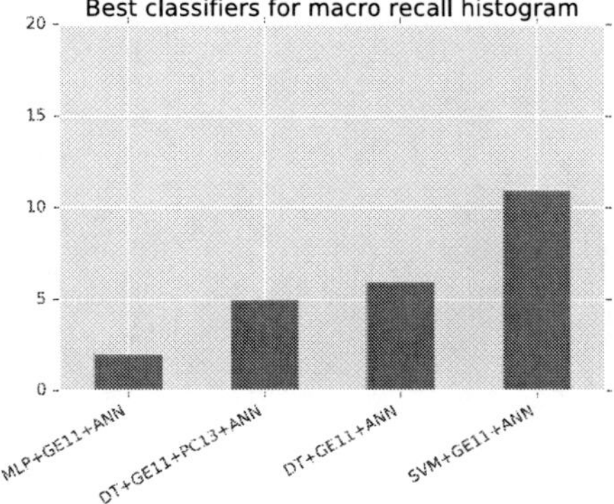

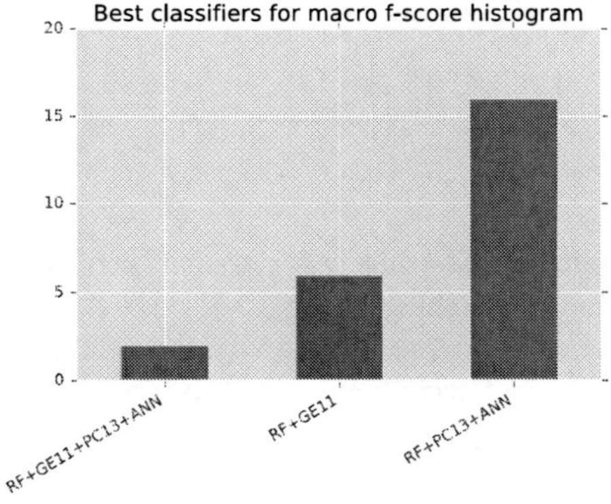

Figure 2: Histogram of best classifiers. This histogram is generated by counting how often a classifier is the best for a particular subgraph matching strategy.

RF+PC13+ANN is the most frequent best classifier w.r.t precision. RF+GE11 and RF+GE11+PC13+ANN also performing comparably. Compared to recall this means that RF wins F-score because they are best in precision.

No RF classifier performs best in recall. Results show that MLP, DT and SVM all perform well for certain subgraph overlap strategies with SVM be-

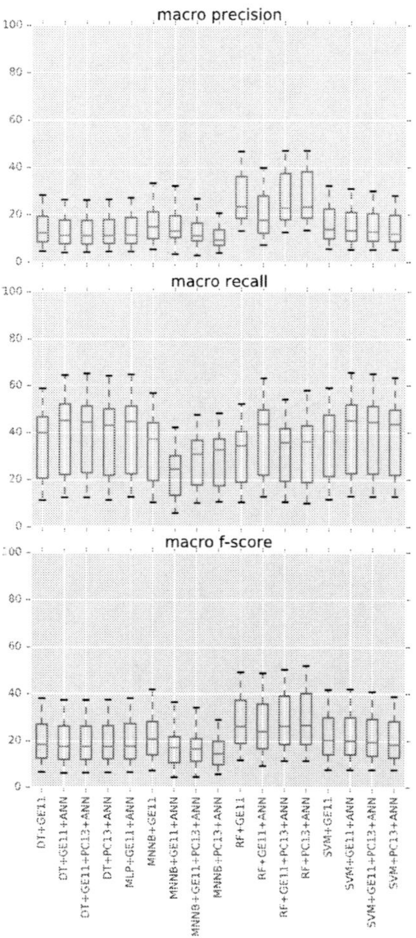

Figure 3: Statistics of classifier performance across all matching strategies. X-axis - classifiers. Y-axis - macro precision top, macro recall middle and macro f-score bottom (with 100 being perfect score).

ing most often the best classifier, followed by various DT-based classifiers and MLP.

Figure 3 gives results for all classifiers across all matching strategies. Looser strategies give the max and strict matching strategies the min data points. We can see that performance is primarily driven by the choice of classifier as the F-score mostly varies with the type of classifier used (even though there are a few outliers). Situation is a bit more varied for precision and recall. Interestingly choice of dataset seems to have less impact. Generally speaking MNNB are the least successful. RF clearly dominate precision on average but close enough to DT and SVM on recall.

7 Conclusion

This paper continues the current trend of extending NLP systems for APC and building more complete systems that allow evaluation with respect to some external standard - here the hand curated mTOR pathway.

We measured the impact of different classifiers on retrieval performance and showed that certain classifiers have the potential to increase retrieval performance. Especially Random Forest classifiers perform much better on mTOR than previously tried Support Vector Machines. On the other hand, the training data choice seems to have little impact (at least for the tested ANN, GE11 and PC13 training datasets).

Spranger et al. (2016) argue that not all of the problems of APC can be overcome by using more training data on event extraction systems. They argue that additions such as complex species recognition, co-reference resolution and pathway construction are needed to ultimately solve the problem posed by APC. This certainly remains true and is not directly questioned by results in this paper. The system described here does not automatically compose single pathway maps from the extracted data. Nevertheless, our results suggest that a lot of progress can be made by improving on the event extraction part of the pipeline.

This paper focuses on evaluating current machine learning techniques for event extraction. We are currently in the process of evaluating other systems including rule-based ones.

Acknowledgments

We would like to thank the authors of the Turku Event Extraction System for providing such an excellent system to the community. We also thank SBI for making the mTOR dataset available and their collaboration on evaluation and measurements.

References

B. B Aldridge, J. M Burke, D. A Lauffenburger, and P. K Sorger. 2006. Physicochemical modelling of cell signalling pathways. *Nature cell biology* 8(11):1195–1203.

J. Allen, W. de Beaumont, L. Galescu, and C. M. Teng. 2015. Complex event extraction using drum. *BioNLP 2015 Workshop on Biomedical Natural Language Processing* pages 1–11.

S. Ananiadou, S. Pyysalo, J. Tsujii, and D. Kell. 2010. Event extraction for systems biology by text mining the literature. *Trends in biotechnology* 28(7):381–90.

J. Björne, F. Ginter, S. Pyysalo, J. Tsujii, and T. Salakoski. 2010. Complex event extraction at pubmed scale. *Bioinformatics* 26(12):i382–i390.

J. Björne, F. Ginter, and T. Salakoski. 2012. University of turku in the bionlp'11 shared task. *BMC bioinformatics* 13(11):1.

J. Björne, J. Heimonen, F. Ginter, A. Airola, T. Pahikkala, and T. Salakoski. 2011. Extracting contextualized complex biological events with rich graph-based feature sets. *Computational Intelligence* 27(4):541–557.

J. Björne and T. Salakoski. 2015. Tees 2.2: Biomedical event extraction for diverse corpora. *BMC Bioinformatics* 16(16):S4.

E. Caron, S. Ghosh, Y. Matsuoka, et al. 2010. A comprehensive map of the mtor signaling network. *Molecular systems biology* 6(1).

P. Creixell, J. Reimand, S. Haider, et al. 2015. Pathway and network analysis of cancer genomes. *Nature methods* 12(7):615.

A. Efeyan and D. Sabatini. 2010. mtor and cancer: many loops in one pathway. *Current opinion in cell biology* 22(2):169–176.

A. Funahashi, Y. Matsuoka, A. Jouraku, et al. 2008. Celldesigner 3.5: a versatile modeling tool for biochemical networks. *Proceedings of the IEEE* 96(8):1254–1265.

J. Hakenberg, M. Gerner, M. Haeussler, I. Solt, C. Plake, M. Schroeder, G. Gonzalez, G. Nenadic, and C. M Bergman. 2011. The gnat library for local and remote gene mention normalization. *Bioinformatics* 27(19):2769–2771.

M. Hucka, A. Finney, H. Sauro, et al. 2003. The systems biology markup language (sbml): a medium for representation and exchange of biochemical network models. *Bioinformatics* 19(4):524–531.

T. Joachims. 1999. Making large-scale SVM learning practical. In B. Schölkopf, C. Burges, and A. Smola, editors, *Advances in Kernel Methods - Support Vector Learning*, MIT Press, Cambridge, MA, chapter 11, pages 169–184.

S. Katiyar, E. Liu, C. Knutzen, et al. 2009. Reddl, an inhibitor of mtor signalling, is regulated by the cul4a–ddb1 ubiquitin ligase. *EMBO reports* 10(8):866–872.

P. Khatri, M. Sirota, and A. Butte. 2012. Ten years of pathway analysis: current approaches and outstanding challenges. *PLoS Comput Biol* 8(2).

J.-D. Kim, N. Nguyen, Y. Wang, J. Tsujii, T. Takagi, and A. Yonezawa. 2012. The genia event and protein coreference tasks of the bionlp shared task 2011. *BMC Bioinformatics* 13(11):S1.

H. Kitano. 2002. Systems biology: a brief overview. *Science* 295(5560):1662–1664.

N. Le Novère. 2006. Model storage, exchange and integration. *BMC neuroscience* .

V. I Levenshtein. 1966. Binary codes capable of correcting deletions, insertions, and reversals. In *Soviet physics doklady*. volume 10, pages 707–710.

D. Maglott, J. Ostell, K. D Pruitt, and T. Tatusova. 2005. Entrez gene: gene-centered information at ncbi. *Nucleic acids research* 33(suppl 1):D54–D58.

S. Mizuno, R. Iijima, S. Ogishima, et al. 2012. Alzpathway: a comprehensive map of signaling pathways of alzheimer's disease. *BMC systems biology* 6(1):52.

K. Oda and H. Kitano. 2006. A comprehensive map of the toll-like receptor signaling network. *Molecular systems biology* 2(1).

T. Ohta, S. Pyysalo, R. Rak, et al. 2013. Overview of the pathway curation (pc) task of bionlp shared task 2013. In *Proceedings of the BioNLP Shared Task 2013 Workshop*. ACL, pages 67–75.

T. Ohta, S. Pyysalo, and J. Tsujii. 2011. From pathways to biomolecular events: opportunities and challenges. In *Proceedings of BioNLP 2011 Workshop*. ACL, pages 105–113.

F. Pedregosa, G. Varoquaux, A. Gramfort, V. Michel, B. Thirion, O. Grisel, M. Blondel, P. Prettenhofer, R. Weiss, V. Dubourg, J. Vanderplas, A. Passos, D. Cournapeau, M. Brucher, M. Perrot, and E. Duchesnay. 2011. Scikit-learn: Machine learning in Python. *Journal of Machine Learning Research* 12:2825–2830.

M. Sokolova and G. Lapalme. 2009. A systematic analysis of performance measures for classification tasks. *Information Processing & Management* 45(4):427–437.

M. Spranger, S. Palaniappan, and S. Ghosh. 2015. Extracting biological pathway models from nlp event representations. In *Proceedings of the 2015 Workshop on Biomedical Natural Language Processing (BioNLP 2015)*. ACL, pages 42—51.

M. Spranger, S. K. Palaniappan, and S. Ghosh. 2016. Measuring the State of the Art of Automated Pathway Curation Using Graph Algorithms - A Case Study of the mTOR Pathway. *ArXiv e-prints* .

J. Szostak, S. Ansari, S. Madan, J. Fluck, M. Talikka, A. Iskandar, H. De Leon, M. Hofmann-Apitius, M. C Peitsch, and J. Hoeng. 2015. Construction of biological networks from unstructured information based on a semi-automated curation workflow. *Database* 2015:bav057.

	DT+GE11	DT+GE11+ANN	DT+GE11+PC13+ANN	DT+PC13+ANN	MLP+GE11+ANN	MNNB+GE11	MNNB+GE11+ANN	MNNB+GE11+PC13+ANN	MNNB+PC13+ANN	RF+GE11	RF+GE11+ANN	RF+GE11+PC13+ANN	RF+PC13+ANN	SVM+GE11	SVM+GE11+ANN	SVM+GE11+PC13+ANN	SVM+PC13+ANN
nmeq, sboeq	6.5	6.1	6.3	6.4	6.6	7.2	4.3	4.5	5.7	**11.7**	9.3	11.5	11.5	7.6	7.5	7.5	7.5
nmeq, sboov	10.1	9.5	9.8	9.6	9.7	12.0	10.1	11.0	9.5	**15.3**	13.8	15.1	14.6	11.4	11.4	11.1	10.2
nmeq, sboisa	11.9	11.4	11.7	11.7	11.6	14.2	12.2	12.3	10.2	**18.1**	16.1	17.6	17.7	13.5	13.5	13.2	12.3
appeq, sboeq	7.0	6.6	6.8	7.0	7.1	7.5	4.5	4.6	6.0	**12.5**	10.0	12.4	12.4	8.0	7.9	8.0	8.0
appeq, sboov	10.8	10.1	10.4	10.4	10.3	12.5	10.8	11.5	10.0	**16.3**	14.5	16.1	15.7	12.0	12.0	11.7	10.8
appeq, sboisa	12.8	12.2	12.5	12.6	12.5	14.9	13.1	13.0	10.7	**19.4**	17.1	18.9	19.1	14.4	14.4	14.1	13.1
appeq/enteq, sboeq	10.2	9.9	10.2	10.4	10.5	10.6	6.0	6.0	8.3	16.8	14.1	**16.9**	16.7	11.7	11.5	11.7	11.8
appeq/enteq, sboov	15.2	14.7	15.0	14.9	15.0	17.2	14.1	14.9	13.6	21.6	20.0	**21.7**	21.2	17.0	16.9	16.6	15.7
appeq/enteq, sboisa	18.1	17.6	18.0	18.0	17.9	20.6	17.4	17.1	14.7	25.9	23.5	25.4	**26.0**	20.2	20.1	19.8	18.9
appeq/entov, sboeq	23.6	23.0	22.9	23.3	24.2	24.1	17.2	16.3	17.2	33.8	31.6	35.2	**36.2**	26.4	26.2	25.7	25.3
appeq/entov, sboov	31.0	30.1	30.2	30.2	31.0	34.2	29.6	29.2	25.9	39.9	40.6	41.6	**41.9**	34.1	34.2	33.2	31.3
appeq/entov, sboisa	35.5	34.6	34.9	35.0	35.6	39.0	33.9	32.3	27.4	46.4	45.9	47.2	**48.6**	38.9	39.2	38.2	36.4
nmeq/wc, sboeq	13.4	12.4	12.2	12.5	12.9	13.6	6.9	6.4	8.4	22.0	17.9	22.3	**23.3**	14.9	14.4	13.9	14.0
nmeq/wc, sboov	17.9	16.9	16.6	16.5	16.9	20.1	16.8	16.4	14.1	25.4	23.5	25.8	**26.0**	19.5	19.3	18.3	17.2
nmeq/wc, sboisa	20.1	19.1	18.8	19.0	19.1	22.5	18.5	17.6	14.7	28.5	26.2	28.5	**29.1**	22.0	21.8	20.8	19.7
appeq/wc, sboeq	14.1	13.1	12.9	13.2	13.7	14.2	7.1	6.6	8.8	23.2	18.9	23.4	**24.6**	15.6	15.0	14.7	14.8
appeq/wc, sboov	18.7	17.6	17.3	17.3	17.7	20.8	17.3	16.9	14.6	26.7	24.5	27.0	**27.4**	20.3	20.0	19.2	18.1
appeq/wc, sboisa	21.1	20.1	19.8	20.1	20.2	23.5	19.3	18.3	15.2	30.3	27.5	30.1	**30.9**	23.1	22.8	22.0	20.8
appeq/enteq/wc, sboeq	25.7	25.1	24.8	25.3	26.3	26.4	18.3	16.9	18.0	36.7	34.1	38.3	**39.7**	28.7	28.6	28.0	27.5
appeq/enteq/wc, sboov	33.5	32.7	32.5	32.4	33.5	37.1	32.3	31.2	27.6	42.7	43.4	44.7	**45.3**	36.7	37.0	35.8	33.6
appeq/enteq/wc, sboisa	38.2	37.4	37.4	37.6	38.2	42.0	36.6	34.3	29.1	49.4	48.9	50.4	**52.0**	41.7	42.1	41.0	39.0
appeq/entov/wc, sboeq	25.7	25.1	24.8	25.3	26.3	26.4	18.3	16.9	18.0	36.7	34.1	38.3	**39.7**	28.7	28.6	28.0	27.5
appeq/entov/wc, sboov	33.5	32.7	32.5	32.4	33.5	37.1	32.3	31.2	27.6	42.7	43.4	44.7	**45.3**	36.7	37.0	35.8	33.6
appeq/entov/wc, sboisa	38.2	37.4	37.4	37.6	38.2	42.0	36.6	34.3	29.1	49.4	48.9	50.4	**52.0**	41.7	42.1	41.0	39.0

Table 5: Supplementary materials: Macro F-score all results

Toward Automated Early Sepsis Alerting: Identifying Infection Patients from Nursing Notes

Emilia Apostolova
Language.ai
Chicago, IL, USA
emilia@language.ai

Tom Velez
Vivace Health Solutions
Cardiff, CA, USA
tom.velez@cta.com

Abstract

Severe sepsis and septic shock are conditions that affect millions of patients and have close to 50% mortality rate. Early identification of at-risk patients significantly improves outcomes. Electronic surveillance tools have been developed to monitor structured Electronic Medical Records and automatically recognize early signs of sepsis. However, many sepsis risk factors (e.g. symptoms and signs of infection) are often captured only in free text clinical notes. In this study, we developed a method for automatic monitoring of nursing notes for signs and symptoms of infection. We utilized a creative approach to automatically generate an annotated dataset. The dataset was used to create a Machine Learning model that achieved an F1-score ranging from 79 to 96%.

1 Introduction

Severe sepsis and septic shock are rapidly progressive, life-threatening conditions caused by complications from an infection. They are major healthcare problems that affect millions of patients globally each year (Kim and Hong, 2016). The mortality rate for severe sepsis and septic shock is approaching 50% (Nguyen et al., 2006).

A key goal in critical care medicine is the early identification and treatment of infected patients with early stage sepsis. The most recent guidelines for the management of severe sepsis and septic shock include early recognition and management of these conditions as medical emergencies, immediate administration of resuscitative fluids, frequent reassessment, and empiric antibiotics as soon as possible following recognition (Dellinger et al., 2008).

Early recognition of infections that can lead to sepsis, severe sepsis and/or septic shock can be challenging for several reasons: 1) these conditions can quickly develop from any form of common infections (bacterial, viral or fungal) and can be localized or generalized; 2) culture-dependent diagnosis of infection is commonly slow and prior use of antibiotics may make cultures falsely negative (Vincent, 2016); 3) systemic inflammatory response syndrome, traditionally associated with sepsis, may be the result of other noninfectious disease processes (Bone et al., 1992). Consequently, clinicians frequently rely on a myriad of non-specific symptoms of infections and physiological signs of systemic inflammatory response for rapid diagnosis. Each hour of delay in the administration of recommended therapy is associated with a linear increase in the risk of mortality rate (Kumar et al., 2006; Han et al., 2003), driving the need for automation of early sepsis recognition.

In response to this need, electronic surveillance tools have been developed to monitor for the arrival of new patient electronic medical record (EMR) data, automatically recognize early signs of sepsis risk in specific patients, and trigger alerts to clinicians to help guide timely, effective interventions (Herasevich et al., 2011; Azzam et al., 2009; Koenig et al., 2011). The automated decision logic used in many existing sepsis screening tools, for example (Nguyen et al., 2014; Hooper et al., 2012; Nelson et al., 2011), relies on consensus criteria-based rules.

Structured EMR data, such as diagnostic codes, vital signs and orders for tests, imaging and medications, can be a reliable source of sepsis criteria. However, many sepsis risk factors (e.g. symptoms and signs of infection) are routinely captured solely in free text clinical notes. The aim of this

Proceedings of the BioNLP 2017 workshop, pages 257–262,
Vancouver, Canada, August 4, 2017. ©2017 Association for Computational Linguistics

study is to develop a system for the detection of signs and symptoms of infection from free-text nursing notes. The output of the system is later used, in conjunction with available structured data, as an input to an electronic surveillance tool for an early detection of sepsis.

2 Task Definition and Dataset

Depending on the infection source and the specifics of the patient history, signs and symptoms of infection can vary widely. In addition, similar symptoms can be stated using a number of synonymous expressions, complicated by the presence of abbreviations, variant spellings and misspellings. Table 1 lists nursing note snippets indicating a possible presence of infection with various degrees of certainty. For example, line items 1, 8, and 12 indicate increased temperature; line items 1, 3, 6, 7, and 13 indicate the use of infection-treating antibiotics; line items 4, 7, 9, 11, and 13 mention specific infectious diseases. A number of examples mention additional infection symptoms or infection detecting tests.

1	Afebrile on antibiotics.
2	Very large copious amount of secretions,sputum
3	... medicated with iv cefazolin dose 2 of 3
4	UA positive for UTI.
5	Blood culture pos for gram neg organisms.
6	Elevated WBC count, on clindamycin IV.
7	... continues on clindamycin(D6), pen-G(D5) and doxycycline(D4) for LLL pneumonia
8	84 year old male with h/o throat cancer who presented on [DATE] to [LOCATION] with fever, diffuse rash, renal failure and altered mental status.
9	Pt had a positive sputum specs for GBS and GPC, and he has HSV on lips.
10	... blood, urine and sputum culture sent today ...
11	PEG tube to gravity, episodes of vomitting on day shift,NPO. Contact precautions MRSA.
12	Pt had temp spike of 102.4
13	Levaquin started for pnuemonia.
14	Had infected knee prosthesis which led to wash out of joint yesterday.

Table 1: MIMIC-III nursing note snippets indicating a presence of infection with various degrees of certainty. Abbreviations and misspellings are preserved to demonstrate the task challenges.

In the literature of identifying patient phenotype cohorts using electronic health records, most studies map textual elements to standard vocabularies, such as the Unified Medical Language System (UMLS) (Shivade et al., 2014). The standard vocabulary concepts are later used in rule-based, and, in some cases, Machine Learning (ML) approaches to identify patient cohorts.

In the context of identifying infection from clin-

ical notes, however, such an approach poses a number of challenges. Symptoms can vary widely depending on the source of infection, for example, *redness, sputum, swelling, pus, phlegm, vomiting, increased white blood cell count*, etc. The same symptom can also be expressed in a large number of ways, for example, *afebrile, temp spike of 102.4, fever*, etc. There is a large number of conditions indicating infections, for example *UTI, strep throat, hepatitis, HIV/AIDS*, etc. In addition, abbreviations and misspellings are quite common in the context of ICU care, for example, *pneumonia, PNA, pnuemonia, pneu*, etc.

Due to their nature, the dataset and task are better suited for ML approaches that are not relying on standard vocabularies or a structured set of features. As with most ML tasks in the clinical domain, the challenge in this approach is obtaining a sufficient amount of training data (Chapman et al., 2011).

To address these challenges, we utilized the MIMIC-III dataset (Johnson et al., 2016) and developed a creative solution to automatically generate training data as described in section 3. MIMIC (Medical Information Mart for Intensive Care) is a large, freely-available database comprising deidentified health-related data associated with over 40,000 patients who stayed in critical care units of the Beth Israel Deaconess Medical Center between 2001 and 2012. The dataset contains over 2 million free-text clinical notes. We focused only on nursing notes for adult patients, and our dataset consists of a total of 634,369 nursing notes.

3 Rule-based Training Dataset Creation

To obtain a sizable training dataset we explored the use of available MIMIC-III structured data, such as test orders and results, prescribed medications, and diagnosis codes. However, this approach did not translate to accurately identifying nursing notes suggesting infection for a number of reasons[1]. Instead, we utilized a simple heuristic. We observed that whenever there is an existing infection or a suspicion of infection, the nursing notes describe the fact that the patient is taking or is prescribed infection-treating antibiotics. Thus, identifying nursing notes describing the use of antibiotics will, in most cases, also identify nursing notes describing signs and symptoms of infection.

---

[1]Challenges include missing or incorrect data, discontinuous or disordered EMR data entry timestamps, etc.

To identify positive mentions of administered antibiotics, we used a list of the 60 most commonly administered infection-treating antibiotics in the MIMIC dataset (Misquitta, 2013). This initial list was then extended to include additional antibiotic names, brands, abbreviations, spelling variations, and common misspellings. We semi-automated this laborious task by utilizing *word embeddings* (Mikolov et al., 2013). Word embeddings were generated utilizing all available MIMIC-III nursing notes[2]. The initial set of antibiotics was then extended using the closest *word embeddings* in terms of cosine distance. For example, the closest words to the antibiotic *amoxicillin* are *amox, amoxacillin, amoxycillin, cefixime, suprax, amoxcillin, amoxicilin*. As shown, this includes misspellings, abbreviations, similar drugs and brand names. The extended list was then manually reviewed. The final infection-treating antibiotic list consists of 402 unambiguous expressions indicating antibiotics.

Antibiotics, however, are sometimes negated and are often mentioned in the context of allergies (e.g. *allergic to penicillin*). To distinguish between affirmed, negated, and speculated mentions of administered antibiotics, we also developed a set of rules in the form of keyword triggers. Similarly to the NegEx algorithm (Chapman et al., 2001), we identified phrases that indicate uncertain or negated mentions of antibiotics that precede or follow a list of hand-crafted expression at the sentence and clause levels. Word embeddings were again used to extend the list of triggers with synonyms, spelling variations, abbreviations, and misspellings. For example, the words *allergic, anaphylaxis, anaphalaxis, allerg,* and *anaphylaxsis* are all used as triggers indicating the negation of an antibiotic use. The full list of keywords indicating antibiotics, negation/speculation triggers and conjunctions is available online[3].

The described approach identified 186,158 nursing notes suggesting the unambiguous presence of infection (29%) and 3,262 notes suggesting possible infection. The remaining 448,211 notes (70%) were considered to comprise our negative dataset, i.e. not suggesting infection.

4 Machine Learning Results

We modeled the task as a binary classification of free-form clinical notes. It has been shown that Support Vector Machines (Cortes and Vapnik, 1995) achieve superior results in most text classifications tasks and were selected as a sensible first choice. The individual nursing notes were represented as a bag-of-words (1-grams). The tokens were all converted to lower case and non-alphanumeric characters were discarded. Tokens that are present in more than 60% of all samples or less than 6 times were also discarded. The tokens were weighted using the tf-idf scheme (Salton and McGill, 1986). We trained the model using linear kernel SVMs[4] (Chang and Lin, 2011). We set the positive class weight to 2 to address the unbalanced dataset. 70% of the automatically generated dataset was used for training and the remaining 30% for testing. This resulted in a precision of 93.12 and a recall of 99.04 as shown in Table 2.

	Precision	**Recall**	**F1-score**
SVM^{auto}	93.12	99.04	95.99
SVM^{gold}	92.10	68.46	78.53

Table 2: Classification Results. SVM^{auto}=Results from applying the SVM model on an automatically generated test set of 190,000 nursing notes; SVM^{gold}=Results from applying the SVM model on a manually reviewed dataset of 200 nursing notes.

As the training dataset was automatically created, the above results do not truly reflect the model performance. To evaluate the model on the ground truth, a qualified professional manually reviewed 200 randomly selected nursing notes. These results are also shown in Table 2. While the model precision remained high (92.10), the recall dropped significantly to 68.46.

The drop in recall can be partially attributed to the manner in which the testing data was created. Nursing notes describing signs of infection but failing to mention the use of antibiotics were considered (incorrectly) negative examples. However, an error analysis revealed that the majority of the false negatives (contributing to the low recall) were actually all indicating low level of suspicion of infection. For example, the human annotator considered the following snippets sufficient to indicate a possible infection *afebrile, bld cx's sent; monitor temp, wbc's, await stool cx results; lungs coarse, thick yellow secretions suctioned from ett;*

---

[2]We used vector size 200, window size 7, and continuous bag-of-words model.

[3]https://github.com/ema-/antibiotic-dictionary

[4]We used the LibSVM library with both the cost and gamma parameters set to 2 (obtained via grid-search parameter estimation).

awaiting results of CT, malignancy vs pneumonia.
In all cases, the note expresses only a suspicion for
infection, pending further tests.

We further attempted to improve the system performance by utilizing Paragraph Vectors (Le and
Mikolov, 2014). Unsupervised algorithms have
been used to represent variable pieces of texts such
as paragraphs and documents as fixed-length feature representations (Paragraph Vectors). Studies have shown that Paragraph Vectors outperform
bag-of-words models on some text classification
tasks. We used the text from all nursing notes
to create Paragraph Vectors. We generated document embeddings using a distributed memory
model and distributed bag-of-words model, each
of size 300 with a window size of 7. Combining the vectors of the distributed memory model
and the distributed bag-of-words model, we represented each document as a vector of size 600.
The paragraph vectors of the training instances
were then fed to a logistic regression, K-nearest
neighbors, and an SVM classifier. Results significantly under-performed the SVM bag-of-words
model and we were able to achieve a maximum
precision and recall of 63% and 77% respectively.

5   Related Work

A review of approaches to identifying patient phenotype cohorts using EMR data (Shivade et al.,
2014) describes a number of studies using clinical
notes, most often in combination with additional
structured information, such as diagnosis codes.
The study asserts that clinical notes are often the
only source of information from which to infer important phenotypic characteristics.

Demner-Fushman et al. (2009) note that clinical
events monitoring is one of the most common and
essential tasks of Clinical Decision Support systems. The task is in many respects similar to the
task of identifying patient phenotype cohorts and
it has been observed that free text clinical notes are
again the best source of information. For example,
Murff et al. (2003) found the electronic discharge
summaries to be an excellent source for detecting
adverse events. They also note that simple keywords and triggers are not sufficient to detect such
events.

In the context of identifying infection from clinical text, most studies map textual elements to
standard vocabularies, such as UMLS. For example, Matheny et al. (2012) develop a system for detecting infectious symptoms from emergency department and primary care clinical documentation,
utilizing keywords and SNOMED-CT concepts.
Bejan et al. (2012) describe a system for pneumonia identification from narrative reports using
n-grams and UMLS concepts. Similarly, Elkin
et al. (2008) encoded radiology reports using
SNOMED-CT concepts and developed a set of
rules to identify pneumonia cases.

Horng et al. (2017) develop an automated trigger for sepsis clinical decision support at emergency department triage. They utilize machine
learning and establish that free text drastically improves the discriminatory ability of identifying
infection (increase in AUC from 0.67 to 0.86).
Arnold et al. (2014) develop an EHR screening tool to identify sepsis patients. They utilize
NLP applied to clinical documentation, providing
greater clinical context than laboratory and vital
sign screening alone. DeLisle et al. (2010) used
a combination of structured EMR parameters and
text analysis to detect acute respiratory infections.
Murff et al. (2011) develop a natural language
processing search approach to identify postoperative surgical complications within a comprehensive electronic medical record.

Halpern et al. (2014) describe a system for
learning to estimate and predict clinical state variables without labeled data. Similar to our approach, they use a combination of domain expertise and vast amounts of unlabeled data, without
requiring labor-intensive manual labeling. In their
system, an expert encodes a certain amount of
domain knowledge (identifying anchor variables)
which is later used to train classifiers. Elkan and
Noto (2008) show that a classifier trained on positive and unlabeled examples predicts probabilities
that differ by only a constant factor from the true
conditional probabilities of being positive.

6   Discussion

We presented an approach to identifying nursing
notes describing the suspicion or presence of an
infection. We utilized the MIMIC-III dataset and
a creative approach to obtain an ample amount of
annotated data. We then applied ML methods to
the task and achieved performance sufficient for
practical applications. The ultimate goal of this
study is to utilize free-text notes, in combination
with structured EMR data, to build an automated
surveillance system for early detection of patients
at risk of sepsis.

References

R Arnold, J Isserman, S Smola, and E Jackson. 2014. Comprehensive assessment of the true sepsis burden using electronic health record screening augmented by natural language processing. *Critical Care* 18(1):P244.

Helen C Azzam, Satjeet S Khalsa, Richard Urbani, Chirag V Shah, Jason D Christie, Paul N Lanken, and Barry D Fuchs. 2009. Validation study of an automated electronic acute lung injury screening tool. *Journal of the American Medical Informatics Association* 16(4):503–508.

Cosmin Adrian Bejan, Fei Xia, Lucy Vanderwende, Mark M Wurfel, and Meliha Yetisgen-Yildiz. 2012. Pneumonia identification using statistical feature selection. *Journal of the American Medical Informatics Association* 19(5):817–823.

Roger C Bone, William J Sibbald, and Charles L Sprung. 1992. The accp-sccm consensus conference on sepsis and organ failure. *CHEST journal* 101(6):1481–1483.

Chih-Chung Chang and Chih-Jen Lin. 2011. LIB-SVM: A library for support vector machines. *ACM Transactions on Intelligent Systems and Technology* 2:27:1–27:27. Software available at http://www.csie.ntu.edu.tw/~cjlin/libsvm.

Wendy W Chapman, Will Bridewell, Paul Hanbury, Gregory F Cooper, and Bruce G Buchanan. 2001. A simple algorithm for identifying negated findings and diseases in discharge summaries. *Journal of biomedical informatics* 34(5):301–310.

Wendy W Chapman, Prakash M Nadkarni, Lynette Hirschman, Leonard W D'avolio, Guergana K Savova, and Ozlem Uzuner. 2011. Overcoming barriers to nlp for clinical text: the role of shared tasks and the need for additional creative solutions.

Corinna Cortes and Vladimir Vapnik. 1995. Support-vector networks. *Machine learning* 20(3):273–297.

Sylvain DeLisle, Brett South, Jill A Anthony, Ericka Kalp, Adi Gundlapallli, Frank C Curriero, Greg E Glass, Matthew Samore, and Trish M Perl. 2010. Combining free text and structured electronic medical record entries to detect acute respiratory infections. *PloS one* 5(10):e13377.

R Phillip Dellinger, Mitchell M Levy, Jean M Carlet, Julian Bion, Margaret M Parker, Roman Jaeschke, Konrad Reinhart, Derek C Angus, Christian Brun-Buisson, Richard Beale, et al. 2008. Surviving sepsis campaign: international guidelines for management of severe sepsis and septic shock: 2008. *Intensive care medicine* 34(1):17–60.

Dina Demner-Fushman, Wendy W Chapman, and Clement J McDonald. 2009. What can natural language processing do for clinical decision support? *Journal of biomedical informatics* 42(5):760–772.

Charles Elkan and Keith Noto. 2008. Learning classifiers from only positive and unlabeled data. In *Proceedings of the 14th ACM SIGKDD international conference on Knowledge discovery and data mining*. ACM, pages 213–220.

Peter L Elkin, David Froehling, Dietlind Wahner-Roedler, Brett E Trusko, Gail Welsh, Haobo Ma, Armen X Asatryan, Jerome I Tokars, S Trent Rosenbloom, and Steven H Brown. 2008. Nlp-based identification of pneumonia cases from free-text radiological reports. In *AMIA*.

Yoni Halpern, Youngduck Choi, Steven Horng, and David Sontag. 2014. Using anchors to estimate clinical state without labeled data. In *AMIA Annual Symposium Proceedings*. American Medical Informatics Association, volume 2014, page 606.

Yong Y Han, Joseph A Carcillo, Michelle A Dragotta, Debra M Bills, R Scott Watson, Mark E Westerman, and Richard A Orr. 2003. Early reversal of pediatric-neonatal septic shock by community physicians is associated with improved outcome. *Pediatrics* 112(4):793–799.

Vitaly Herasevich, Mykola Tsapenko, Marija Kojicic, Adil Ahmed, Rachul Kashyap, Chakradhar Venkata, Khurram Shahjehan, Sweta J Thakur, Brian W Pickering, Jiajie Zhang, et al. 2011. Limiting ventilator-induced lung injury through individual electronic medical record surveillance. *Critical care medicine* 39(1):34–39.

Michael H Hooper, Lisa Weavind, Arthur P Wheeler, Jason B Martin, Supriya Srinivasa Gowda, Matthew W Semler, Rachel M Hayes, Daniel W Albert, Norment B Deane, Hui Nian, et al. 2012. Randomized trial of automated, electronic monitoring to facilitate early detection of sepsis in the intensive care unit. *Critical care medicine* 40(7):2096.

Steven Horng, David A Sontag, Yoni Halpern, Yacine Jernite, Nathan I Shapiro, and Larry A Nathanson. 2017. Creating an automated trigger for sepsis clinical decision support at emergency department triage using machine learning. *PloS one* 12(4):e0174708.

Alistair EW Johnson, Tom J Pollard, Lu Shen, Li-wei H Lehman, Mengling Feng, Mohammad Ghassemi, Benjamin Moody, Peter Szolovits, Leo Anthony Celi, and Roger G Mark. 2016. Mimic-iii, a freely accessible critical care database. *Scientific data* 3.

Won-Young Kim and Sang-Bum Hong. 2016. Sepsis and acute respiratory distress syndrome: recent update. *Tuberculosis and respiratory diseases* 79(2):53–57.

Helen C Koenig, Barbara B Finkel, Satjeet S Khalsa, Paul N Lanken, Meeta Prasad, Richard Urbani, and Barry D Fuchs. 2011. Performance of an automated electronic acute lung injury screening system in intensive care unit patients. *Critical care medicine* 39(1):98–104.

Anand Kumar, Daniel Roberts, Kenneth E Wood, Bruce Light, Joseph E Parrillo, Satendra Sharma, Robert Suppes, Daniel Feinstein, Sergio Zanotti, Leo Taiberg, et al. 2006. Duration of hypotension before initiation of effective antimicrobial therapy is the critical determinant of survival in human septic shock. *Critical care medicine* 34(6):1589–1596.

Quoc V Le and Tomas Mikolov. 2014. Distributed representations of sentences and documents. In *ICML*. volume 14, pages 1188–1196.

Michael E Matheny, Fern FitzHenry, Theodore Speroff, Jennifer K Green, Michelle L Griffith, Eduard E Vasilevskis, Elliot M Fielstein, Peter L Elkin, and Steven H Brown. 2012. Detection of infectious symptoms from va emergency department and primary care clinical documentation. *International journal of medical informatics* 81(3):143–156.

Tomas Mikolov, Ilya Sutskever, Kai Chen, Greg S Corrado, and Jeff Dean. 2013. Distributed representations of words and phrases and their compositionality. In *Advances in neural information processing systems*. pages 3111–3119.

Donald Misquitta. 2013. *Early Prediction of Antibiotics in Intensive Care Unit Patients*. Ph.D. thesis, The Center for Biomedical Informatics at the Harvard Medical School.

Harvey J Murff, Fern FitzHenry, Michael E Matheny, Nancy Gentry, Kristen L Kotter, Kimberly Crimin, Robert S Dittus, Amy K Rosen, Peter L Elkin, Steven H Brown, et al. 2011. Automated identification of postoperative complications within an electronic medical record using natural language processing. *Jama* 306(8):848–855.

Harvey J Murff, Vimla L Patel, George Hripcsak, and David W Bates. 2003. Detecting adverse events for patient safety research: a review of current methodologies. *Journal of biomedical informatics* 36(1):131–143.

Jessica L Nelson, Barbara L Smith, Jeremy D Jared, and John G Younger. 2011. Prospective trial of real-time electronic surveillance to expedite early care of severe sepsis. *Annals of emergency medicine* 57(5):500–504.

H Bryant Nguyen, Emanuel P Rivers, Fredrick M Abrahamian, Gregory J Moran, Edward Abraham, Stephen Trzeciak, David T Huang, Tiffany Osborn, Dennis Stevens, David A Talan, et al. 2006. Severe sepsis and septic shock: review of the literature and emergency department management guidelines. *Annals of emergency medicine* 48(1):54–e1.

Su Q Nguyen, Edwin Mwakalindile, James S Booth, Vicki Hogan, Jordan Morgan, Charles T Prickett, John P Donnelly, and Henry E Wang. 2014. Automated electronic medical record sepsis detection in the emergency department. *PeerJ* 2:e343.

Gerard Salton and Michael J McGill. 1986. Introduction to modern information retrieval .

Chaitanya Shivade, Preethi Raghavan, Eric Fosler-Lussier, Peter J Embi, Noemie Elhadad, Stephen B Johnson, and Albert M Lai. 2014. A review of approaches to identifying patient phenotype cohorts using electronic health records. *Journal of the American Medical Informatics Association* 21(2):221–230.

Jean-Louis Vincent. 2016. The clinical challenge of sepsis identification and monitoring. *PLoS Med* 13(5):e1002022.

Enhancing Automatic ICD-9-CM Code Assignment for Medical Texts with PubMed

Danchen Zhang[1], Daqing He[1], Sanqiang Zhao[1], Lei Li[2]
[1]School of Information Sciences, University of Pittsburgh
[2]School of Economics and Management, Nanjing University of Science and Technology
{daz45, dah44, saz31}@pitt.edu, leili@njust.edu.cn

Abstract

Assigning a standard ICD-9-CM code to disease symptoms in medical texts is an important task in the medical domain. Automating this process could greatly reduce the costs. However, the effectiveness of an automatic ICD-9-CM code classifier faces a serious problem, which can be triggered by unbalanced training data. Frequent diseases often have more training data, which helps its classification to perform better than that of an infrequent disease. However, a diseases frequency does not necessarily reflect its importance. To resolve this training data shortage problem, we propose to strategically draw data from PubMed to enrich the training data when there is such need. We validate our method on the CMC dataset, and the evaluation results indicate that our method can significantly improve the code assignment classifiers' performance at the macro-averaging level.

1 Introduction and Background

The rapid computerization of medical content such as electronic medical records (EMRs), doctors notes and death certificates, drives a crucial need to apply automatic techniques to better assist medical professionals in creating and managing medical information. A standard procedure in hospital is to assign the International Classification of Diseases (ICD) codes to diseases appearing in medical texts by professional coders. As a result, several recent studies have been devoted to automatically extracting ICD code from medical texts to help manual coders (Crammer et al., 2007; Farkas and Szarvas, 2008; Aronson et al., 2007; Kavuluru et al., 2015, 2013; Zuccon and Nguyen,

Radiology report
Clinical History: Ten year old with chest pain x two weeks.   **Impression**: The lungs are well expanded and clear. There is no focal infiltrate or pleural effusion. The cardiac and mediastinal silhouette is normal. No bony abnormalities are appreciated. There is no evidence of pneumothorax or pleural disease to explain chest pain.
Code assignment
ICD-9-CM code: 786.2 (cough)

Figure 1: An example radiology report with manually labeled ICD-9-CM code from CMC dataset.

2013; Koopman et al., 2015).

In this paper, we focus on ICD-9-CM (the 9th version ICD, Clinical Modification), although our work is portable to ICD-10-CM (the 10th version ICD). The reason to conduct our study on ICD-9-CM is to compare with the state-of-art methods, whose evaluations have mostly conducted on ICD-9-CM code (Aronson et al., 2007; Kavuluru et al., 2015, 2013; Patrick et al., 2007; Ira et al., 2007; Zhang, 2008). ICD-9-CM codes are organized hierarchically, and each code corresponds to a textual description, such as "786.2, cough". Multiple codes can be assigned to a medical text, and a specific ICD-9-CM code is preferred than a more generic one when both are suitable (Pestian et al., 2007). Figure 1 shows a code assignment example where a radiology report is labeled with "786.2, cough".

Existing methods for automatic ICD-9-CM assignment have been mostly supervised methods because of the effectiveness the training; however, classification performance heavily relies on the sufficiency of training data (He and Garcia, 2009). To certain degree, micro-average measures, commonly used to evaluate the classification performance of existing algorithms, pays at-

Proceedings of the BioNLP 2017 workshop, pages 263–271,
Vancouver, Canada, August 4, 2017. ©2017 Association for Computational Linguistics

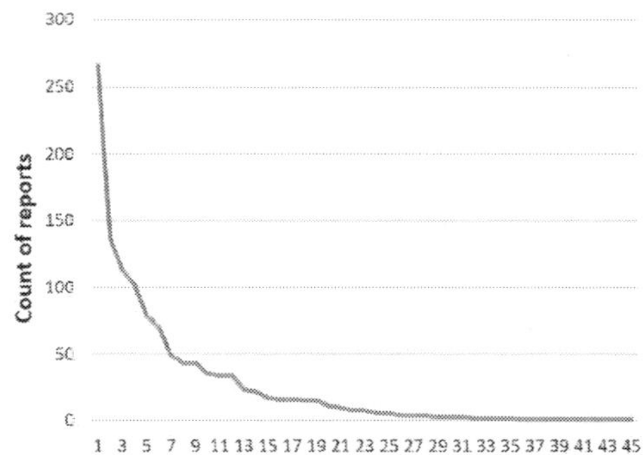

Figure 2: The distribution of radiology reports for 45 ICD-9-CM codes in the CMC dataset.

tention to the correctness of the code assignment to each EHR (individual case), which helps to hide the impact of unbalanced training data. However, a useful classification system should perform consistently across all ICD-9-CM codes regardless of the popularity of the codes (Jackson and Moulinier, 2007). This motivated us to examine the imbalanced training data and its impacts to the classifier. Specifically, we pay more attention to macro-average measures, which helps to examine the consistency across all codes.

Unfortunately, in a real dataset for studying ICD-9-CM code classification, the data available for each code is highly imbalanced. For example, Figure 2 shows the count of available radiology reports for each of the 45 ICD-9-CM codes in the CMC dataset (Pestian et al., 2007). Common diseases like "786.2, cough", can have 266 reports as the training data, whereas unpopular disease "758.6, Gonadal dysgenesis" only has one. Similarly, Kavuluru et al. (2015) found that 874 of 1,231 ICD-9-CM code in their UKLarge dataset have less than 350 supporting data, and only 92 codes have more than 1,430 supporting data. In another example, Koopman et al. (2015) found that 85% of the whole death certificate dataset are related to top 20 common cancers, and only rest 15% is associated with 65 rarer cancers. These long tail supporting data problems are very common, which makes data imbalance an noticeable problem.

Our approach for resolving this problem is to introduce additional information resources. Furthermore, due to the privacy concern of medical-related content, this study is particularly interested in obtaining extra relevant training data from pub-

licly available medical datasets. PubMed[1], as a vast and broad medical literature dataset, covers a great number of disease related information and imposes few restrictions on data access. Therefore, it is a perfect starting point to explore our approach. The hypothesis is that training data can be obtained from PubMed articles that talk about a disease corresponding to a ICD-9-CM code. With the abundant PubMed articles, we would be able to alleviate the training data imbalance problem.

There are several contributions in our study. Firstly, we examine the data imbalance problem in ICD-9-CM code assignment. Secondly, we propose and compare several methods to resolve the data imbalance problem. Thirdly, we give a comprehensive discussion on the current classification challenges. Finally, our method can be adapted to ICD-10-CM code assignment task with minor modifications.

The rest of this paper is organized as follows. In Section 2, we will discuss related research. Our methods and experiments will appear in Section 3 and 4. Limitations are discussed in Section 5 and the conclusion is provided in Section 6.

2 Related Work

The existing studies of automating ICD-9 code assignment can be classified into two groups. Through examining how professional coders assigning ICD codes, the first one used rule-based approaches. Ira et al. (2007) developed a rule-based system considering factors such as uncertainty, negation, synonymy, and lexical elements. Farkas and Szarvas (2008) used Decision Tree (DT) and Maximum Entropy (ME) to automatically generate a rule-based coding system. Crammer et al. (2007) composed a hybrid system consisting of a machine learning system with natural language features, a rule-based system based on the overlap between the reports and code descriptions, and an automatic policy system. Their results showed better performance than each single system.

The second group employed supervised machine learning methods for the assignment task, and their performance has been being equivalent or even better than those rule-based systems that need experts manually crafting knowledge. Aronson et al. (2007) used a stacked model to combine the results of four modules: Support Vector Ma-

---

[1] https://www.ncbi.nlm.nih.gov/pubmed/

chine (SVM), K-Nearest Neighbors (KNN), Pattern Matching (PM) and a hybrid Medical Text Indexer (MTI) system. Patrick et al. (2007) used ME and SVM classifiers, enhanced by a feature-engineering module that explores the best combination of several types of features. Zhang (2008) proposed a hierarchical text categorization method utilizing the ICD-9-CM codes structure. Zuccon and Nguyen (2013) conducted a comparison study on four classifiers (SVM, Adaboost, DT, and Naive Bayes) and different features on a 5,000 free-text death certificate dataset, and found that SVM with a stemmed unigram feature performed the best.

Along with the introduction of supervised methods, many past studies indicated that data imbalance problem can severely affect the classifier's performance. For example, Kavuluru et al. (2015) found that 874 of 1,231 ICD-9-CM codes in UK-Large dataset have less than 350 supporting data, whereas only 92 codes have more than 1,430 supporting data. The former group has macro F1 value of 51.3%, but the latter group only has 16.1%. To resolve data imbalance problem, they used optimal training set (OTS) selection approach to sample negative instance subset that provides best performance on validation set. However, OTS did not work on UKLarge dataset because several codes have so few training examples that even carefully selecting negative instances could not help. When Koopman et al. (2015) found that 85% of the whole death certificate dataset is associated with only top 20 common cancers, whereas the other 65 rarer cancers only have the rest 15% of the dataset, they tried to construct the balanced training set by randomly sampling a static number of negative examples for each class. Their results reflected the benefits of having more training data in improving the classifiers' performance. Since result of original model learned with imbalanced data is not provided, we cannot know the actual improvement. In addition, to deal with codes that only appear once in the dataset, Patrick et al. (2007) used a rule-based module to supplement ME and SVM classifiers.

Consistent to the existing works, our approach utilizes supervised methods for automatic ICD-9-CM code assignment, and our focus is on addressing the training data imbalance problem. But our work tries to solve the data imbalance problem by adding extra positive instances, which is not limited to the existing training data distribution or expert's knowledge. Adding positive instances have been proven to be effective in supervised machine learning in other domains(Caruana, 2000; He and Garcia, 2009), and we are first to find open source dataset as supplementary data for improving ICD-9-CM assignment performance.

3 Methods

In this section, we will first introduce the dataset on which our methods will be evaluated, then we propose two methods of collecting supplementary training data from PubMed dataset.

3.1 Dataset

We validate our methods on the official dataset of the 2007 Computational Medicine Challenge (CMC dataset), collected by Cincinnati Children's Hospital Medical Center (Pestian et al., 2007), which is frequently used by researchers working on the ICD-9-CM code assignment task. The CMC dataset consists of training and testing dataset, but only training dataset is accessible for us. Fortunately, most studies publish their system performance on both training and testing dataset, and then we can compare our methods with state-of-art methods. This corpus consists of 978 radiological reports taken from real medical records, and each report has been manually labeled with ICD-9-CM codes by professional companies. The example in Figure 1 comes from this dataset. In total, there are 45 ICD-9-CM codes appearing in the CMC dataset, and each report is labeled with one or more ICD-9-CM codes. This is a very imbalanced collection, with around half codes having less than 10 training data (see Figure 2).

3.2 Method I: Retrieving PubMed articles using ICD-9-CM code official description

Through examining the reports available to us, and also based on the discussions in previous work (Farkas and Szarvas, 2008; Ira et al., 2007; Crammer et al., 2007; Farkas and Szarvas, 2008), we hypothesize that the text description part of ICD-9-CM code can play important role for code assigners to build up the connection between a medical text and a ICD-9 code. Therefore, this motivated us to view the identifying extra training data in PubMed for an ICD-9-CM code as a retrieval problem where the text description part of an ICD-9-CM code can act as the query, and the whole PubMed dataset as a document collection. For example, based on ICD-9-CM code "786.2, cough", we can retrieve PubMed articles with a query "cough". Our initial informal testing confirmed our hypothesis.

To avoid bring back too much noise, we restricted the PubMed retrieval to only search on the article title field. Our motivation is that the title generally introduces the main topic of the whole paper. For the same reason, we also only utilized the title and abstract of top returned articles as the supplementary training data. In case of empty retrieval result, certain ICD-9-CM description terms that would not appear in PubMed article titles, such as "other", "unspecified", "specified", "NOS", and "nonspecific", are removed from the query. For example, ICD-9-CM code "599.0", whose description is "urinary tract infection, site not specified", will generate a cleaned query "urinary tract infection", and ICD-9-CM code "596.54", whose description is "neurogenic bladder NOS", will generate a cleaned query "neurogenic bladder".

3.3 Method II: Retrieving PubMed articles with both official and synonyms ICD-9-CM code description

Despite great overlap among them, ICD-9-CM code descriptions and the radiology reports in the CMC collection are written by different groups of people with different purposes. Therefore, there could be term mis-match problems between them. When this happens, it is actually better to not use the terms in the ICD-9-CM official description as the query for finding relevant PubMed articles, but actually to use the related terms in the CMC dataset as the query terms instead. This would enable the model trained on these returned PubMed articles can be more effectively classifying CMC reports. For example, the description "Anorexia" of code "783.0" does not appear in CMC dataset. Instead, "loss of appetite" exists in the radiology reports labeled with "783.0", while according to data in ICD9Data.com, "loss of appetite" is the synonym of "Anorexia". Therefore, in this case, it is better to use "loss of appetite" rather than "Anorexia" to be the query when search for training data in PubMed.

ICD9Data.com is an online website, providing rich and free ICD-9-CM coding information. It contains code definition, hierarchy structure, ap-

proximate synonyms, etc. We crawled the 45 codes' synonyms from the website. In method II, besides the queries from the official description, we also conducted PubMed searches with queries based on the synonyms of the descriptions. Each synonym is an individual PubMed query, and only when all its terms appear in CMC dataset, the query is considered. If one ICDcode has n queries and totally needs m supplementary documents for training, only top m/n retrieved PubMed articles from each query are considered.

4 EXPERIMENTS

4.1 Evaluation metrics

Following the past studies (Pestian et al., 2007; Kavuluru et al., 2015), we evaluate the classification performance through a micro F1 score (i.e., sum of the individual classification performance and divided by the individual amount) and a macro F1 score (i.e., sum of the classifiers performance and divided by the classifiers amount). We expect that by alleviating the data imbalance problem, macro F1 scores can increase significantly. All experiments in this study have gone through 10-fold cross validation, because it can provide a reliable result when data is limited (Witten et al., 2016).

4.2 Pre-process and Features

Following the past studies (Crammer et al., 2007; Aronson et al., 2007; Kavuluru et al., 2015, 2013; Koopman et al., 2015; Patrick et al., 2007; Ira et al., 2007), the CMC dataset is preprocessed with following steps:

- Full name restoration. Medical abbreviation restoration is a hard topic, which is not explored in this study. We manually generate a list of full names for abbreviations appearing in CMC dataset [2].

- Word lemmatization. Lemmatization of words are restored with WordNet 3.0 (Miller, 1995).

- Negation detection and removal. Negex (Chapman et al., 2001) is used to detect the negation expression, and negation target terms are removed after detection.

[2]https://github.com/daz45/CMC-data-set-abbreviations/tree/master

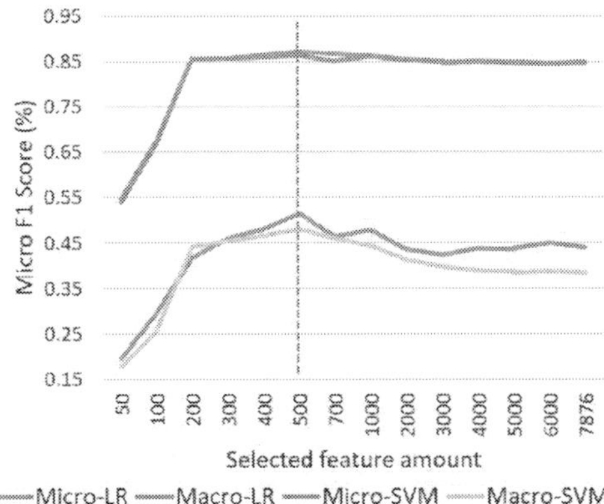

Figure 3: Feature selection on LR and SVM.

- Phrase recognition. MetaMap (Aronson and Lang, 2010) is utilized to extract the medical concept phrase appearing in the text, which is appended to the text.

After preprocessing, the example radiology report in Figure 1 will be "ten year old with chest pain x two week. the lung be well expand and clear. there be. the cardiac and mediastinal silhouette be normal. there be. chest_pain". Supplementary data collected from PubMed will be preprocessed in the same way.

4.3 Baselines

According to the past studies(Farkas and Szarvas, 2008; Aronson et al., 2007; Kavuluru et al., 2015), Support Vector Machine (SVM) and Logistic Regression (LR) are the most effective and commonly used classification models in this task. Therefore, we selected them as the two baselines. Each ICD-9-CM code has one binary classifier implemented using Scikit-Learn(Pedregosa et al., 2011). We name these two sets of baselines as Baseline_LR and Baseline_SVM.

Features consist of unigrams and bigrams appearing in preprocessed radiology reports, and the feature vector values are binary, indicating the appearance or absence of the word in text.

We performed feature selection on two baselines to avoid over-fit and extra computation cost. χ^2 based feature selection was employed for feature selection (Liu and Setiono, 1995). As shown in Figure 3, We find that 500 features can provide stable micro F1 performance and best macro F1 performance for Baseline_LR and Baseline_SVM. In all the following experiments, all classifiers are trained on these 500 selected features.

Our baseline performance were compared with the state-of-art methods in Table 1. Stacking is a stacked model combining four classification models (Aronson et al., 2007). Hybrid rule-based+MaxEnt is a hybrid system combining rule-based method with MaxEnt (Aronson et al., 2007). Although Table 1 shows that their performance is significantly better than our baselines, for the purpose of studying the methods for addressing imbalanced training data, we have to use the two current baselines since these advanced and complicated systems would hide the effects that we want to observe. In addition, any improvement we achieve in single classifier can be later integrated into these systems, which could be an interesting future work. These methods concentrated on micro averaging performance, while in this study we will explore the macro averaging performance.

Method	Micro F1
Baseline_LR	86.51%
Baseline_SVM	87.26%
Stacking	89.00%
Hybrid rule-based+MaxEnt	90.26%

Table 1: Baseline performance and existing best performed methods from related work.

Figure 4 shows the individual classification performance of 45 classifiers, and we can find an unstable performance across 45 classifiers. We use Macro F1 score as the split line, and we can find that, for both baseline system, there are 21 classifiers having a below-average performance, and all of them have relatively less training data than the classifiers with above-average performance. This indicates that the data imbalance leads to the performance instability across all classes.

With the Macro F1 score, we separate 45 classifiers into two groups: Group 1 consists of 24 classifiers, union set of classifiers with below-average performance in two baselines, and Group 2 consists of rest 21 classifiers with above-average classification performance. Though Group 1 has 24 classifier, radiological reports labeled with them only takes 11.56% of 978 reports. To deal with this data imbalance problem, we will introduce supplementary training data from PubMed dataset. Through adding additional data, we expect that classification performance of the whole system, especially Group 1, will be improved.

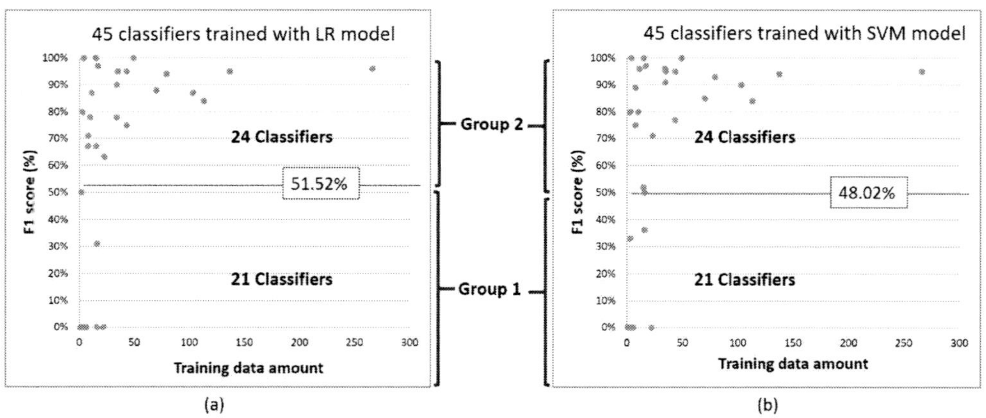

Figure 4: Individual classification performance of 45 classifiers trained with LR and SVM model in Baseline.

4.4 Experiment I: retrieving PubMed articles with ICD code official description

In the first experiment, supplementary data is collected based on ICD-9-CM code official description, as described in method I. The supplementing document size is set to be 10, 20, 40 and 60. Supplementary training data is added to 24 classifiers in Group 1. We name these two new runs as Group_1_Description_LR (G1_desc_LR) and Group_1_Description_SVM (G1_desc_SVM), appended with supplementary data size. The results in Table 2 also show that supplementing 10 documents can generate best performance, and with more documents added, both macro and micro F1 will decrease.

Method	Micro F1	Macro F1
Baseline_LR	86.51%	51.52%
G1_desc_LR_10	86.68%	55.78%
G1_desc_LR_20	86.07%	55.23%
G1_desc_LR_40	85.18%	52.01%
G1_desc_LR_60	84.97%	51.40%
Baseline_SVM	87.26%	48.03%
G1_desc_SVM_10	86.96%	57.09%
G1_desc_SVM_20	86.67%	55.43%
G1_desc_SVM_40	85.87%	57.61%
G1_desc_SVM_60	86.25%	54.77%

Table 2: Enhance classifiers in Group 1 with supplementary data collected with method I, while the evaluation is performed on all classes.

Through Wilcoxon Signed Ranks test, there is no significant difference between G1_desc_LR_10 and Baseline_LR. Nor does G1_desc_SVM_10. Further, we compare both methods against baseline on Group 1 and Group 2 separately. However, there is still no significant difference existing. Take G1_desc_SVM_10 for example, from Figure 5, we can see that 11 classes still have F1=0%, while 5 classes' performance decrease, and only 8 got F1 improved. It indicates that the method I is ineffective.

After exploring the results, we find sometimes the supplementary data does not help training. For example, for ICD-9-CM code "783.0 Anorexia", the classification performance stays 0%. The corresponding radiology report doesn't have term "Anorexia", making the supplementary data useless. It implies we need to collect PubMed articles containing same features with the radiology reports in CMC dataset.

4.5 Experiment II: Retrieving PubMed articles with ICD code official and synonyms descriptions

In this second experiment, we collect PubMed data through the ICD-9-CM code's both official and synonyms description that appears in CMC dataset. We name these two runs as Group_1_Synonym_LR (G1_syn_LR) and Group_1_Synonym_SVM (G1_syn_SVM). Due to the paper size limitation, here we only show the best results with supplementary document size being 10 in Table 3.

Through Wilcoxon Signed Ranks test, G1_syn_SVM_10 significantly outperforms baseline ($p - value < 0.01$), but has no significant difference compared with G1_desc_SVM_10. However, if only classifiers in Group 1 are considered, G1_syn_SVM_10 significantly outperforms G1_desc_SVM_10 ($p - value < 0.01$). This indicates that our propose method II can generate effective supplementary training data. On the other hand, G1_syn_LR_10 is found to outperform Baseline_LR significantly only on Group 1 classes ($p - value < 0.01$).

Method	Micro F1	Macro F1
Baseline_LR	86.51%	51.52%
G1_desc_LR_10	86.68%	55.78%
G1_syn_LR_10	86.30%	62.85%‡
All_syn_LR_10	86.60%	62.19%‡
Baseline_SVM	87.26%	48.03%
G1_desc_SVM_10	86.96%	57.09%
G1_syn_SVM_10	87.22%	67.43%†‡
All_syn_SVM_10	87.88%	64.54%†‡

Table 3: Experiment results. †means significantly outperform baseline. ‡means significant outperform baseline on Group 1.

It shows that on SVM model, PubMed data collected with ICD-9-CM code descriptions synonyms works better in solving the data imbalance problem than with the official descriptions. After data supplementation, there are still 6 classifiers with F1 score being 0%, which will be further discussed in Section 5.

4.6 Experiment III: adding supplementary training data to all classifiers

In the third experiment, we add supplementary data to all 45 classifiers to explore whether adding supplementary data to the classifiers that originally have sufficient training data still can gain performance improvement. We name these two runs as All_Synonym_LR (All_syn_LR) and All_Synonym_SVM (All_syn_SVM). Also, only the best results with supplementary document size being 10 is shown in Table 3. Through Wilcoxon Signed Ranks test, All_syn_SVM_10 significantly outperforms the baseline, and All_syn_LR_10 significantly outperforms the baseline only on Group 1 ($p - value < 0.01$), but both have no significant difference with G1_syn and G1_desc. These means that adding supplementary training data is effective on solving data imbalance problem, but for the classifiers that originally have sufficient training data, extra training data seems have no

significant effect.

5 Discussion

Experiment results indicates that our proposed supplementing training data method can help the classifiers to reach to a relatively balanced performance. Such improvement mainly comes from changing the word weight ranking so that important words rank higher. For example, for code "758.6 turner syndrome", in the baseline_LR (F1=0%), top 3 features with highest weights are "duplicate left, partially, turner_syndrome". But in G1_syn_LR_10 (F1=67%), top 3 features are "turner, turner syndrome, syndrome". Supplementary data trains term in "turner syndrome" a higher weight in LR model, explaining this code's classification performance increase.

In addition, supplementary data will improve classification through boosting the weight of the features. For example, the top features for code "786.59, Other chest pain" are basically similar in both baseline_LR (F1=0%) and G1_syn_LR_10 (F1=40%), including "tightness, chest_tightness, chest pain". However, the weight differs a lot. For baseline_LR, weights are all under 1.5, while in G1_syn_LR_10, top 5 features are all above 1.5, indicating the classification model have much higher confidence on these features.

Finally, supplementary data mainly support code assignment effectively in Group 1, and we find that classification performance in Group 2 basically has no significant difference across all experiments. Meanwhile, 978 reports, dominated by Group 2 classes, also show no significant difference across all experiments. Therefore, extra training data does not improve Group 2's performance, and hence supplementary data is not suggested for classes having sufficient training data.

Besides, our proposed methods can be directly used in ICD-10-CM classification with little modification. Just update the PubMed query with the ICD-10-CM textual descriptions and synonyms.

Though data imbalance problem has been largely alleviated, there are still a few classifiers in Group 1 have poor performance. After exploring, we think there are mainly four reasons:

- Word level feature matching limitation. For example, description of code "V72.5 Radiological examination" does not appear in the collection, and it has no synonyms. Radiological examination actually means a variety

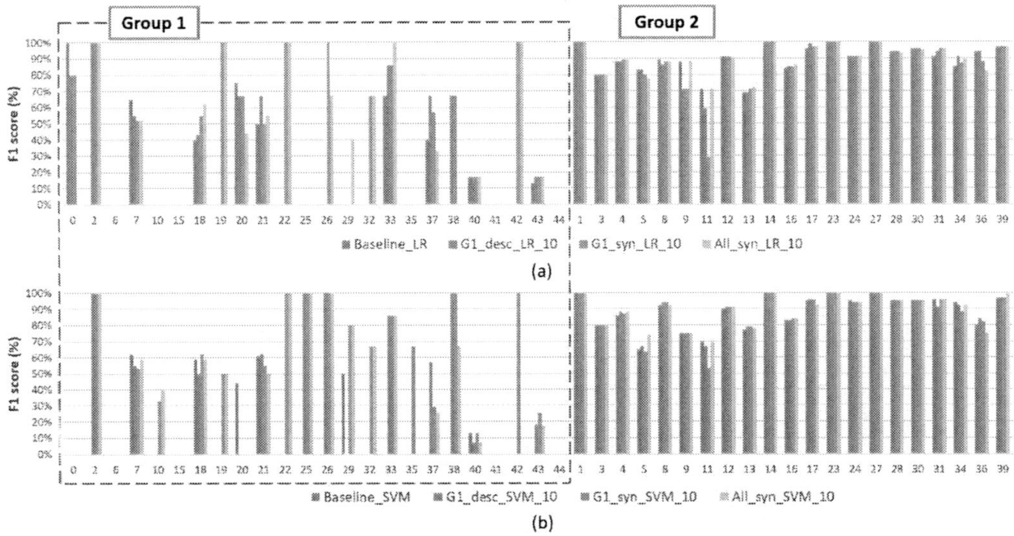

Figure 5: Individual classification performance of 45 classifiers on Baseline and three experiments.

of imaging techniques, and such word level feature matching cannot help classification.

- "History of" ICD-9-CM codes. For codes "V13.02, Personal history, urinary tract infection" and "V13.09 Personal history of other specified urinary system disorders", adding supplementary data doesn't help their performance. We find their radiology reports are basically classified to "599.0 urinary tract infection" and "593.70 vesicoureteral reflux". "history of" feature is ignored. Extra training data has no effect on this problem.

- Speculative expression. In prepossessing procedure, negation terms are removed, but speculative expressions are kept. It results in that when doctor is not sure whether a patient may get a disease, but write it down to reports, classification results will rely on these speculative terms, and cause false positive. For example, code "518.0" has a low F1 score because in many reports labeled with other codes, doctors write that the patient may have disease "atelectasis", while "atelectasis" is a very important word to recognize "518.0".

- Data missing due to expert disagreement. In CMC dataset, three experts manually assign codes to 978 radiology reports. Only when two or more experts agree, code is approved. However, sometimes the conflict opinions re-

sults in code assignment failure. For example, reports 99619963 and 99803917 should be labeled with "741.90 Spina bifida". However, one expert assigned "741.90", another assigned "741.9", and the third expert miss this code at all. This led to "741.90 Spina bifida" was not assigned to these two reports. However, with the supplementary data added into the training, our method correctly assigns "741.90 Spina bifida" to these two reports, but this assignment was counted as wrong since the ground truth does not have this code due to expert disagreements.

6 Conclusion and Future Work

In this study, we studied to address the data imbalance problem in ICD-9-CM code automatic assignment task. Using ICD-9-CM codes synonyms can accurately search medical texts relevant documents from PubMed. Collected data, used as supplementary training data, can significantly boost systems macro averaging performance as the data imbalance problem is largely alleviated. However, for the classifiers that originally have sufficient training data, additional data basically has no significant effect. As future work, we will modify the Context algorithm (Harkema et al., 2009) to detect the historical mentions and speculative expressions in the radiology reports. Also, we would explore the difference of same features extracted from different field of radiology reports.

References

Alan R Aronson, Olivier Bodenreider, Dina Demner-Fushman, Kin Wah Fung, Vivian K Lee, James G Mork, Aurélie Névéol, Lee Peters, and Willie J Rogers. 2007. From indexing the biomedical literature to coding clinical text: experience with mti and machine learning approaches. In *Proceedings of the Workshop on BioNLP 2007: Biological, Translational, and Clinical Language Processing*. Association for Computational Linguistics, pages 105–112.

Alan R Aronson and François-Michel Lang. 2010. An overview of metamap: historical perspective and recent advances. *Journal of the American Medical Informatics Association* 17(3):229–236.

Rich Caruana. 2000. Learning from imbalanced data: Rank metrics and extra tasks. In *Proc. Am. Assoc. for Artificial Intelligence (AAAI) Conf.* pages 51–57.

Wendy W Chapman, Will Bridewell, Paul Hanbury, Gregory F Cooper, and Bruce G Buchanan. 2001. A simple algorithm for identifying negated findings and diseases in discharge summaries. *Journal of biomedical informatics* 34(5):301–310.

Koby Crammer, Mark Dredze, Kuzman Ganchev, Partha Pratim Talukdar, and Steven Carroll. 2007. Automatic code assignment to medical text. In *Proceedings of the Workshop on BioNLP 2007: Biological, Translational, and Clinical Language Processing*. Association for Computational Linguistics, pages 129–136.

Richárd Farkas and György Szarvas. 2008. Automatic construction of rule-based icd-9-cm coding systems. *BMC bioinformatics* 9(3):S10.

Henk Harkema, John N Dowling, Tyler Thornblade, and Wendy W Chapman. 2009. Context: an algorithm for determining negation, experiencer, and temporal status from clinical reports. *Journal of biomedical informatics* 42(5):839–851.

Haibo He and Edwardo A Garcia. 2009. Learning from imbalanced data. *IEEE Transactions on knowledge and data engineering* 21(9):1263–1284.

Goldstein Ira, Arzumtsyan Anna, and Uzuner Ozlem. 2007. Three approaches to automatic assignment of icd-9-cm codes to radiology reports .

Peter Jackson and Isabelle Moulinier. 2007. *Natural language processing for online applications: Text retrieval, extraction and categorization*, volume 5. John Benjamins Publishing.

Ramakanth Kavuluru, Sifei Han, and Daniel Harris. 2013. Unsupervised extraction of diagnosis codes from emrs using knowledge-based and extractive text summarization techniques. In *Canadian Conference on Artificial Intelligence*. Springer, pages 77–88.

Ramakanth Kavuluru, Anthony Rios, and Yuan Lu. 2015. An empirical evaluation of supervised learning approaches in assigning diagnosis codes to electronic medical records. *Artificial intelligence in medicine* 65(2):155–166.

Bevan Koopman, Guido Zuccon, Anthony Nguyen, Anton Bergheim, and Narelle Grayson. 2015. Automatic icd-10 classification of cancers from free-text death certificates. *International journal of medical informatics* 84(11):956–965.

Huan Liu and Rudy Setiono. 1995. Chi2: Feature selection and discretization of numeric attributes. In *Tools with artificial intelligence, 1995. proceedings., seventh international conference on*. IEEE, pages 388–391.

George A Miller. 1995. Wordnet: a lexical database for english. *Communications of the ACM* 38(11):39–41.

Jon Patrick, Yitao Zhang, and Yefeng Wang. 2007. Developing feature types for classifying clinical notes. In *Proceedings of the Workshop on BioNLP 2007: Biological, Translational, and Clinical Language Processing*. Association for Computational Linguistics, pages 191–192.

Fabian Pedregosa, Gaël Varoquaux, Alexandre Gramfort, Vincent Michel, Bertrand Thirion, Olivier Grisel, Mathieu Blondel, Peter Prettenhofer, Ron Weiss, Vincent Dubourg, et al. 2011. Scikit-learn: Machine learning in python. *Journal of Machine Learning Research* 12(Oct):2825–2830.

John P Pestian, Christopher Brew, Paweł Matykiewicz, Dj J Hovermale, Neil Johnson, K Bretonnel Cohen, and Włodzisław Duch. 2007. A shared task involving multi-label classification of clinical free text. In *Proceedings of the Workshop on BioNLP 2007: Biological, Translational, and Clinical Language Processing*. Association for Computational Linguistics, pages 97–104.

Ian H Witten, Eibe Frank, Mark A Hall, and Christopher J Pal. 2016. *Data Mining: Practical machine learning tools and techniques*. Morgan Kaufmann.

Yitao Zhang. 2008. A hierarchical approach to encoding medical concepts for clinical notes. In *Proceedings of the 46th Annual Meeting of the Association for Computational Linguistics on Human Language Technologies: Student Research Workshop*. Association for Computational Linguistics, pages 67–72.

Guido Zuccon and Anthony Nguyen. 2013. Classification of cancer-related death certificates using machine learning .

Evaluating Feature Extraction Methods for Knowledge-based Biomedical Word Sense Disambiguation

Sam Henry **Clint Cuffy** **Bridget T. McInnes**

Department of Computer Science
Virginia Commonwealth University
Richmond, VA 23284 USA

henryst@vcu.edu cuffyca@vcu.edu btmcinnes@vcu.edu

Abstract

In this paper, we present an analysis of feature extraction methods via dimensionality reduction for the task of biomedical Word Sense Disambiguation (WSD). We modify the vector representations in the 2-MRD WSD algorithm, and evaluate four dimensionality reduction methods: Word Embeddings using Continuous Bag of Words and Skip Gram, Singular Value Decomposition (SVD), and Principal Component Analysis (PCA). We also evaluate the effects of vector size on the performance of each of these methods. Results are evaluated on five standard evaluation datasets (Abbrev.100, Abbrev.200, Abbrev.300, NLM-WSD, and MSH-WSD). We find that vector sizes of 100 are sufficient for all techniques except SVD, for which a vector size of 1500 is preferred. We also show that SVD performs on par with Word Embeddings for all but one dataset.

1 Introduction

Word Sense Disambiguation (WSD) is the task of automatically identifying the intended sense (or concept) of an ambiguous word based on the context in which the word is used. Automatically identifying the intended sense of ambiguous words improves the performance of clinical and biomedical applications such as medical coding and indexing for quality assessment, cohort discovery (Plaza et al., 2011; Preiss and Stevenson, 2015), and other secondary uses of data such as information retrieval and extraction (Stokoe et al., 2003), and question answering systems (Ferrández et al., 2006). These capabilities are becoming essential tasks due to the growing amount of information available to researchers, the transition of health care documentation towards electronic health records, and the push for quality and efficiency in health care.

Previous methods using distributional context vectors have been shown to perform well for the task of WSD. One problem with distributional vectors is the sparseness of the vectors and noise (defined here as information that does not aid in the discrimination between word senses). Word embeddings have become an increasingly popular method to reduce the dimensionality of vector representations, and have been shown to be a valuable resource for NLP tasks including WSD (Sabbir et al., 2016).

Prior to word embeddings, (Deerwester et al., 1990) proposed Latent Semantic Indexing (LSI) which reduces dimensionality using the factor analysis technique, singular value decomposition (SVD). When performing SVD, some information is lost. Intuitively the lost information is noise, and its removal causes the similarity and non-similarity between words to be more discernible (Pedersen, 2006).

Similar to SVD is principal component analysis (PCA). PCA transforms the vectors into a new basis of *principal components*, which are created by orthogonal linear combinations of the original features. Each principal component captures as much variance in the data as possible while maintaining orthogonality. Dimensionality reduction is performed by removing principal components that capture little variance.

In this paper, we evaluate the performance of word embeddings, SVD, and PCA for dimensionality reduction for the task of knowledge-based WSD. Explicit vectors are trained on Medline abstracts and performance is evaluated on five reference standards. Specifically, the contributions of this paper are an analysis of:

Proceedings of the BioNLP 2017 workshop, pages 272–281,
Vancouver, Canada, August 4, 2017. ©2017 Association for Computational Linguistics

- Vector Representation: SVD, PCA, and word embeddings using continuous bag of words (CBOW) and skip-gram are evaluated as dimensionality reduction techniques applied to the task of knowledge-based WSD. Evaluation is performed on several standard evaluation datasets, and compared against explicit co-occurrence vectors as a baseline.

- Dimensionality: the dimensionality of the reduced vectors is a parameter, and the value can effect performance. We evaluate each vector representation's performance at dimensionalities of 100, 200, 500, 1000, and 1500.

2 Related Work

Existing biomedical WSD methods can be classified into three groups: unsupervised (Brody and Lapata, 2009; Pedersen, 2010), supervised (Zhong and Ng, 2010; Stevenson et al., 2008), and knowledge-based methods (Navigli et al., 2011). Unsupervised methods use the distributional characteristics of an outside corpus and do not rely on sense information or a knowledge source (Pedersen, 2006).

Supervised methods use machine learning algorithms to assign senses to instances containing the ambiguous word. Although supervised methods have the best performance, they require training data for each target word to be disambiguated. Whether this is done manually or automatically, it is infeasible to create such data on a large scale. Recently, (Sugawara et al., 2015) created a supervised system that uses word2vec word embeddings as input to a support vector machine classifier. They compare the word vectors generated by word2vec with the word vectors generated by SVD, and show that word2vec slightly outperforms SVD with vector dimensionality of 300.

Knowledge-based methods do not use manually or automatically generated training data, but instead use information from an external knowledge source (e.g. taxonomy). These knowledge-based methods can be classified into two categories, graph-based and vector-based approaches. Here, we focus on vector-based approaches as it relates to this research.

(Humphrey et al., 2006) introduce a vector-based method that assigns a sense to a target word by first identifying its semantic type with the assumption that each possible sense has a distinct semantic type. In this method, semantic type (st-) vectors are created for each possible semantic type. The st-vectors consist of binary values for each one word term in the United Medical Language System (UMLS); a one if that word has a sense of the semantic type, else a zero. A target word (tw-) vector is created using the words surrounding the target word. The cosine of the angle between the tw-vector and each of the st-vectors is calculated and the sense whose st-vector is closest to the tw-vector is assigned to the target word. The limitation of this method is that two possible senses may have the same semantic type. For example, the term *cortices* can refer to either the cerebral cortex (C0007776) or the kidney cortex (C0022655), both of which have the same semantic type, "Body Part, Organ, or Organ Component". Analysis of the 2009 Medline data [1] shows that there are 1,072,902 terms in Medline that exist in the UMLS of which 35,013 are ambiguous and 2,979 have two or more senses with the same semantic type. This indicates that approximately 12% of the ambiguous words cannot be disambiguated using the knowledge-based methods discussed above, and another method is required.

(Jimeno-Yepes et al., 2011) attempt to address this limitation by introducing two methods, MRD and 2-MRD. In these methods a sense vector (s-vector) is created for each possible sense of a target word using the definition information from the UMLS. A target word (tw-) vector is created using the words surrounding the target word. The cosine of the angle between the tw-vector and each of the s-vectors is calculated and the sense whose s-vector is closest to the tw-vector is assigned to the target word. The MRD method uses the words within the definition weighted based on their occurrence statistics across definitions in the UMLS. The 2-MRD method (discussed more fully in Section 3) uses second-order context vectors to represent the concept's definition.

(Pakhomov et al., 2016) and (Tulkens et al., 2016) explore using the 2-MRD method in conjunction with word embeddings, and evaluate their performance with varying training corpora. Their results are promising, however evaluation is limited to a single dataset (MSH-WSD), vector size is not varied, and they do not compare performance with different word2vec models.

---

[1]http://mbr.nlm.nih.gov/index.shtml

3 Method

We modify the vector representations of the 2-MRD WSD algorithm using four different vector representations: SVD, PCA, and word embeddings using continuous bag of words (CBOW) and skip-gram. Explicit vectors are word-by-word co-occurrence vectors, and are used as a baseline. The disadvantage of explicit vectors is that the word-by-word co-occurrence matrix is sparse and subject to noise introduced by features that do not distinguish between the different senses of a word. The goal of the dimensionality reduction techniques is to generate vector representations that reduce this type of noise. Each method is described in detail here.

3.1 2-MRD Algorithm

In this section we describe the 2-MRD WSD algorithm at a high level: a vector is created for each possible sense of an ambiguous word, and the ambiguous word itself. The appropriate sense is then determined by computing the cosine similarity between the vector representing the ambiguous word and each of the vectors representing the possible senses. The sense whose vector has the smallest angle between it and the vector of the ambiguous word is chosen as the most likely sense.

To create a vector for a possible sense, we first obtain a textual description of sense from the UMLS, which we refer to as the *extended definition*. Each sense, from our evaluation set, was mapped to a concept in the UMLS, therefore, we use the sense's definition plus the definition of its parent/children and narrow/broader relations and associated synonymous terms. After the extended definition is obtained, we create the second-order vector by first creating a word by word co-occurrence matrix in which the rows represent the content words in the extended definition, and the columns represent words that co-occur in Medline abstracts with the words in the definition. Each word in the extended definition is replaced by its corresponding vector, as given in the co-occurrence matrix. The centroid of these vectors constitutes the second order co-occurrence vector that is used to represent the sense.

The second-order co-occurrence vector for the ambiguous word is created in a similar fashion, only rather than using words in the extended definition, we use the words surrounding the word in the instance. Second-order co-occurrence vectors

were first described by (Schütze, 1998) and extended by (Purandare and Pedersen, 2004) and (Patwardhan and Pedersen, 2006) for the task of word sense discrimination. Later, (McInnes et al., 2011; Jimeno-Yepes et al., 2011) adapted these vectors for the task of disambiguation rather than discrimination.

3.2 Singular Value Decomposition

Singular Value Decomposition (SVD), used in Latent Semantic Indexing, is a factor analysis technique to decompose a matrix, M into a product of three simpler matrices, such that $M = U \cdot \Sigma \cdot V^T$. The matrices U and V are orthonormal and Σ is a diagonal matrix of eigenvalues in decreasing order. Limiting the eigenvalues to d, we can reduce the dimensionality of our matrix to $M_d = U_d \cdot \Sigma_d \cdot V_d^T$. The columns of U_d correspond to the eigenvectors of M_d. Typically this decomposition is achieved without any loss of information. Here though, SVD reduces a word-by-word co-occurrence matrix from thousands of dimensions to hundreds, and therefore the original matrix cannot be perfectly reconstructed from the three decomposed matrices. The intuition is that any information lost is noise, the removal of which causes the similarity and non-similarity between words to be more discernible (Pedersen, 2006).

3.3 Principal Component Analysis

Principal Component Analysis (PCA) is similar to SVD, and is commonly used for dimensionality reduction. The goal of PCA is to map data to a new basis of orthogonal principal components. These principal components are linear combinations of the original features, and are ordered by their variance. Therefore, the first principal components capture the most variance in the data. Under the assumption that the dimensions with the most variance are the most discriminative, dimensions with low variance (the last principal components) can safely be removed with little information loss.

PCA may be performed in a variety of ways, however the implementation we chose makes the parallels between PCA and SVD clear. First the co-occurrence matrix, M is centered to produce the matrix C. Centering consists of subtracting the mean of each column from values in that column. PCA is sensitive to scale, and this prevents the variance of features with higher absolute counts from dominating. Mathematically, this allows us to compute the principal components us-

"

ing SVD on C. This is because $C^T C$ is proportional to the covariance matrix of M, and is used in the calculation of SVD. Applying SVD to C, such that $C = U \cdot \Sigma \cdot V^T$, the principal components are obtained by the product of U and Σ (e.g. $M_{PCA} = U \cdot \Sigma$). For dimensionality reduction all but the first d columns of M_{PCA} are removed. This captures as much variation in the data with the fewest possible dimensions.

3.4 Word embeddings

The word embeddings method, proposed by (Mikolov et al., 2013), is a neural network based approach that learns a representation of a word-word co-occurrence matrix. The basic idea is that a neural network is used to learn a series of weights (hidden layer with in the neural network) that either maximizes the probability of a word given the surrounding context, referred to as the continuous bag of words (CBOW) approach, or to maximize the probability of the context given a word, referred to as the Skip-gram approach;

For either approach, the resulting hidden layer consists of a matrix where each row represents a word in the vocabulary and columns a word embedding. The basic intuition behind this method is that words closer in meaning will have vectors closer to each other in this reduced space.

4 Data

4.1 Training Data

We develop our vectors using co-occurrence information from Medline [2]. Medline is a bibliographic database containing around 23 million citations to journal articles in the biomedical domain and is maintained by National Library of Medicine. The 2015 Medline Baseline encompasses approximately 5,600 journals starting from 1948, and contains 22,775,609 citations, of which 13,835,206 contain abstracts. In this work, we use Medline titles and abstracts from 1975 to present day to generate word embeddings, and to generate the co-occurrence matrix of explicit vectors that is the input into SVD and PCA. Prior to 1975, only 2% of the citations contained an abstract.

4.2 Evaluation Data

We evaluate using several standard WSD evaluation datasets which include the following.

Abbrev. The Abbrev dataset [3] developed by Stevenson, et al. (Stevenson et al., 2009) contains examples of 300 ambiguous abbreviations found in MEDLINE that were initially presented by (Liu et al., 2001). The data set was automatically re-created by identifying the abbreviations and long-forms (unabbreviated terms) in MEDLINE abstracts, and replacing the long-form in the abstract with its abbreviation. The abbreviations' long-forms were manually mapped to concepts in the UMLS by Stevenson, et al. Each abstract contains approximately 216 words. The datasets consist of a set of 21 different ambiguous abbreviations for which the number of labeled instances of those abbreviations varies. Abbrev.100 contains 100 instances, Abbrev.200 contains 200, and Abbrev.300 contains 300 labeled instances. Two abbreviations contain less than 200 instances, and three abbreviations contain less than 300 instances, and are omitted from Abbrev.200 and Abbrev.300 respectively. The average number of long-forms per abbreviation is 2.6 and the average majority sense across all subsets is 70%.

NLM-WSD. The National Library of Medicine's Word Sense Disambiguation (NLM-WSD) dataset [4] developed by (Weeber et al., 2001) contains 50 frequently occurring ambiguous words from the 1998 MEDLINE baseline. Each ambiguous word in the NLM-WSD dataset contains 100 ambiguous instances randomly selected from the abstracts totaling to 5,000 instances. The instances were manually disambiguated by 11 evaluators who assigned the ambiguous word to a concept (CUI) in the UMLS, or assigned the concept as "None" if none of the possible concepts described the term. The average number of senses per term is 2.3, and the average majority sense is 78%.

MSH-WSD. The National Library of Medicine's MSH Word Sense Disambiguation (MSH-WSD) dataset [5] developed by (Jimeno-Yepes et al., 2011) contains 203 ambiguous terms and abbreviations from the 2010 MEDLINE baseline. Each target word contains approximately 187 instances, has 2.08 possible senses, and has a 54.5% majority sense. Out of 203 target words, 106 are terms, 88 are abbreviations, and 9 have possible senses that are both abbreviations and

---

[2]http://mbr.nlm.nih.gov/Download/index.shtml

[3]http://nlp.shef.ac.uk/BioWSD/downloads/corpora
[4]http://wsd.nlm.nih.gov
[5]http://wsd.nlm.nih.gov

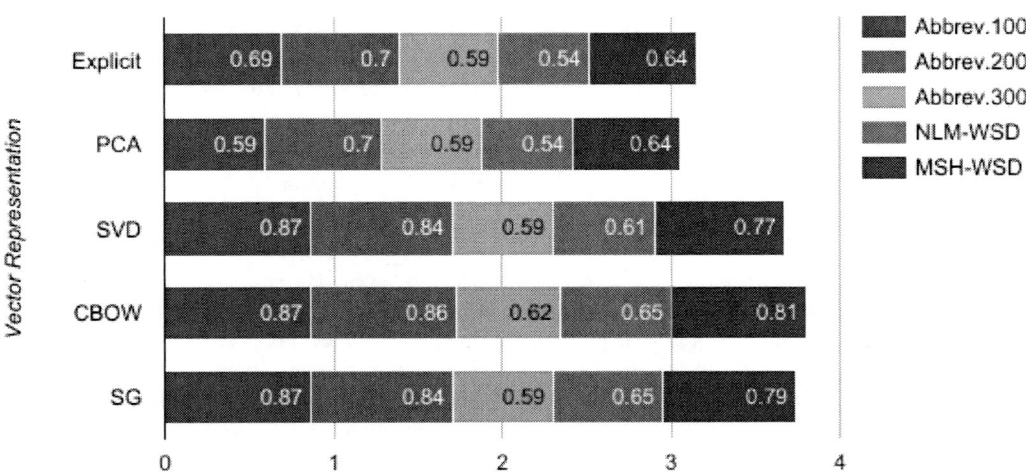

Figure 1: Comparison between accuracy of vector representations on WSD datasets

terms. For example, the target word *cold* has the abbreviation *Chronic Obstructive Lung Disease* as a possible sense, as well as the term *Cold Temperature*. The total number of instances is 37,888.

5 Experimental Framework

We used the following packages to obtain our vector representations:

[1] Explicit Representation: We used the Text::NSP packaged developed by (Pedersen et al., 2011). We used a windows size of 8, a frequency cutoff of 5, and removed stopwords.

[2] Singular Value Decomposition: We ran the MATLAB R2016b implementation of sparse matrix SVD (svds) on the explicit representation matrix, and used each row of the resulting U matrix as a reduced vector.

[3] Principal Component Analysis: We centered the explicit representation matrix, and used the MATLAB R2016b implementation of sparse matrix SVD (svds) on the centered matrix to obtain the U and Σ matrices. The reduced vectors are obtained from the product of U and Σ.

[4] Word Embeddings: We used the *word2vec* package developed by (Mikolov et al., 2013)

for the continuous-bag-of-words (CBOW) and skip-gram word embedding models with a window size of 8, a frequency cutoff of 5, and default settings for all other parameters.

We use the Word2vec::Interface package [6] version 0.03 to obtain the disambiguation accuracy for each of the WSD datasets. The differences between the means of disambiguation accuracy were tested for statistical significance using pair-wise Students t-test.

6 Results and Analysis

6.1 Results

Figure 1 compares the performance of each vector representation technique, and shows the best results (best among all dimensionalities tested) of each of the vector representations on the WSD datasets. Explicit refers to the co-occurrence vector without dimensionality reduction, PCA refers to the principal component analysis representation, SVD refers to singular value decomposition representation, CBOW refers to the word embeddings continuous bag of words representation and SG refers to the word embeddings skip gram representation. The colored bars show results for individual datasets, and the total length shows the sum of accuracies for all datasets.

---

[6]http://search.cpan.org/dist/Word2vec-Interface/

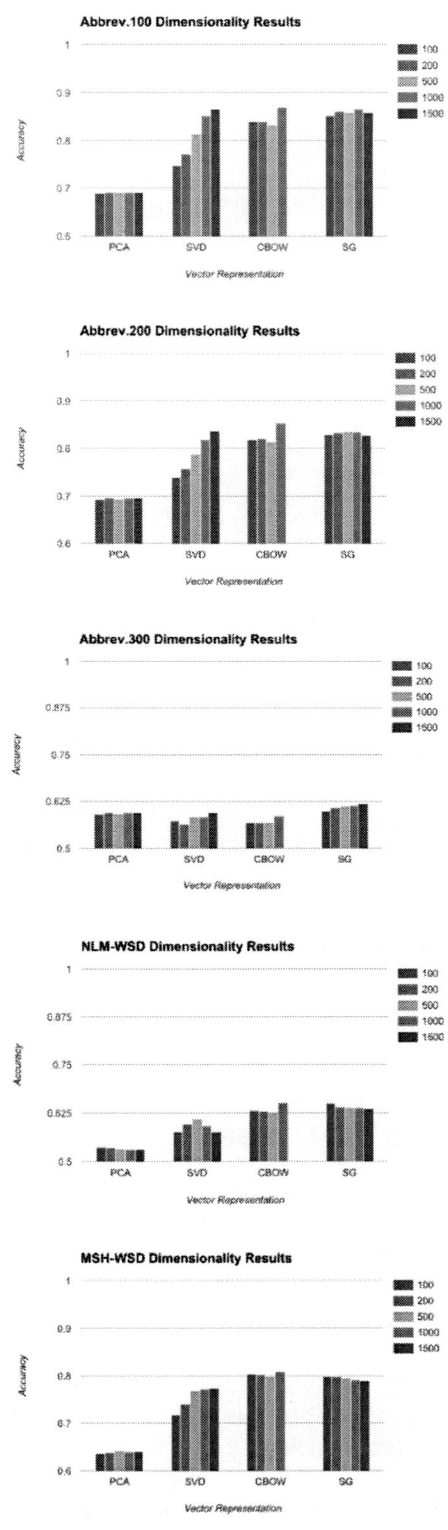

Figure 2: Effect of dimensionality on accuracy

Abbrev.100			
	PCA	SVD	CBOW	SG
explicit	0.65	0.0008	0.0015	0.0006
PCA		0.0007	0.0013	0.0005
SVD			0.94	0.97
CBOW				0.93

Abbrev.200			
	PCA	SVD	CBOW	SG
explicit	0.29	0.006	0.0047	0.0045
PCA		0.005	0.0042	0.0037
SVD			0.56	0.93
CBOW				0.60

Abbrev.300			
	PCA	SVD	CBOW	SG
explicit	1.0	1.0	0.41	0.63
PCA		1.0	0.41	0.63
SVD			0.29	0.21
CBOW				0.08

NLM-WSD			
	PCA	SVD	CBOW	SG)
explicit	0.35	0.10	0.0062	0.0127
PCA		0.087	0.0042	0.009
SVD			0.2489	0.2993
CBOW				0.66

MSH-WSD			
	PCA	SVD	CBOW	SG
explicit	0.37	0.0001	0.0001	0.0001
PCA		0.0356	0.0005	0.0001
SVD			0.0005	0.0346
CBOW				0.056

Table 1: The p-values using Student's pairwise t-$test$. Each table corresponds to a different dataset, each row and column a different dimensionality reduction technique.

The Abbrev.100, Abbrev.200, and Abbrev.300 results show that SVD (0.87/0.84/0.62), CBOW (0.87/0.86/0.62), and SG (0.87/0.84/0.59) obtained a statistically higher overall disambiguation accuracy ($p \leq 0.05$) than explicit (0.69/0.70/0.59) and PCA (0.59/0.70/0.59), while the difference between their respective results was not statistically significant. The NLM-WSD results also show that SVD (0.61), CBOW (0.65), and SG (0.65) obtained a statistically higher disambiguation accuracy than explicit (0.54) and PCA (0.54), while the difference between their respective results was not statistically significant. The MSH-WSD results show a statistically significant difference ($p \leq 0.05$) between explicit (0.64), PCA (0.64), SVD (0.77), CBOW (0.81), and SG (0.79) except for Explicit and PCA. Table 1 shows the p-$values$ between the vector representations for each of the datasets.

Figure 2 shows the effects of dimensionality on

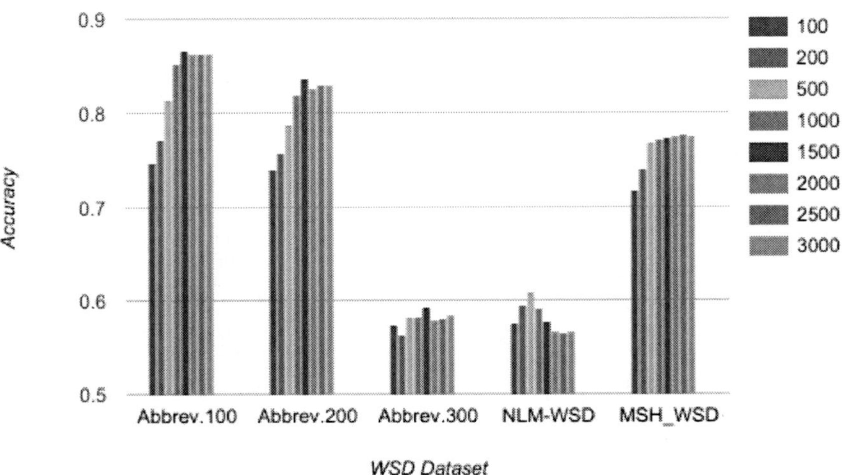

Figure 3: Effect of dimensionality on the accuracy of SVD

disambiguation accuracy of PCA, SVD, CBOW [7], and SG for each of the datasets for dimensionality reduction of d = 100, 200, 500, 1000 and 1500. The PCA, CBOW, and SG all show little change in accuracy as the dimensions vary. This indicates lower dimensional vector representations are sufficient for these techniques. SVD on the other hand shows for all of the datasets except NLM-WSD, an increase in accuracy as dimensionality increases. To discover an upper bound on dimensionality and performance, we continued to increase the dimensions of SVD up to 3000. Results are shown in Figure 3, and indicate that after $d = 1500$ there are not significant gains in accuracy, indicating that a dimensionality of 1500 is sufficient for SVD.

6.2 Analysis

This study indicates that SVD performs on par with word embeddings for most datasets. This is exciting because the co-occurrence matrix that is the input for SVD can be easily modified to hopefully increase performance. The word embeddings algorithms use a neural network approach which can approximate any function, but does not provide any insights about the features being approximated; instead accuracy gains are often achieved by increasing the amount of training data.

One disadvantage of SVD is that it, unlike word embeddings, may not be scalable to massive cor-

pora. Since we are using the majority of MED-LINE, we feel that SVD is sufficient, and previous studies (Pakhomov et al., 2016; Pedersen et al., 2007) have shown that beyond 100 million tokens little performance gains can be achieved.

Surprisingly the results showed that PCA did not obtain a higher accuracy than the explicit co-occurrence vector. We believe this is a result of centering the matrix, and believe that in language absolute counts are important. When the matrix is centered, only relative counts are considered. This could create a situation where infrequently used words have distributions similar to commonly used words, adversely effecting results.

With respect to dimensionality, we found that low vector dimensionality ($d = 100$) is sufficient for CBOW and SG, but that a higher dimensionality ($d = 1500$) obtained better results with SVD. In addition, we found that although PCA is commonly used for dimensionality reduction in many fields, it does not improve results for WSD.

We found that CBOW and SG achieve approximately the same accuracy which is important because SG takes much longer to compute (our rough estimates indicate that SG takes between 5 and 9 times as long to train).

6.3 Comparison with previous work

Recently, word embeddings have been used for word sense disambiguation in the biomedical do-

---

[7]CBOW crashed due to memory constraints for $d = 1500$

Table 2: Comparison with Previous Work on MSH-WSD

Method	Medline	MIMIC-III	BioASQ	Fairview	PMC
(Pakhomov et al., 2016) (CBOW)				0.72	0.78
(Tulkens et al., 2016) (SG)	0.80	0.69	0.84		
SG	0.81				
CBOW	0.79				
SVD	0.77				
PCA	0.64				
Explicit	0.64				

main. (Tulkens et al., 2016) evaluated the skip gram model on the MSH-WSD dataset with three different sets of training data: a subset of Medline abstracts, the MIMIC-III corpus of clinical notes, and BioASQ Medline abstracts. (Pakhomov et al., 2016) evaluated CBOW on the MSH-WSD dataset using two different types of training data: clinical (clinical notes from the Fairview Health System) and biomedical (PMC corpus).

Table 2 shows the comparison between the previous works' reported results and our current results. The table shows that our skip gram and CBOW results are similar to those reported by both (Tulkens et al., 2016) (0.80 versus 0.81) and (Pakhomov et al., 2016) (0.78 versus 0.79) respectively. We believe that the small variations in accuracy are due to the difference in training data. The table also shows that SVD performs on par with previous word embeddings results.

6.4 Limitations

This study focused on comparing vector representations and the effects of dimensionality for WSD. We did not experiment with other parameters, such as window size, cut-off level, and sampling parameters. We also limited our technique to the 2-MRD WSD algorithm. This is a well known algorithm that has been shown to perform well in the past, and allows comparison between similar papers. These vector representations can be used for other WSD algorithms as well, including supervised or "distantly supervised" approaches (Sabbir et al., 2016) which may achieve higher accuracies, but are limited to pre-labeled or preprocessed datasets.

7 Conclusions and Future Work

In this study we analyzed the performance of vector representations using the dimensionality reduction techniques of word embeddings (continuous bag of words and skip-gram), singular value decomposition (SVD), and principal component analysis (PCA) on five evaluation standards (Abbrev.100, Abbrev.200, Abbrev.300, NLM-WSD, MSH-WSD). We used explicit co-occurrence vectors as the baseline. The results show that word embeddings and SVD outperform PCA and explicit representations for all datasets. PCA does not increase performance over explicit, and word embeddings are significantly different from SVD on just a single dataset (MSH-WSD). The method (CBOW versus SG) in which word embeddings are generated makes no statistically significant difference in WSD results. We also varied the dimensionality of the vectors to 100, 300, 500, 1000, and 1500. We found that the smallest dimensionality of 100 is sufficient for all vector representations except SVD. For SVD we found that increasing dimensionality does increase performance, and continued to increase the dimensionality to 2000, 2500, and 3000. Accuracy stopped increasing at 1500, indicating that a dimensionality of 1500 is sufficient for SVD.

An interesting result of this research is that SVD performs essentially on par with word embeddings. In the future we hope to increase the accuracy of SVD by modifying the co-occurrence matrix that is input into SVD to include incorporating knowledge sources (such as the UMLS) for term expansion by capturing co-occurrences with synonymous terms, and creating a UMLS concept (CUI) co-occurrence matrix. Additionally, this concept co-occurrence matrix can then be augmented to exploit the hierarchical structure of the UMLS. Using a matrix of similarities or association scores may also be interesting. Independent from how vectors are generated, we could use similarity metrics other than cosine, similar to those from (Sabbir et al., 2016) that incorporate both magnitude and orientation.

References

S. Brody and M. Lapata. 2009. Bayesian word sense induction. In *Proceedings of the 12th Conference of the European Chapter of the Association for Computational Linguistics*. Association for Computational Linguistics, pages 103–111.

S. Deerwester, S.T. Dumais, G.W. Furnas, T.K. Landauer, and R. Harshman. 1990. Indexing by latent semantic analysis. *Journal of the American society for information science* 41(6):391.

S. Ferrández, S. Roger, A. Ferrández, A. Aguilar, and P. López-Moreno. 2006. A new proposal of word sense disambiguation for nouns on a question answering system. *Advances in Natural Language Processing. Research in Computing Science* 18:83–92.

S.M. Humphrey, W.J. Rogers, H. Kilicoglu, D. Demner-Fushman, and T.C. Rindflesch. 2006. Word sense disambiguation by selecting the best semantic type based on journal descriptor indexing: Preliminary experiment. *Journal of the American Society for Information Science and Technolology* 57(1):96–113.

A. Jimeno-Yepes, B.T. McInnes, and A. Aronson. 2011. Exploiting mesh indexing in medline to generate a data set for word sense disambiguation. *BMC Bioinformatics* .

H. Liu, Y.A. Lussier, and C. Friedman. 2001. Disambiguating ambiguous biomedical terms in biomedical narrative text: an unsupervised method. *Journal of Biomedical Informatics* 34(4):249–261.

B.T. McInnes, T. Pedersen, Y. Liu, S.V. Pakhomov, and G.B Melton. 2011. Using second-order vectors in a knowledge-based method for acronym disambiguation. In *Proceedings of the Fifteenth Conference on Computational Natural Language Learning*. Association for Computational Linguistics, pages 145–153.

T. Mikolov, I. Sutskever, Kai Chen, G.S. Corrado, and J. Dean. 2013. Distributed representations of words and phrases and their compositionality. In *Advances in neural information processing systems*. pages 3111–3119.

R. Navigli, S. Faralli, A. Soroa, O. de Lacalle, and E. Agirre. 2011. Two birds with one stone: learning semantic models for text categorization and word sense disambiguation. In *Proceedings of the 20th ACM international conference on Information and knowledge management*. ACM, pages 2317–2320.

S.V. Pakhomov, G. Finley, R. McEwan, Y. Wang, and G.B. Melton. 2016. Corpus domain effects on distributional semantic modeling of medical terms. *Bioinformatics* 32(23):3635–3644.

S. Patwardhan and T. Pedersen. 2006. Using WordNet-based Context Vectors to Estimate the Semantic Relatedness of Concepts. In *Proceedings of the EACL 2006 Workshop Making Sense of Sense - Bringing Computational Linguistics and Psycholinguistics Together*. Trento, Italy, pages 1–8.

T. Pedersen. 2006. Unsupervised corpus-based methods for wsd. *Word sense disambiguation: algorithms and applications* pages 133–166.

T. Pedersen. 2010. The Effect of Different Context Representations on Word Sense Discrimination in Biomedical Texts. In *Proceedings of the 1st ACM International Health Informatics Symposium*. pages 56–65.

T. Pedersen, S. Banerjee, B.T. McInnes, S. Kohli, M. Joshi, and Y. Liu. 2011. The Ngram statistics package (Text::NSP): A flexible tool for identifying ngrams, collocations, and word associations. In *Proceedings of the Workshop on Multiword Expressions: from Parsing and Generation to the Real World*. Association for Computational Linguistics, pages 131–133.

T. Pedersen, S.V. Pakhomov, S. Patwardhan, and C.G. Chute. 2007. Measures of semantic similarity and relatedness in the biomedical domain. *Journal of Biomedical Informatics* 40(3):288–299.

L. Plaza, A.and Díaz A. Jimeno-Yepes, and A.R. Aronson. 2011. Studying the correlation between different word sense disambiguation methods and summarization effectiveness in biomedical texts. *BMC bioinformatics* 12(1):355.

J. Preiss and M. Stevenson. 2015. The effect of word sense disambiguation accuracy on literature based discovery. In *Proceedings of the ACM Ninth International Workshop on Data and Text Mining in Biomedical Informatics*. ACM, pages 1–1.

A. Purandare and T. Pedersen. 2004. Word sense discrimination by clustering contexts in vector and similarity spaces. In *Proceedings of the Conference on Natural Language Learning*. Boston, MA, pages 41–48.

AKM Sabbir, A.J. Yepes, and R. Kavuluru. 2016. Knowledge-based biomedical word sense disambiguation with neural concept embeddings and distant supervision. *arXiv preprint arXiv:1610.08557*

H. Schütze. 1998. Automatic word sense discrimination. *Computational Linguistics* 24(1):97–123.

M. Stevenson, Y. Guo, A. Al Amri, and R. Gaizauskas. 2009. Disambiguation of biomedical abbreviations. In *Proceedings of the ACL BioNLP Workshop*. pages 71–79.

M. Stevenson, Y. Guo, R. Gaizauskas, and D. Martinez. 2008. Disambiguation of biomedical text using diverse sources of information. *BMC Bioinformatics* 9(Suppl 11):11.

C. Stokoe, M.P. Oakes, and J. Tait. 2003. Word sense disambiguation in information retrieval revisited. In *Proceedings of the 26th annual international ACM SIGIR conference on Research and development in informaion retrieval*. ACM, pages 159–166.

H. Sugawara, H. Takamura, R. Sasano, and M. Okumura. 2015. Context representation with word embeddings for wsd. In *International Conference of the Pacific Association for Computational Linguistics*. Springer, pages 108–119.

S. Tulkens, S. Šuster, and W. Daelemans. 2016. Using distributed representations to disambiguate biomedical and clinical concepts. *arXiv preprint arXiv:1608.05605* .

M. Weeber, J.G. Mork, and A.R. Aronson. 2001. Developing a test collection for biomedical word sense disambiguation. In *Proceedings of the American Medical Informatics Association Symposium*. Washington, DC, pages 746–750.

Z. Zhong and H.T. Ng. 2010. It makes sense: A wide-coverage word sense disambiguation system for free text. In *Proceedings of the ACL 2010 System Demonstrations*. Association for Computational Linguistics, pages 78–83.

Investigating the Documentation of Electronic Cigarette Use in the Veteran Affairs Electronic Health Record: A Pilot Study

Danielle L. Mowery
Dept. of Biomedical Informatics
University of Utah
Salt Lake City, UT 84108, USA
danielle.mowery@utah.edu

Brett R. South
Dept. of Biomedical Informatics
University of Utah
Salt Lake City, UT 84108, USA
Brett.South@hsc.utah.edu

Shu-Hong Zhu
Dept. of Family Medicine & Public Health
University of California San Diego
La Jolla, CA 92093, USA
szhu@ucsd.edu

Olga V. Patterson
Division of Epidemiology
University of Utah
Salt Lake City, UT 84132, USA
olga.patterson@utah.edu

Mike Conway
Dept. of Biomedical Informatics
University of Utah
Salt Lake City, UT 84108, USA
mike.conway@utah.edu

Abstract

In this paper, we present pilot work on characterising the documentation of electronic cigarettes (e-cigarettes) in the United States Veterans Administration Electronic Health Record. The Veterans Health Administration is the largest health care system in the United States with 1,233 health care facilities nationwide, serving 8.9 million veterans per year. We identified a random sample of 2000 Veterans Administration patients, coded as current tobacco users, from 2008 to 2014. Using simple keyword matching techniques combined with qualitative analysis, we investigated the prevalence and distribution of e-cigarette terms in these clinical notes, discovering that for current smokers, 11.9% of patient records contain an e-cigarette related term.

1 Introduction

Electronic cigarettes — e-cigarettes — were developed in China in the early 2000s and first introduced to the US market in 2007. Once established in the US, the product experienced explosive growth, with the number of e-cigarette users doubling every year between 2008 and 2012 (Grana et al., 2014). In 2012 it was estimated that 75% of US adults had heard of e-cigarettes, and 8.1% had tried them (Zhu et al., 2013). By 2014, the proportion of adult Americans who had tried e-cigarettes increased to 12.6% (Schoenborn and Gindi, 2015).

Public health practitioners, government regulatory authorities, professional associations, the media, as well as individual clinicians and health workers are divided as to whether e-cigarettes represent an exciting new smoking cessation opportunity (Green et al., 2016; McNeill et al., 2015; Caponnetto et al., 2013) or are an untested, potentially dangerous technology that risks undermining recent successes in "denormalising" smoking (Choi et al., 2012; Etter et al., 2011; Gornall, 2015; U.S. Department of Health and Human Services, 2016; Department of Health and Human Services, 2014).

Currently, little is known about how clinicians "on-the-ground" advise patients who use, or are considering using, e-cigarettes. While Winden et al. (2015) has gone some way to describing e-cigarette Electronic Health Record (EHR) documentation behaviour in the context of a medical system in Vermont, national patterns in e-cigarette documentation have not been explored. In this paper, we present pilot work on characterising the documentation of e-cigarettes in the United States Veterans Administration Electronic Health Record. The Veterans Health Administration (VA) is the largest health care system in the

282

Proceedings of the BioNLP 2017 workshop, pages 282–286,
Vancouver, Canada, August 4, 2017. ©2017 Association for Computational Linguistics

United States with 1,233 health care facilities nationwide, serving 8.9 million veterans per year. VA EHR data provides the opportunity for nationwide population-health surveillance of e-cigarette use.

The remainder of this document consists of five sections. Following a discussion of related work in Section 2, Section 3 describes both our cohort selection procedure, and our method of identifying e-cigarette documentation in clinical notes, while Sections 4 and 5 present the results of out analysis, and some discussion of those results. The final section outlines some broad conclusions.

2 Background

The VA collects data about patient smoking history and status using several approaches at the time of a patient encounter. Most patient clinical encounters have an associated *health factor* (i.e. semi-structured data that describes patient smoking status or smoking history (Barnett et al., 2014)). In addition, if the veteran has received dental care, the VA dental data contains descriptions of patient smoking status as a coded database field. However, neither of these data sources can be used to define what type of tobacco the patient uses and more specifically, if the patient uses e-cigarettes. This information is only found embedded in clinical text.

Given the rapid rise in popularity of e-cigarettes, and the lack of adequate public health surveillance systems currently focussing on these novel tobacco products, various methods and data sources have been used to understand changes in e-cigarette prevalence and usage patterns, including analysing search engine queries relevant to e-cigarettes (Ayers et al., 2011), mining social media data (Myslín et al., 2013; Chen et al., 2015), and — the focus of this paper — analysing EHR data for e-cigarette related documentation (Winden et al., 2015).

Previous work on smoking status identification in the EHR context has focussed on *structured data* (e.g. Wiley et al. (2013) used ICD-9 codes successfully to identify current smokers in the Vanderbilt Medical Center EHR), *semi-structured* data (e.g. McGinnis et al. (2011) used VA EHR *health factors* to reliably identify current smokers), and *unstructured* data (e.g. Clark et al. (2008); Savova et al. (2008); Da Silva et al. (2011) applied natural language processing methods to EHR clinical notes to identify smoking status).

EHR corpus analysis has been the focus of several research efforts in the tobacco domain. For example, Chen et al. (2014) investigated the documentation of general tobacco use in clinical notes from Vermont's Fletcher Allen Health Center, discovering that free-text clinical notes are frequently used to document amount of tobacco used, tobacco use frequency, and start and end dates of tobacco use (i.e. important clinical information that is difficult to represent with structured data). In follow-up work focussing specifically on e-cigarettes rather than general tobacco use, Winden et al. (2015), again using EHR data from Fletcher Allen Health Center, developed a sophisticated annotation scheme to code e-cigarette documentation, with categories including *dose, device type, frequency*, and *use for smoking cessation*. One result of particular note garnered from this research is the observation that less than 1% of patients had e-cigarette mentions in their note.

In this pilot study, our aim is to complete an initial corpus analysis of VA patient record data with the goal of quantifying the frequency with which e-cigarette usage is documented within the VA patient record.

3 Materials and Methods

We queried the VA dental record data found in the VA Corporate Data Warehouse to identify a national cohort of all Veterans Affairs patients with a coded history of current (or current and past) smoking between the years 2008-2014. Dental records were chosen as a data source as they are believed to be the most reliable indicators of smoking status in the VA context. From these data we identified 87,392 unique patients (77,491 current smokers, 9,901 current and past smokers). We then selected a random sample of 2,000 patients and extracted their associated clinical notes yielding 154,991 clinical notes. Note types include progress notes, consultation notes, consent documents, instructions, triage notes, history and physical notes, amongst others.

Based on an iterative process of corpus exploration, along with insights gleaned from previous work on e-cigarette related natural language processing (Myslín et al., 2013; Winden et al., 2015), we identified twenty e-cigarette related terms (listed in Table 1), and — using these terms — performed a keyword search within the patient clinical notes. We reviewed each e-cigarette term instance in its context to ascertain whether the e-cigarette term instance actually referred to e-cigarette usage.

We report the precision of each e-cigarette term defined as the proportion of term match instances actually referencing e-cigarette usage of all term matches.

4 Results

Term	Total	TP	FP	Precision
ecig	14	14	0	100.0
electronic cig	10	10	0	100.0
liquid nicotine	5	5	0	100.0
Ecig	4	4	0	100.0
E CIG	4	4	0	100.0
electric cig	1	1	0	100.0
ecigarette	1	1	0	100.0
Ecigarette	1	1	0	100.0
E Cig	1	1	0	100.0
vape	19	18	1	0.947
e cig	7	6	1	0.857
Vape	6	5	1	0.833
VAPOR	9	7	2	0.778
VAPE	2	1	1	0.500
vapor	241	81	160	0.336
vaporizer	192	36	156	0.188
Vapor	73	4	69	0.055
Vaporizer	3	0	3	0.000
VAPORIZER	3	0	3	0.000
ECIG	5	0	5	0.000
Total	601	199	402	–

Table 1:   Proposed e-cigarette related terms ranked by precision

We analysed notes from 2,000 VA patients. From these notes, we observed 238 patients (11.9%) with one or more e-cigarette mentions within their notes (see **Figure 1**). In total, there were 601 mentions, with 436 notes containing more than one mention. Of these 601 mentions, 199 (33.1%) mentions described true e-cigarette usage (**Table 1**) as ascertained by manual inspection. The most frequent e-cigarette term matches included variants of the term *vapor* (*vapor*: 241, *vaporizer*: 192, *Vapor*: 73). These terms were also the most frequent sources of false positives (*vapor*: 160, *vaporizer*: 156, and *Vapor*: 69). Thirteen of the twenty terms yielded precision scores greater than 0.500. Of these high-precision terms,

the most prevalent terms included *vape*: 19, *ecig*: 14, and *electronic cig*: 10.

5 Discussion

We observed a variety of linguistic contexts describing e-cigarette usage. Patients report use of e-cigarettes with other tobacco products (e.g., "smokes 10 tobacco cigs per day and uses *vape*"). Similar to tobacco cessation, clinicians report providing encouragement and counselling for patients to stop e-cigarette use. Patients often contemplate e-cigarettes as an alternative to tobacco usage (e.g., "thinking about switching to *ecig*") or as an approach to tobacco cessation (e.g., "uses *nicotine vaporizer* and hasn't smoked tobacco in 6 mos"). This was not a surprising finding given that, according to the Centers for Disease Control, "among current cigarette smokers who had tried to quit smoking in the past year, more than one-half had ever tried an e-cigarette and 20.3% were current e-cigarette users" (Schoenborn and Gindi, 2015). Patients reported differing experiences of using e-cigarettes as a smoking cessation aid, with one patient stating directly that e-cigarettes were an ineffective tool in his struggle to quit smoking. Consistent with current uncertainty regarding the safety of e-cigarettes and their utility as a smoking cessation aid, not all clinicians support the use of e-cigarettes as a safe alternative to tobacco usage (e.g., "I do not recommend *ecig/vapor*").

Analogous to the "packs-per-day" metric used by clinicians to document volume of combustable tobacco use, patients report their frequency of e-cigarette use in volume over time (e.g., "6mg/day"). E-cigarette usage goals are often set by both clinicians ("reducing consumption from 9 grams to 3 with goal of quitting") and patients ("using *e cig* and cutting back by half") alike. One clinician reported a patient's use of e-cigarettes with "no side effects with current meds" suggesting that clinicians are aware that known side effects with medication use is a possibility.

Although most of the twenty e-cigarette terms used in this study yielded precision scores greater than 0.500, we also observed a substantial proportion of term matches that did not indicate actual e-cigarette usage. Many false positives occurred due to the ambiguous nature of the word *vaporizer* and its variants. For example, the domestic use of a vaporizer to increase room humidity, the treatment of patients with over-the-counter sinus

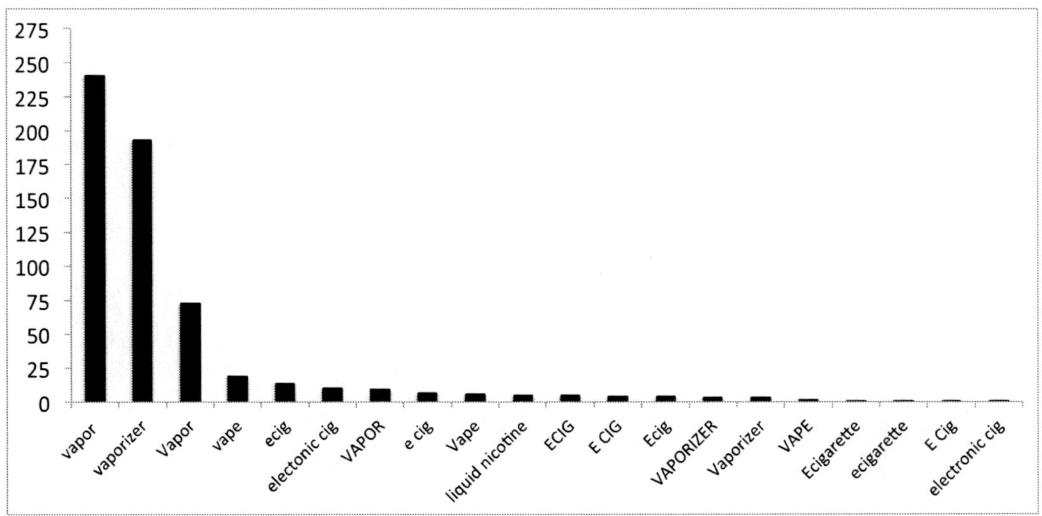

Figure 1: Frequency of e-cigarette terms in the clinical notes of a cohort of 2,000 VA smokers

relief ("Vicks *vapor* rub"), or the use of vaporisers to ingest medical marijuana ("MJ via *vaporizer*"). These non-relevant mentions of e-cigarette related terms are most frequently found in anaesthesia consent notes (n=185 mentions).

From notes containing matched e-cigarette variants, we discovered several co-occurring terms which could improve the term's precision, with examples including *nicotine vaporizer*, *vapor-nicotine*, *vapor cig*, *vapor cigarettes*, *vapor pens*, *vapor cigarets*, *methonol vapor*, and *vapor nicotine*.

The pilot work described in this short paper has several limitations. First, our list of e-cigarette related keywords was limited to twenty. As indicated above, there may well be additional high precision e-cigarette related terms that we did not use in this work. Second, unlike Winden et al. (2015) we have not conducted a large scale annotation effort or mapped to an annotation scheme. Finally, while the VA is the largest integrated medical system in the United States, and the only nationwide system, VA patients are not necessarily representative of the general population. It is particularly important to note that approximately 92% of veterans are male (National Center for Veterans Analysis and Statistics, 2013).

6 Conclusion

In conclusion, we have demonstrated that for current smokers, e-cigarette terms are present in 11.9% (238) of VA patient records. Of this 11.9%

of patients, it is estimated that around two thirds of e-cigarette mentions are false positives, suggesting that around 4% of smokers have e-cigarette use documented in their clinical notes.

Acknowledgments

This research was supported by the National Library of Medicine of the National Institutes of Health under award numbers R00LM011393 & K99LM011393, and University of Utah funds.

Ethics Statement

This study was approved by the University of Utah Institutional Review Board (IRB_00088382).

References

John W. Ayers, Kurt M. Ribisl, and John S. Brownstein. 2011. Tracking the rise in popularity of electronic nicotine delivery systems (electronic cigarettes) using search query surveillance. *American Journal of Preventive Medicine* 40(4):448–453.

Paul Barnett, Adam Chow, and Nicole Flores. 2014. Using health factors data for VA health service research. Technical Report 28, Health Economic Resource Center. https://www.herc.research.va.gov/files/RPRT_768.pdf.

Pasquale Caponnetto, Cristina Russo, Cosimo Bruno, Angela Alamo, Maria Amaradio, and Ricardo Polosa. 2013. Electronic cigarette: a possible substitute for cigarette dependence. *Monaldi Arch Chest Dis* 79(1):12–9.

Annie T Chen, Shu-Hong Zhu, and Mike Conway. 2015. What online communities can tell us about electronic cigarettes and hookah use: A study using text mining and visualization techniques. *J Med Internet Res* 17(9):e220.

Elizabeth S Chen, Elizabeth W Carter, Indra Neil Sarkar, Tamara J Winden, and Genevieve B Melton. 2014. Examining the use, contents, and quality of free-text tobacco use documentation in the electronic health record. *AMIA Annu Symp Proc* 2014:366–74.

Kelvin Choi, Lindsey Fabian, Neli Mottey, Amanda Corbett, and Jean Forster. 2012. Young adults' favorable perceptions of snus, dissolvable tobacco products, and electronic cigarettes: findings from a focus group study. *Am J Public Health* 102(11):2088–93.

Cheryl Clark, Kathleen Good, Lesley Jezierny, Melissa Macpherson, Brian Wilson, and Urszula Chajewska. 2008. Identifying smokers with a medical extraction system. *J Am Med Inform Assoc* 15(1):36–9.

Lalindra Da Silva, Thomas Ginter, Tyler Forbus, Neil Nokes, Brian Fay, Ted Mikuls, Grant Cannon, and Scott DuVall. 2011. Extraction and quantification of pack-years and classification of smoker information in semi-structured medical records. In *Proceedings of the 28th International Conference on Machine Learning*, Bellevue, WA.

Department of Health and Human Services. 2014. The health consequences of smoking – 50 years of progress. Technical report, Surgeon General.

Jean-François Etter, Chris Bullen, Andreas D Flouris, Murray Laugesen, and Thomas Eissenberg. 2011. Electronic nicotine delivery systems: a research agenda. *Tob Control* 20(3):243–8.

Jonathan Gornall. 2015. Public Health England's troubled trail. *BMJ* 351:h5826.

Rachel Grana, Neal Benowitz, and Stanton A Glantz. 2014. E-cigarettes: a scientific review. *Circulation* 129(19):1972–86.

Sharon H Green, Ronald Bayer, and Amy L Fairchild. 2016. Evidence, policy, and e-cigarettes–will England reframe the debate? *N Engl J Med* 374(14):1301–3.

Kathleen A McGinnis, Cynthia A Brandt, Melissa Skanderson, Amy C Justice, Shahida Shahrir, Adeel A Butt, Sheldon T Brown, Matthew S Freiberg, Cynthia L Gibert, Matthew Bidwell Goetz, Joon Woo Kim, Margaret A Pisani, David Rimland, Maria C Rodriguez-Barradas, Jason J Sico, Hilary A Tindle, and Kristina Crothers. 2011. Validating smoking data from the Veteran's Affairs health factors dataset, an electronic data source. *Nicotine Tob Res* 13(12):1233–9.

Ann McNeill, Leonie Brose, Robert Calder, and Sara Hitchman. 2015. E-cigarettes: an evidence update - report commissioned by Public Health England. Technical report, Public Health England.

Mark Myslín, Shu-Hong Zhu, Wendy Chapman, and Mike Conway. 2013. Using Twitter to examine smoking behavior and perceptions of emerging tobacco products. *J Med Internet Res* 15(8):e174.

National Center for Veterans Analysis and Statistics. 2013. Women veteran profile. Technical report, United States Department of Veterans Affairs. https://www.va.gov/vetdata/docs/SpecialReports/ Women_Veteran_Profile5.pdf.

Guergana K Savova, Philip V Ogren, Patrick H Duffy, James D Buntrock, and Christopher G Chute. 2008. Mayo clinic NLP system for patient smoking status identification. *J Am Med Inform Assoc* 15(1):25–8.

Charlotte Schoenborn and Renee Gindi. 2015. Electronic cigarette use among adults: United States, 2014. Technical Report 217, NCHS. https://www.cdc.gov/nchs/data/databriefs/db217.pdf.

U.S. Department of Health and Human Services. 2016. E-cigarette use among youth and young adults: a report of the Surgeon General. Technical report.

Laura K Wiley, Anushi Shah, Hua Xu, and William S Bush. 2013. ICD-9 tobacco use codes are effective identifiers of smoking status. *J Am Med Inform Assoc* 20(4):652–8.

Tamara J Winden, Elizabeth S Chen, Yan Wang, Indra Neil Sarkar, Elizabeth W Carter, and Genevieve B Melton. 2015. Towards the standardized documentation of e-cigarette use in the electronic health record for population health surveillance and research. *AMIA Jt Summits Transl Sci Proc* 2015:199–203.

Shu-Hong Zhu, Anthony Gamst, Madeleine Lee, Sharon Cummins, Lu Yin, and Leslie Zoref. 2013. The use and perception of electronic cigarettes and snus among the US population. *PloS One* 8(10):e79332

Automated Preamble Detection in Dictated Medical Reports

Wael Salloum, Greg Finley, Erik Edwards, Mark Miller, and David Suendermann-Oeft
EMR.AI Inc
90 New Montgomery St #400
San Francisco, CA 94105, USA
`david@emr.ai`

Abstract

Dictated medical reports very often feature a preamble containing metainformation about the report such as patient and physician names, location and name of the clinic, date of procedure, and so on. In the medical transcription process, the preamble is usually omitted from the final report, as it contains information already available in the electronic medical record. We present a method which is able to automatically identify preambles in medical dictations. The method makes use of state-of-the-art NLP techniques including word embeddings and Bi-LSTMs and achieves preamble detection performance superior to humans.

1 Introduction

For decades, medical dictation and transcription has been used as a convenient and cost-effective way to document patient-physician encounters and procedures and bring reports into a form which can be stored in an electronic medical record (EMR) system, formatted as an out-patient letter, etc (Häyrinen et al., 2008; Johnson et al., 2008; Meystre et al., 2008; Holroyd-Leduc et al., 2011; Kalra et al., 2012; Logan, 2012; Hyppönen et al., 2014; Campanella et al., 2015; Moreno-Conde et al., 2015; Alkureishi et al., 2016; Ford et al., 2016). While dictated speech has traditionally been transcribed by humans (such as clinical assistants or professional transcription personnel), sometimes in multiple stages, it is common nowadays for speech recognition technology to be deployed in the *first stage* to increase transcription speed and cope with the enormous amount of dictated episodes in the clinical context (Hammana et al., 2015; Hodgson and Coiera, 2016; Edwards et al., 2017).

In its purest form, a speech recognizer transforms spoken into written words, as exemplified in Figure 1. Obviously, this raw output will have to undergo multiple transformation steps to format it in a way it can be stored in an EMR or sent out as a letter to the patient, including: formatting numbers, dates, units, etc.; punctuation restoration (Salloum et al., 2017b); and processing physician normals.

Furthermore, dictated reports often contain metadata in a preamble containing information not intended to be copied into the letter, such as patient and physician names, location and name of the clinic, date of procedure, and so on. Rather, the metadata serves the sole purpose of enabling realigning dictations with a particular record or file, in case this alignment is not otherwise possible (usually, metadata in medical transcription systems is automatically retrieved from the EMR system and inserted into the outpatient letter). See Figure 2 for the same text sample as Figure 1 with the preamble highlighted and the above post-processing rules applied.

In a *second stage*, medical transcriptionists take the speech recognizer output and perform a post-editing exercise and quality check before entering the final report into the EMR or sending it off as an outpatient letter. This stage usually involves the removal of metadata, i.e. the preamble, from the dictation's main text body. To facilitate this procedure, this paper explores techniques to automatically mark preambles.

It is worth noting that the accurate detection of preambles in dictated reports is a non-trivial task, even for humans. Clinical dictations may (a) contain metadata at multiple places throughout the report (see Figure 3 for an example), (b) or no such data at all, (c) feature sentences convolving metadata and general narrative, or (d) have grammati-

Proceedings of the BioNLP 2017 workshop, pages 287–295,
Vancouver, Canada, August 4, 2017. ©2017 Association for Computational Linguistics

```
this is doctor mike miller dictating
a maximum medical improvement slash
impairment rating evaluation for
john j o h n doe d o e social one
two three four five six seven eight
nine service i_d one two three four
five six seven eight nine service
date august eight two thousand and
sixteen subjective and treatment to
date the examinee is a thirty-nine
year-old golf course maintenance
worker with the apache harding park
who was injured on eight seven two
thousand sixteen
```

Figure 1: Raw output of a medical speech recognizer.

This is Dr Mike Miller dictating a Maximum Medical Improvement/Impairment Rating Evaluation for John Doe.
SSN: 123-45-6789
Service ID: 123 456 789
Service Date: 08/08/16

Subjective and Treatment:
To date, the examinee is a 39 year-old golf course maintenance worker with the Apache Harding Park who was injured on 08/07/16.

Figure 2: Output of post-processor with preamble highlighted.

cal inaccuracies and lack overall structure caused by the spontaneous nature of dictated speech, including the total absence of punctuations. To systematically quantify the task's complexity, we also determined the human baseline performance of detecting the preamble in clinical dictation.

This paper is structured as follows: After discussing related work in Section 2, we describe the corpus and determine the human baseline in Section 3.3. Section 4 provides details on the techniques we used for the automated detection of preambles, followed by evaluation results and discussion in Section 5. We conclude the paper and provide an outlook on future work in Section 6.

This is Dr Mike Miller.
The patient is a baking associate over at Backwerk.
Today's date is 03/10/2016.
The patient noted he strained his back while he was helping his mother move some household items.

Figure 3: Example of a report intertwining preamble and main body. Physician name and date of the visit are commonly considered preamble, whereas the patient's profession and employer are not. When spontaneously dictating, physicians sometimes remember to mention preamble statements only after they have already started the main body narrative, such as the date of visit in this example.

2 Related Work

To our knowledge, the problem of automated preamble detection in medical transcriptions has not been addressed before. That said, we do build upon classic methods in NLP: specifically, our system is a generalization of sequence tagging, which has seen use in other tasks such as part-of-speech tagging, shallow parsing or chunking, named entity recognition, and semantic role labeling. Traditionally, sequential tagging has been handled using either generative methods, such as hidden Markov models (Kupiec, 1992), or sequence-based discriminative methods, such as conditional random fields (Lafferty et al., 2001; Sha and Pereira, 2003).

More modern approaches have shown performance gains and increased generalizability with neural networks (NNs). Collobert and colleagues (Collobert and Weston, 2008; Collobert et al., 2011) successfully apply NNs to several sequential NLP tasks without the need for separate feature engineering for each task. Their networks feature concatenated windowed word vectors as inputs or, in the case of sentence-level tasks, a convolutional architecture to allow interaction over the entire sentence.

However, this approach still does not cleanly capture nonlocal information. In recent years, recurrent NN architectures, often using gated recurrent (Cho et al., 2014; Tang et al., 2016; Dey and Salem, 2017) or long short-term memory (LSTM) units (Hochreiter and Schmidhuber, 1997; Hammerton, 2003), have been applied with excellent results to various sequence labeling problems. Many linguistic problems feature dependencies at longer distances, which LSTMs are better able to capture than convolutional or plain recurrent approaches. Bidirection LSTM (Bi-LSTM) networks (Graves and Schmidhuber, 2005; Graves et al., 2005; Wöllmer et al., 2010) also use future context, and recent work has shown advantages of Bi-LSTM networks for sequence labeling and named entity recognition (Huang et al., 2015; Chiu and Nichols, 2015; Wang et al., 2015; Lample et al., 2016; Ma and Hovy, 2016; Plank et al., 2016).

In some approaches, tag labels from NN outputs are combined in a final step, such as conditional random fields, especially when the goal is to apply a single label to a continuous sequence of tags. Our architecture, as described in Section 4, also utilizes a post-tagging step to define a clear preamble endpoint.

3 Corpus and Inter-Annotator Agreement

In this section we report on the corpus used for this study, the methodology for computing inter-annotator agreement, and we analyze the preamble split positions in more detail.

3.1 The Data

A total of 10,517 dictated medical reports were transcribed by a team of professional medical transcriptionists (MTs) organized in a private crowd as described in (Salloum et al., 2017a). The produced transcriptions were raw, i.e., only lowercase alphabetic characters, dash, and underscore were permitted, resulting in output as shown in Figure 1.

In a separate round, we sent these transcribed reports to a private crowd of MTs to acquire a total of five annotation jobs per file. Since we cannot specify all types of information that are expected to be found in preambles ahead of time, we let the MTs, who are well experienced in transcribing medical dictations, determine the exact split position that, in their opinion, separated preamble text from main report. This approach allows us to harvest the wisdom of the crowd and define what they agree on as the ground truth, which we can then learn automatically.

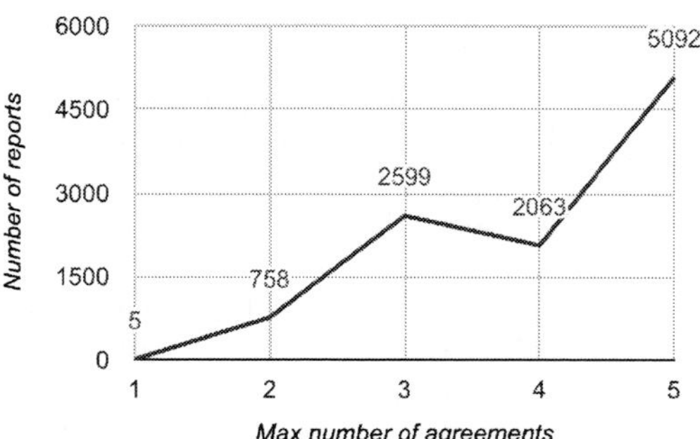

Figure 4: Histogram of the maximum number of exact agreements obtained for the annotated reports

3.2 Inter-Annotator Agreement

In order to establish a corpus with reliable labels which subsequently can be used to measure human accuracy and train and test the automatic preamble detector, we defined a gold-standard annotation to be one where at least three annotators agreed on the exact split between preamble and main body. Figure 4 shows a histogram of the frequency of number of agreements. For example, out of the 10,517 reports, 5,092 have all annotators agree on the split position while only 5 reports have 5 different annotations. By reducing the corpus to only those reports with at least three annotators in agreement about the split position, we ended up with a total of 9,754 reports, or 92.75% of the original body of data. 4.4% of the reports were *not* annotated by all five annotators, constituting the majority of omitted files. The lack of annotations is presumably due to annotators not being sure how to split, or due to oversight. Missing annotations makes it harder for such files to match the three-agreement threshold.

Overall, it became clear that the lack of guidelines on specific types of phenomena featured in the preamble, such as including or excluding a patient's employer, led to disagreements that ultimately caused the exclusion of reports—although note that nearly half of included reports do have at least one dissenting opinion. This analysis is specifically helpful for designing new guidelines for the next round of annotations, which will lead to cleaner data fed to our system.

We split the 9,754 reports randomly into training and test sets. Table 1 shows some statistics about the data split. The test set out-of-vocabulary

Set	# reports	# tokens	# types
Training Set	8,711	3,335,588	30,707
Test Set	1,043	415,491	13,507

Table 1: Corpus statistics.

(OOV) rate against the training set is 10.76% (1,454 types).

In order to quantify the inter-annotator agreement, we compared each annotator against the majority vote, resulting in the following annotator split accuracy scores: 83.22%, 86.09%, 86.09%, 86.58%, 88.20%. The average inter-annotator agreement score, 86.04%, will serve as standard of comparison in this paper.

3.3 Analysis of Preamble Split Positions

As motivated in the introduction, the use of preambles in medical dictations is not very consistent. E.g., a good amount of dictations do not contain a preamble at all, whereas others contain multiple, even others convolve preamble and main text so much that it is very hard to determine the exact split position. In this work, annotators were required to provide a single split tag at the location were they found the boundary to be most appropriate. If annotators did not find any preamble in the dictation, the tag was placed in front of the first token of the dictation.

Figure 5 displays a histogram of the split position in reports. The vast majority of split positions are below 100 tokens into the dictation (compared to the average total token count for the dictations in our corpus of 385; see Table 1 for exact statistics). There are 319 reports (3.3%) with no preamble and, hence, split position 0.

If we define the problem as a sequence tagging problem where every token in a preamble is tagged with I-P (Inside Preamble) and every token in the main report is tagged with I-M (Inside Main), we get the histogram in Figure 6.

4 Approach

Although the training data contains 3.3 M tokens, the evaluation is at the level of reports, of which we have only 8.7 K examples. We determined from preliminary experiments that this limited amount of examples is not enough to train an end-to-end neural network to predict the split position. Therefore, we use a two-step approach to preamble detection:

1. A sequence **tagger** that labels every word in the input sequence with one of two tags: I-P (Inside Preamble) and I-M (Inside Main). This tagger leverages the large number of tokens in our data, as opposed to the small number of example reports, which leads to near perfect tagging accuracy.

2. A report **splitter** that determines heuristically at what position to split the tagged report into preamble and main. This splitter attempts to correct the tagger's mistakes.

4.1 The Tagging Model

Like other recent work, our model is based on LSTM NNs. We experimented with both unidirectional and bidirectional networks. The stack consists of an embedding layer (see Section 4.3 for details), a (Bi-)LSTM layer, and a time-distributed dense layer with softmax activation (illustrated in Figure 7). For the present study, we used Keras with TensorFlow backend (Chollet, 2015; Abadi et al., 2016; Chollet, 2017). We applied a categorical cross-entropy cost function and Adam optimization (Kingma and Ba, 2014).

In addition to word meaning and context, the analysis we did in Section 3.3 motivates that the correct prediction of tags depends on the location of words in the report as well (Figure 5 and Figure 6). Therefore, instead of tagging the input sequence using a sliding window like many taggers do, we have a fixed size input to the network comprising the first 512 tokens of the report. Words after this limit are truncated. We add padding for reports with less than 512 tokens. Informal experiments showed that varying the window length to 256 or 1024 tokens deteriorated preamble detection performance.

Since the data we have is limited in size, we use word vectors pretrained on large amounts of unlabeled text collected from medical reports and medical dictation transcriptions. This transfer learning technique is often used in deep learning approaches to NLP since the vectors learned from massive amounts of unlabeled text can be transfered to another NLP task where labeled data is limited and might not be enough to train the embedding layer.

4.2 The Heuristic Splitter

The training examples of the tagging model always have preamble tags (I-P) preceding main re-

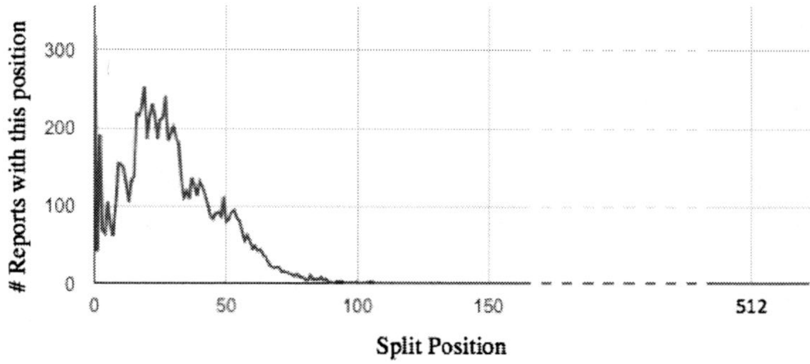

Split Position

Figure 5: A histogram of the split position in our training set. A point of interest is the split at position 0, which indicates that 319 reports have no preamble text. The longest preamble text ends at position 131, after that the curve stays on 0.

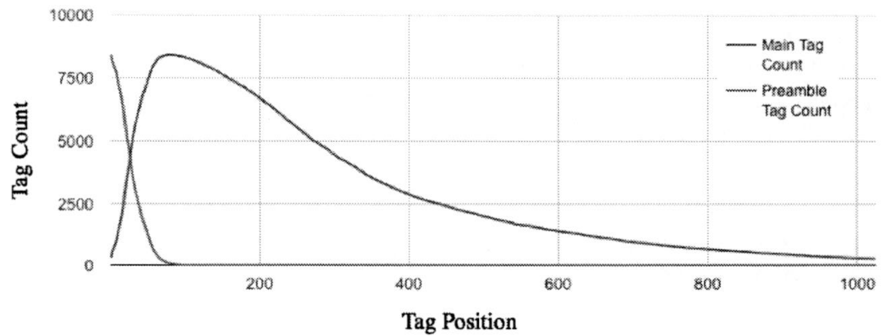

Tag Position

Figure 6: Frequency of the two tags at positions in the first 1000 words of reports. The last preamble tag, I-P, appears at position 131, after that the red curve stays on 0. The main tag, I-M, starts at position 1 with a value 319, and goes up as the report grows longer. The main tag curve then falls down as longer reports are less frequent than shorter ones.

port tags (I-M). Nevertheless, the neural network sometimes produces mixed sequences of I-P and I-M. An example of such output starts with I-P, switches briefly to I-M, then back to I-P, and then to I-M. This situation requires another system to find the exact position in which we need to split preamble from main report. We use simple heuristics to determine the split position as explained in Algorithm 1.

The algorithm looks for concentrations of preamble and main tag sequences. It initializes the split position it is trying to predict, $splitPos$, and a sequence counter, $counter$, to 0. While scanning the tagged sequence, it increases $counter$ if it sees an I-P (Line 6) and decreases it if it sees an I-M (Line 11). $counter > 0$ means that we have seen a long enough I-P tag sequence since the last I-M tag to consider the text so far to be preamble and the previous I-M tags to be errors. However,

the next I-M tag will set restart the counter (Line 9) and set $splitPos$ to the previous position (Line 10). Lines 12-13 handle the edge case where the sequence ends while $counter > 0$, which means that the whole report is preamble.

It is important to point out that our splitter is biased, by design, to vote in favor of including more words in main (i.e., shorter preambles). The reason for this bias is that in applications where the main text is more valued than preamble (e.g., to create a formatted note), we take the safe option not to omit content words.

4.3 Pretrained Word Embeddings

Word embeddings were trained offline using the original implementation of the word2vec package (Mikolov et al., 2013b,a). All vectors are 200 dimensions and trained using 15 iterations of the continuous bag-of-words model over a window of 8 words, with no word count minimum.

291

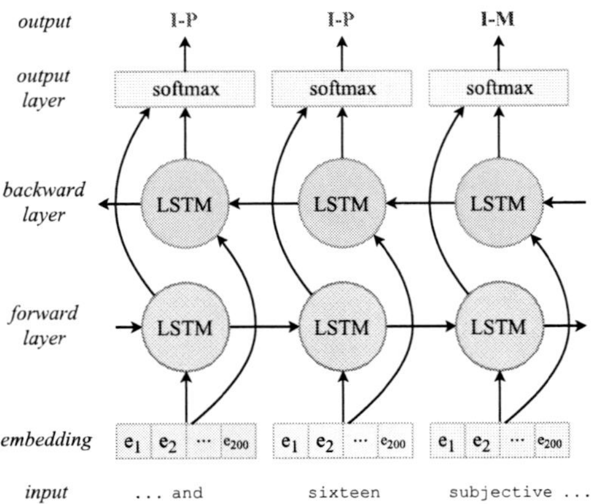

Figure 7: The NN stack using Bi-LSTM. An embedding at each word step is fed into forward and backward LSTM layers, which are fully connected to a softmax-activated output layer. (For the unidirectional LSTM, the backward layer is omitted.)

We experimented with three sets of embeddings, each trained on cumulatively more text:

- "SplitEmb" was trained on the same transcriptions as the tagging model (plus those on which only two annotators agreed on the split), with the insertion of a line break at the split between the preamble and main text. This break causes word2vec not to train on co-occurrences of tokens on either side of the split, hypothetically leading to decreased similarity between words typically found inside and outside of preambles. (3.7 M tokens total.)

- "SplitTransEmb" added more transcribed medical dictations which were not part of the preamble-annotated set. (8.3 M tokens.)

- "SplitTransRepEmb" added formatted medical reports processed to look like transcriptions—numerals spelled out, punctuation removed, etc. (60 M tokens.)

5   Evaluation

As a first sanity check, we measured the preamble tagging accuracy on the token level. In other words, we determined how many of the tokens in the test set were correctly tagged as being either part of the preamble or the main body. In this task, our system achieved an accuracy of 99.80%, with only 816 mismatches among the total of 415,491 tokens in the test set.

As motivated in Section 3.2, the ultimate performance measure we are using counts how many perfect splits the preamble detector found, i.e. the split accuracy. Table 2 shows detailed results of the systems introduced in Section 4, comparing all pre-trained word embedding models across two embedding schemes (trainable vs. frozen) and for both Uni- and Bi-LSTM. The best overall system uses Bi-LSTMs and frozen embeddings, performing at 89.84% split accuracy. In comparison, as calculated earlier, the human split accuracy on our corpus was determined to be 86.04% which constitutes a statistically significant difference. The fact that our automated preamble detection system outperforms humans demonstrates the strength of the presented methods in exploiting synergistic effects across a crowd of annotators.

We were also interested in the effectiveness of the heuristic splitter introduced in Section 4.2. We therefore determined results for both Uni-LSTM (75.74%) and Bi-LSTM (87.44%) when leaving out the splitter. Compared to the individual best results for Uni- and Bi-LSTMs in Table 2, this constitutes a difference of 8.25% and 2.4%, demonstrating a clear positive impact of the heuristic splitter.

Algorithm 1 The Heuristic Splitter.

1: $splitPos \leftarrow 0$ // predicted split position.
2: $counter \leftarrow 0$ // sequence counter.

3: **for** $pos := 1 \rightarrow length(tags)$ **do**
4:     **switch** $tags[pos]$ **do**

        // ... padding is ignored.

5:         **case** I-P
6:           $counter$++
7:         **case** I-M
8:           **if** $counter > 0$ **then**
9:             $counter \leftarrow 0$ // reset
10:             $splitPos \leftarrow pos - 1$
11:           $counter$--
12: **if** $counter > 0$ **then**
13:     **return** $length(predictedTags)$
14: **else**
15:     **return** $splitPos$

	Test OOVs (first 512)		LSTM				Bi-LSTM			
			Trainable Emb.		Frozen Emb.		Trainable Emb.		Frozen Emb.	
	# types	*rate*	*# PS*	*%*	*# PS*	*%*	*# PS*	*%*	*# PS*	*%*
Fully-trained Emb.	n/a	n/a	754	72.29%	n/a	n/a	863	82.74%	n/a	n/a
SplitEmb	1133	8.59%	809	77.56%	839	80.44%	911	87.34%	916	87.82%
SplitTransEmb	905	6.86%	809	77.56%	846	81.11%	899	86.19%	**937**	**89.84%**
SplitTransRepEmb	230	1.74%	798	76.51%	876	83.99%	907	86.96%	925	88.69%

Table 2: Evaluation of our LSTM and Bi-LSTM models across all pretrained word embedding models. The first column shows the different pretrained word embedding models we used. The "Test OOVs" column shows the OOV count and rate against each pretrained embedding model. This only includes types in the first 512 words of the report (that are passed to the NN) which contain 13,186 types out of the 13,507. Columns with title "Trainable Emb." report results where backpropagation is allowed to update the pretrained embedding layer after it is loaded, while columns with title "Frozen Emb." does not allow such updates. # PS is the number of Perfect Splits.

6 Conclusion and Future Work

The work presented in this paper shows yet again that careful design and execution of state-of-the-art NLP techniques when applied to traditionally manual tasks (in this case, the detection of preambles in medical dictations) can approach or even surpass human performance. We assume that the presented NLP stack with Bi-LSTMs makes use of the wisdom of the crowd: it exploits the fact that, even though the annotators working on this task were professional MTs, the provided guidelines on how to tell preambles from main body were not very detailed.

In future investigations, we would like to see how more elaborate annotation guidelines can improve human performance and what impact the improved annotations have on the performance of an automated preamble detector. It is specifically interesting to investigate how situations of intertwined preamble and main body, as exemplified in Figure 3, can be resolved by clearer guidelines or, alternatively, by an annotation scheme allowing for more than a single hard split.

We are also interested to further enhance the automatic preamble detector by combining the tagger and splitter into a joint neural network model, or by implementing a transfer learning step which reuses the learned tagger weight in a neural-network-based splitter.

References

M Abadi, A Agarwal, P Barham, E Brevdo, Z Chen, C Citro, GS Corrado, A Davis, J Dean, M Devin, S Ghemawat, I Goodfellow, A Harp, G Irving, M Isard, Y Jia, R Jozefowicz, L Kaiser, M Kudlur, J Levenberg, D Mané, R Monga, S Moore, D Murray, C Olah, M Schuster, J Shlens, B Steiner, I Sutskever, K Talwar, P Tucker, V Vanhoucke, V Vasudevan, F Viégas, O Vinyals, P Warden, M Wattenberg, M Wicke, Y Yu, and X Zheng. 2016. Tensorflow: large-scale machine learning on heterogeneous distributed systems. *arXiv* 1603(04467):1–19.

MA Alkureishi, WW Lee, M Lyons, VG Press, S Imam, A Nkansah-Amankra, D Werner, and VM Arora. 2016. Impact of electronic medical record use on the patient-doctor relationship and communication: a systematic review. *J Gen Intern Med* 31(5):548–560.

P Campanella, E Lovato, C Marone, L Fallacara, A Mancuso, W Ricciardi, and ML Specchia. 2015. The impact of electronic health records on healthcare quality: a systematic review and meta-analysis. *Eur J Public Health* 26(1):60–64.

JPC Chiu and E Nichols. 2015. Named entity recognition with bidirectional lstm-cnns. *arXiv* 1511(08308):1–14.

K Cho, B van Merriënboer, D Bahdanau, and Y Bengio. 2014. On the properties of neural machine translation: encoder-decoder approaches. *arXiv* 1409(1259):1–9.

F Chollet. 2015. Keras: deep learning library for theano and tensorflow. https://keras.io/.

F Chollet. 2017. *Deep learning with Python*. Manning, Shelter Island, NY.

R Collobert and J Weston. 2008. A unified architecture for natural language processing: deep neural networks with multitask learning. In *Proc ICML*. ACM, pages 160–167.

R Collobert, J Weston, L Bottou, M Karlen, K Kavukcuoglu, and P Kuksa. 2011. Natural language processing (almost) from scratch. *J Mach Learn Res* 12(8):2493–2537.

R Dey and FM Salem. 2017. Gate-variants of gated recurrent unit (gru) neural networks. *arXiv* 1701(05923):1–5.

E Edwards, W Salloum, GP Finley, J Fone, G Cardiff, M Miller, and D Suendermann-Oeft. 2017. Medical speech recognition: reaching parity with humans. In *SPECOM*. page 10 p.

E Ford, JA Carroll, HE Smith, D Scott, and JA Cassell. 2016. Extracting information from the text of electronic medical records to improve case detection: a systematic review. *J Am Med Inform Assoc* 23(5):1007–1015.

A Graves, S Fernández, and J Schmidhuber. 2005. Bidirectional lstm networks for improved phoneme classification and recognition. In *Proc ICANN*. Springer, pages 799–804.

A Graves and J Schmidhuber. 2005. Framewise phoneme classification with bidirectional lstm networks. In *Proc IJCNN*. IEEE, volume 4, pages 2047–2052.

I Hammana, L Lepanto, T Poder, C Bellemare, and M-S Ly. 2015. Speech recognition in the radiology department: a systematic review. *HIM J* 44(2):4–10.

J Hammerton. 2003. Named entity recognition with long short-term memory. In *Proc HLT-NAACL*. ACL, volume 4, pages 172–175.

K Häyrinen, K Saranto, and P Nykänen. 2008. Definition, structure, content, use and impacts of electronic health records: a review of the research literature. *Int J Med Inform* 77(5):291–304.

S Hochreiter and J Schmidhuber. 1997. Long short-term memory. *Neural Comput* 9(8):1735–1780.

T Hodgson and EW Coiera. 2016. Risks and benefits of speech recognition for clinical documentation: a systematic review. *J Am Med Inform Assoc* 23(e1):e169–e179.

JM Holroyd-Leduc, D Lorenzetti, SE Straus, L Sykes, and H Quan. 2011. The impact of the electronic medical record on structure, process, and outcomes within primary care: a systematic review of the evidence. *J Am Med Inform Assoc* 18(6):732–737.

Z Huang, W Xu, and K Yu. 2015. Bidirectional lstm-crf models for sequence tagging. *arXiv* 1508(01991):1–10.

H Hyppönen, K Saranto, R Vuokko, P Mäkelä-Bengs, P Doupi, M Lindqvist, and M Mäkelä. 2014. Impacts of structuring the electronic health record: a systematic review protocol and results of previous reviews. *Int J Med Inform* 83(3):159–169.

SB Johnson, S Bakken, D Dine, S Hyun, E Mendonca, F Morrison, T Bright, T Van Vleck, J Wrenn, and P Stetson. 2008. An electronic health record based on structured narrative. *J Am Med Inform Assoc* 15(1):54–64.

D Kalra, B Fernando, Z Morrison, and A Sheikh. 2012. A review of the empirical evidence of the value of structuring and coding of clinical information within electronic health records for direct patient care. *Inform Prim Care* 20(3):171–180.

DP Kingma and J Ba. 2014. Adam: a method for stochastic optimization. *arXiv* 1412(6980):1–9.

J Kupiec. 1992. Robust part-of-speech tagging using a hidden markov model. *Comput Speech Lang* 6(3):225–242.

JD Lafferty, A McCallum, and FCN Pereira. 2001. Conditional random fields: probabilistic models for segmenting and labeling sequence data. In *Proc ICML*. ACM, pages 282–289.

G Lample, M Ballesteros, S Subramanian, K Kawakami, and C Dyer. 2016. Neural architectures for named entity recognition. *arXiv* 1603(01360):1–11.

J Logan. 2012. Electronic health information system implementation models a review. *Stud Health Technol Inform* 178:117–123.

X Ma and EH Hovy. 2016. End-to-end sequence labeling via bi-directional lstm-cnns-crf. *arXiv* 1603(01354):1–12.

SM Meystre, GK Savova, KC Kipper-Schuler, and JF Hurdle. 2008. Extracting information from textual documents in the electronic health record: a review of recent research. *Yearb Med Inform* 2008:128–144.

T Mikolov, K Chen, GS Corrado, and J Dean. 2013a. Efficient estimation of word representations in vector space. *arXiv* 1301(3781):1–13.

T Mikolov, W-t Yih, and G Zweig. 2013b. Linguistic regularities in continuous space word representations. In *Proc HLT-NAACL*. ACL, volume 13, pages 746–751.

A Moreno-Conde, D Moner, WD da Cruz, MR Santos, JA Maldonado, M Robles, and D Kalra. 2015. Clinical information modeling processes for semantic interoperability of electronic health records: systematic review and inductive analysis. *J Am Med Inform Assoc* 22(4):925–934.

B Plank, A Søgaard, and Y Goldberg. 2016. Multilingual part-of-speech tagging with bidirectional long short-term memory models and auxiliary loss. *arXiv* 1604(05529):1–7.

W Salloum, E Edwards, S Ghaffarzadegan, D Suendermann-Oeft, and M Miller. 2017a. Crowdsourced continuous improvement of medical speech recognition. In *Proc AAAI Wrkshp Crowdsourcing*. AAAI, San Francisco, CA.

Wael Salloum, Greg Finley, Erik Edwards, Mark Miller, and David Suendermann-Oeft. 2017b. Deep learning for punctuation restoration in medical reports. In *Proceedings of the ACL BioNLP Workshop*. Association for Computational Linguistics.

F Sha and F Pereira. 2003. Shallow parsing with conditional random fields. In *Proc HLT-NAACL*. ACL, volume 1, pages 134–141.

Y Tang, Z Wu, H Meng, M Xu, and L Cai. 2016. Analysis on gated recurrent unit based question detection approach. In *Proc Interspeech*. ISCA, San Francisco, CA, pages 735–739.

P Wang, Y Qian, FK Soong, L He, and H Zhao. 2015. A unified tagging solution: bidirectional lstm recurrent neural network with word embedding. *arXiv* 1511(00215):1–10.

M Wöllmer, F Eyben, A Graves, B Schuller, and G Rigoll. 2010. Bidirectional lstm networks for context-sensitive keyword detection in a cognitive virtual agent framework. *Cognit Comput* 2(3):180–190.

A Biomedical Question Answering System in BioASQ 2017

Mourad Sarrouti, Said Ouatik El Alaoui
Laboratory of Computer Science and Modeling
Faculty of Sciences Dhar El Mahraz
Sidi Mohammed Ben Abdellah University
Fez, Morocco
`mourad.sarrouti@usmba.ac.ma`

Abstract

Question answering, the identification of short accurate answers to users questions, is a longstanding challenge widely studied over the last decades in the open-domain. However, it still requires further efforts in the biomedical domain. In this paper, we describe our participation in phase B of task 5b in the 2017 BioASQ challenge using our biomedical question answering system. Our system, dealing with four types of questions (i.e., yes/no, factoid, list, and summary), is based on (1) a dictionary-based approach for generating the exact answers of yes/no questions, (2) UMLS metathesaurus and term frequency metric for extracting the exact answers of factoid and list questions, and (3) the BM25 model and UMLS concepts for retrieving the ideal answers (i.e., paragraph-sized summaries). Preliminary results show that our system achieves good and competitive results in both exact and ideal answers extraction tasks as compared with the participating systems.

1 Introduction

Finding accurate answers to biomedical questions written in natural language from the biomedical literature is the key to creating high-quality systematic reviews that support the practice of evidence-based medicine (Kropf et al., 2017; Wang et al., 2017; Sarrouti and Lachkar, 2017) and improve the quality of patient care (Sarrouti and Alaoui, 2017b). However, with the large and increasing volume of textual data in the biomedical domain makes it difficult to absorb all relevant information (Sarrouti and Alaoui, 2017a). Since time and quality are of the essence in finding an-

swers to biomedical questions, developing and improving question answering systems are desirable. Question answering (QA) systems aim at directly producing and providing short precise answers to users questions by automatically analyzing thousands of articles using information extraction and natural language processing methods.

Although different types of QA systems have different architectures, most of them, especially in the biomedical domain, follow a framework in which (1) question classification and query formulation, (2) document retrieval, (3) passage retrieval, and (4) answer extraction components play a vital role (Athenikos and Han, 2010; Neves and Leser, 2015; Abacha and Zweigenbaum, 2015).

Question answering in the open-domain is a longstanding challenge widely studied over the last decades (Green et al., 1961; Katz et al., 2002). However, it still remains a real challenge in the biomedical domain. As has been extensively documented in the recent research literature (Athenikos and Han, 2010), open-domain QA is concerned with questions which were not restricted to any domain, while in restricted-domain QA such as the biomedical one, the domain of application provides a context for the QA process. Additionally, Athenikos and Han (2010) report the following characteristics for QA in the biomedical domain: (1) large-sized textual corpora, (2) highly complex domain-specific terminology, and (3) domain specific format and typology of questions.

Since the launch of the biomedical QA track at the BioASQ[1] challenge (Tsatsaronis et al., 2015), various approaches in biomedical QA have been presented. The BioASQ challenge, within 2017 edition, comprised three tasks: (1) task 5a on large-scale online biomedical semantic indexing, (2) task 5b on biomedical semantic QA, and (3)

---

[1] `http://bioasq.org/`

Proceedings of the BioNLP 2017 workshop, pages 296–301,
Vancouver, Canada, August 4, 2017. ©2017 Association for Computational Linguistics

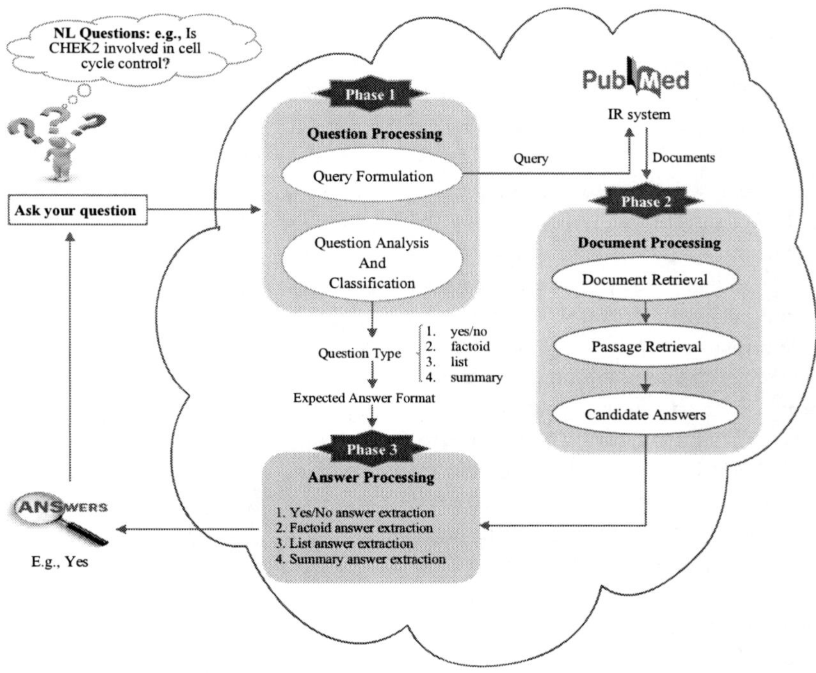

Figure 1: Overall architecture of the proposed biomedical question-answering system

task 5c on funding information extraction from biomedical literature. Task 5b consists of two phases: In phase A, BioASQ released questions in English from benchmark datasets. There were four types of questions: yes/no, factoid, list and summary questions (Balikas et al., 2013). Participants had to respond with relevant concepts, relevant documents, relevant snippets retrieved from the relevant documents, and relevant RDF triples. In phase B, the released questions contained the golden answers for the required elements (documents and snippets) of the first phase. The participants had to answer with exact answers (e.g., biomedical entity, number, list of biomedical entities, yes, no, etc.) as well as with ideal answers (i.e., paragraph-sized summaries) (Krithara et al., 2016). In this paper, we describe our participation in the phase B (i.e., exact and ideal answers) of task5b in the 2017 BioASQ challenge. In our biomedical QA system, we have used (1) a dictionary-based approach to generate the exact answers to yes/no questions, (2) the unified medical language system (UMLS) metathesaurus and term frequency metrics for extracting the exact answers of factoid and list questions, and (3) the BM25 model and UMLS concepts for retrieving

the ideal answers. Figure 1 illustrates the generic architecture of our biomedical QA system.

The remainder of the paper is organized as follows. Section 2 introduces related work and discussion about the main biomedical QA approaches with a particular focus on BioASQ participants. Section 3 describes the answer extraction methods used in our biomedical QA system. Section 4 presents the preliminary results we obtained in the 2017 BioASQ challenge. Finally, the conclusion and future work are made in Section 5.

2 Related work

Since the launch of the BioASQ challenge (Tsatsaronis et al., 2015), QA in the biomedical domain has received much attention from the research community. The BioASQ challenge, which takes place regularly every year since 2013, is an EU-funded support action to set up a challenge on biomedical semantic indexing and QA. Yenala et al. (2015) have presented IIITH biomedical QA system in BioASQ 2015 based on PubMed search engine, leverage web search results, and domain words. The authors have relied on the PubMed search engine to retrieve relevant documents and then applied their own snippet extraction methods, which is based on number of common domain words of the top 10 sentences of the retrieved documents and the question. Zhang et al. (2015) have described USTB biomedical QA system in the 2015 BioASQ challenge. They have built a generic retrieval model based on the sequential dependence model, word embedding and ranking model for document retrieval. After splitting the top-ranked documents into sentences, the authors then have applied the same approach for snippets retrieval. Yang et al. (2016) have described the OAQA system in BioASQ 4b based on NLP annotators, machine learning algorithms for search result scoring, collective answer re-ranking, and yes/no answer prediction. Schulze et al. (2016) have presented HPI biomedical QA system based on NLP functionality from a in-memory database (IMDB). The authors have participated in phase A and B of BioASQ 4b. They have used the LexRank algorithm and biomedical entity names for generating ideal answers. Lee et al. (2016) have described KSAnswer biomedical QA system that returns relevant documents and snippets in BioASQ 4b. KSAnswer, which is participated in phase A of task 4b in the 2016 BioASQ challenge, retrieves candidate snippets using a cluster-based language model. Then, it reranks the retrieved top-N snippets using five independent similarity models based on shallow semantic analysis.

3   Methods

In this section, we describe the answer extraction module of our biomedical QA system. Although our biomedical QA system is composed of many components (cf. Figure 1) which are included in three main phases, i.e., question processing, document processing, and answer processing, we have only used its answer extraction module since we have participated only in phase B (i.e., exact and ideal answers) of task 5b in BioASQ 2017.

During phase B, BioASQ organizers released the test set of biomedical questions along with their relevant documents, relevant snippets, and questions types, i.e., whether yes/no, factoid, list or summary. For each question, each participating system may return an ideal answer, i.e., a paragraph-sized summary of relevant information. In the case of yes/no, factoid, and list questions, the systems may also return exact answers; for summary questions, no exact answers will be returned. In the following sections (cf. Sections 3.1

and 3.2), we will provide a detailed description of the proposed methods used to extract exact and ideal answers for yes/no, factoid, list and summary questions.

3.1   Exact answers

As it has already been described by the BioASQ challenge, the participating systems in phase B of task 5b may return exact answers for yes/no, factoid, and list questions, while no exact answers will be returned for summary questions.

Yes/No questions: For each yes/no question, the exact answer of each participating system will have be either "yes" or "no". The decision for the answers "yes" or "no" in our system is obtained by a sentiment analysis-based approach. Indeed, we first have used the Stanford CoreNLP (Manning et al., 2014) for tokenization and part-of-speech tagging one by one the N retrieved snippets $(s_1, s_2, ..., s_n)$ from benchmark datatsets. We then have assigned a sentiment score using the SentiWordNet (Baccianella et al., 2010) lexical resource to each word in the set of retrieved snippets. Finally, the decision for the answers "yes" or "no" is based on the number of positive and negative snippets.

Factoid questions:

For each factoid question, each participating system will have to return a list of up to 5 entity names (e.g., up to 5 names of drugs), numbers, or similar short expressions, ordered by decreasing confidence. To answer a factoid question in our biomedical QA system, we have first mapped both the given question and its N relevant snippets retrieved from benchmark datasets to the UMLS metathesaurus in order to extract a set of biomedical entity names. To do so, the MetaMap[2] program was used (Aronson, 2001). We then re-ranked the obtained set of biomedical entity names based on term frequency metrics, i.e., the number of times an entity name appeared in the set of biomedical entity names. Indeed, the biomedical entity names appeared in the question are ignored. We finally kept the 5 top-ranked biomedical entity names as answers. A factoid question has one correct answer, but up to five candidate answers are allowed in BioASQ 2017.

List questions: For each list question, each participating system will have to return a single list of entity names, numbers, or similar short expres-

---

[2] https://metamap.nlm.nih.gov/

sions. The proposed method used to answer list questions in our system is similar to the one described for factoid questions. Indeed, the exact answer is the same of factoid questions, only the interpretation is different for list questions: All N top-ranked biomedical entities are considered part of the same answer for the list question, not as candidates. In this work, we have used the five top-ranked ($N = 5$) entities as answers for list questions.

3.2 Ideal answers

To formulate and generate the ideal answers for a given yes/no, factoid, list or summary question, we have used the proposed retrieval model presented in (Sarrouti and Alaoui, 2017b). More specifically, after retrieving the N relevant snippets from benchmark datasets to a given biomedical question, we have re-ranked them based on the BM25 model as retrieval model, stemmed words and UMLS concepts as features. First, we have preprocessed the retrieved set of snippets, including tokenization using the Stanford CoreNLP (Manning et al., 2014), removing stop words[3], and applying Porter' stemmer (Porter, 1980) to extract stemmed words. Additionally, we have used MetaMap to map both questions and snippets to UMLS metathesaurus concepts so as to extract UMLS concepts. Then, we have re-ranked the set of snippets using stemmed words and UMLS concepts as features for the BM25 model. Finally, the ideal answer is obtained by concatenating the two top-ranked snippets.

4 Experimental results and discussion

In this section, we present the preliminary results we obtained in BioASQ 2017. We first introduce the evaluation metrics, then give the experimental results, and finally discuss the results.

4.1 Evaluation metrics

The evaluation metrics used for the exact answers in phase B of task 5b are accuracy, strict accuracy and lenient accuracy, mean reciprocal rank (MRR), mean precision, mean recall, and mean F-measure. Accuracy, MRR and F-measure are the official measures used for evaluating the exact answers of yes/no, factoid and list questions, respectively. ROUGE-2 and ROUGE-SU4, on the

other hand, are the main measures for an automatic evaluation of ideal answers. Details of these evaluations metrics appear in (Balikas et al., 2013).

4.2 Results and discussion

Table 1 highlights the preliminary results of our system in phase B (i.e., exact and ideal answers) of BioASQ task 5b. More details on the results can be found in the BioASQ web site[4].

Our system performed well in the challenge ranking as compared with the participating systems. In batch 1, it achieved the third and the fifth position within the 15 participating systems for extracting the exact answers of list and factoid questions respectively. More specifically, our system obtained the second and the third position when considering results by teams, instead of each individual run. In batch 2, considering results by teams, our system obtained the second and the fourth position for extracting the exact answers of list and factoid questions respectively. For yes/no questions, our system achieved the first and the second position respectively in batch 3 and batch 4, while it obtained the fifteenth position in batch 5.

On the other hand, for the ideal answers, our system in terms of ROUGE-2 achieved the fourth position as compared to the 15 and 21 participating systems in batch 1, batch 2 and batch 3 respectively. While in terms of ROUGE-SU4, our system obtained the third position in batch 1 and the fourth position in batch 2. In batch 4 and batch 5, our systems achieved respectively the second and third position within the 27 participating systems in terms of ROUGE-2 and ROUGE-SU4 when considering results by teams, instead of each individual run. This proves that the proposed method could effectively identify the ideal answers to a given biomedical question.

Overall, from the results and analysis on fives batches of testing data of BioASQ task 5b, we can draw a conclusion that our system is very competitive as compared with the participating systems in both exact and ideal answers tasks.

5 Conclusion and future work

In this paper, we presented the obtained results for the answer extraction module of our biomedical QA system that participated in task 5b of

---

[3]http://www.textfixer.com/resources/
common-english-words.txt

[4]http://participants-area.bioasq.org/
results/5b/phaseB/

Datasets	Exact answers					Idial answers	
	Yes/No	Factoid	List				
	Accuracy	MRR	P	R	F	ROUGE-2	ROUGE-SU4
Batch 1	0.7647	0.2033 (5/15)	0.1909	0.2658	0.2129 (3/15)	0.4943 (4/15)	0.5108 (3/15)
Batch 2	0.7778	0.0887 (10/21)	0.2400	0.3922	0.2920 (6/21)	0.4579 (4/21)	0.4583 (4/21)
Batch 3	0.8387 (1/21)	0.2212 (9/21)	0.2000	0.4151	0.2640 (6/21)	0.5566 (4/21)	0.5656 (4/21)
Batch 4	0.6207 (2/27)	0.0970 (13/27)	0.1077	0.2013	0.1369 (12/27)	0.5895 (4/27)	0.5832 (4/27)
Batch 5	0.4615 (15/25)	0.2071 (9/25)	0.2091	0.3087	0.2438 (11/25)	0.5772 (7/25)	0.5756 (7/25)

Table 1: The primarily results of our system in phase B of BioASQ task 5b. P, R, and F indicate precision, recall, and F-measure, respectively. The values inside parameters indicate our current rank and the total number of submissions for the batch.

the 2017 BioASQ challenge. The proposed approach is based on (1) the SentiWordNet lexical resource to generate the exact answers for yes/questions, (2) UMLS metathesaurus and term frequency metrics for answering factoid and list questions, (3) our retrieval model based on UMLS concepts and the BM25 model for generating the ideal answers. The preliminary results show that our system achieved good performances and is very competitive as compared with the participating systems.

In future research, we intend to present the end-to-end evaluations of our biomedical QA system which includes question classification, document retrieval, passage retrieval, and answer extraction components.

References

Asma Ben Abacha and Pierre Zweigenbaum. 2015. MEANS: A medical question-answering system combining NLP techniques and semantic Web technologies. *Information Processing & Management* 51(5):570–594. https://doi.org/10.1016/j.ipm.2015.04.006.

Alan R Aronson. 2001. Effective mapping of biomedical text to the UMLS metathesaurus: the MetaMap program. In *Proceedings of the AMIA Symposium*. American Medical Informatics Association, page 17.

Sofia J Athenikos and Hyoil Han. 2010. Biomedical question answering: A survey. *Computer methods and programs in biomedicine* 99(1):1–24. https://doi.org/10.1016/j.cmpb.2009.10.003.

Stefano Baccianella, Andrea Esuli, and Fabrizio Sebastiani. 2010. SentiWordNet 3.0: An enhanced lexical resource for sentiment analysis and opinion mining. In *LREC*. volume 10, pages 2200–2204.

Georgios Balikas, Ioannis Partalas, Aris Kosmopoulos, Sergios Petridis, Prodromos Malakasiotis, Ioannis Pavlopoulos, Ion Androutsopoulos, Nicolas Baskiotis, Eric Gaussier, Thierry Artiere, and Patrick Gallinari. 2013. Evaluation framework specifications. project deliverable d4.1, 05/2013.

Bert F. Green, Alice K. Wolf, Carol Chomsky, and Kenneth Laughery. 1961. Baseball. In *western joint IRE-AIEE-ACM computer conference on - IRE-AIEE-ACM 61 (Western)*. Association for Computing Machinery (ACM). https://doi.org/10.1145/1460690.1460714.

Boris Katz, Sue Felshin, Deniz Yuret, Ali Ibrahim, Jimmy Lin, Gregory Marton, Alton Jerome McFarland, and Baris Temelkuran. 2002. Omnibase: Uniform access to heterogeneous data for question answering. In *Natural Language Processing and Information Systems*, Springer Nature, pages 230–234. https://doi.org/10.1007/3-540-36271-1_23.

Anastasia Krithara, Anastasios Nentidis, George Paliouras, and Ioannis Kakadiaris. 2016. Results of the 4th edition of BioASQ challenge. In *Proceedings of the Fourth BioASQ workshop at the Conference of the Association for Computational Linguistics*. pages 1–7.

S. Kropf, P. Krcken, W. Mueller, and K. Denecke. 2017. Structuring legacy pathology reports by openEHR archetypes to enable semantic querying. *Methods of Information in Medicine* 56(2). https://doi.org/10.3414/me16-01-0073.

Hyeon-gu Lee, Minkyoung Kim, Harksoo Kim, Juae Kim, Sunjae Kwon, Jungyun Seo, Jungkyu Choi, and Yi-reun Kim. 2016. KSAnswer: Question-answering system of kangwon national university and sogang university in the 2016 BioASQ challenge. *ACL 2016* page 45.

Christopher Manning, Mihai Surdeanu, John Bauer, Jenny Finkel, Steven Bethard, and David McClosky. 2014. The stanford CoreNLP natural language processing toolkit. In *Proceedings of 52nd Annual Meeting of the Association for Computational Linguistics: System Demonstrations.* https://doi.org/10.3115/v1/p14-5010.

Mariana Neves and Ulf Leser. 2015. Question answering for biology. *Methods* 74:36–46. https://doi.org/10.1016/j.ymeth.2014.10.023.

M.F. Porter. 1980. An algorithm for suffix stripping. *Program: electronic library and information systems* 14(3):130–137. https://doi.org/10.1108/eb046814.

Mourad Sarrouti and Said Ouatik El Alaoui. 2017a. A machine learning-based method for question type classification in biomedical question answering. *Methods of Information in Medicine* 56(3). https://doi.org/10.3414/me16-01-0116.

Mourad Sarrouti and Said Ouatik El Alaoui. 2017b. A passage retrieval method based on probabilistic information retrieval and UMLS concepts in biomedical question answering. *Journal of Biomedical Informatics* 68:96–103. https://doi.org/10.1016/j.jbi.2017.03.001.

Mourad Sarrouti and Abdelmonaime Lachkar. 2017. A new and efficient method based on syntactic dependency relations features for ad hoc clinical question classification. *International Journal of Bioinformatics Research and Applications* 13(2):161–177. https://doi.org/10.1504/ijbra.2017.10003490.

Frederik Schulze, Ricarda Schüler, Tim Draeger, Daniel Dummer, Alexander Ernst, Pedro Flemming, Cindy Perscheid, and Mariana Neves. 2016. HPI question answering system in bioasq 2016. In *Proceedings of the Fourth BioASQ workshop at the Conference of the Association for Computational Linguistics.* pages 38–44.

George Tsatsaronis, Georgios Balikas, Prodromos Malakasiotis, Ioannis Partalas, Matthias Zschunke, Michael R. Alvers, Dirk Weissenborn, Anastasia Krithara, Sergios Petridis, Dimitris Polychronopoulos, Yannis Almirantis, John Pavlopoulos, Nicolas Baskiotis, Patrick Gallinari, Thierry Artiéres, Axel-Cyrille Ngonga Ngomo, Norman Heino, Eric Gaussier, Liliana Barrio-Alvers, Michael Schroeder, Ion Androutsopoulos, and Georgios Paliouras. 2015. An overview of the BIOASQ large-scale biomedical semantic indexing and question answering competition. *BMC Bioinformatics* 16(1):1–28. https://doi.org/10.1186/s12859-015-0564-6.

Liqin Wang, Guilherme Del Fiol, Bruce E. Bray, and Peter J. Haug. 2017. Generating disease-pertinent treatment vocabularies from MEDLINE citations. *Journal of Biomedical Informatics* 65:46–57. https://doi.org/10.1016/j.jbi.2016.11.004.

Zi Yang, Yue Zhou, and Eric Nyberg. 2016. Learning to answer biomedical questions: OAQA at BioASQ 4b. *ACL 2016* page 23.

Harish Yenala, Avinash Kamineni, Manish Shrivastava, and Manoj Kumar Chinnakotla. 2015. IIITH at BioASQ challange 2015 task 3b: Bio-medical question answering system. In *CLEF 2015.*

Zhijuan Zhang, Tiantian Liu, Bo-Wen Zhang, Yan Li, Chun Hua Zhao, Shao-Hui Feng, Xu-Cheng Yin, and Fang Zhou. 2015. A generic retrieval system for biomedical literatures: USTB at BioASQ 2015 question answering task. In *CLEF 2015.*

Adapting Pre-trained Word Embeddings For Use In Medical Coding

Kevin Patel[1], Divya Patel[2], Mansi Golakiya[2], Pushpak Bhattacharyya[1], Nilesh Birari[3]
[1]Indian Institute of Technology Bombay, India
[2]Dharmsinh Desai University, India, [3]ezDI Inc, India
[1]{kevin.patel,pb}@cse.iitb.ac.in, [3]nilesh.b@ezdi.us
[2]{divya.patel.8796,golkiya.mansi}@gmail.com

Abstract

Word embeddings are a crucial component in modern NLP. Pre-trained embeddings released by different groups have been a major reason for their popularity. However, they are trained on generic corpora, which limits their direct use for domain specific tasks. In this paper, we propose a method to add task specific information to pre-trained word embeddings. Such information can improve their utility. We add information from medical coding data, as well as the first level from the hierarchy of ICD-10 medical code set to different pre-trained word embeddings. We adapt CBOW algorithm from the word2vec package for our purpose. We evaluated our approach on five different pre-trained word embeddings. Both the original word embeddings, and their modified versions (the ones with added information) were used for automated review of medical coding. The modified word embeddings give an improvement in f-score by 1% on the 5-fold evaluation on a private medical claims dataset. Our results show that adding extra information is possible and beneficial for the task at hand.

1 Introduction

Word embeddings are a recent addition to an NLP researcher's toolkit. They are dense, real-valued vector representations of words that capture interesting properties among them. Word embeddings are learned from raw corpora. Usually, the larger the corpora, the better is the quality of the embeddings learned. However, the larger the corpora, the larger is the amount of resources and time needed for their training. Thus, different groups release their learned embeddings publicly. Such pre-trained embeddings is a primary reason for the inclusion of word embeddings in mainstream NLP. However, such pre-trained embeddings are usually learned on generic corpora. Using such embeddings in a particular domain such as medical domain leads to following problems:

- No embeddings for domain-specific words. For example, *phenacetin* is not present in pre-trained vectors released by Google.

- Even those words that do have embeddings, may have a poor quality of the embedding, due to different senses of the words, some of which belonging to different domains.

It is difficult to obtain large amounts of domain-specific data. However, many NLP applications have benefited from the addition of information from small domain-specific corpus to that obtained from a large generic corpus (Ito et al., 1997). This raises the following questions:

- *Can we use additional domain-specific data to learn the missing embeddings?*

- *Can we use additional domain-specific data to improve the quality of already available embeddings?*

In this paper, we address the second question: Given pre-trained word embeddings, and domain specific data, we tune the pre-trained word embeddings such that they can achieve better performance. We tune the embeddings for and evaluate them on an automated review of medical coding.

The rest of the paper is organized as follows: Section 2 provides some background on different notions used later in the paper. Section 3 motivates our approach through examples. Section 4 explains our approach in detail. Section 5 enlists the experimental setup. Section 6 details the results and analysis, followed by conclusion and future work.

302

Proceedings of the BioNLP 2017 workshop, pages 302–306,
Vancouver, Canada, August 4, 2017. ©2017 Association for Computational Linguistics

2 Background

2.1 Word Embeddings

Word embeddings are a crucial component of modern NLP. They are learned in an unsupervised manner from large amounts of raw corpora. Bengio et al. (2003) were the first to propose neural word embeddings. Many word embedding models have been proposed since then (Collobert and Weston, 2008; Huang et al., 2012; Mikolov et al., 2013; Levy and Goldberg, 2014). The central idea behind word embeddings is the distributional hypothesis, which states that *words which are similar in meaning occur in similar contexts* (Rubenstein and Goodenough, 1965). Consider the Continuous Bag of Words model by (Mikolov et al., 2013), where the following problem is poised to a neural network: given the context, predict the word that comes in between. The weights of the network are the word embeddings. Training the model over running text brings embeddings of words with similar meaning closer.

2.2 Medical Coding

Medical coding is the process of assigning predefined alphanumeric medical codes to information contained in patient medical records.

Babre et al. (2010) shows a typical medical coding pipeline. Note that the coding (both automatic and/or manual) is followed by a manual review. This is due to the critical nature of the coding process, and the high cost incurred due to any errors. However, any human involvement increases cost both in terms of time and money. Thus, in order to reduce human involvement in the review process, an automatic review component can be inserted just before the human review. Automated reviewing is a binary classification problem. Those instances that are rejected by the automated review component can be directly sent back for recoding, whereas those instances that are accepted by the automated review component should be sent to human reviewers for further checking. Such a modification decreases the load on the human reviewer, thereby reducing the cost of overall pipeline.

Given the textual nature of medical data, many natural language processing challenges manifest themselves while performing either automated medical coding or automated review of medical coding. Common challenges include, but are not limited to:

- Synonymy: Multiple words can have same meaning (Synonym). For instance, *High Blood Sugar* and *Diabetes* have the same meaning.

- Abbreviation: Medical staff, in their hurry, often abbreviate words and sentences. For instance, *hypertension* can be written as *HTN*. The automated system needs to understand that both these strings ultimately mean the same thing.

One can note that both in case of synonym and abbreviations, the context will be almost same. Thus, word embeddings are well suited to handle both these challenges.

3 Motivation

Consider the following medical terms (the abbreviations in parentheses will be used to refer to the terms later):
- High Blood Pressure (HBP)
- Low Blood Pressure (LBP)
- High Blood Sugar (HBS)
- Liver Failure (LF)
- Diabetes (D)
- Hypertension
- HTN

We would ideally like the embeddings of the terms to be learned such that the following constraints hold:

- Similarity (HBP, HBS) should be higher than Similarity (HBP, LBP), which in turn, should be higher than Similarity (HBP, LF) (as per medical knowledge).

- Similarity (HBS, D) should be high (as they are synonyms).

- Similarity (Hypertension, HTN) should be high (as HTN is abbreviation of hypertension).

Information about such relations might not be available in generic corpus on which most pretrained embeddings are trained. However, it might be available in domain specific corpora, or even labeled data, such as those used in medical claims. Approaches that can add that information to pretrained embeddings will definitely improve their utility.

4 Approach

We adapt the Continuous Bag Of Words (CBOW) approach (Mikolov et al., 2013) for our situation. Given labeled medical claims data, we consider the terms in the transcripts as context words, and the corresponding codes as target word. We have both positive and negative samples in our data. Thus we have both normal samples as well as negative samples needed for applying negative sampling.

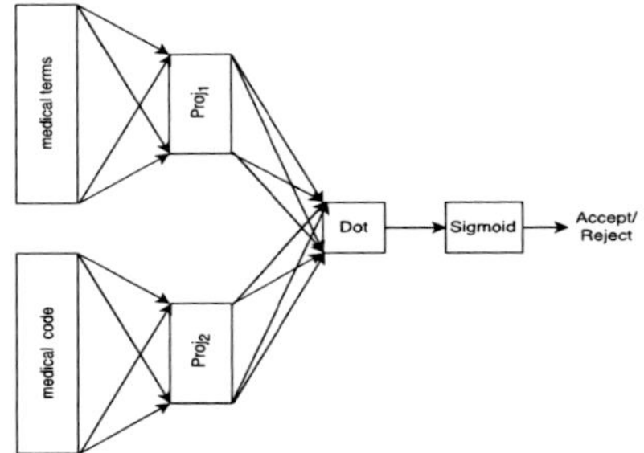

Figure 1: Network architecture of our approach

Figure 1 shows the network of our approach. The inputs to the network are a bag of words representation of medical terms, and a one-hot representation of the corresponding code. The output of the network is a binary value indicating whether the input code is accepted for the corresponding input medical terms.

Exploiting ICD10 Code hierarchy

Another information that can be included is the hierarchical nature of the ICD10 code set. Currently, the network considers the error of misclassifying codes in same subcategory, say F32.9 and F11.20, the same as the error of misclassifying codes belonging to different subcategories, say F32.9 and 30233N1. Ideally, error(F32.9, F11.20) should be less than error(F32.9, E87.1), which in turn should be less than error(F32.9, 30233N1). Such hierarchical information can be encoded by a network like the one in figure 2. Due to resource and time constraints, we have currently considered only the top level hierarchy, *i.e.* whether the code is ICD-10 Diagnosis or ICD-10 Procedural.

The learned weights between Proj$_1$ and codes input in hierarchy network (figure 2) are used to initialize the weights between Proj$_2$ and codes in the original network (figure 1). Then the original network is trained as usual. The weights between

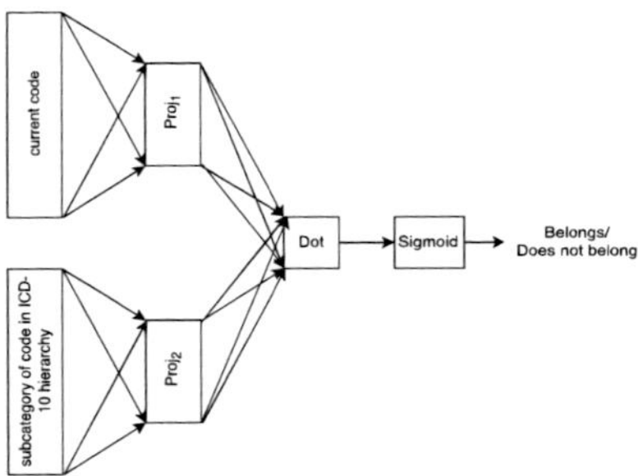

Figure 2: Encoding hierarchy information

Proj$_1$ and medical terms in the original network are the modified word embeddings.

5 Experimental Setup

5.1 Dataset

We used a private medical claims review dataset, which we cannot release publicly due to privacy concerns. The dataset consists of 280k records, consisting of medical terms along with a code. Each entry is labeled as *accept* or *reject*, depending on whether the entry has correct code, or whether it was sent for recoding.

5.2 Pre-trained word embeddings

We used 5 different pre-trained word embeddings. The first one is the one released along with Google's word2vec toolkit. The remaining four are medical domain specific, and were released by (Pyysalo et al., 2013). They are as follows:

- PMC: Trained on 4 million PubMed Central's full articles

- PubMed: Trained on 26 million abstracts and citations in PubMed.

- PubMed_PMC: Trained on combination of previous two resources

- Wikipedia_PubMed_PMC: Trained on combination of Wikipedia, PubMed and PMC resources.

5.3 Classifiers

Once we tune the embeddings, we use them to learn a binary classifier. For our experiments, we report the results we got by using logistic regression..

		Medical Knowledge			Synonym	Abbreviation
		HBP,HBS	HBP,LBP	HBP,LF	HBS,Diabetes	Hypertension,HTN
Google	Orig	0.534	0.895	0.181	0.293	0
	Mod	0.549	0.640	0.089	0.350	-0.004
PMC	Orig	0.599	0.980	0.173	0.141	0.608
	Mod	0.638	0.477	-0.054	0.221	0.947
PubMed	Orig	0.529	0.970	0.006	0.091	0.465
	Mod	0.636	0.474	-0.090	0.188	0.952
PubMed_PMC	Orig	0.592	0.976	0.116	0.141	0.575
	Mod	0.641	0.450	-0.039	0.241	0.952
Wikipedia_ PubMed_PMC	Orig	0.595	0.976	0.158	0.156	0.617
	Mod	0.653	0.474	-0.061	0.190	0.950

Table 1: Cosine similarities of pairs of examples from Section 3

Pre-trained Embeddings	Original Embeddings	Modified Embeddings
Google	82.78	**83.37**
PMC	82.93	**83.96**
PubMed	83.18	**84.00**
PubMed_PMC	82.88	**83.92**
Wikipedia_ PubMed_PMC	83.12	**83.91**

Table 2: Average 5-fold cross validation F-score on automated review of medical coding

6 Results and Analysis

Table 2 shows the results of 5-fold evaluation on automated review of medical coding. Note that the modified embeddings consistently outperform the original ones for all pre-trained embeddings that we used. The reason behind this improvement is evident from the analysis table 1 where we show how the constraints are better modeled by the modified embeddings (Mod) as compared to the original embeddings (Orig).

7 Related Work

Word embeddings have proved to be useful for various tasks, such as Part of Speech Tagging (Collobert and Weston, 2008), Named Entity Recognition Sentence Classification (Kim, 2014), Sentiment Analysis (Liu et al., 2015), Sarcasm Detection (Joshi et al., 2016). Medical domain specific pre-trained word embeddings were released by different groups, such as Pyysalo et al. (2013), Brokos et al. (2016), *etc.* Wu et al. (2015) apply word embeddings for clinical abbreviation disambiguation.

8 Conclusion and Future Work

In this paper, we proposed a modification of the CBOW algorithm to add task and domain specific information to pre-trained word embeddings. We added information from a medical claims dataset and the ICD-10 code hierarchy to improve the utility of the pre-trained word embeddings. We obtained an improvement of approximately 1% using the modified word embeddings as compared to using the original word embeddings. Such improvement was achieved by including only the top level hierarchy. We hypothesize that using the full hierarchy will lead to better improvements, which we shall investigate in the future.

References

Deven Babre et al. 2010. Medical coding in clinical trials. Perspectives in clinical research 1(1):29.

Yoshua Bengio, Réjean Ducharme, Pascal Vincent, and Christian Janvin. 2003. A neural probabilistic language model. J. Mach. Learn. Res. 3:1137–1155.

Georgios-Ioannis Brokos, Prodromos Malakasiotis, and Ion Androutsopoulos. 2016. Using centroids of word embeddings and word mover's distance for biomedical document retrieval in question answering. In Proceedings of 15th Workshop on Biomedical Natural Language Processing (BioNLP 2016), at the 54th Annual Meeting of the Association for Computational Linguistics (ACL 2016).

Ronan Collobert and Jason Weston. 2008. A unified architecture for natural language processing: deep neural networks with multitask learning. In William W. Cohen, Andrew McCallum, and Sam T. Roweis, editors, ICML. ACM, volume 307 of ACM International Conference Proceeding Series, pages 160–167.

Eric H. Huang, Richard Socher, Christopher D. Manning, and Andrew Y. Ng. 2012. Improving Word Representations via Global Context and Multiple Word Prototypes. In Annual Meeting of the Association for Computational Linguistics (ACL).

Akinori Ito, Hideyuki Saitoh, Masaharu Katoh, and Masaki Kohda. 1997. N-gram language model adaptation using small corpus for spoken dialog recognition. In ASJ. volume 3000, page 96779.

Aditya Joshi, Vaibhav Tripathi, Kevin Patel, Pushpak Bhattacharyya, and Mark Carman. 2016. Are word embedding-based features useful for sarcasm detection? In Proceedings of the 2016 Conference on Empirical Methods in Natural Language Processing. Association for Computational Linguistics, Austin, Texas, pages 1006–1011.

Yoon Kim. 2014. Convolutional neural networks for sentence classification. In Proceedings of the 2014 Conference on Empirical Methods in Natural Language Processing (EMNLP). Association for Computational Linguistics, Doha, Qatar, pages 1746–1751.

Omer Levy and Yoav Goldberg. 2014. Dependency-based word embeddings. In Proceedings of the 52nd Annual Meeting of the Association for Computational Linguistics, ACL 2014, June 22-27, 2014, Baltimore, MD, USA, Volume 2: Short Papers. pages 302–308.

Pengfei Liu, Shafiq R Joty, and Helen M Meng. 2015. Fine-grained opinion mining with recurrent neural networks and word embeddings. In EMNLP. pages 1433–1443.

Tomas Mikolov, Ilya Sutskever, Kai Chen, Greg S Corrado, and Jeff Dean. 2013. Distributed representations of words and phrases and their compositionality. In Advances in neural information processing systems. pages 3111–3119.

S Pyysalo, F Ginter, H Moen, T Salakoski, and S Ananiadou. 2013. Distributional semantics resources for biomedical text processing. In Proceedings of LBM 2013. pages 39–44.

Herbert Rubenstein and John B. Goodenough. 1965. Contextual correlates of synonymy. Commun. ACM 8(10):627–633. https://doi.org/10.1145/365628.365657.

Yonghui Wu, Jun Xu, Yaoyun Zhang, and Hua Xu. 2015. Clinical abbreviation disambiguation using neural word embeddings. In Proceedings of 14th Workshop on Biomedical Natural Language Processing (BioNLP 2016), at the 53th Annual Meeting of the Association for Computational Linguistics (ACL 2015). page 171.

Initializing neural networks for hierarchical multi-label text classification

Simon Baker [1,2]   **Anna Korhonen** [2]

[1]Computer Laboratory, 15 JJ Thomson Avenue
[2] Language Technology Lab, DTAL
University of Cambridge, UK
`simon.baker@cl.cam.ac.uk, alk23@cam.ac.uk`

Abstract

Many tasks in the biomedical domain require the assignment of one or more predefined labels to input text, where the labels are a part of a hierarchical structure (such as a taxonomy). The conventional approach is to use a one-*vs.*-rest (OVR) classification setup, where a binary classifier is trained for each label in the taxonomy or ontology where all instances not belonging to the class are considered negative examples. The main drawbacks to this approach are that dependencies between classes are not leveraged in the training and classification process, and the additional computational cost of training parallel classifiers. In this paper, we apply a new method for hierarchical multi-label text classification that initializes a neural network model final hidden layer such that it leverages label co-occurrence relations such as hypernymy. This approach elegantly lends itself to hierarchical classification. We evaluated this approach using two hierarchical multi-label text classification tasks in the biomedical domain using both sentence- and document-level classification. Our evaluation shows promising results for this approach.

1   Introduction

Many tasks in biomedical natural language processing require the assignment of one or more labels to input text, where there exists some structure (such as a taxonomy or ontology) between the labels: for example, the assignment of Medical Subject Headings (MeSH) to PubMed abstracts (Lipscomb, 2000).

A typical approach to classifying multi-label documents is to construct a binary classifier for each label in the taxonomy or ontology where all documents not belonging to the class are considered negative examples, *i.e.* one-*vs.*-rest (OVR) classification (Hong and Cho, 2008). This approach has two major drawbacks: first, it makes the hard assumption that the classes are independent which often does not reflect reality; second, it is more computationally expensive (albeit by a constant factor): if there are a very large number of classes, the approach becomes computationally unrealistic.

In this paper, we investigate a simple and computationally fast approach for multi-label classification with a focus on labels that share a structure, such as a hierarchy (taxonomy). This approach can work with established neural network architectures such as a convolutional neural network (CNN) by simply initializing the final output layer to leverage the co-occurrences between the labels in the training data.

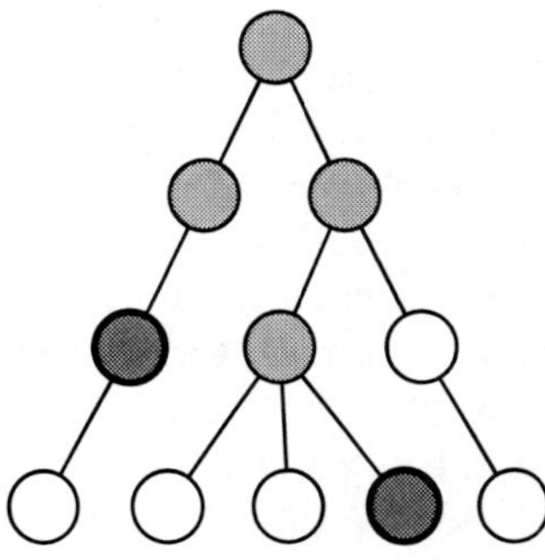

Figure 1: Hierarchical multi-label classification. Nodes represent possible labels that can be assigned to text: a dark grey node denotes an explicit label assignment and light grey denotes implicit assignment due to a hypernymy relationship with the explicitly assigned label.

Proceedings of the BioNLP 2017 workshop, pages 307–315,
Vancouver, Canada, August 4, 2017. ©2017 Association for Computational Linguistics

First, we need to define hierarchical multi-label classification. In multi-label text classification, input text can be associated with multiple labels (label co-occurrence). When the labels form a hierarchy, they share a hypernym–hyponym relation (Figure 1). When multiple labels are assigned to the text, if it is explicitly labeled by a subclass it must also implicitly include all of the its superclasses.

The co-occurrence between subclasses and superclasses as labels for the input text contains information we would like to leverage to improve multi-label classification using a neural network.

In this paper we experiment with this approach using two hierarchical multi-label text classification tasks in the biomedical domain, using both document- and sentence-level classification.

We first briefly summarize related literature on the topic of multi-label classification using neural networks, we then describe our methodology and evaluation procedure, and then we present and discuss our results.

2 Related work

There have been numerous works that focus on solving hierarchical text classification. Sun and Lim (2001) proposed top-down level-based SVM classification. More recently, Sokolov and Ben-Hur (2010); Sokolov et al. (2013) predict ontology terms by explicitly modeling the structure hierarchy using kernel methods for structured output space. Clark and Radivojac (2013) use a Bayesian network, structured according to the underlying ontology to model the prior probability.

Within the context of neural networks, Kurata et al. (2016) propose a scheme for initializing neural networks hidden output layers by taking into account multi-label co-occurrence. Their method treats some of the neurons in the final hidden layer as dedicated neurons for each pattern of label co-occurrence. These dedicated neurons are initialized to connect to the corresponding co-occurring labels with stronger weights than to others. They evaluated their approach on the *RCV1-v2* dataset (Lewis et al., 2004) from the general domain, containing only flat labels. Their evaluation shows promising results. However, their applicability to the biomedical domain with more a complex set of labels that share a hierarchy remains an open question.

Chen et al. (2017) propose a convolutional neural network (CNN) and recurrent neural network (RNN) ensemble method that is capable of efficiently representing text features and modeling high-order label correlation (including co-occurrence). However, they show that their method is susceptible to overfitting with small datasets.

Cerri et al. (2014) propose a method for hierarchical multi-label text classification that incrementally trains a multi-layer perceptron for each level of the classification hierarchy. Predictions made by a neural network in a given level are used as inputs to the neural network responsible for the prediction in the next level. Their method was evaluated against several datasets with convincing results.

There are also several relevant works that propose the inclusion of multi-label co-occurrence into loss functions such as pairwise ranking loss (Zhang and Zhou, 2006) and more recent work by Nam et al. (2014), who report that binary cross-entropy can outperform the pairwise ranking loss by leveraging rectified linear units (ReLUs) for nonlinearity.

3 Method

In this section, we describe the approach of initializing a neural network for multi-label classification. We base our CNN architecture on the model of Kim (2014), which has been used widely in text classification tasks, but this approach can be applied to any other architecture.

Briefly, this model consists of an initial embedding layer that maps input texts into matrices, followed by convolutions of different filter sizes and 1-max pooling, and finally a fully connected layer. The architecture is illustrated in Figure 2.

To perform multi-label classification using this architecture, the final output layer uses logistic (sigmoid) activation function σ:

$$\sigma(x) = \frac{1}{1 + \mathrm{e}^{-x}} \tag{1}$$

where x is the input signal. The output range of the function is between zero and one; if it is above a cut-off threshold T_σ (which is tuned by grid search on the development dataset) then the prediction y'_k for label y_k is positive. We use a binary cross-entropy loss function L:

$$L(\theta, (x, y)) = -\sum_{k=1}^{K} y_k \log(y_k') + (1 - y_k) \log(1 - y_k') \tag{2}$$

where θ is the model parameters and K is the number of classes.

As shown in Figure 2, the multi-label initialization happens in output layer of the network. Figure 3 illustrates the initialization process. The rows represent the units in the final hidden layer, while the columns represent the output classes.

The idea is to initialize the final hidden layer with rows that map to co-occurrence of labels in the training data. This can be implicit hypernymy relations between the labels, or explicit co-occurrence in the annotation. For each co-occurrence, the value ω is assigned to the associated classes and a value of zero is assigned to the rest. The value ω is the upper bound of the normalized initialization proposed by Glorot and Bengio (2010), which is calculated as follows:

$$\omega = \frac{\sqrt{6}}{\sqrt{n_h + n_k}} \tag{3}$$

where n_h is the number of units in the final hidden layer and n_k is the number of units in the output layer (*i.e.* classes). This value was also successfully used by Kurata et al. (2016) in their initialization procedure.

The motivation for this initialization is to incline units in the hidden layer to be dedicated to representing co-occurrence of labels by triggering only the corresponding label nodes in the output layer when they are active.

The number of units in the final hidden layer can exceed the number of label co-occurrences in the training data. We must therefore decide what to do with the remaining hidden units. Kurata et al. (2016) assign random values to these units (shown in Figure 3 (B)). We will also use this scheme, but in addition we propose another variant: we assign the value zero for these neurons, so that the hidden layer will only be initialized with nodes that represent label co-occurrence.

We implement the neural network and the initialization using Keras (Chollet, 2015). the hyperparameters for our model and baselines are those of Kim (2014), summarized in Table 1.

We use word2vec embeddings trained on PubMed by Chiu et al. (2016).

Hyperparameter	Value
Word vector size	300
Filter sizes	3, 4, and 5
Number of filters	300 (100 of each size)
Dropout probability	0.5
Minibatch size	50
Input size (in tokens)	500 (documents), 100 (sentences)

Table 1: Our baseline model, based on Kim (2014) model hyperparameters.

4 Data

We investigate our approach using two multi-label classification tasks. In this section, we describe the nature of these tasks and the annotated gold standard data.

Task 1: The Hallmarks of Cancer The Hallmarks of Cancer describe a set of interrelated biological properties and behaviors that enable cancer to thrive in the body. Introduced in the seminal paper by Hanahan and Weinberg (2000)— the most cited paper in the journal *Cell*—the hallmarks of cancer have seen widespread use in BioNLP for many systems and works, including the BioNLP Shared Task 2013, 'Cancer Genetics task' (Pyysalo et al., 2013), which involved the extraction of events (*i.e.* biological processes) from cancer-domain texts. Baker et al. (2016) have released an expert-annotated dataset for cancer hallmark classification for both sentences and documents from PubMed. The data consists of multi-labelled documents and sentences using a taxonomy of 37 classes.

Task 2: The exposure taxonomy Larsson et al. (2017) introduce a new task and an associated annotated dataset for the classification of text (documents or sentences) for chemical risk assessment: more specifically, the assessment of exposure routes (such as ingestion, inhalation, or dermal absorption) and human biomonitoring (the measurement of exposure biomarkers). The taxonomy of 32 classes is divided into two branches: Biomonitoring and Exposure routes.

We split both datasets (by documents) into train, development (dev), and test splits in order to evaluate our methodology. Table 4 summarizes key statistics for these splits.

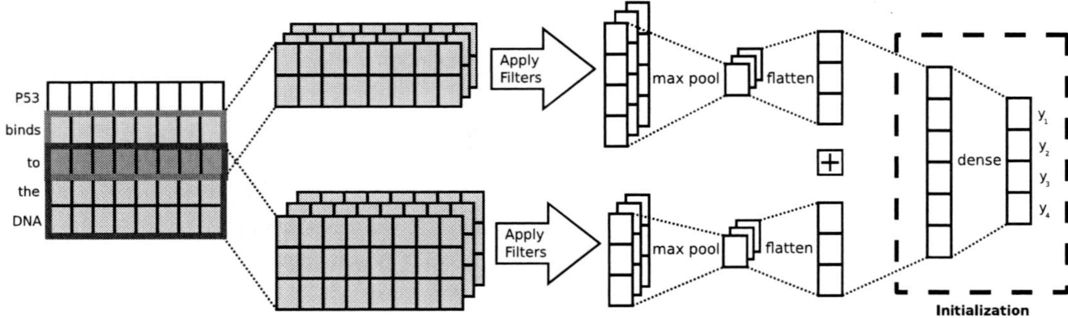

Figure 2: Convolutional Neural Network (CNN) architecture with the initialization layer outlined. The CNN architecture is based on the model of Kim (2014).

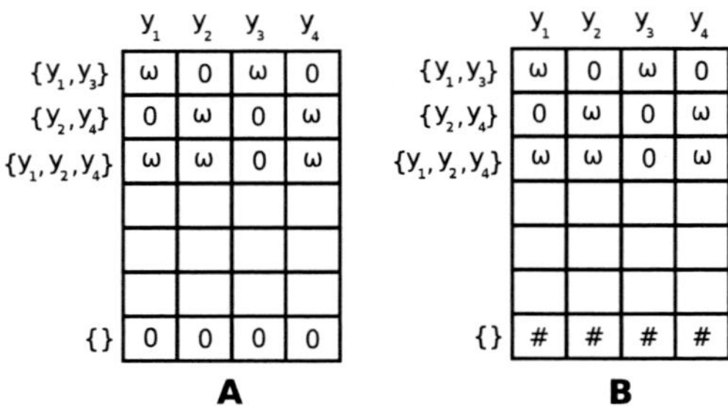

Figure 3: The two initialization schemes: (A) initializing non label co-occurrence nodes with zero, (B) initializing non label co-occurrence with a random value (#) drawn from a uniform distribution.

	Task 1		Task 2	
	Document	Sentence	Document	Sentence
Train	1,303	12,279	2,555	25,307
Dev	183	1,775	384	3,770
Test	366	3,410	722	7,100
Total	1,852	17,464	3,661	36,177

Table 2: Summary statistics for Tasks 1 and 2.

We also measure the overlap in the data between pairs of labels. We use Jaccard similarity J to measure this overlap using the following equation:

$$J(A, B) = \frac{A \cap B}{A \cup B} \qquad (4)$$

Where A and B are sets of instances labelled with these classes. Table 4 summarizes the average and maximum pairwise Jaccard similarity between the labels in both tasks.

Table 4 shows that Task 1 labels have slightly more overlap than those of Task 2.

	Task 1		Task 2	
	Document	Sentence	Document	Sentence
Avg	4.1	2.3	5.7	3.0
Max	49.3	49.4	45.7	42.5

Table 3: Jaccard similarity scores (expressed as percentages) between label pairs.

The large difference in values between document and sentence label overlap is due to the fact that documents have more labels per instance than sentences. The average score is much lower as most pair combinations would not have overlaps; where there is overlap it is typically significant (as shown by the Max row in Table 4).

5 Evaluation

In this section, we describe our experimental setup and our baselines.

5.1 Experimental setup

We ascertain the performance of our approach under a controlled experimental setup. We compare two baseline models (described in the next section), and two variants of the initialization models corresponding to the two initialization schemes described in Figure 3. We will refer to the first scheme (allocating all units in the final hidden layer to representing label co-occurrences and zeroing all other units) as INIT-A, and the second scheme (allocating a random value drawn from a uniform distribution for non co-occurrence hidden units) as INIT-B. We use the hyperparameters in Table 1 and data splits in Table 4 for all models.

We check the model's performance (F_1-score) on development data at the end of every epoch. We select the model from the best-performing epoch and train it until its performance does not improve for ten epochs.

5.2 Baselines

We compare two baselines in our setup: one-*vs*.-rest (OVR) and multi-label baseline (MULTI-BASIC)

One-*vs*.-rest (OVR) We train and evaluate K independent binary CNN classifiers (*i.e.* a single classifier per class with the instances of that class as positive samples and all other instances as negatives).

Multi-label baseline (MULTI-BASIC) We train and evaluate a multi-label baseline based on Figure 2 without initialization of the final hidden layer. This enables us to directly compare the effect of the initialization step. As with the initialization models (INIT-A and INIT-B), we grid search the sigmoid cut-off parameter T_σ on the development data at the end of each epoch to be used with the selected best model on the test split.

5.3 Post-processing label correction

The predicted output labels from all of our models can be inconsistent with respect to the label hierarchy: a subclass label might be positive while its superclass is negative, thereby contradicting the hypernmy relation (illustrated in Figure 4 (A)).

We can apply two kinds of post-processing corrections to the predicted labels in order for them to be well-formed. We call the first *transitive correction* (Figure 4 (B)), wherein we correct all superclass labels (transitively) to be positive. The alternative is *retractive correction* (Figure 4 (C)), where we ignore the positive classification of the subclass label, and accept only the chain of superclass labels (from the root), as long as they are well-formed.

We apply both of these post-processing correction policies to all of the models, and observe the effect on their performance.

6 Results

In this section, we describe the results for the evaluation setup described in the last section. We assess the performance of the models by measuring the precision (P), recall (R), and F_1-scores of the labels in the model using the one-*vs*.-rest setup.

	Document			Sentence		
	P	R	F_1	P	R	F_1
Task 1						
OVR	77.8	51.7	62.1	56.8	30.7	39.9
MULTI-BASIC	71.0	71.6	71.3	42.0	71.9	53.0
INIT-A	73.4	76.9	75.1	42.7	70.6	53.2
INIT-B	68.3	83.4	75.1	40.1	72.2	51.6
Task 2						
OVR	89.5	87.1	88.3	66.2	62.8	64.5
MULTI-BASIC	86.0	90.0	88.0	51.7	75.6	61.4
INIT-A	86.7	91.1	88.9	49.5	80.7	61.4
INIT-B	75.7	91.3	82.8	47.0	83.2	60.1

Table 4: Performance results for Tasks 1 and 2. All figures are micro-averages expressed as percentages.

Table 6 shows the micro-averaged scores across all labels for both tasks.

The results show that for Task 1, all multi-labeled models significantly outperform the OVR model in F_1-score, which is explained by a very substantial improvement in recall. INIT-A outperforms all models in this task, particularly at the document level where there is 5 point improvement over MULTI-BASIC.

The results for Task 2 on are more mixed. Overall, all models achieve a similar F_1-score at the document level. However, there is a clear improvement in recall at the cost of lower precision when compared to OVR. The best performing model at the document level is INIT-A. On the sentence level, OVR seems to outperform all multi-label models by a good margin. This indicates that the multi-label approach did not aid sentence-level classification in this particular task.

The figures in Table 6 do not show a complete picture as the interactions between the labels are not taken into account.

We can observe the proportion of the number of labels assigned to each instance by the classifiers, and compare these proportions to the annotated gold standard test data. Figure 5 shows this distribution for each classifier. We can see in Figure 5 that the overall distributions for all sentence-level classifiers (for both tasks) are closer to the gold standard distribution (compared to document level). This is due to the fact that most sentences have no assigned labels. For Task 2, the classifiers tend to assign more labels than the gold standard.

Document-level classification shows two out-

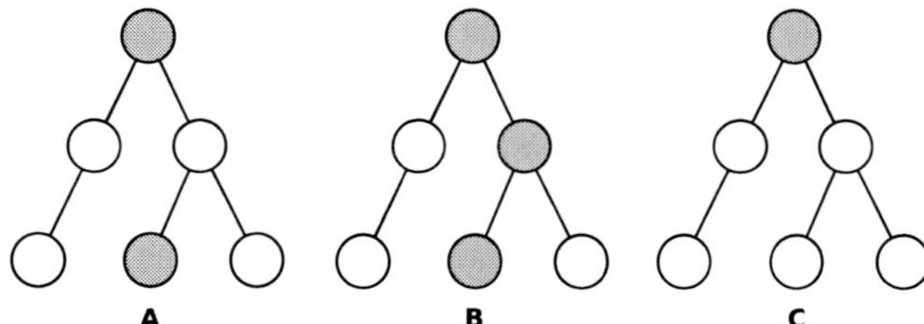

Figure 4: Illustrating post-processing label correction, with (A) showing the output prediction from the neural network model, (B) applying transitive correction, (C) applying retractive correction.

liers. For Task 1, we observe that OVR disproportionately assigns exactly one label per document compared to gold standard (where documents have two to three labels on average). In Task 2, INIT-B assigns more labels per document than the gold standard (and every other model).

In addition to looking at the number of labels per class, we also measure the proportion of exact label matches that each model predicts as shown in Table 6.

	Task 1		Task 2	
	Doc.	Sent.	Doc.	Sent.
OVR	18.0	67.9	43.4	61.7
MULTI-BASIC	26.2	59.3	40.9	54.2
INIT-A	33.9	65.9	45.6	53.1
INIT-B	31.3	62.6	12.7	49.7

Table 5: The proportion (%) of exact matches.

For document classification in Task 1, INIT-A outperforms all models, while OVR significantly underperforms. However, OVR performs significantly better than all other models when classifying sentences when considering exact matches only.

Finally, we look at how consistent (well-formed) the predictions output by each model are. We do this by running the post-processing label correction policies described in Section 5.3. Table 6 summarizes these results.

For Task 1, OVR shows the largest variance after the application of any method of correction, whereas no multi-labeled model shows this variation. This indicates that the post-processing corrections had little effect on the predicted results as they were already well-formed. For Task 2, there is very little variance for all multi-labeled models, with only a slight change for OVR.

	Document			Sentence		
	O	T	R	O	T	R
Task 1						
OVR	62.1	63.9	60.6	39.9	42.2	37.5
MULTI-BASIC	71.3	71.3	71.2	53.0	53.0	53.0
INIT-A	75.1	75.0	75.2	53.2	53.2	53.3
INIT-B	75.1	74.9	75.3	51.6	51.5	51.6
Task 2						
OVR	88.3	88.4	88.2	64.5	65.3	63.3
MULTI-BASIC	88.0	87.7	88.1	61.4	61.3	61.7
INIT-A	88.9	88.7	89.0	61.4	61.3	61.5
INIT-B	82.8	82.8	82.8	60.1	59.8	60.4

Table 6: Post-processing label correction. O is the predicted output, T is transitive correction, and R is retractive correction. All figures are micro-averaged F_1-scores expressed as percentages.

7   Discussion

The strength of using the hidden-layer initialization for multi-label classification lies in leveraging the co-occurrence between labels. Naturally, if such co-occurrences are relatively rare in the dataset, then this approach becomes less effective. This implies that this approach is especially attractive for hierarchical multi-label classification, because of the implicit hypernym–hyponym relations between the labels, which by definition guarantees co-occurrence of labels in the datasets. The superclass labels must be included when labeling a given example in order to model the hierarchical nature of the labels.

Another key strength of this approach is its low computational cost, which is only proportional to the size of the input text, and the number of label co-occurrences.

However, when there is a large amount of training data, the number of label co-occurrences can be larger than the number of the hidden units. In such a case, one possible option is to select an ap-

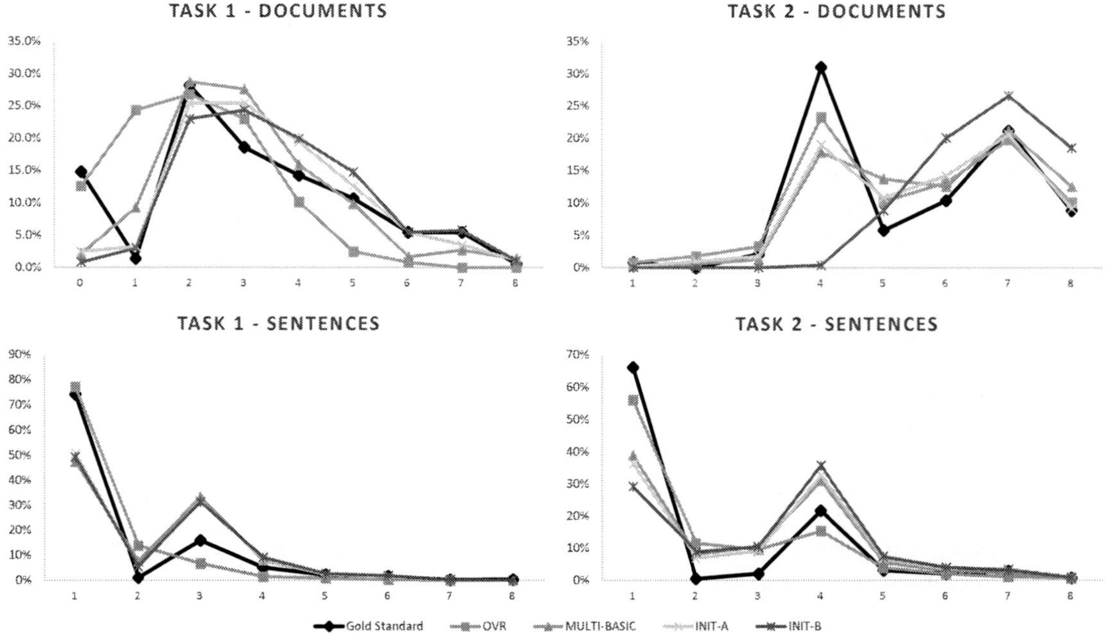

Figure 5: The distribution of instances according to the number labels per instance. The number of labels per instance (x-axis), and y-axis is the proportion of instances in the test dataset that have that number of labels. The black line indicates the distribution of the gold standard annotation (*i.e.* ground truth).

propriate subset of label co-occurrences using a certain criteria such as the frequency in the training data. For the datasets used in this paper, this was not necessary.

Overall, the results of the evaluation show that initializing the model using only label co-occurrences (INIT-A) generally produced a higher performance compared to the other models, including the random initialization of remaining hidden units in the final hidden layer (the INIT-B model) as proposed by Kurata et al. (2016). However, there was one key exception in Task 2 sentence level classification, where the one-*vs.*-rest OVR model achieved the best results.

Both variants of the initialization models investigated here achieved generally positive results when the scope of text is larger (*i.e.* documents), where there are more labels assigned per text instance. However, due to time and computational constraints, this initialization method was not fully utilized as we could only investigate its performance under a closed set of hyperparamaters for the CNN model.

It may be possible for this approach to yield even better results if further parameters are in-

cluded in the CNN models (*e.g.* more filters and filter sizes). It is also important to note that collectively the one-*vs.*-rest models have much more parameters than any of the the multi-label models in our experiment setup, and therefore they have a higher capacity to capture correlations. In spite of this, the multi-label models have largely outperformed the OVR model.

8 Conclusions

There are many tasks in the biomedical domain that require the assignment of one or more labels to input text. These labels often exists within some hierarchical structure (such as a taxonomy).

The conventional approach is to use a one-*vs.*-rest classification setup: a binary classifier is trained for each label in the taxonomy or ontology where all instances not belonging to the class are considered negative examples. The main drawbacks to this approach are that dependencies between classes are not leveraged in the training and classification process, and the additional computational cost of training a classifier for each class.

We applied a new method for multi-label classification that initializes a neural network model final hidden layer to leverage label co-occurrence. This approach elegantly lends itself to hierarchical classification.

We evaluated this approach using two hierarchical multi-label classification tasks using both sentence and document level classification. We use a baseline CNN model with a sigmoid output for each class, and a binary cross-entropy loss function. We investigated two variants of the initialization procedure. One used only co-occurrence (and hierarchical information), while the other randomly assigned random values to the remaining hidden units in the final hidden layer as proposed by Kurata et al. (2016). The experimental results for both tasks show that overall, our proposed initialization procedure (INIT-A) achieved better results than all of the the other models, with the exception of sentence-level classification in Task 2, where one-*vs*.-rest classification attained the best result. We believe that this approach shows promising potential for improving the performance for hierarchical multi-label text classification tasks.

For future work, we plan to try different initialization schemes in addition to the upper bound parameter by Glorot and Bengio (2010) that was used in the paper, and the application of this approach to other tasks and datasets such as Medical Subject headings (MeSH) text classification.

Acknowledgements

The first author is funded by the Commonwealth Scholarship Commission and the Cambridge Trust. This work is supported by Medical Research Council grant MR/M013049/1 and the Google Faculty Award. We thank Tyler Griffiths for his help in proofreading and editing this paper.

References

Simon Baker, Ilona Silins, Yufan Guo, Imran Ali, Johan Högberg, Ulla Stenius, and Anna Korhonen. 2016. Automatic semantic classification of scientific literature according to the hallmarks of cancer. *Bioinformatics* 32(3):432–440.

Ricardo Cerri, Rodrigo C Barros, and André CPLF De Carvalho. 2014. Hierarchical multi-label classification using local neural networks. *Journal of Computer and System Sciences* 80(1):39–56.

Guibin Chen, Deheng Ye, Erik Cambria, Jieshan Chen, and Zhenchang Xing. 2017. Ensemble application of convolutional and recurrent neural networks for multi-label text categorization. IJCNN.

Billy Chiu, Gamal Crichton, Anna Korhonen, and Sampo Pyysalo. 2016. How to train good word embeddings for biomedical NLP. In *Proceedings of BioNLP*.

François Chollet. 2015. Keras. `https://github.com/fchollet/keras`.

Wyatt T Clark and Predrag Radivojac. 2013. Information-theoretic evaluation of predicted ontological annotations. *Bioinformatics* 29(13):i53–i61.

Xavier Glorot and Yoshua Bengio. 2010. Understanding the difficulty of training deep feedforward neural networks. In *Aistats*. volume 9, pages 249–256.

Douglas Hanahan and Robert A Weinberg. 2000. The hallmarks of cancer. *Cell* 100(1):57–70.

Jin-Hyuk Hong and Sung-Bae Cho. 2008. A probabilistic multi-class strategy of one-vs.-rest support vector machines for cancer classification. *Neurocomputing* 71(16):3275–3281.

Yoon Kim. 2014. Convolutional neural networks for sentence classification. *arXiv preprint arXiv:1408.5882* .

Gakuto Kurata, Bing Xiang, and Bowen Zhou. 2016. Improved neural network-based multi-label classification with better initialization leveraging label co-occurrence. In *Proceedings of NAACL-HLT*. pages 521–526.

Kristin Larsson, Simon Baker, Ilona Silins, Yufan Guo, Ulla Stenius, Anna Korhonen, and Marika Berglund. 2017. Text mining for improved exposure assessment. *PloS one* 12(3):e0173132.

David D Lewis, Yiming Yang, Tony G Rose, and Fan Li. 2004. Rcv1: A new benchmark collection for text categorization research. *Journal of machine learning research* 5(Apr):361–397.

Carolyn E Lipscomb. 2000. Medical subject headings (mesh). *Bulletin of the Medical Library Association* 88(3):265.

Jinseok Nam, Jungi Kim, Eneldo Loza Mencía, Iryna Gurevych, and Johannes Fürnkranz. 2014. Large-scale multi-label text classificationrevisiting neural networks. In *Joint European Conference on Machine Learning and Knowledge Discovery in Databases*. Springer, pages 437–452.

Sampo Pyysalo, Tomoko Ohta, and Sophia Ananiadou. 2013. Overview of the cancer genetics (CG) task of BioNLP Shared Task 2013. In *BioNLP Shared Task 2013 Workshop*.

Artem Sokolov and Asa Ben-Hur. 2010. Hierarchical classification of gene ontology terms using the gostruct method. *Journal of bioinformatics and computational biology* 8(02):357–376.

Artem Sokolov, Christopher Funk, Kiley Graim, Karin Verspoor, and Asa Ben-Hur. 2013. Combining heterogeneous data sources for accurate functional annotation of proteins. *BMC bioinformatics* 14(3):S10.

Aixin Sun and Ee-Peng Lim. 2001. Hierarchical text classification and evaluation. In *Data Mining, 2001. ICDM 2001, Proceedings IEEE International Conference on*. IEEE, pages 521–528.

Min-Ling Zhang and Zhi-Hua Zhou. 2006. Multilabel neural networks with applications to functional genomics and text categorization. *IEEE transactions on Knowledge and Data Engineering* 18(10):1338–1351.

Biomedical Event Trigger Identification Using Bidirectional Recurrent Neural Network Based Models

Patchigolla V S S Rahul, Sunil Kumar Sahu, Ashish Anand
Department of Computer Science and Engineering
Indian Institute of Technology Guwahati, Assam, India
`rahul.rahul.pvss@gmail.com`
`{sunil.sahu, anand.ashish}@iitg.ernet.in`

Abstract

Biomedical events describe complex interactions between various biomedical entities. Event trigger is a word or a phrase which typically signifies the occurrence of an event. Event trigger identification is an important first step in all event extraction methods. However many of the current approaches either rely on complex hand-crafted features or consider features only within a window. In this paper we propose a method that takes the advantage of recurrent neural network (RNN) to extract higher level features present across the sentence. Thus hidden state representation of RNN along with word and entity type embedding as features avoid relying on the complex hand-crafted features generated using various NLP toolkits. Our experiments have shown to achieve state-of-art F1-score on Multi Level Event Extraction (MLEE) corpus. We have also performed category-wise analysis of the result and discussed the importance of various features in trigger identification task.

1 Introduction

Biomedical events play an important role in improving biomedical research in many ways. Some of its applications include pathway curation (Ohta et al., 2013) and development of domain specific semantic search engine (Ananiadou et al., 2015). So as to gain attraction among researchers many challenges such as BioNLP'09 (Kim et al., 2009), BioNLP'11 (Kim et al., 2011), BioNLP'13 (Nédellec et al., 2013) have been organized and many novel methods have also been proposed addressing these tasks.

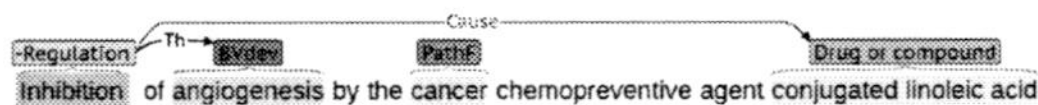

Figure 1: Example of a complex biomedical event

An event can be defined as a combination of a trigger word and arbitrary number of arguments. Figure 1 shows two events with trigger words as "*Inhibition*" and "*Angiogenesis*" of trigger types "*Negative Regulation*" and "*Blood Vessel Development*" respectively. Pipelined based approaches for biomedical event extraction include event trigger identification followed by event argument identification. Analysis in multiple studies (Wang et al., 2016b; Zhou et al., 2014) reveal that more than 60% of event extraction errors are caused due to incorrect trigger identification.

Existing event trigger identification models can be broadly categorized in two ways: *rule based approaches* and *machine learning based approaches*. Rule based approaches use various strategies including pattern matching and regular expression to define rules (Vlachos et al., 2009). However, defining these rules are very difficult, time consuming and requires domain knowledge. The overall performance of the task depends on the quality of rules defined. These approaches often fail to generalize for new datasets when compared with machine learning based approaches. Machine learning based approaches treat the trigger identification problem as a word level classification problem, where many features from the data are extracted using various NLP toolkits (Pyysalo et al., 2012; Zhou et al., 2014) or learned automatically (Wang et al., 2016a,b).

In this paper, we propose an approach using RNN to learn higher level features without the requirement of complex feature engineering. We thoroughly evaluate our proposed approach on

Proceedings of the BioNLP 2017 workshop, pages 316–321,
Vancouver, Canada, August 4, 2017. ©2017 Association for Computational Linguistics

MLEE corpus. We also have performed category-wise analysis and investigate the importance of different features in trigger identification task.

2   Related Work

Many approaches have been proposed to address the problem of event trigger identification. Pyysalo et al. (2012) proposed a model where various hand-crafted features are extracted from the processed data and fed into a Support Vector Machine (SVM) to perform final classification. Zhou et al. (2014) proposed a novel framework for trigger identification where embedding features of the word combined with hand-crafted features are fed to SVM for final classification using multiple kernel learning. Wei et al. (2015) proposed a pipeline method on BioNLP'13 corpus based on Conditional Random Field (CRF) and Support vector machine (SVM) where the CRF is used to tag valid triggers and SVM is finally used to identify the trigger type. The above methods rely on various NLP toolkits to extract the hand-crafted features which leads to error propagation thus affecting the classifier's performance. These methods often need to tailor different features for different tasks, thus not making them generalizable. Most of the hand-crafted features are also traditionally sparse one-hot features vector which fail to capture the semantic information.

Wang et al. (2016b) proposed a neural network model where dependency based word embeddings (Levy and Goldberg, 2014) within a window around the word are fed into a feed forward neural network (FFNN) (Collobert et al., 2011) to perform final classification. Wang et al. (2016a) proposed another model based on convolutional neural network (CNN) where word and entity mention features of words within a window around the word are fed to a CNN to perform final classification. Although both of the methods have achieved good performance they fail to capture features outside the window.

3   Model Architecture

We present our model based on bidirectional RNN as shown in Figure 2 for the trigger identification task. The proposed model detects trigger word as well as their type. Our model uses embedding features of words in the input layer and learns higher level representations in the subsequent layers and

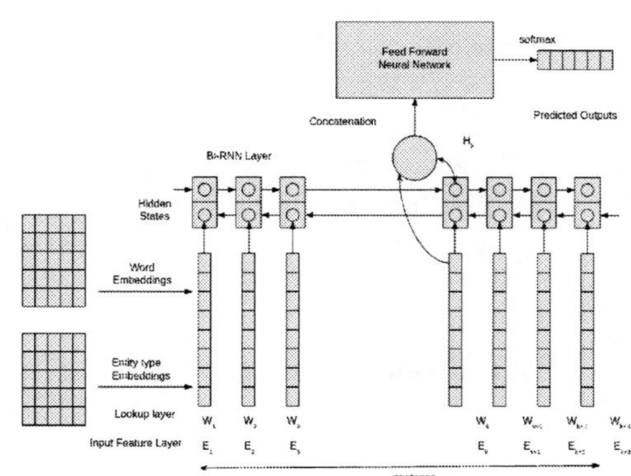

Figure 2: Model Architecture

makes use of both the input layer and higher level features to perform the final classification. We now briefly explain about each component of our model.

3.1   Input Feature Layer

For every word in the sentence we extract two features, exact word $w \in W$ and entity type $e \in E$. Here W refers the dictionary of words and E refers to dictionary of entities. Apart from all the entities, E also contains a $None$ entity type which indicates absence of an entity. In some cases the entity might span through multiple words, in that case we assign every word spanned by that entity the same entity type.

3.2   Embedding or Lookup Layer

In this layer every input feature is mapped to a dense feature vector. Let us say that E_w and E_e be the embedding matrices of W and E respectively. The features obtained from these embedding matrices are concatenated and treated as the final word-level feature (l) of the model.

The $E_w \in \mathbb{R}^{n_w \times d_w}$ embedding matrix is initialized with pre-trained word embeddings and $E_e \in \mathbb{R}^{n_e \times d_e}$ embedding matrix is initialized with random values. Here n_w, n_e refer to length of the word dictionary and entity type dictionary respectively and d_w, d_e refer to dimension of word and entity type embedding respectively.

3.3   Bidirectional RNN Layer

RNN is a powerful model for learning features from sequential data. We use both LSTM (Hochreiter and Schmidhuber, 1997) and GRU (Chung et al., 2014) variants of RNN in our experiments as they handle the vanishing and exploding gradient problem (Pascanu et al., 2012) in a better way. We use bidirectional version of RNN (Graves, 2013) where for every word forward RNN captures features from the past and the backward RNN captures features from future, inherently each word has information about whole sentence.

3.4 Feed Forward Neural Network

The hidden state of the bidirectional RNN layer acts as sentence-level feature (g), the word and entity type embeddings (l) act as a word-level features, are both concatenated (1) and passed through a series of hidden layers (2), (3) with dropout (Srivastava et al., 2014) and an output layer. In the output layer, the number of neurons are equal to the number of trigger labels. Finally we use $Softmax$ function (4) to obtain probability score for each class.

$$f = g^k \oplus l^k \tag{1}$$
$$h_0 = \tanh(W_0 f + b_0) \tag{2}$$
$$h_i = \tanh(W_i h_{i-1} + b_i) \tag{3}$$
$$p(y|x) = Softmax(W_o h_i + b_o) \tag{4}$$

Here k refers to the kth word of the sentence, i refers to the ith hidden layer in the network and $\oplus$ refers to concatenation operation. W_i, W_o and b_i, b_o are parameters of the hidden and output layers of the network respectively.

3.5 Training and Hyperparameters

We use cross entropy loss function and the model is trained using stochastic gradient descent. The implementation[1] of the model is done in python language using $Theano$ (Bergstra et al., 2010) library. We use pre-trained word embeddings obtained by Moen et al. (2013) using $word2vec$ tool (Mikolov et al., 2013).

We use training and development set for hyperparameter selection. We use word embeddings of 200 dimension, entity type embeddings of 50 dimension, RNN hidden state dimension of 250 and 2 hidden layers with dimension 150 and 100. In both the hidden layers we use dropout of 0.2.

---

[1]Implementation is available at `https://github.com/rahulpatchigolla/EventTriggerDetection`

4 Experiments and discussion

4.1 Dataset Description

We use MLEE (Pyysalo et al., 2012) corpus for performing our trigger identification experiments. Unlike other corpora on event extraction it covers events across various levels from molecular to organism level. The events in this corpus are broadly divided into 4 categories namely *"Anatomical"*, *"Molecular"*, *"General"*, *"Planned"* which are further divided into 19 sub-categories as shown in Table 1. Here our task is to identify correct sub-category of an event. The entity types associated with the dataset are summarized in Table 2.

Category	Trigger label	Train count	Test count
Anatomical	Cell Proliferation (CELLP)	82	43
	Development (DEV)	202	98
	Blood Vessel Development (BVD)	540	305
	Death (DTH)	57	36
	Breakdown (BRK)	44	23
	Remodeling (REMDL)	22	10
	Growth (GRO)	107	56
Molecular	Synthesis (SYN)	13	4
	Gene Expression (GENEXP)	210	132
	Transcription (TRANS)	16	7
	Catabolism (CATA)	20	4
	Phosphorylation (PHO)	26	3
	Dephosphorylation (DEPHO)	2	1
General	Localization (LOC)	282	133
	Binding (BIND)	102	56
	Regulation (REG)	362	178
	Positive Regulation (PREG)	654	312
	Negative Regulation (NREG)	450	233
Planned	Planned Process (PLP)	407	175

Table 1: Statistics of event triggers in MLEE corpus

Category	Entity label	Train count	Test count
Molecule	Drug or Compound	637	307
	Gene or Gene Product	1961	1001
Anatomy	Organism Subdivision	27	22
	Anatomical System	10	8
	Organ	123	53
	Multi-tissue Structure	348	166
	Tissue	304	122
	Cell	866	332
	Cellular Component	105	40
	Developing Anatomical Structure	4	2
	Organism Substance	82	60
	Immaterial Anatomical Entity	11	4
	Pathological Formation	553	357
Organism	Organism	485	237

Table 2: Statistics of entities in MLEE corpus

4.2 Experimental Design

The data is provided in three parts as training, development and test sets. Hyperparameters are tuned using development set and then final model is trained on the combined set of training and development sets using the selected hyperparameters. The final results reported here are the best results over 5 runs.

318

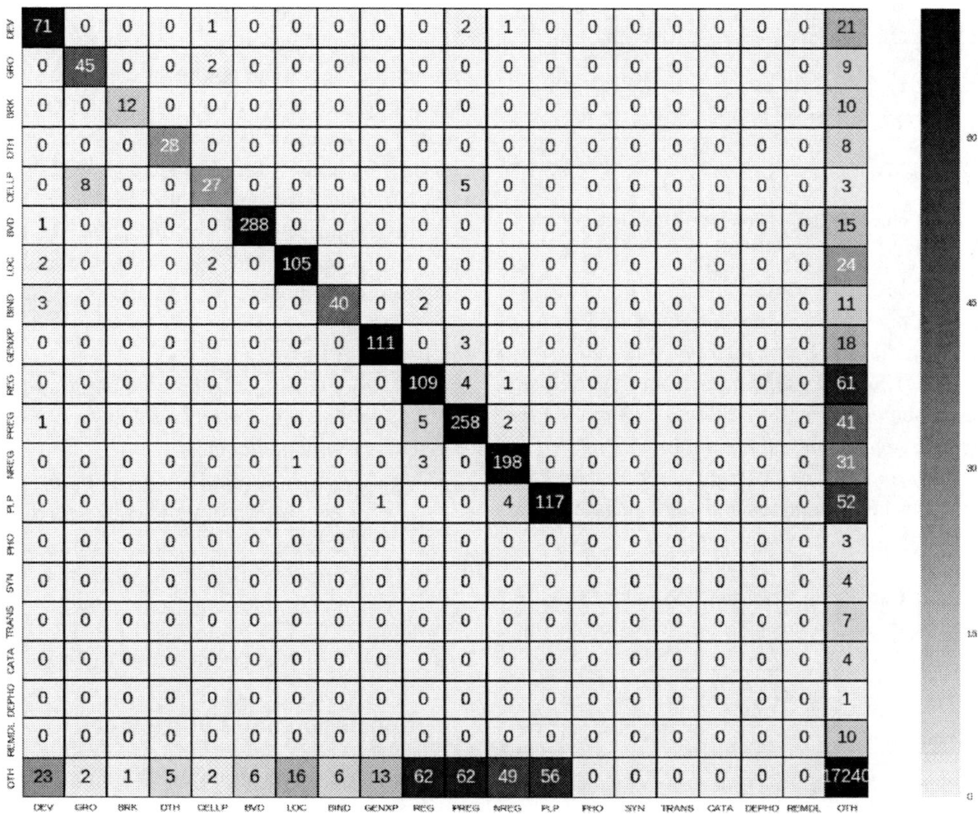

Figure 3: Confusion matrix of trigger classes with abbreviations mentioned in Table 1

We have used micro averaged F1-score as the evaluation metric and evaluated the performance of the model by ignoring the trigger classes with counts ≤ 10 in test set while training and considered them directly as false-negative while testing.

4.3 Performance comparison with Baseline Models

We compare our results with baseline models shown in Table 3. Pyysalo et al. (2012) defined a SVM based classifier with hand-crafted features. Zhou et al. (2014) also defined a SVM based classifier with word embeddings and hand-crafted features. Wang et al. (2016a) defined window based CNN classifier. Apart from the proposed models we also compare our results with two more baseline methods FFNN and CNN^{ψ} which are our implementations. Here FFNN is a window based feed forward neural network where embedding features of words within the window are used to predict the trigger label (Collobert et al., 2011). We chose window size as 3 (one word from left and one word from right) after tuning it in validation set. CNN^{ψ} is our implementation of window based CNN classifier proposed by Wang et al. (2016a) due to unavailability of their code in public domain. Our proposed model have shown slight improvement in F1-score when compared with baseline models. The proposed model's ability to capture the context of the whole sentence is likely to be one of the reasons of improvement in performance.

We perform one-side t-test over 5 runs of F1-Scores to verify our proposed model's performance when compared with FFNN and CNN^{Ψ}. The p value of the proposed model (GRU) when compared with FFNN and CNN^{ψ} are 8.57×10^{-07} and 1.178×10^{-10} respectively. This indicates statistically superior performance of the proposed model.

4.4 Category Wise Performance Analysis

The category wise performance of the proposed model is shown in Table 4. It can be observed that

Method	Precision	Recall	F1-Score
SVM (Pyysalo et al., 2012)	81.44	69.48	75.67
SVM+W_e (Zhou et al., 2014)	80.60	74.23	77.82
CNN (Wang et al., 2016a)	80.67	76.76	78.67
FFNN	77.53	75.55	76.53
CNN$^\psi$	80.75	69.36	74.62
Proposed (LSTM)	78.58	78.84	**78.71**
Proposed (GRU)	79.78	78.45	**79.11**

Table 3: Comparison of performance of our model with baseline models

model's performance in *anatomical* and *molecular* categories are better than *general* and *planned* categories. We can also infer from the confusion matrix shown in Figure 3 that *positive regulation*, *negative regulation* and *regulation* among *general* category and *planned* category triggers are causing many false positives and false negatives thus degrading the model's performance.

Trigger Category	Precision	Recall	F1-Score
Anatomical	88.86	83.06	85.87
Molecular	88.80	73.51	80.43
General	75.69	78.53	77.09
Planned	67.63	67.24	67.43
Overall	79.78	78.45	79.11

Table 4: Category wise performance of the model

4.5 Further Analysis

In this section we investigate the importance of various features and model variants as shown in Table 5. Here E_w and E_e refer to using word and entity type embedding as a feature in the model, l and g refer to using word-level and sentence-level feature respectively for the final prediction. For example, $E_w + E_e$ and g means using both word and entity type embedding as the input feature for the model and g means only using the global feature (hidden state of RNN) for final prediction.

Index	Method	F1-Score
1	E_w and g	76.52
2	E_w and $l + g$	77.59
3	$E_w + E_e$ and g	78.70
4	$E_w + E_e$ and $l + g$	79.11

Table 5: Affect on F1-Score based on feature analysis and model variants

Examples in Table 6 illustrate importance of features used in best performing models. In phrase 1 the word "*knockdown*", is a part of an entity namely "*round about knockdown endothelial*

cells" of type "*Cell*" and in phrase 2 it is trigger word of type "*Planned Process*", methods 1 and 2 failed to differentiate both of them because of no knowledge about the entity type. In phrase 3 "*impaired*" is a trigger word of type "*Negative Regulation*" methods 1 and 3 failed to correctly identify but when reinforced with word-level feature the model succeeded in identification. So, we can say that E_e feature and $l + g$ model variant help in improving the model's performance.

Index	Phrase
1	silencing of directional migration in *round about knockdown endothelial cells*
2	we show that PSMA inhibition *knockdown* or deficiency decrease
3	display altered maternal hormone concentrations indicative of an *impaired* trophoblast capacity

Table 6: Example phrases for Further Analysis

5 Conclusion and Future Work

In this paper we have proposed a novel approach for trigger identification by learning higher level features using RNN. Our experiments have shown to achieve state-of-art results on MLEE corpus. In future we would like to perform complete event extraction using deep learning techniques.

References

Sophia Ananiadou, Paul Thompson, Raheel Nawaz, John McNaught, and Douglas B Kell. 2015. Event-based text mining for biology and functional genomics. *Briefings in functional genomics* 14(3):213–230.

James Bergstra, Olivier Breuleux, Frédéric Bastien, Pascal Lamblin, Razvan Pascanu, Guillaume Desjardins, Joseph Turian, David Warde-Farley, and Yoshua Bengio. 2010. Theano: A cpu and gpu math compiler in python. In *Proc. 9th Python in Science Conf.* pages 1–7.

Junyoung Chung, Çaglar Gülçehre, KyungHyun Cho, and Yoshua Bengio. 2014. Empirical evaluation of gated recurrent neural networks on sequence modeling. *CoRR* abs/1412.3555.

Ronan Collobert, Jason Weston, Léon Bottou, Michael Karlen, Koray Kavukcuoglu, and Pavel Kuksa. 2011. Natural language processing (almost) from scratch. *J. Mach. Learn. Res.* 12:2493–2537.

Alex Graves. 2013. Generating sequences with recurrent neural networks. *CoRR* abs/1308.0850.

Sepp Hochreiter and Jürgen Schmidhuber. 1997. Long short-term memory. *Neural Comput.* pages 1735–1780.

Jin-Dong Kim, Tomoko Ohta, Sampo Pyysalo, Yoshinobu Kano, and Jun'ichi Tsujii. 2009. Overview of bionlp'09 shared task on event extraction. In *Proceedings of the Workshop on Current Trends in Biomedical Natural Language Processing: Shared Task*. Association for Computational Linguistics, pages 1–9.

Jin-Dong Kim, Sampo Pyysalo, Tomoko Ohta, Robert Bossy, Ngan Nguyen, and Jun'ichi Tsujii. 2011. Overview of bionlp shared task 2011. In *Proceedings of the BioNLP Shared Task 2011 Workshop*. Association for Computational Linguistics, pages 1–6.

Omer Levy and Yoav Goldberg. 2014. Dependency-based word embeddings. In *ACL 2014*. pages 302–308.

Tomas Mikolov, Ilya Sutskever, Kai Chen, Greg S Corrado, and Jeff Dean. 2013. Distributed representations of words and phrases and their compositionality. In *Advances in neural information processing systems*. pages 3111–3119.

Hans Moen, Sampo Pyysalo, Filip Ginter, Tapio Salakoski, and Sophia Ananiadou. 2013. Distributional semantics resources for biomedical text processing.

Claire Nédellec, Robert Bossy, Jin-Dong Kim, Jung-Jae Kim, Tomoko Ohta, Sampo Pyysalo, and Pierre Zweigenbaum. 2013. Overview of bionlp shared task 2013. In *Proceedings of the BioNLP Shared Task 2013 Workshop*. Association for Computational Linguistics, pages 1–7.

Tomoko Ohta, Sampo Pyysalo, Rafal Rak, Andrew Rowley, Hong-Woo Chun, Sung-Jae Jung, Changhoo Jeong, Sung-pil Choi, and Sophia Ananiadou. 2013. Overview of the pathway curation (pc) task of bionlp shared task 2013 .

Razvan Pascanu, Tomas Mikolov, and Yoshua Bengio. 2012. Understanding the exploding gradient problem. *CoRR, abs/1211.5063* .

Sampo Pyysalo, Tomoko Ohta, Makoto Miwa, Han-Cheol Cho, Jun'ichi Tsujii, and Sophia Ananiadou. 2012. Event extraction across multiple levels of biological organization. *Bioinformatics* 28(18):575–581.

Nitish Srivastava, Geoffrey E Hinton, Alex Krizhevsky, Ilya Sutskever, and Ruslan Salakhutdinov. 2014. Dropout: a simple way to prevent neural networks from overfitting. *Journal of Machine Learning Research* 15(1):1929–1958.

Andreas Vlachos, Paula Buttery, Diarmuid O Séaghdha, and Ted Briscoe. 2009. Biomedical event extraction without training data. In *Proceedings of the Workshop on Current Trends in Biomedical Natural Language Processing: Shared Task*. Association for Computational Linguistics, pages 37–40.

Jian Wang, Honglei Li, Yuan An, Hongfei Lin, and Zhihao Yang. 2016a. Biomedical event trigger detection based on convolutional neural network. *International Journal of Data Mining and Bioinformatics* 15(3):195–213.

Jian Wang, Jianhai Zhang, Yuan An, Hongfei Lin, Zhihao Yang, Yijia Zhang, and Yuanyuan Sun. 2016b. Biomedical event trigger detection by dependency-based word embedding. *BMC Medical Genomics* 9(2):45.

Xiaomei Wei, Qin Zhu, Chen Lyu, Kai Ren, and Bo Chen. 2015. A hybrid method to extract triggers in biomedical events. *Journal of Digital Information Management* 13(4):299.

Deyu Zhou, Dayou Zhong, and Yulan He. 2014. Event trigger identification for biomedical events extraction using domain knowledge. *Bioinformatics* 30(11):1587–1594.

Representations of Time Expressions for Temporal Relation Extraction with Convolutional Neural Networks

Chen Lin[1], Timothy Miller[1], Dmitriy Dligach[2], Steven Bethard[3] and Guergana Savova[1]

[1]Boston Children's Hospital and Harvard Medical School
[2]Loyola University Chicago
[3]University of Arizona
[1]{first.last}@childrens.harvard.edu
[2]ddligach@luc.edu
[3]bethard@email.arizona.edu

Abstract

Token sequences are often used as the input for Convolutional Neural Networks (CNNs) in natural language processing. However, they might not be an ideal representation for time expressions, which are long, highly varied, and semantically complex. We describe a method for representing time expressions with single pseudo-tokens for CNNs. With this method, we establish a new state-of-the-art result for a clinical temporal relation extraction task.

1 Introduction

Convolutional Neural Networks (CNNs) utilize convolving filters and pooling layers for exploring and subsampling a feature space, and show excellent results in tasks such as semantic parsing (Yih et al., 2014), search query retrieval (Shen et al., 2014), sentence modeling (Kalchbrenner et al., 2014), and many other natural language processing (NLP) tasks (Collobert et al., 2011).

Token sequences are often used as the input for a CNN model in NLP. Each token is represented as a vector. Such vectors could be either word embeddings trained on the fly (Kalchbrenner et al., 2014), pre-trained on a corpus (Pennington et al., 2014; Mikolov et al., 2013), or one-hot vectors that index the token into a vocabulary (Johnson and Zhang, 2014). CNN filters then act as n-grams over continuous representations. Subsequent network layers learn to combine these n-gram filters to detect patterns in the input sequence.

This token vector sequence representation has worked for many NLP tasks, but has not been well-studied for temporal relation extraction. Time expressions are complex linguistic expressions that are challenging to represent because of their length and variety. For example, for the time expressions in the THYME (Styler IV et al., 2014) colon cancer training corpus, there are 3,833 occurrences of

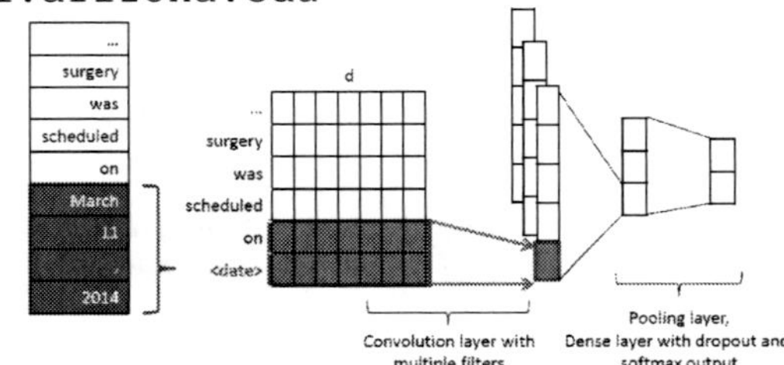

Figure 1: CNN with encoded timex

2,014 unique expressions of which 1,624 (80.6%) are multi-token, 1,104 span three or more tokens, and some span as many as 10 tokens. CNNs, which represent meaning through fragments of word sequences, might struggle to compose these fragments to represent the meaning of time expressions. For example, can a CNN properly generalize that *May 7* as a date is closer to *April 30* than *May 20*? Can it embed years like *2012* and *2040* to recognize that the former was in the past, while the latter is in the future? Time normalization systems can handle such phenomena, but they are complex and language-specific, and often require significant manual effort to re-engineer for a new domain (Strötgen and Gertz, 2013; Bethard, 2013).

Fortunately, not all tasks require full time normalization, so if the CNN can at least embed a meaningful subset of the time expression semantics, it may still be helpful in such tasks. An open question then, is how to best feed time expressions to the CNN so that it can usefully generalize over them as part of its solution to a larger task.

We propose representing time expressions as single pseudo-tokens, with single vector representations (as in Figure 1), that encode easily extractable information about the time expression that is valuable for the task of temporal relation extraction. The benefits are two-fold: 1) Only minimal linguistic preprocessing is required: off-the-shelf time expression identifiers are available with low over-

Proceedings of the BioNLP 2017 workshop, pages 322–327,
Vancouver, Canada, August 4, 2017. ©2017 Association for Computational Linguistics

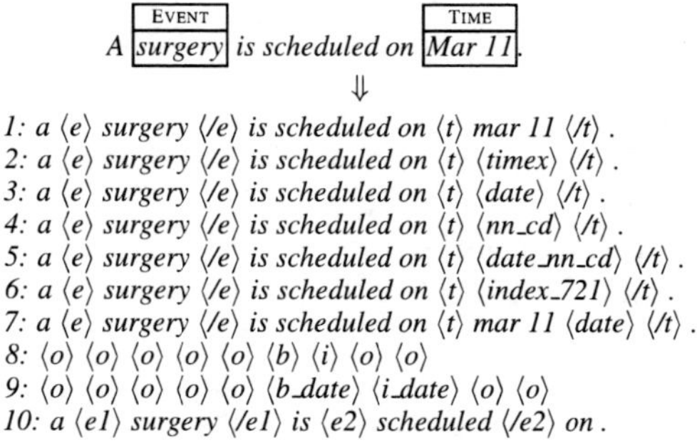

1: a ⟨e⟩ surgery ⟨/e⟩ is scheduled on ⟨t⟩ mar 11 ⟨/t⟩ .
2: a ⟨e⟩ surgery ⟨/e⟩ is scheduled on ⟨t⟩ ⟨timex⟩ ⟨/t⟩ .
3: a ⟨e⟩ surgery ⟨/e⟩ is scheduled on ⟨t⟩ ⟨date⟩ ⟨/t⟩ .
4: a ⟨e⟩ surgery ⟨/e⟩ is scheduled on ⟨t⟩ ⟨nn_cd⟩ ⟨/t⟩ .
5: a ⟨e⟩ surgery ⟨/e⟩ is scheduled on ⟨t⟩ ⟨date_nn_cd⟩ ⟨/t⟩ .
6: a ⟨e⟩ surgery ⟨/e⟩ is scheduled on ⟨t⟩ ⟨index_721⟩ ⟨/t⟩ .
7: a ⟨e⟩ surgery ⟨/e⟩ is scheduled on ⟨t⟩ mar 11 ⟨date⟩ ⟨/t⟩ .
8: ⟨o⟩ ⟨o⟩ ⟨o⟩ ⟨o⟩ ⟨o⟩ ⟨b⟩ ⟨i⟩ ⟨o⟩ ⟨o⟩
9: ⟨o⟩ ⟨o⟩ ⟨o⟩ ⟨o⟩ ⟨o⟩ ⟨b_date⟩ ⟨i_date⟩ ⟨o⟩ ⟨o⟩
10: a ⟨e1⟩ surgery ⟨/e1⟩ is ⟨e2⟩ scheduled ⟨/e2⟩ on .

Figure 2: Representations of an input sequence

head and high accuracy (Miller et al., 2015). 2) CNN filters are more effective because they operate over the time expression as one unit. The filter process can thus focus on the informative surrounding context to catch generalizable patterns instead of being trapped within lengthy time expressions.

We explored a variety of one-tag representations for time expressions, from very specific to very general. We also experimented with other ways to inject temporal information into the CNN models and compared them with our one-tag representations. We picked a challenging learning task where time expressions are critical cues for evaluating our proposed representation: clinical temporal relation extraction. The identification of temporal relations in medical text has been drawing growing attention because of its potential to dramatically increase the understanding of many medical phenomena such as disease progression, longitudinal effects of medications, a patient's clinical course, and its many clinical applications such as question answering (Das and Musen, 1995; Kahn et al., 1990), clinical outcomes prediction (Schmidt et al., 2005), and the recognition of temporal patterns and timelines (Zhou and Hripcsak, 2007; Lin et al., 2014).

Through experiments, we not only demonstrate the usefulness of one-tag representations for time expressions, but also establish a new state-of-the-art result for clinical temporal relation extraction.

2 Methods

We trained two CNN-based classifiers for recognizing two types of within-sentence temporal relations, event-event and event-time relations, as they usually call for different temporal cues (Lin et al., 2016a). The input to our classifiers was manually annotated (gold) events and time expressions during both training and testing stages. That

way we isolated the task of time expression representation for temporal relation extraction from the tasks of event and time expression recognition. We adopted the same xml-tag marked-up token sequence representation and model setup as (Dligach et al., 2017). Figure 2(1) illustrates the marked-up token sequence for an event-time instance, in which the event is marked by ⟨e⟩ and ⟨/e⟩ and the time expression is marked by ⟨t⟩ and ⟨/t⟩. Event-event instances are handled similarly, e.g. *a ⟨e1⟩ surgery ⟨/e1⟩ is ⟨e2⟩ scheduled ⟨/e2⟩ on march 11.*

We tried different ways of representing a time expression as a one-token tag. The most coarse option would be to represent all time expressions with one universal tag, ⟨timex⟩, as in Figure 2(2). For more granular options, we experimented with these additional representations: 1) The time class[1] of a time expression, as in Figure 2(3), where the time expression, *Mar 11*, is represented by its class, ⟨date⟩. 2) The Penn Treebank POS tags of the tokens in a time expression, as in Figure 2(4), where the time expression, *Mar 11*, is represented by concatenating two POS tags, ⟨nn_cd⟩. 3) The combination of time class and POS tags, as in Figure 2(5), where the time expression is represented by ⟨date_nn_cd⟩. 4) A fine-grained representation that assigns an index to each unique time expression, as in Figure 2(6), where the time expression is represented by ⟨index_721⟩, the index used every time the time expression *Mar 11* appears. For event-event relations, where time expressions are not part of the relational arguments, we tried removing the time expressions altogether, as in Figure 2(10), where *Mar 11* has been removed.

To show the contribution of one-tag representations versus adding new information to the system, we explored incorporating temporal information by adding time-class tags to the original token sequences (Figure 2(7)) and adding BIO tags with/without time classes for time expression (Figure 2(8,9)) alongside the original token sequences.

We used the same CNN architecture as the CNN used in (Dligach et al., 2017), and focused on extracting the *contains* relation. The word embeddings were randomly initialized[2] and

---

[1] We used the standard clinical domain classification (Styler IV et al., 2014), where the classes are date (e.g., *next Friday, this month*), time (e.g. *3:00 pm*), duration (e.g., *five years*), quantifier (e.g. *twice, four times*), prepostexp (e.g., *preoperative, post-surgery*), and set (e.g., *twice monthly*).

[2] Our preliminary experiments showed better results for randomly-initialized embeddings than several pre-trained embeddings. One-hot vectors were too slow for processing.

Model	Event-time relations			Event-event relations		
	P	R	F1	P	R	F1
THYME system	0.583	0.810	0.678	0.569	0.574	**0.572**
1. CNN tokens	0.660	0.775	0.713	0.566	0.522	0.543
2. CNN <timex>	0.697	0.710	0.703	0.681	0.397	0.501
3. CNN time class tags	0.705	0.759	**0.731**	0.582	0.495	0.535
4. CNN POS tags	0.727	0.710	0.719	0.619	0.462	0.529
5. CNN time class+ POS tags	0.709	0.747	0.727	0.553	0.521	0.537
6. CNN indexed time expressions	0.692	0.727	0.709	0.645	0.429	0.516
7. CNN token + time class tags	0.749	0.626	0.682	0.437	0.589	0.502
8. CNN token + BIO tags	0.691	0.708	0.700	0.570	0.423	0.486
9. CNN token + BIO-time class tags	0.713	0.726	0.719	0.428	0.542	0.478
10. CNN remove all time expressions	n/a	n/a	n/a	0.635	0.446	0.524

Table 1: Event-time and event-event *contains* relation on the dev set (all notes included)

learned through training. For the combined token and BIO sequence input, we used two embedding/convolutional branches: one for the token sequence, and one for the BIO sequence; the resulting vectors were concatenated into the same dense, dropout and final softmax layers. All models were implemented in Keras 1.0.4 (Chollet, 2015) with Theano (Theano Development Team, 2016) backend. Models were trained with a batch size of 50, a dropout rate of 0.25, RMSprop optimizer, and a learning rate of 0.0001, on a GTX Titan X GPU. Our code will be made publicly available.

3 Evaluation Methodology and Results

We tested our new representations of time expressions on the THYME corpus (Styler IV et al., 2014). We followed the evaluation setup of Clinical Temp-Eval 2016 (Bethard et al., 2016). The THYME corpus contains a colon cancer set and a brain cancer set. The colon cancer set was our main focus. Models were trained on the colon cancer training set, hyper-parameters were tuned on the colon cancer development set. Finally, the best models were re-trained using the best hyper-parameters on the combined training and development sets, tested and compared on the colon cancer test set.

As a secondary validation set, we also considered the brain cancer portion of the THYME corpus. The models were re-trained on the brain cancer training and development sets (using the best hyper-parameters found for colon cancer) and tested on the brain cancer test set.

For results on the test sets, we used the official Clinical TempEval evaluation scripts (with closure-enhanced precision, recall, and F1-score).

Table 1 shows performance on the colon development set for the THYME system and the various methods of representing time expressions to CNN models. The order of representation settings is identical to that in Figure 2. For event-time relations, all our neural models outperformed the state-of-the-art THYME system's F1. Three one-tag temporal representations with moderate granularity, time class (Table 1(3)), POS tags (Table 1(4)), and time class plus POS tags (Table 1(5)), performed better than the token sequence CNN baseline (Table 1(1)), with the time class tag representation achieving the highest score (Table 1(3)). CNNs were better able to leverage time class information in our tag-based representation (Table 1(3)), than adding time class information to the original token sequence (Table 1(7)) or adding a separate time-class neural embedding (Table 1(9)).

For event-event relations, none of the neural models performed as well as the state-of-the-art THYME system. The CNN token-based model had similar performance as some of the one-tag temporal representations (Table 1(3,4,5)). Removing the time expression entirely (Table 1(10)) did not hurt performance much, confirming that time expressions were not critical cues for within-sentence event-event relation reasoning (Xu et al., 2013). Thus, on the colon test set, we evaluated the contribution of encoding time expressions on the event-time CNN model only. For the event-event part, we used the THYME event-event system, so that our results were directly comparable with the outcomes of Clinical TempEval 2016 (Bethard et al., 2016) and the performance of the THYME system (Lin et al., 2016a,b). As for the Brain cancer data, we

Corpus	Model	*contains* relations			
		P	R	F1	*p*-value
Colon cancer	Top Clinical TempEval 2016 system	0.588	0.559	0.573	
	THYME system	0.669	0.534	0.594	
	CNN (tokens) event-time + THYME event-event	0.654	0.576	0.612	
	CNN (encode) event-time + THYME event-event	0.662	0.585	**0.621**	0.03
Brain cancer	CNN (tokens) event-time	0.765	0.371	0.500	
	CNN (encode) event-time	0.726	0.429	**0.539**	0.0002

Table 2: Performance on both Colon and Brain test sets with the Clinical TempEval evaluation.

only evaluated on the event-time CNN models, so that we could directly assess the contribution of encoding time expressions as time class tags.

The top 4 rows of Table 2 show performance on the colon cancer test set for the best model from Clinical TempEval 2016, the THYME system, our CNN model with tokens only, and our CNN model where time expressions are encoded with time class tags. (To allow comparison with prior work, the event-time relation predictions made by our CNN models were coupled with the event-event relation predictions from the THYME system.) The bottom two rows of Table 2 show performance on the brain cancer test set. On both colon and brain corpora, the encoded CNN model outperformed the regular CNN model significantly, based on a Wilcoxon signed-rank test over document-by-document comparisons, as in (Cherry et al., 2013).

4 Discussion

The CNN filters in the first layers are designed to detect the presence of highly discriminative patterns. For the event-time relation extraction task, one such pattern signaling a *contains* relation is "*on Mar 11, 2014*" as in Figure 1. However, a more generalizable pattern should be – "*on DATE*". Our time-class tag representation provided such information and contributed towards generalizability. A size-two filter can easily capture such a useful pattern, instead of picking up less generalizable patterns like "on March" or "11 ," (shown in Figure 1). For a time-sensitive learning task, especially the event-time relation extraction, our time encoding technique has been proved effective on two corpora. We hypothesize the contribution is from generalizability and efficient filter computation.

Our method did not work for event-event relations because time expressions are not critical cues for such relations. CNN models as a whole did not outperform the conventional THYME event-event system, as confirmed by Dligach et al. (2017). Event-event relations have lower inter-annotator agreement and usually leverage more of the syntactic information and event properties (Xu et al., 2013), which are not perfectly captured by token sequences. The class imbalance issues are more severe for event-event relations than for event-time relations as well (Dligach et al., 2017). These likely lead to a lower performance for event-event CNNs. In the future, we will investigate methods to improve the event-event model including incorporating syntactic information and event properties into a deep neural framework, and positive instance augmentation Yu and Jiang (2016).

Word embeddings trained by conventional methods such as word2vec and GloVe did not prove to be useful in our preliminary experiments. This is likely due to (1) lack of sufficiently large publicly available domain-specific corpora, and (2) inability of the conventional methods to capture the semantic properties of events that are key for the relation extraction task (such as event durations).

Currently, when we combined our encoded CNN-based event-time model with the THYME event-event model, we achieved the state-of-the-art performance (0.621F) on the colon cancer data. The best 2016 Clinical TempEval system achieved 0.573F (Bethard et al. (2016); row 1 of Table 2), the result of the THYME system was 0.594F (Lin et al. (2016b); row 2 of Table 2), while our best combined model reached 0.621F, significantly higher (p=0.03) than the 0.612F of the combination of a regular CNN event-time model and the THYME event-event model. Note that the number of gold event-time *contains* relation instances is similar to the number of gold event-event *contains* relations (Lin et al., 2016a). Having a better event-time model indeed made the difference.

The conventional machine learning world has focused on heavy feature engineering, while the new deep learning world has called for minimalistic pre-processing as input to powerful learners. We propose a new direction to combine the best of both worlds – infusing some knowledge into the learner input. For CNN models, multi-word time expressions are imperfectly represented in the token sequence representation. With a little engineering, we can encapsulate the time expressions in one tag with different granularities. Our experiments show that this small change still takes minimum linguistic preprocessing but delivers a significant performance boost for a temporal relation extraction task. There are other multi-token named entities (locations, organizations, etc.) where it may be hard to generalize over their multiple tokens. We believe our encoding strategy is likely to benefit tasks where critical linguistic information resides in phrases or multi-word units.

Acknowledgments

The study was funded by R01LM10090 (THYME), R01GM103859 (iPGx), and U24CA184407 (Deep-Phe). The content is solely the responsibility of the authors and does not necessarily represent the official views of the National Institutes of Health. The Titan X GPU used for this research was donated by the NVIDIA Corporation.

References

Steven Bethard. 2013. A synchronous context free grammar for time normalization. In *Proceedings of the 2013 Conference on Empirical Methods in Natural Language Processing*. Association for Computational Linguistics, Seattle, Washington, USA, pages 821–826. http://www.aclweb.org/anthology/D13-1078.

Steven Bethard, Guergana Savova, Wei-Te Chen, Leon Derczynski, James Pustejovsky, and Marc Verhagen. 2016. Semeval-2016 task 12: Clinical tempeval. *Proceedings of SemEval* pages 1052–1062.

Colin Cherry, Xiaodan Zhu, Joel Martin, and Berry de Bruijn. 2013. la recherche du temps perdu: extracting temporal relations from medical text in the 2012 i2b2 nlp challenge. *Journal of the American Medical Informatics Association* 20(5):843–848. https://doi.org/10.1136/amiajnl-2013-001624.

François Chollet. 2015. Keras. https://github.com/fchollet/keras.

Ronan Collobert, Jason Weston, Léon Bottou, Michael Karlen, Koray Kavukcuoglu, and Pavel Kuksa. 2011. Natural language processing (almost) from scratch. *Journal of Machine Learning Research* 12(Aug):2493–2537.

Amar K Das and Mark A Musen. 1995. A comparison of the temporal expressiveness of three database query methods. In *Proceedings of the Annual Symposium on Computer Application in Medical Care*. American Medical Informatics Association, page 331.

Dmitriy Dligach, Timothy Miller, Chen Lin, Steven Bethard, and Guergana Savova. 2017. Neural temporal relation extraction. *EACL 2017* page 746.

Rie Johnson and Tong Zhang. 2014. Effective use of word order for text categorization with convolutional neural networks. *arXiv preprint arXiv:1412.1058* .

Michael G Kahn, Larry M Fagan, and Samson Tu. 1990. Extensions to the time-oriented database model to support temporal reasoning in medical expert systems. *Methods of information in medicine* 30(1):4–14.

Nal Kalchbrenner, Edward Grefenstette, and Phil Blunsom. 2014. A convolutional neural network for modelling sentences. *arXiv preprint arXiv:1404.2188* .

Chen Lin, Dmitriy Dligach, Timothy A Miller, Steven Bethard, and Guergana K Savova. 2016a. Multilayered temporal modeling for the clinical domain. *Journal of the American Medical Informatics Association* 23(2):387–395.

Chen Lin, Elizabeth W Karlson, Dmitriy Dligach, Monica P Ramirez, Timothy A Miller, Huan Mo, Natalie S Braggs, Andrew Cagan, Vivian Gainer, Joshua C Denny, and Guergana K Savova. 2014. Automatic identification of methotrexate-induced liver toxicity in patients with rheumatoid arthritis from the electronic medical record. *Journal of the American Medical Informatics Association* https://doi.org/10.1136/amiajnl-2014-002642.

Chen Lin, Timothy Miller, Dmitriy Dligach, Steven Bethard, and Guergana Savova. 2016b. Improving temporal relation extraction with training instance augmentation. In *Proceedings of the 15th Workshop on Biomedical Natural Language Processing*. Association for Computational Linguistics, pages 108–113.

Tomas Mikolov, Ilya Sutskever, Kai Chen, Greg S Corrado, and Jeff Dean. 2013. Distributed representations of words and phrases and their compositionality. In *Advances in neural information processing systems*. pages 3111–3119.

Timothy A Miller, Steven Bethard, Dmitriy Dligach, Chen Lin, and Guergana K Savova. 2015. Extracting time expressions from clinical text. In *Proceedings of the 2015 Workshop on Biomedical Natural Language Processing (BioNLP 2015)*. Association for Computational Linguistics, pages 81–91.

Jeffrey Pennington, Richard Socher, and Christopher D Manning. 2014. Glove: Global vectors for word representation. In *EMNLP*. volume 14, pages 1532–43.

Reinhold Schmidt, Stefan Ropele, Christian Enzinger, Katja Petrovic, Stephen Smith, Helena Schmidt, Paul M Matthews, and Franz Fazekas. 2005. White matter lesion progression, brain atrophy, and cognitive decline: the austrian stroke prevention study. *Annals of neurology* 58(4):610–616.

Yelong Shen, Xiaodong He, Jianfeng Gao, Li Deng, and Grégoire Mesnil. 2014. Learning semantic representations using convolutional neural networks for web search. In *Proceedings of the 23rd International Conference on World Wide Web*. ACM, pages 373–374.

Jannik Strötgen and Michael Gertz. 2013. Multilingual and cross-domain temporal tagging. *Language Resources and Evaluation* 47(2):269–298. https://doi.org/10.1007/s10579-012-9179-y.

William F Styler IV, Steven Bethard, Sean Finan, Martha Palmer, Sameer Pradhan, Piet C de Groen, Brad Erickson, Timothy Miller, Chen Lin, Guergana Savova, et al. 2014. Temporal annotation in the clinical domain. *Transactions of the Association for Computational Linguistics* 2:143–154.

Theano Development Team. 2016. Theano: A Python framework for fast computation of mathematical expressions. *arXiv e-prints* abs/1605.02688. http://arxiv.org/abs/1605.02688.

Yan Xu, Yining Wang, Tianren Liu, Junichi Tsujii, I Eric, and Chao Chang. 2013. An end-to-end system to identify temporal relation in discharge summaries: 2012 i2b2 challenge. *Journal of the American Medical Informatics Association* 20(5):849–858.

Wen-tau Yih, Xiaodong He, and Christopher Meek. 2014. Semantic parsing for single-relation question answering. In *ACL (2)*. Citeseer, pages 643–648.

Jianfei Yu and Jing Jiang. 2016. Pairwise relation classification with mirror instances and a combined convolutional neural network. *Proceedings of COLING 2016*.

Li Zhou and George Hripcsak. 2007. Temporal reasoning with medical dataa review with emphasis on medical natural language processing. *Journal of biomedical informatics* 40(2):183–202.

Automatic Diagnosis Coding of Radiology Reports: A Comparison of Deep Learning and Conventional Classification Methods

Sarvnaz Karimi[1], Xiang Dai[1,2], Hamed Hassanzadeh[3], and Anthony Nguyen[3]

[1]Data61, CSIRO, Sydney, Australia
[2]School of Information Technologies, University of Sydney, Sydney, Australia
[3]The Australian e-Health Research Centre, CSIRO, Brisbane, Australia

Abstract

Diagnosis autocoding is intended to both improve the productivity of clinical coders and the accuracy of the coding. We investigate the applicability of deep learning at autocoding of radiology reports using International Classification of Diseases (ICD). Deep learning methods are known to require large training data. Our goal is to explore how to use these methods when the training data is sparse, skewed and relatively small, and how their effectiveness compares to conventional methods. We identify optimal parameters for setting up a convolutional neural network for autocoding with comparable results to that of conventional methods.

1 Introduction

Hospitals and other medical clinics invest in clinical coders to abstract relevant information from patients' medical records and decide which diagnoses and procedures meet the criteria for coding, as per coding standards such as International Statistical Classification of Diseases referred to as ICD Code. For example, *Multiple fractures of foot* is represented by the ICD-10 code 'S92.7'. These codes are used to find statistics on diseases and treatments as well as for billing purposes. Clinical coding is a specialized skill requiring excellent knowledge of medical terminology, disease processes, and coding rules, as well as attention to detail, and analytical skills. Apart from high costs of labor, human errors could lead to over and undercoding resulting in misleading statistics.

To alleviate the costs and increase the accuracy of coding, autocoding has been studied by the Natural Language Processing (NLP) community. It has been studied for a variety of clinical texts such as radiology reports (Crammer et al., 2007; Perotte et al., 2014; Kavuluru et al., 2015; Scheurwegs et al., 2016), surveillance of diseases or type of cancer from death certificates (Koopman et al., 2015a,b), and coding of cancer sites and morphologies (Nguyen et al., 2015).

Text classification using deep learning is relatively recent with promises to reduce the load of domain or application specific feature engineering. Conventional classifiers such as SVMs with well-engineered features have long shown high performance in different domains. We investigate if deep learning methods can further improve clinical text classification. Specifically, we investigate how and in what setting some of the most popular neural architectures such as Convolutional Neural Networks (CNNs) can be applied to the autocoding of radiology reports. The outcomes of our work can inform similar tasks with decision making on type and settings of text classifiers.

2 Related Work

In 2007 Pestian et al. (2007) organised a shared task which introduced a dataset of radiology reports to be autocoded with ICD9 codes. This multi-label classification task attracted a large body of research over the years—e.g., (Farkas and Szarvas, 2008; Suominen et al., 2008)—which tackled the problem with methods such as rule-based, decision trees, entropy and SVM classifiers. Text classification using SVM has long been known to be state-of-the-art.

Proceedings of the BioNLP 2017 workshop, pages 328–332,
Vancouver, Canada, August 4, 2017. ©2017 Association for Computational Linguistics

| | | | | Best Values | | |
Parameter	Definition	Range	Default	ICD9	rICD9	IMDB
Batch size	Number of samples that will be propagated through the network at each point of time	8–256	8	16	16	32
Number of epochs	Epoch is one forward pass and one backward pass of all training data	1–40	30	30	30	3
Activation function on convolution layer	Non-linearity function applied on the output of convolution layer neurons	linear, tanh, sigmoid, softplus, softsign, relu, relu6, leaky_relu, prelu, elu	relu	leaky_relu	relu6	relu
Activation function on fully connected layer	Non-linearity function applied on the output of neurons in the fully-connected layer	linear, tanh, sigmoid, softmax, softplus, softsign, relu, relu6, leaky_relu, prelu, elu	softmax	softmax	softplus	softmax
Dropout rate	At each training stage, node can be dropped out of the network with probability $1 - p$. The reduced network is then trained on the data in that stage	0.1 - 1.0	0.5	0.7	0.5	0.3
Filter size	Receptive field of each neuron also known as local connectivity	1–7 and all combinations	(3, 4, 5)	(2, 3, 4)	(2;3;4;5;6)	(3;4;5)
Depth	Number of filters per filter size	40 - 5000	100	800	100	200
Learning rate	Controls the size of weight and bias changes during training	0.0001 - 0.03	0.001	0.001	0.001	0.001
Word representation	How words in text are represented as input to the network	See Table 3	random	Medline (300)	Medline (100)	Medline (40)
Vector size	Size of input vectors. When word embeddings are used, this represents the embedding size	32–512	128	128	128	128
Stride	Size of sliding window for moving filter over input	1	1	1	1	1

Table 1: Hyperpameters, range of grid search for finding optimal values, initial and best values for three datasets.

Recently, neural network based learning methods have been investigated in generic NLP as well as domain-specific applications. For text classification, two dominant methods are: (1) Convolutional Neural Networks (CNNs) from the category of feed-forward neural networks; and (2) Long Short-Term Memory (LSTM) with a recurrent neural network (RNN) architecture. Also the use of word embeddings (Le and Mikolov, 2014)—which are to capture semantic representations of words in text—has been investigated in a variety of applications to replace one-hot (vector space) models which is the traditional method of text representation.

Text classification using CNNs has been increasingly studied in recent years (Kalchbrenner et al., 2014; Kim, 2014; Rios and Kavuluru, 2015). For example, Rios and Kavuluru (2015) applied CNN to classify biomedical articles for indexing, and Kavuluru et al. (2016) on suicide watch forums.

3 Method

We build a CNN network with the architecture proposed by (Kim, 2014). It consists of one convolutional layer using multiple filters and filter sizes followed by a max pooling and fully-connected layer to assign a label. This model is chosen based on its success in other tasks. This will set a base for what is achievable using this set of algorithms without using a very deep network or more complicated architecture.

Input text to the network is represented using two different settings: (1) a matrix of random vectors representing all the words in a document; or (2) word embeddings. We refer to word embeddings created from a corpus of medical text such as Medline citations as *in-domain*, and *out-of-domain* otherwise (i.e., using Wikipedia). We also experimented with *static* and *dynamic* embeddings. In static setting, the embedding vector values were pre-fixed based on the collection they were created on, whereas dynamic embeddings changed values during the training.

One goal of this work is to quantify the impact of CNN hyperparameters. Tuning hyperparameters can be considered equivalent of feature engineering in conventional machine learning tasks. We list some of the main hyperparameters to be set in a CNN in Table 1 (first two columns). Our experiments are focused on tuning these and investigate how they differ for different datasets.

Classifier	Accuracy	Precision	Recall	F1-score
SVM	80.52	66.02	67.69	65.63
Random Forests	68.22	50.85	49.38	48.66
Logistic Regression	79.43	66.08	66.15	64.63
CNN (default)	81.55	78.93	81.55	79.05
CNN (optimal)	**83.84**	**81.44**	**83.84**	**81.55**

Table 2: Comparison of conventional classifiers with CNN on ICD9.

4 Datasets

We experiment on two different datasets, *in-domain* and *out-of-domain*, in order to find common characteristics and domain specific properties of these datasets for text classification. These datasets are: (a) ICD9, a dataset of radiology reports, and (b) IMDB, a sentiment analysis dataset. These corpora are publicly available and are explained below.

ICD9 dataset is an open challenge dataset published by the Computational Medicine Center in 2007 (Pestian et al., 2007). The dataset consists of clinical free text which is a set of 978 anonymized radiology reports and their corresponding ICD-9-CM codes.[1] There are 38 unique ICD-9 codes present in the dataset. Given the imbalance of different disease categories in the dataset with some categories only having one or two instances, we created a revised subset rICD9. In rICD9 those codes with less than 15 instances are removed. This subset contains 894 documents with 16 unique codes. To measure how our grid search for hyper-parameters are robust and how much they are task and dataset dependent, we use an out-of-domain dataset. IMDB movie review dataset is a sentiment analysis dataset provided by Maas et al. (2011). It contains $100,000$ movie reviews from IMDB.

5 Experimental Setup

We treat this task as a multi-label classification problem. Our implementations use Tensorflow and Scikit-learn. For word embeddings we use Word2Vec (Mikolov et al., 2013). For SVM and other conventional methods, we used normalized tf-idf features similar to (Wang and Manning, 2012).

Evaluation For evaluations on ICD9 and its variant rICD9, we use stratified 10-fold cross-validation. We measure classification accu-

---

[1] Testing data for this dataset is no longer available.

Word embedding	Vector Size	Dynamic	ICD9	rICD9	IMDB
Random embedding			81.93	86.69	87.75
Word2Vec Wikipedia	40	Yes	81.03	86.24	88.79
	40	No	69.75	74.90	85.02
	100	Yes	82.04	86.86	88.55
	100	No	75.93	81.40	86.98
	300	Yes	82.41	87.22	88.21
	300	No	79.34	84.94	88.14
	400	Yes	82.60	87.24	88.10
	400	No	80.03	85.53	88.19
Word2Vec Medline	40	Yes	81.59	87.06	**89.00**
	40	No	72.31	78.05	82.11
	100	Yes	82.55	**87.76**	89.00
	100	No	78.66	84.06	85.70
	300	Yes	**83.84**	87.45	88.58
	300	No	80.88	86.30	87.10
	400	Yes	82.55	87.56	88.62
	400	No	81.39	86.66	87.21

Table 3: Impact of methods of generating word embeddings on classification accuracy.

racy, precision, recall, and F-score by macro-averaging. Stratified cross-validation is used to make label distribution in each training and validation fold as consistent as possible. IMDB dataset has been divided into training data and testing data by its providers. We therefore train the model on the training data and evaluate the results on the test data. For all datasets, all experiments are run for 50 times, and reported results are averaged over repeated experiments.

Hyperparameter Tuning Effect of varying different hyperparameters on classification accuracy is examined by a grid search method that incrementally changes the values of hyperparameters. We start from a default setting as shown in Table 1 as a baseline. We also change one parameter at a time, according to a wide range given in column three, and analyze the results to find the optimal hyperparameter values. Based on the optimal parameter values, all experiments are repeated to measure the effects.

6 Experiments and Results

CNN versus Conventional Classifiers

Classification accuracy was calculated varying values of different hyperparameters. Based on the best results we chose the optimal values for each hyper parameter as listed in columns 5 to 7 of Table 1. Table 2 compares three conventional classifiers, including SVM, Random Forests and logistics regression to CNNs. The results for CNN with default values as well as accuracy- optimized values on ICD9 dataset shows comparable results to all the three conventional classifiers. That means the two sets of algorithms can achieve similar baselines with minimal feature engineering or parameter tuning.

Effect of Pre-trained Word Vectors
Pre-trained word vectors can be considered as prior knowledge on meaning of words in a dataset. That is, instead of random values, the embedding layer can be initialized to values obtained from word embeddings. We investigated whether using word embeddings would improve classification accuracy in our coding task. Therefore, we created different word vectors trained using both Wikipedia and Medline with various vector sizes. We then compared the accuracy of random embeddings with these pre-trained embeddings. Our results, shown in Table 3, can be summarized as below: (1) Pre-trained word vectors improve the classification accuracy: The best accuracy achieved on all three datasets come from using pre-trained word vectors. It shows that pre-trained word vectors did improve the effectiveness of our model (t-test, p-value < 0.05); (2) Dynamic word vectors are better than static ones: Almost all dynamic word vectors achieve better accuracy than their corresponding static word vectors; (3) In-domain word vectors are better than generic ones: On ICD9 and its variant dataset, word vectors trained using Medline which is a collection of medical articles outperformed the word vectors trained using Wikipedia. It shows in-domain word vectors can better capture the meaning of medical terminology. On the other hand, for IMDB dataset, word vectors trained using Wikipedia were more effective than word vectors trained using Medline, but that was only if the word vectors were static. We believe that a dynamic word vector, regardless of what source it is built on, eventually leads to more accurate classification; and (4) Larger embedding size does not always lead to higher accuracy: For all three datasets, once the vector size was set to 100, the accuracy leveled with higher vector sizes. It means that the computation load associated with bigger vectors may not be necessary.

Error Analysis To identify how accurate our CNN classifier was and what mistakes it makes, we manually inspected some of these classification mistakes. We found two major sources of mistakes as below: (1) Not all the documents in ICD9 dataset have exactly one target label. 212 out of 978 documents (22%) have two target labels, and 14 documents have three. These multi-label annotations imply that even human experts cannot have full consensus on some of these coding tasks; and, (2) Companion diseases: Human experts may focus on different symptoms present on a patient report and therefore reach to different conclusions. For example, based on the following text: *"UTI with fever. Bilateral hydroureteronephrosis. Diffuse scarring lower pole right kidney."*, one expert labeled the instance as '591' (Hydronephrosis), and a second expert labeled it as both '591' and '599.0' (Urinary tract infection, site not specified), and a third expert labeled it as '591', '599.0' and '780.6' (Fever and other physiologic disturbances of temperature regulation). In this case, '591' is a majority vote, however, '599.0' may also be a reasonable target, since two of the experts agreed on that. Based on our experiments, accommodating this increases the overall accuracy on ICD9 by approximately 4%.

7 Conclusion

We explored the potential of machine learning methods using neural networks to compete with conventional classification methods. We used ICD9 coding of radiology reports. Our experiments showed that some of CNN hyperparameters such as depth are specific to a dataset or task and should be tuned, whereas some of the parameters (e.g., learning rate or vector size) can be set in advance without sacrificing the results. Our results also showed the value of using dynamic word embeddings. Our best classification results achieved comparable or superior results to SVM and logistic regression classifiers for autocoding of radiology reports. Our work is continuing in two major directions: (1) quantifying the relationships between hyperparameters using linear-regression analysis; and (2) applying CNN and LSTM models for ICD-10 autocoding of patient encounters in hospital settings.

References

K. Crammer, M. Dredze, K. Ganchev, P. Pratim Talukdar, and S. Carroll. 2007. Automatic code assignment to medical text. In *Proceedings of the Workshop on BioNLP*. Prague, Czech Republic, pages 129–136.

R. Farkas and G. Szarvas. 2008. Automatic construction of rule-based ICD-9-CM coding systems. *BMC bioinformatics* 9(3):S10.

N. Kalchbrenner, E. Grefenstette, and P. Blunsom. 2014. A convolutional neural network for modelling sentences. In *Proceedings of the 52nd Annual Meeting of the Association for Computational Linguistics*. Baltimore, Maryland, pages 655–665.

R. Kavuluru, M. Ramos-Morales, T. Holaday, A. Williams, L. Haye, and J. Cerel. 2016. Classification of helpful comments on online suicide watch forums. In *Proceedings of the 7th ACM International Conference on Bioinformatics, Computational Biology, and Health Informatics*. Seattle, WA, pages 32–40.

R. Kavuluru, A. Rios, and Y. Lu. 2015. An empirical evaluation of supervised learning approaches in assigning diagnosis codes to electronic medical records. *Artificial Intelligence in Medicine* 65(2):155–166.

Y. Kim. 2014. Convolutional neural networks for sentence classification. In *Proceedings of the Conference on Empirical Methods in Natural Language Processing*. Doha, Qatar, pages 1746–1751.

B. Koopman, S. Karimi, A. N. Nguyen, R. McGuire, D. Muscatello, M. Kemp, D. Truran, M. Zhang, and S. Thackway. 2015a. Automatic classification of diseases from free-text death certificates for real-time surveillance. *BMC Mededical Informatics & Decision Making* 15:53.

B. Koopman, G. Zuccon, A. N. Nguyen, A. Bergheim, and N. Grayson. 2015b. Automatic ICD-10 classification of cancers from free-text death certificates. *International Journal of Medical Informatics* 84(11):956–965.

Q. Le and T. Mikolov. 2014. Distributed representations of sentences and documents. In *Proceedings of The 31st International Conference on Machine Learning*. Beijing, China, pages 1188–1196.

A.L. Maas, R.E. Daly, P.T. Pham, D. Huang, A.Y. Ng, and C. Potts. 2011. Learning word vectors for sentiment analysis. In *Proceedings of the 49th Annual Meeting of the Association for Computational Linguistics: Human Language Technologies*. Portland, Oregon, pages 142–150.

T. Mikolov, I. Sutskever, K. Chen, G. Corrado, and J. Dean. 2013. Distributed representations of words and phrases and their compositionality. pages 3111–3119.

A. Nguyen, J. Moore, J. O'Dwyer, and S. Philpot. 2015. Assessing the utility of automatic cancer registry notifications data extraction from free-text pathology reports. In *American Medical Informatics Association Annual Symposium*. pages 953–962.

A. Perotte, R. Pivovarov, K. Natarajan, N. Weiskopf, F. Wood, and N. Elhadad. 2014. Diagnosis code assignment: models and evaluation metrics. *Journal of the American Medical Informatics Association* 21(2):231–237.

J. Pestian, C. Brew, P. Matykiewicz, D. Hovermale, N. Johnson, K.B. Cohen, and W. Duch. 2007. A shared task involving multi-label classification of clinical free text. In *Proceedings of the Workshop on BioNLP 2007: Biological, Translational, and Clinical Language Processing*. Prague, Czech Republic, pages 97–104.

A. Rios and R. Kavuluru. 2015. Convolutional neural networks for biomedical text classification: Application in indexing biomedical articles. In *Proceedings of the 6th ACM Conference on Bioinformatics, Computational Biology and Health Informatics*. Atlanta, Georgia, pages 258–267.

E. Scheurwegs, K. Luyckx, L. Luyten, W. Daelemans, and T. Van den Bulcke. 2016. Data integration of structured and unstructured sources for assigning clinical codes to patient stays. *Journal of the American Medical Informatics Association* 23(e1).

H. Suominen, F. Ginter, S. Pyysalo, A. Airola, T. Pahikkala, S. Salantera, and T. Salakoski. 2008. Machine learning to automate the assignment of diagnosis codes to free-text radiology reports: a method description. In *Proceedings of the ICML/UAI/COLT 2008 Workshop on Machine Learning for Health-Care Applications*. Helsinki, Finland.

S. Wang and C. Manning. 2012. Baselines and bigrams: Simple, good sentiment and topic classification. In *Proceedings of the 50th Annual Meeting of the Association for Computational Linguistics*, Jeju Island, Korea, pages 90–94.

Automatic classification of doctor-patient questions
for a virtual patient record query task

Leonardo Campillos Llanos   **Sophie Rosset**   **Pierre Zweigenbaum**
LIMSI, CNRS,
Université Paris-Saclay, Orsay, France
`{campillos|rosset|pz}@limsi.fr`

Abstract

We present the work-in-progress of automating the classification of doctor-patient questions in the context of a simulated consultation with a virtual patient. We classify questions according to the computational strategy (rule-based or other) needed for looking up data in the clinical record. We compare 'traditional' machine learning methods (Gaussian and Multinomial Naive Bayes, and Support Vector Machines) and a neural network classifier (FastText). We obtained the best results with the SVM using semantic annotations, but the neural classifier achieved promising results without it.

1 Introduction

Previous work on question classification has mostly been undertaken within the framework of question answering (hereafter, QA) tasks, where classification is but one step of the overall process. Other steps are linguistic/semantic question processing, answer retrieval and generation by integrating data; indeed, these make QA a different task to that of standard information retrieval. Biomedical QA (Zweigenbaum, 2003) has mostly focused on questions that aim to obtain knowledge to help diagnose or cure diseases, by medical doctors (Demner-Fushman and Lin, 2007) or by patients (Roberts et al., 2014b), or to obtain knowledge on biology (Neves and Leser, 2015). Clinical questions to obtain data from patient records have also been addressed (Patrick and Li, 2012).

Herein, we address a question classification task from a different perspective to existing research. Our task is set in a simulated consultation scenario where a user (a medical doctor trainee) asks questions to a virtual patient (hereafter, VP) (Jaffe

et al., 2015; Talbot et al., 2016) during the anamnesis stage, i.e. the interview to the patient to obtain diagnostic information. Question types need accurate classification to search the data in the clinical record.

In this context, question classification has aimed at identifying detailed question types (Jaffe et al., 2015). In contrast, we consider a situation where we already have a rule-based question analysis system that classifies questions according to the semantic function or content (in order to restrict the search for data in the patient record and reply coherently). This strategy works well as long as questions remain within its specifications; other questions should be handled by a different strategy. What is needed in this context is a way to determine whether a given question should be transmitted to the rule-based system or to a fallback strategy. This is the goal of the present research, which is tackled as a binary classification task. Figure 1 is a schema of the processing steps we address in this work (note that we do not represent other stages such as dialogue management).

Guiding the processing of input questions is a common step in QA systems. Questions may be filtered through an upfront classifier based on machine-learning techniques, parsing (Hermjakob, 2001), regular expressions and syntactic rules, or hybrid methods (Lally et al., 2012). To achieve that, a question analysis process might precede, which may involve detecting lexical answer types, question targets or the question focus.

Our VP system relies on named entity recognition and domain semantic labels in the question analysis. The results we report seem to show that leveraging this semantic information was beneficial for the classification step. We also tested a neural method without the semantic information, and indeed did not achieve the best performance (despite having promising results). We suggest

333

Proceedings of the BioNLP 2017 workshop, pages 333–341,
Vancouver, Canada, August 4, 2017. ©2017 Association for Computational Linguistics

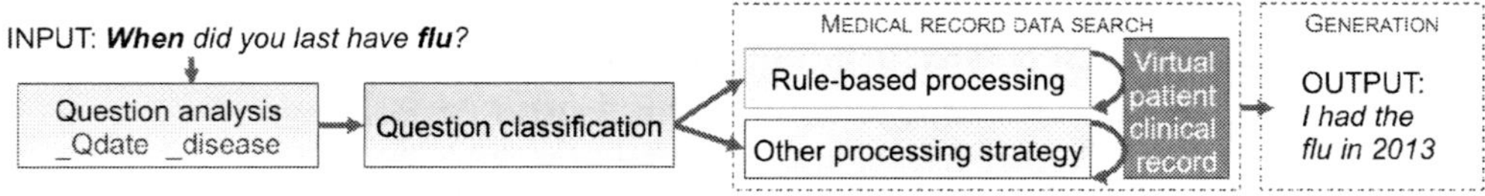

Figure 1: Schema of the question processing and search for data in the virtual patient record

that using a linear SVM classifier with the semantic information defined for the task (together with features such as token frequency and 3-grams) is a reliable technique for question triage in a rule-based system similar to the one we present.

We report results of the classification task and compare traditional machine-learning and a neural-network supervised classifiers (Bojanowski et al., 2016). We briefly review approaches to question classification (§2) and outline our task (§3). Then, we explain the sources of our data and describe them (§4). We present our methods (§5) and give our results (§6) then conclude (§7).

2 Related work

2.1 Question classification in medical QA

QA in medicine has extensively been researched. Approaches have addressed doctor questions on clinical record data (Patrick and Li, 2012), with the purpose of, among others, improving clinical decision support (Roberts et al., 2015; Goodwin and Harabagiu, 2016) or meeting the information needs of evidence-based medicine (EBM) practitioners (Demner-Fushman and Lin, 2007). EBM-focused approaches have relied on a specific knowledge framework, decomposing question topics into Problem/Population, Intervention, Comparison, and Outcome (PICO). Taxonomies of clinical question types already exist (Ely et al., 2000). (Patrick and Li, 2012) report an ontology and classification model for clinical QA applied to electronic patient notes.

Consumer health questions are another area of interest (Roberts et al., 2014b). Research has focused on classifiying the question according to the user (consumer or clinician) and question type (e.g focusing on the cause of a condition or the affected anatomic entity (Roberts et al., 2014a), or how consumer queries differ at the lexical, syntactic and/or semantic level (Slaughter et al., 2006; Roberts and Demner-Fushman, 2016).

We refer to (Athenikos and Han, 2010; Neves and Leser, 2015), respectively, for state-of-the-art reviews of QA for biomedicine and biology. Questions are generally classified into *Yes/No*, *Factoid/List* and *Definition/summary*.

Questions to a virtual patient have been addressed by mapping the user input to a set of predefined questions (Jaffe et al., 2015), as is done in a large subset of recent general-domain QA work which queries lists of frequently asked questions (FAQs) and returns their associated predetermined answers (Leuski et al., 2006; Nakov et al., 2016). Our setting is different in two ways: first, we do not rely on a FAQ but instead generate answers based on the question and on the contents of the virtual patient's record; second, we already perform fine-grained question classification with a rule-based system (Campillos et al., 2015), and aim to determine whether a given question should be referred to this rule-based strategy or deserves to be handled by a fallback strategy.

2.2 Approaches

Across the mentioned tasks, machine-learning methods for classifying questions range from hierarchical classifiers (Li and Roth, 2002) to linear support vector machines (SVM, hereafter) (Zhang and Lee, 2003). The benefit of using semantic features to improve question classification varies across experiments. For example, (Roberts et al., 2014a) reported improvements when classifying a dataset of consumer-related topics. They used an SVM with combinations of features including semantic information, namely Unified Medical Language System® (Bodenreider, 2004) Semantic Types and Concept Unique Identifiers. For their part, (Patrick and Li, 2012) used SNOMED categories. They reported improvements in classification through models including this type of feature, but not systematically. The type of the semantic information used in each task might explain these results. The impact of using semantic features is a point we explore in the present work in the context of questions to a virtual patient.

Neural network representations and classifiers are more and more applied to natural language processing (Bengio et al., 2003; Collobert et al., 2011). Word embeddings—i.e. vector representations of words—allow the prediction of a word according to the surrounding context, and vice-versa. New research questions are being raised with regard to current architectures (Mikolov et al., 2013; Pennington et al., 2014; Goldberg, 2016), parameters (e.g. vector dimension or window size), hyperparameters or the effect of input data.

The latest models include subword information in word embeddings, encoding both n-grams of characters and the standard occurrence of words (Bojanowski et al., 2016). There is a growing interest in research on word embeddings for sentence classification (Kim, 2014; Zhang et al., 2016) and question classification (Mou et al., 2015). However, a far as we know, a neural network classifier using subword information has not yet been tested on a medical question classification task. This is another point we explore herein.

3 Task description

We classify questions into those that a rule-based dialogue system can process, and those needing a supplementary method. Table 1 gives examples of these two classes of questions, and shows the semantic annotation performed in our task. A rule-based system is to be favoured to maximize precision, but developing rules for any question type is not feasible in the long term. Thus, we need a classifier to distinguish which questions should be processed through our rules and which should resort to another strategy. Those *rule-based process-*

Example of questions	Strategy	Semantic annotation
Do you cough every day ?	Rule-based	SYMPTOM, FREQUENCY
Are your parents still alive ?	Other	FAMILYMEMBER

Table 1: Examples of questions and classes

ing strategy (*RBPS* hereafter) types of question are thought to have specific patterns (e.g. recurrent n-grams, question roots or domain semantic labels), which make it possible to formalise rules.

In our system, rules are formalised based on the semantic annotation of questions.[1] For example, a rule processing the combination of SYMPTOM and FREQUENCY labels interprets the input as a query on the frequency of a symptom. Accordingly, the VP agent will answer with a fixed type

of reply instantiated with the corresponding data in the record. We hypothesize that questions not fitting this scheme will require some *other processing strategy* (*OPS* hereafter), be it statistical, neural or machine-learning-based techniques, to search data in the record (or to reply adequately when data are not available).

4 Data sources and preparation

4.1 Data sources

We collected French language questions from books aimed at medical consultation and clinical examination (Bates and Bickley, 2014; Epstein et al., 2015), as well as resources for medical translation (Coudé et al., 2011; Pastore, 2015).[2] We also collected questions from 25 transcribed doctor-patient interactions performed by human standardized patients (i.e. actors simulating medical consultations).

4.2 Additional data creation

The purpose of collecting the corpus is to train health dialogue systems aimed at simulating a consultation with virtual patients. There is a growing interest of research groups towards integrating Natural Language Interaction (NLI) features in medical education simulation systems (Hubal et al., 2003; Stevens et al., 2006; Kenny et al., 2008; Jaffe et al., 2015; Talbot et al., 2016).

Due to the lack of availability of questions, a subset of data was generated automatically by using question formation templates, semantic labels and resources from the UMLS. An example of template is *Do you suffer from* SYMPTOM *in your* ANATOMY*?*. There, the label SYMPTOM is replaced automatically with symptom terms (e.g. *pain* or *tingling*), and ANATOMY, with anatomic entities (e.g. *leg* or *arm*). We also generated automatically paraphrases of questions through a list of paraphrase patterns (e.g. *can you → are you able to*). These procedures allowed us to increase the corpus data, making up around 25% of the total number of questions. Of note is that we did not increase the corpus with more generated questions in order to avoid getting a too artificial dataset. Table 2 provides statistics on the experimental data.

4.3 Data preparation

We processed each question with our VP dialogue system (Campillos et al., 2015). Then, we manu-

---

[1] The semantic labels we use encode domain data (DISEASE), miscellanea (e.g. time or quantity) and question type or tense: e.g. QPASTYESNO (Campillos et al., 2016).

Questions	RBPS	OPS	Total
Original	1,607	825	2,432
Generated	510	328	838
Total	2,117	1,153	**3,270**

		RBPS	OPS	Total
Words	Tokens	15,276	10,299	**25,575**
	Types	3,470	2,624	**4,985**
	Mean	7.21	8.93	**7.82**
	Stdev	2.68	3.35	**3.04**
	Minimum	1	2	**1**
	Maximum	20	27	**27**
Sem. labels	Tokens	6,816	3,375	**10,291**
	Types	111	90	**119**
	Mean	3.22	3.01	**3.15**
	Stdev	1.30	1.59	**1.41**
	Minimum	0	0	**0**
	Maximum	11	11	**11**

Table 2: Distribution of experimental data (*stdev* = standard deviation)

ally labelled the output of question analysis, based on our knowledge of the dialogue system, into questions that should be processed by rule-based processing (*RBPS*) and questions requiring some other processing strategy (*OPS*). Specifically, we labelled as *RBPS* those questions with correct replies through the rule-based dialogue manager, or those questions for which the system has rules, but did not understand the questions or produced incorrect replies due to processing errors. We labelled as *OPS* the remaining questions that were not understood by the system or had wrong replies.

We split our corpus into 80% training and 20% test data (respectively, 2616 and 654 questions of both types). We performed 10-fold cross-validation on the training set for the non-neural classifiers, then applied the model to the test set.

5   Methods

We carried out tests with a linear support vector machine classifier and two Naive Bayes classifiers (Gaussian and Multinomial; from here on, respectively, Gaussian NB and Multinomial NB). We used Scikit-learn v0.18 (Pedregosa et al., 2011); the SVM used the LinearSVC implementation based on liblinear, one versus the rest scheme.

The combination of features used were inspired by (Roberts et al., 2014a). We used four sources of information:

1. The question Q itself, i.e., morphological and lexical features:

- Token and frequency in Q (TK)

- Question root (QR): the three first words of Q

- Three-character-grams (3CG) and frequency

- Three-grams (3G) and their frequency

- Number of words in Q (WC)

- Minimum, maximum and average word length in Q (WL)

2. The relation of Q to system knowledge, i.e., the term is found in the core system lexicon:

- Out-of-vocabulary words (NIL): terms in Q not found in system lexicon

3. Word representations computed from an external corpus:

- Average word embeddings of words in Q (WE). We used pre-trained word vectors (see below) with the best combination of parameters we tested (window=10, vector dimension=100, negative samples=10, learning rate=0.1, sampling threshold=1-e4). We only used this feature for the SVM classifier.

4. Annotations produced by the question analysis component of our dialogue system:

- Semantic annotation of Q (SEM)

We also tested the neural method implemented in FastText (Joulin et al., 2016). An extension of word2vec (Mikolov et al. 2013), FastText associates n-grams of words and/or characters to learn word vectors. It is a shallow neural model relying on a hidden layer, where a sentence is represented by averaging the vector representations of each word. This text representation is then input to a linear classifier (a hierarchical softmax, which reduces computational complexity). As our data were scarce, we used word vectors pretrained in a large domain corpus from the European Medicines Agency,[3] which amounts to more than 16 million tokens after tokenization. Several parameter values were tested: window size of 2, 4, 6, 8 and 10, vector dimension of 50, 100 and 300, use of 3-grams or 3-character-grams, number of negative samples (5, 10 or 20), learning rate (0.1 and 0.05) and sub-sampling threshold (1e-3 and 1e-4). We only tested the skip-gram architecture since it has

---

[3]http://opus.lingfil.uu.se/EMEA.php/

been observed to yield better results (Chiu et al., 2016). The minimum word count was fixed to 1, given the scarcity of our labelled data. We did not use semantic annotation to create word vectors.

6 Results and discussion

Table 3 breaks down our results (reported as F1-score) in the training set (top of the table) with different parameter combinations and non-neural classifiers. The weighted average F1-score was computed based upon both F1-scores of classifying *RBPS* and *OPS* types of questions. The best combinations of parameters found in the training set were applied to the test set; their results are placed at the bottom of the table. Note that a baseline method making a majority class decision would categorize each question as *RBPS*: since the proportion of *RBPS* is 0.647, its weighted average F1-score would be $0.647^2 = 0.419$.

The SVM classifier outperformed the other classifiers and the neural classifier. In all combinations of features used with non-neural methods, the use of semantic labels improved question classification. Multinomial NB obtained better results than Gaussian NB. Results with the best combinations of features and Multinomial NB gave similar results to those yielded by the neural method.

In such a small dataset and constrained task, the use of word embeddings as feature did not improve classification performance. This could be due to the data used for pre-training word embeddings. Despite being related to the domain, the nature of texts used for pre-training vectors is different to that of a clinical consultation context. Using the combination of token/frequency and semantic annotation together with another feature provided the highest results (or almost the highest). The use of 3-character-grams, word length or word count contributed to good classification, but their benefit was not strong, nor is it clear which feature was more relevant. Using 3-grams seems to be the exception: the best combination of parameters—as it improved results in all models—is 3-grams, semantic labels and token/frequency. Not shown in the table, when semantic labels are not used, the other features did not improve classification in our task (except 3-grams with Gaussian NB).

We note that the F1-scores obtained on the test set are similar to that obtained by cross-validation on the training set: the system did not overfit the training data.

The fact that we used a subset of generated questions from patterns could be argued as a bias. However, we tested the above models in a subset of 2,282 questions without any generated sentence, and the models and classifiers had similar results (but lower F1-scores). We again obtained the best results (avg. F1=0.81) with Linear SVC, with models using semantic features with or without all other parameters (e.g. QR+TK+WL+WC+SEM+3G+NIL and TK+SEM+QR+WC). We also tested the same combinations of features in Linear SVC with and without computing term frequency-inverse document frequency (tf-idf), and also a Logistic Regression classifier (with and without tf-idf). For each group of parameters, results were similar to those yielded by Linear SVC (which does not use tf-idf).

As for the neural method, Table 4 reports our results. The F1-score was computed based on precision and recall of the top ranked label (precision and recall @1). The best result was an average F-score of 0.812 (window of 10, vector dimension of 100, negative sampling of 10, learning rate of 0.1 and sampling threshold of 1-e4). We achieved similar results by modifying parameters (e.g. window of 6 or 8, vector dimension of 50, or use of 3 grams). Interestingly, using both 3 grams and 3-character-grams tended to lower performance.

We can draw two observations from our results. First, we find it beneficial leveraging the semantic information used for question analysis at the classification step. This could be a hint for developping QAs in a similar task and restricted domain to the one here presented. That is, the question analysis and classification steps for a similar rule-based system would need to build on a comprehensive semantic scheme permeating both rule development, entity type annotation and question triage. This is what seems to explain our lower results obtained when semantic features were not used in with machine-learning classifiers and the neural method. Indeed, (Jaffe et al., 2015) also reported an error reduction in question classification when domain concept-based features were used in the question classifier for their VP system.

Second, we found necessary to complement the neural approaches in this restricted task with natural language processing techniques to raise the classification performance. We trained a large amount of data for generating word embeddings (to use them as features for the LinearSVC classi-

TRAINING			
Parameters	**Linear SVC**	**Gaussian NB**	**Multinomial NB**
TK	**0.798**	0.573	0.783
3G	0.766	**0.678**	**0.787**
3CG	0.752	0.539	0.751
SEM	0.746	0.389	0.676
QR	0.679	0.427	0.670
WE	0.616	-	-
WC	0.611	0.554	0.612
WL	0.519	0.467	0.576
TK+SEM	**0.839**	0.595	**0.805**
TK+3G	0.814	**0.729**	0.794
TK+SEM+3G	**0.861**	**0.741**	**0.815**
TK+SEM+QR	0.844	0.657	0.802
TK+SEM+WC	0.841	0.596	0.803
TK+SEM+NIL	0.839	0.595	0.805
TK+SEM+3G+NIL	**0.862**	0.741	**0.815**
TK+SEM+3G+WC	**0.858**	**0.742**	0.809
TK+SEM+QR+WC	0.843	0.659	0.797
TK+SEM+3G+3CG	0.834	0.756	0.796
TK+SEM+QR+WC+WL	0.844	0.693	0.800
TK+SEM+QR+WC+WL+NIL	0.844	0.693	0.800
TK+SEM+3G+QR+WC+WL+NIL	**0.860**	**0.763**	**0.811**
TK+SEM+3CG+QR+WC+WL+NIL	0.816	0.701	0.781
TK+SEM+3G+QR+WC+WE+WL+NIL	**0.862**	-	-
TK+SEM+3G+QR+WC+WL+3CG+NIL	0.840	0.764	0.795
TEST			
TK+SEM+3G	**0.866**	**0.765**	**0.817**
TK+SEM+3G+WC	**0.871**	**0.766**	0.806
TK+SEM+3G+NIL	**0.866**	0.765	**0.817**
TK+SEM+3G+QR+WC+WL+NIL	**0.870**	0.759	**0.810**

TK: token; SEM: semantic labels; WL: maximum, minimum and average word length; WC: word count; QR: question root (3 first words); 3G: 3-grams; 3CG: 3-character-grams; NIL: word not in lexicon

Table 3: Avg. F1 of non-neural classifiers with the best tested features in training and test sets

WS	DIM	GR	CHGR	NEG	LR	SAMP	Avg F1
10	100	0	0	10	0.1	1-e4	**0.812**
8	50	3	0	20	0.1	1-e4	0.804
8	100	0	3	20	0.1	1-e4	0.803
6	50	0	3	10	0.1	1-e4	0.803
10	50	0	3	10	0.1	1-e4	0.800
2	50	3	3	20	0.1	1-e4	0.800
4	300	0	0	20	0.05	1-e3	0.792
10	300	0	3	10	0.05	1-e4	0.789

WS: window size; DIM: vector dimension; GR: n-grams; CHGR: character-grams; NEG: number of negative samples; LR: learning rate; SAMP: sampling threshold

Table 4: Results of the best tested models (neural approach)

fier) and also used a neural model to classify questions. However, our results agree with the observation that restricted-domain QA is less affected by data-intensive methods, but depend on refined language processing methods (Mollá and Vicedo, 2007)—in this type of system, accurate semantic annotation. On the other hand, the neural method seems promising in this kind of classification task, and how to use domain semantic information with it requires further exploration, in line with current works (Yu et al., 2016). We also need to pretrain vectors on domain data of different nature (e.g. clinical records) to confirm our results. Finally, other methods for computing vector representations of sentences deserve to be explored.

7   Conclusions

For the task of optimizing question processing in a VP natural language system, we reported the improvement of using the semantic information in the question analysis step as a feature for question classification. This is likely due to the idiosyncrasy of our task, where the dialogue system makes use of semantic rules for processing input questions. We are nonetheless interested in confirming to which extent reusing semantic information from the question analysis would benefit the classification step in QA systems for other tasks and domains. Anyhow, the neural method here tested yielded promising results for similar classification tasks. Other approaches to test might be including semantic annotation to generate vector representations of questions, pretraining word vectors on clinical record data, and using information from the VP clinical record as another source of features for classification.

Acknowledgments

The Société d'Accélération de Transfert Technologique (SATT) Paris-Saclay funded this research. The authors kindly appreciate the anonymous reviewers' comments and thank Julien Tourille for the helpful discussions on ScikitLearn.

References

Sofia J Athenikos and Hyoil Han. 2010. Biomedical question answering: A survey. *Computer methods and programs in biomedicine* 99(1):1–24.

Barbara Bates and Lynn S Bickley. 2014. *Guide de l'examen clinique-Nouvelle édition 2014*. Arnette-John Libbey Eurotext.

Yoshua Bengio, Réjean Ducharme, Pascal Vincent, and Christian Jauvin. 2003. A neural probabilistic language model. *Journal of machine learning research* 3(Feb):1137–1155.

Olivier Bodenreider. 2004. The Unified Medical Language System (UMLS): Integrating biomedical terminology. *Nucleic Acids Research* 32(Database issue):D267–270.

Piotr Bojanowski, Edouard Grave, Armand Joulin, and Tomas Mikolov. 2016. Enriching word vectors with subword information. *CoRR* abs/1607.04606. http://arxiv.org/abs/1607.04606.

Leonardo Campillos, Dhouha Bouamor, Éric Bilinski, Anne-Laure Ligozat, Pierre Zweigenbaum, and Sophie Rosset. 2015. Description of the PatientGenesys dialogue system. In *Proceedings of the SIGDIAL 2015 Conference*. Association for Computational Linguistics, pages 438–440.

Leonardo Campillos, Dhouha Bouamor, Pierre Zweigenbaum, and Sophie Rosset. 2016. Managing linguistic and terminological variation in a medical dialogue system. In *Proceedings of LREC 2016, Portoroz, Slovenia, 24-27 May 2016*. pages 3167–3173.

Billy Chiu, Gamal Crichton, Anna Korhonen, and Sampo Pyysalo. 2016. How to train good word embeddings for biomedical NLP. *ACL 2016* pages 166–174.

Ronan Collobert, Jason Weston, Léon Bottou, Michael Karlen, Koray Kavukcuoglu, and Pavel Kuksa. 2011. Natural language processing (almost) from scratch. *Journal of Machine Learning Research* 12(Aug):2493–2537.

Claire Coudé, Franois-Xavier Coudé, and Kai Kassmann. 2011. *Guide de conversation médicale - français-anglais-allemand*. Lavoisier, Médecine Sciences Publications.

Dina Demner-Fushman and Jimmy Lin. 2007. Answering clinical questions with knowledge-based and statistical techniques. *Computational Linguistics* 33(1):63–103.

John W Ely, Jerome A Osheroff, Paul N Gorman, Mark H Ebell, M Lee Chambliss, Eric A Pifer, and P Zoe Stavri. 2000. A taxonomy of generic clinical questions: classification study. *British Medical Journal* 321(7258):429–432.

Owen Epstein, David Perkin, John Cookson, and David P. de Bono. 2015. *Guide pratique de l'examen clinique*. Elsevier Masson, Paris.

Yoav Goldberg. 2016. A primer on neural network models for natural language processing. *Journal of Artificial Intelligence Research* 57:345–420.

Travis R Goodwin and Sanda M Harabagiu. 2016. Medical question answering for clinical decision support. In *Proceedings of the 25th ACM International on Conference on Information and Knowledge Management*. ACM, pages 297–306.

Ulf Hermjakob. 2001. Parsing and question classification for question answering. In *Proceedings of the Workshop on Open-domain Question Answering - Volume 12*. Association for Computational Linguistics, ODQA '01, pages 1–6.

Robert C Hubal, Robin R Deterding, Geoffrey A Frank, Henry F Schwetzke, and Paul N Kizakevich. 2003. Lessons learned in modeling virtual pediatric patients. *Studies in Health Technology and Informatics* pages 127–130.

Evan Jaffe, Michael White, William Schuler, Eric Fosler-Lussier, Alex Rosenfeld, and Douglas Danforth. 2015. Interpreting questions with a log-linear ranking model in a virtual patient dialogue system. In *Proc. of the 10 Workshop on Innovative Use of NLP for Building Educational Applic.*. pages 86–96.

Armand Joulin, Edouard Grave, Piotr Bojanowski, and Tomas Mikolov. 2016. Bag of tricks for efficient text classification. *arXiv preprint arXiv:1607.01759* http://arxiv.org/abs/1607.04606.

Patrick Kenny, Thomas D Parsons, Jonathan Gratch, and Albert A Rizzo. 2008. Evaluation of Justina: a virtual patient with PTSD. In *International Workshop on Intelligent Virtual Agents*. Springer, pages 394–408.

Yoon Kim. 2014. Convolutional neural networks for sentence classification. In *Proceedings of the EMNLP 2014, October 25-29, 2014, Doha, Qatar*. pages 1746–1751.

Adam Lally, John M. Prager, Michael C. McCord, Branimir Boguraev, Siddharth Patwardhan, James Fan, Paul Fodor, and Jennifer Chu-Carroll. 2012. Question analysis: How watson reads a clue. *IBM Journal of Research and Development* 56(2:1):1–14.

Anton Leuski, Ronakkumar Patel, David Traum, and Brandon Kennedy. 2006. Building effective question answering characters. In *Proceedings of the 7th SIGdial Workshop on Discourse and Dialogue*. Association for Computational Linguistics, Stroudsburg, PA, USA, SigDIAL '06, pages 18–27.

Xin Li and Dan Roth. 2002. Learning question classifiers. In *Proceedings of the 19th international conference on Computational linguistics-Volume 1*. Association for Computational Linguistics, pages 1–7.

Tomas Mikolov, Kai Chen, Greg Corrado, and Jeffrey Dean. 2013. Efficient estimation of word representations in vector space. *arXiv preprint arXiv:1301.3781* https://arxiv.org/abs/1301.3781.

Diego Mollá and José Luis Vicedo. 2007. Question answering in restricted domains: An overview. *Computational Linguistics* 33(1):41–61.

Lili Mou, Hao Peng, Ge Li, Yan Xu, Lu Zhang, and Zhi Jin. 2015. Discriminative neural sentence modeling by tree-based convolution. In Lluís Màrquez, Chris Callison-Burch, Jian Su, Daniele Pighin, and Yuval Marton, editors, *EMNLP*. Association for Computational Linguistics, pages 2315–2325.

Preslav Nakov, Lluís Màrquez, Alessandro Moschitti, Walid Magdy, Hamdy Mubarak, Abed Alhakim Freihat, Jim Glass, and Bilal Randeree. 2016. Semeval-2016 task 3: Community question answering. *Proceedings of SemEval* pages 525–545.

Mariana Neves and Ulf Leser. 2015. Question answering for biology. *Methods* 74:36–46.

Flicie Pastore. 2015. *How can I help you today? Guide de la consultation mdicale et paramdicale en anglais*. Ellipses, Paris.

Jon Patrick and Min Li. 2012. An ontology for clinical questions about the contents of patient notes. *J. of Biomedical Informatics* 45(2):292–306.

Fabian Pedregosa, Gaël Varoquaux, Alexandre Gramfort, Vincent Michel, Bertrand Thirion, Olivier Grisel, Mathieu Blondel, Peter Prettenhofer, Ron Weiss, Vincent Dubourg, et al. 2011. Scikit-learn: Machine learning in python. *Journal of Machine Learning Research* 12(Oct):2825–2830.

Jeffrey Pennington, Richard Socher, and Christopher D Manning. 2014. Glove: Global vectors for word representation. In *EMNLP*. volume 14, pages 1532–1543.

Kirk Roberts and Dina Demner-Fushman. 2016. Interactive use of online health resources: a comparison of consumer and professional questions. *Journal of the American Medical Informatics Association* 23(4):802–811.

Kirk Roberts, Halil Kilicoglu, Marcelo Fiszman, and Dina Demner-Fushman. 2014a. Automatically classifying question types for consumer health questions. In *AMIA Annual Symposium*.

Kirk Roberts, Halil Kilicoglu, Marcelo Fiszman, and Dina Demner-Fushman. 2014b. Decomposing consumer health questions. In *BioNLP Workshop*. pages 29–37.

Kirk Roberts, Matthew Simpson, Dina Demner-Fushman, Ellen Voorhees, and William Hersh. 2015. State-of-the-art in biomedical literature retrieval for clinical cases: a survey of the TREC 2014 CDS track. *Information Retrieval Journal* .

Laura A Slaughter, Dagobert Soergel, and Thomas C Rindflesch. 2006. Semantic representation of consumer questions and physician answers. *International journal of medical informatics* 75(7):513–529.

Amy Stevens, Jonathan Hernandez, Kyle Johnsen, Robert Dickerson, Andrew Raij, Cyrus Harrison, Meredith DiPietro, Bryan Allen, Richard Ferdig, Sebastian Foti, et al. 2006. The use of virtual patients to teach medical students history taking and communication skills. *The American Journal of Surgery* 191(6):806–811.

Thomas B Talbot, Nicolai Kalisch, Kelly Christoffersen, Gale Lucas, and Eric Forbell. 2016. Natural language understanding performance & use considerations in virtual medical encounters. *Medicine Meets Virtual Reality 22: NextMed/MMVR22* 220:407.

Zhiguo Yu, Trevor Cohen, Elmer V Bernstam, and Byron C Wallace. 2016. Retrofitting word vectors of MeSH terms to improve semantic similarity measures. *EMNLP 2016* page 43.

Dell Zhang and Wee Sun Lee. 2003. Question classification using support vector machines. In *Proceedings of the 26th annual international ACM SIGIR conference on Research and development in informaion retrieval*. ACM, pages 26–32.

Ye Zhang, Stephen Roller, and Byron C. Wallace. 2016. MGNC-CNN: A simple approach to exploiting multiple word embeddings for sentence classification. In *HLT-NAACL*.

Pierre Zweigenbaum. 2003. Question answering in biomedicine. In Maarten de Rijke and Bonnie Webber, editors, *Proc Workshop on Natural Language Processing for Question Answering, EACL 2003*. ACL, Budapest, pages 1–4.

Assessing the performance of Olelo, a real-time
biomedical question answering application

Mariana Neves, Fabian Eckert, Hendrik Folkerts, Matthias Uflacker
Hasso Plattner Institute at University of Potsdam
August Bebel Strasse 88, Potsdam 14482 Germany
mariana.neves@hpi.de

Abstract

Question answering (QA) can support physicians and biomedical researchers to find answers to their questions in the scientific literature. Such systems process large collections of documents in real time and include many natural language processing (NLP) procedures. We recently developed Olelo, a QA system for biomedicine which includes various NLP components, such as question processing, document and passage retrieval, answer processing and multi-document summarization. In this work, we present an evaluation of our system on the the fifth BioASQ challenge. We participated with the current state of the application and with an extension based on semantic role labeling that we are currently investigating. In addition to the BioASQ evaluation, we compared our system to other on-line biomedical QA systems in terms of the response time and the quality of the answers.

1 Introduction

Question answering (QA) is the task of automatically answering questions posed by users (Jurafsky and Martin, 2013). As opposed to information retrieval (IR), input is in the form of natural language, e.g., English, instead of keywords, and answers are provided as short answers, instead of presenting a list of relevant documents. Therefore, QA systems need to rely on various natural language processing (NLP) components, such as question understanding, named-entity recognition (NER), document and passage retrieval, answer extraction and multi-document summarization, among others. QA systems have been developed for many domains, including

biomedicine (Athenikos and Han, 2010; Neves and Leser, 2015). Given the large collection of biomedical documents, e.g., in PubMed, researchers and physicians need to obtain answers for their various questions in a timely manner.

Much research has been published in the past for biomedical QA (Athenikos and Han, 2010), but focus was previously mainly on clinical documents. QA for biomedicine has recently gained importance owing to the BioASQ challenges (Tsatsaronis et al., 2015), for which the organizers created comprehensive datasets of questions, answers and intermediate results. The BioASQ challenge considers four types of questions: (i) yes/no, (ii) factoid, (iii) list and (iv) summary. For yes/no questions, a system should return either of the two answers, factoid and list questions expect one or more short answers, e.g., a gene name, while a short paragraph should be generated as answer for summary questions. Despite the accessibility of these datasets to support development and evaluation of QA systems for biomedicine, few QA applications are currently available on-line.

We recently developed Olelo[1], a QA system for biomedicine (Kraus et al., 2017). It relies on a local index of the Medline documents, includes domain terminologies and implements algorithms specifically designed for biomedical QA. Previous versions of our system were evaluated in the last three editions of the BioASQ challenges (Schulze et al., 2016; Neves, 2015, 2014).

In this work, we perform a comprehensive evaluation of our application, both automatically, during participation in the fifth edition of the BioASQ challenge, as well as manually, by checking our answers against the gold standard ones from BioASQ benchmarks. We also present re-

---

[1]http://hpi.de/plattner/olelo

342

Proceedings of the BioNLP 2017 workshop, pages 342–350,
Vancouver, Canada, August 4, 2017. ©2017 Association for Computational Linguistics

sults for a new extension based on semantic role labeling (SRL) that we are considering for our application. Finally, we performed a comparison of Olelo to the other on-line biomedical QA tools.

2 Related work

We are only aware of three other biomedical QA services, as surveyed in (Bauer and Berleant, 2012), namely, askHERMES (Cao et al., 2011), EAGLi (Gobeill et al., 2015) and HONQA (Cruchet et al., 2009). However, none of the these systems performs robustly to most question types. Further, as far as we know, they have not been recently evaluated on comprehensive biomedical QA benchmarks, as the ones provided by BioASQ.

askHermes extracts answers from various sources, e.g., PubMed and Wikipedia, and presents answers as a cluster of terms, a ranked list or clustered by content, along with the corresponding relevant passages. However, the result page tends to be very long and contains more information than most users can deal with. The methods behind askHermes include regular expressions for question understanding, classification into 12 topics and keyword identification, both based on machine learning approaches, and the use of the MetaMap system for concept recognition. Document indexing is based on the BM25 model and passage ranking is based on the longest common subsequence (LCS) score.

EAGLi extracts the answers exclusively from PubMed abstracts and returns a list of concepts as answers. When no answer is found, the system returns a list of potential relevant publications, along with selected passages. The system locally indexes Medline with the Terrier information retrieval platform and uses the Okapi BM25 as weighting scheme to rank documents. EAGLi provides answers based on the Gene Ontology (GO) concepts.

Finally, HONQA relies on certified websites from the Health On The Net (HON), from which it extracts the answers, and considers a variety of question types. Questions can also be posed in French and Italian. The system relies on UMLS to detect the type of the expected answer and it follows the typical architecture of QA systems, but no details are presented in the publication.

3 Methods

In this section, we briefly describe the current methods behind Olelo as well as its extension for answer extraction based on SRL.

3.1 Olelo QA application

Olelo relies on the typical three steps of QA workflow (Athenikos and Han, 2010), namely, question processing, document/passage retrieval and answer extraction. Details of our methods have been previously published (Kraus et al., 2017; Schulze and Neves, 2016), but we give an overview of these below.

The application is built on top of an in-memory database (IMDB) that accounts for data storage (question, documents and terminologies) and text analysis. The latter are based both on built-in text processing features from the database, namely, sentence splitting, tokenization and part-of-speech tagging, as well as custom implemented SQL procedures for some QA components, such as question understanding, multi-document summarization and answer extraction. The database also includes an NER component based on custom dictionaries that we compiled based on concepts from MeSH and UMLS. Our document collection currently includes Medline abstracts and full text from PubMed Central Open Access.

When a question is posed to the system, its type (e.g., factoid or summary) is extracted via regular expressions. Further, in the case of factoid or list questions, the expected semantic types are detected, e.g., whether a gene or disease name. A query is then generated for the question based on the detected named entities (from the NER component) and other keywords from the question. Relevant documents and passages are then retrieved based on some simple heuristics that consider keywords and named entities from the question. For the answer extraction, different approaches are considered depending on the question type. For summary questions, a custom summary is generated based on the relevant sentence and corresponding named entities. Our approach is based on a graph-based approach for sentence selection (Schulze and Neves, 2016). In the case of factoid questions, and given the set of potentially relevant sentences, the system returns the corresponding MeSH concepts which belong from the same types of the expected type.

Contrary to BioASQ, our application does not

distinguish between factoid and list questions, thus, more that one exact answer can be returned for a factoid question. Further, it does not yet support yes/no questions. Finally, Olelo supports definition questions, e.g., "What is zika virus?", a type not supported in BioASQ. For these cases, the system returns the respective MeSH definition.

3.2 Semantic role labeling for answer extraction

We currently investigate an extension to our system based on SRL whose goal is to improve both the question understanding and answer extraction steps. Our aim is to find correct answers by identifying semantic conformities between a question and its snippets. As a first investigative step, we experimented with the BioKIT SRL tool (Dahlmeier and Ng, 2010) and used it to label the datasets from the first three years of the BioASQ challenge. We propose an initial rule-based proof-of-concept approach to investigate if SRL could improve Olelo QA system. Therefore, we put focus on finding suitable rules for all question types supported in the BioASQ challenge, i.e., yes/no, factoid, list and summary. In our experiments, we relied only on the gold-standard snippets provided by BioASQ, instead of the ones retrieved by Olelo.

Yes/no questions. When experimenting with yes/no questions, we soon observed problems with the skewed nature of the training data. As more than 4 out of 5 correct answers had to be "yes", the challenge was more about finding out in which rare cases to answer "no", instead of whether the answer was "yes" or "no". The latter approach would regularly lead to worse results than the approach of simply answering "yes" to every question. Initially, we were motivated by the idea that SRL could help us to be more confident when an answer was "yes". This could be achieved by finding a semantically matching answer to a question. However, due to the characteristics of the BioASQ data, being confident of when to output "yes" was not helpful to improve results. Hence, we investigated if we could find out whether an answer is "no" by applying a similar strategy.

We investigated the detection of negation. Looking at specific cases of the training data, we created rules which include negation terms or the occurrence of certain domain-specific terms. If multiple answer snippets matched, we calculated an overall score for taking the yes/no decision. For this score, answer snippets were weighted differently depending on the strength of their match.

Factoid and list questions. Factoid and list questions demand slightly different approaches. For both categories, we implemented a rule-based priority queue on answer candidates. The highest priority was given to answers where question and answer snippet contained the same predicate and for which the argument type of the answer matched the argument type of the question word of the question (e.g. "what"). The next highest priority was given to answers which were somehow related to the matching predicate. Hereby, the argument types "Arg0" and "Arg1" have higher priority. For factoid questions, the top five answers were selected for the submission. For list questions, the maximum number of answers to be listed decreased depending on how low the priority levels got. This should ensure that we do not leave out a high-priority answer with a high probability to be correct in our model. Additionally, too many low-priority answers should be avoided to keep an acceptable precision level. Besides the SRL-based priority queue, we introduced a rule for the list question approach. As of the essence of list questions, we consider enumerations by detecting symbols like commas or the conjunction "and".

Summary questions and ideal answers. We also investigated SRL for the summarization task. For summary questions, out of the given sets of answer snippets, the system selects the ones with the largest semantic conformities. Similar to factoid and list questions, the semantic conformity is determined by the degree to which question and answer snippet contain similar predicate argument structures or vocabulary. The same, previously described priority queue is applied. The ideal answers for yes/no, factoid and list questions were retrieved by selecting the whole answer snippet that included the highest priority answer to the corresponding question. If no answer could be determined, we followed the same procedure as for summary questions.

4 Results

In this section we present an evaluation of Olelo based on two aspects: (a) an automatic evaluation of its QA components and the SRL extension approach on the fifth edition of the BioASQ challenge; and (b) its comparison to other on-line QA

Batch	System	Doc. retr.	Pass. retr.
1	Olelo	0.0465	0.0441
2	Olelo	0.0318	0.0246
3	Olelo	0.0658	0.0386
4	Olelo	0.0449	0.0347
5	Olelo	0.0381	0.0386
top results		[0.0874,0.1157]	[0.0467,0.0898]

Table 1: Results for mean average precision (MAP) for Olelo in BioASQ task 5b phase A, i.e., for document retrieval and passage retrieval. Range of top results in all batches are presented in the last row.

systems, in terms of processing time and quality of the answers.

4.1 BioASQ test sets

We participated in Task 5b (Biomedical Semantic QA) of the fifth edition of the BioASQ challenge. This task is composed of two phases: (a) Phase A, which includes submission of results for relevant concepts, documents, snippets and RDF triples. (b) Phase B, which includes submission of results for exact and ideal answers. A new batch of questions is released every two weeks and participants have 24 hours to submit results. For each batch of Phase A, the organizers release a JSON file which includes 100 questions and their corresponding type and identifier. After the end of phase A (24 hours), phase B starts after the release of an extended version of the JSON file which includes the gold standard concepts, documents, passages and RDF snippets, i.e., the answers for Phase A. Therefore, predictions for phase B can rely on this gold standard information, which we indeed used in some of our runs.

4.2 Evaluation on BioASQ task 5b

In this section we present results for both Olelo and the SRL approach. These are the official results that were made available and based on the official metrics that are described in the guidelines[2].

Table 1 presents the results for phase A based on mean average precision (MAP). For this phase, we provide results only for document and passage retrieval. We simply provide the top 10 documents and passages as returned by Olelo for each question, following the maximum number of documents and snippets which is specified in the BioASQ's guidelines.

[2]http://participants-area.bioasq.org/
general_information/Task5b/

Table 2 presents results of Olelo and the SRL approach for the exact answers of Phase B. We only provide results for yes/no questions using the SRL approach as this question type is not supported by Olelo. For the first batch, we had two submissions for SRL. SRL2 considers the detection of enumerations for list questions and fixes some minor bugs regarding the retrieval of ideal answers. The results in batch 1, SRL2 shows a significant improvement for list questions. Given that SRL2 was an improvement of SRL, we did not submit the latter from the second batch on.

For yes/no questions we did not measure any achievements in comparison to the approach of just saying "yes" to any question. The training data from recent years was very "yes"-biased and subsequently was our system. The results imply that this must have changed for the 4th and 5th batch.

The results for factoid questions based on SRL were constantly lower than the Olelo system, but they both reached a similar magnitude, which indicates a potential for a combination of both.

For list questions, the SRL approach achieved much higher F-Measure scores than Olelo. However, it should be noted that the Olelo QA system was performing its own passage retrieval and was not simply relying on the gold standard snippets provided by the challenge.

Finally, Table 3 presents our results for Olelo and the SRL approach for the ideal answers, i.e., custom summaries. These summaries should be provided for all questions, independent of their type. The difference between the Olelo and the Olelo-GS submissions is that the later relies on the gold standard (GS) snippets, instead of the ones retrieved by the system.

As expected, the Olelo-GS submissions usually obtained a higher score than the Olelo ones, but difference was lower than our expectations. The SRL-based approaches obtained much lower scores than Olelo runs. All Rouge metrics for the SRL approach were below 10%, which can be explained by the fact that it was basically just an answer snippet selection approach.

4.3 Comparison to other on-line QA applications

We compare the time response provided by our system to three other biomedical QA

Batch	System	Yes/No	Factoid	List
1	Olelo	-	0.0400	0.0240
	Olelo-GS	-	0.0400	0.0477
	SRL	0.8824	-	0.0038
	SRL2	0.8824	-	0.1183
2	Olelo	-	0.0430	0.0281
	Olelo-GS	-	0.0323	0.0287
	SRL2	0.9630	0.0129	0.1123
3	Olelo	-	0.0192	0.0408
	Olelo-GS	-	0.0192	0.0549
	SRL2	0.8065	0.0128	0.1715
4	Olelo	-	0.0253	0.0513
	Olelo-GS	-	0.0513	0.0513
	SRL2	0.5517	0.0379	0.0943
5	Olelo	-	-	0.0202
	Olelo-GS	-	-	0.0379
	SRL2	0.4615	0.0286	0.2870
top results		[0.8387,0.9630]	[0.3606,0.5713]	[0.3358,0.5001]

Table 2: Results for Olelo and the SRL approach in the BioASQ task 5b phases B (exact answers). Results for yes/no questions are in terms of accuracy, MRR for factoid questions and f-measure for list questions. Range of top results in all batches are presented in the last row.

Batch	System	Rouge-2	Rouge-SU4
1	Olelo	0.2222	0.2710
	Olelo-GS	0.2958	0.3243
	SRL	0.0467	0.0510
	SRL2	0.0833	0.0870
2	Olelo	0.2751	0.2976
	Olelo-GS	0.2048	0.2500
	SRL2	0.0425	0.0418
3	Olelo	0.3426	0.3604
	Olelo-GS	0.2891	0.3262
	SRL2	0.0411	0.0416
4	Olelo	0.2261	0.2696
	Olelo-GS	0.3460	0.3516
	SRL2	0.0796	0.0740
5	Olelo	0.3418	0.3536
	Olelo-GS	0.2117	0.2626
	SRL2	0.0406	0.0413
top results		[0.5153,0.6891]	[0.5182,0.6789]

Table 3: Results for ideal answers (summaries) in terms of Rouge metrics for Olelo and the SRL approach. Range of top results in all batches are presented in the last row.

systems[3], namely AskHermes (Cao et al., 2011), EAGLi (Gobeill et al., 2015) and HONQA (Cruchet et al., 2009). However, we did not obtain any answer for none of the questions posed to HONQA, instead, only the following message: "A problem has occurred. Try later".

We randomly selected ten factoid questions from the BioASQ dataset and posed these to the three systems - AskHermes, EAGLi and our application. This evaluation was carried out manually, and therefore, we needed to limit the number of questions and types. We decided to limit it to factoid questions given that this type of answer is easier to check manually than summaries. Table 4 shows the list of questions.

We manually recorded the time response using a stopwatch. Time record started when clicking on the search button and stopped when any results was shown. All experiments were carried out from a laptop using the Chrome browser installed in the Ubuntu operating system. Further, it was carried out from home, i.e., not in the network of our institution, in order not to favor lower response times from Olelo. We manually and carefully checked the output provided by each system to look for the gold standard answer as provided by BioASQ. This ranged from short titles, as returned by EAGLi, short summaries returned by Olelo and even three long pages of text, as in the case of AskHermes. Table 5 summarizes the re-

<hr>

[3]respectively, http://www.askhermes. org/; http://eagl.unige.ch/EAGLi/ oldindex.htm; http://www.hon.ch/QA/

Number	Question
1	Which is the protein (antigen) targeted by anti-Vel antibodies in the Vel blood group?
2	Where in the cell do we find the protein Cep135?
3	Which enzyme is involved in the maintenance of DNA (cytosine-5-)-methylation?
4	Which is the most widely used model for the study of multiple sclerosis (MS)?
5	Which medication should be administered when managing patients with suspected acute opioid overdose?
6	What is the lay name of the treatment for CCSVI (chronic cerebro-spinal venous insufficiency) in multiple sclerosis?
7	What is the percentage of responders to tetrabenazine treatment for dystonia in children?
8	Intact macromolecular assemblies are analysed by advanced mass spectrometry. How large complexes (in molecular weight) have been studied?
9	Which is the most important prognosis sub-classification in Chronic Lymphocytic Leukemia?
10	What disease is mirtazapine predominantly used for?

Table 4: List of ten factoid questions that we considered for manual evaluation.

Systems	Output	Answers	Time
AskHermes	7/10	1/10	10.1 [2.09,19.74]
EAGLi	10/10	2/10	58.6 [21.41,107.72]
Olelo	10/10	4/10	8.84 [3.35,28.12]

Table 5: Results in terms of number of correct answers and response time for the on-line QA applications.

sults that we obtained. All output pages (or answers) returned by the systems are available for download[4].

5 Discussion

In this section we discuss our performance in the current edition of the BioASQ challenge and present an error analysis based on datasets from the previous years, given that gold standard results for this year's challenge are not yet available. We also provide a discussion on the comparison of Olelo to other on-line biomedical QA systems.

5.1 Olelo's performance in BioASQ task 5b

Although we have been participating in BioASQ in the last years, the development of our application did not have the challenge as goal. Thus, we still do not use any of past challenge datasets for training data. The system is not tuned to obtain best performance in BioASQ, except for the Olelo-GS submissions. As discussed above (cf. Section 3), Olelo does not distinguish between factoid and list questions, and we might have provided multiple results even for factoid questions.

The methods behind Olelo are constantly being improved. Currently, besides the approach based on SRL that we presented here, we also evaluated a new approach based on neural networks

for the extraction of exact answers (Wiese et al., 2017), which obtained top results for factoid and list questions. We plan to integrate this new component into Olelo soon.

5.2 Error analysis based on previous data

In order to analyze the errors returned by our application, we carried out an evaluation on the test datasets of the two last editions (2015 and 2016) of the BioASQ challenge. We evaluated our exact answers using the BioASQ Oracle system, an on-line system that allows uploading JSON result files and obtaining evaluations at any time. We considered only the questions identified as "factoid" and "list" in the BioASQ dataset. We obtained a MAP that ranged from 0.0000 (no single match) to 0.0909 for factoid questions and a MAP from 0.0010 to 0.1000 for list questions.

This automatic evaluation is based solely on automatic matching procedures and results shown here are for the strict accuracy, i.e., an exact matching should apply. However, as described in our methods, our answers are derived from MeSH terms, while the gold standard answers in BioASQ are mostly based on the text spans as they appear in the document. For instance, for one of the questions, we returned the disease name "Hirschsprung Disease", while the gold standard consists of the text "Aganlionic megacolon or Hirschsprung disease". Indeed, during the BioASQ challenge, the organizers carry out a manual evaluation of all submitted answers, besides performing the automatic evaluation. Finally, our system does have some limitations on the exact answers that it is able to return, given its dependency to the MeSH terms. For instance, it performs particularly bad on questions which require gene/protein names in return, given that these entity types are poorly represented in MeSH. Indeed, almost 30% of the questions in BioASQ expect a gene/proteins in return, as pointed by (Neves and Kraus, 2016).

We manually checked our exact answers for all factoid and list questions. Unfortunately, the BioASQ Oracle system only returns a score for each batch of questions but does not give any information regarding true positives (TP), false negatives (FN) and false positives (FP). In our manual evaluation, we did not simply consider any overlap as a TP. For instance, we did not consider "Receptors, Notch" and "Notch intracellular domain (NICD)" as a match. However, we did record as TP those cases in which our answers were correct, e.g., "Ethambutol" and "Rifampin", even though they did not match exactly the gold standard answer, which is the case of the following very long answer (sentence): "Rifampin 10 mg/kg daily, ciprofloxacin 500 mg twice daily, clofazimine 100 mg every day, and ethambutol 15 mg/kg orally daily for 24 weeks, [...]".

For a total of 502 factoid and list questions, our application was able to provide a total of 116 TPs, and at least one correct answer for a total of 71 questions. However, we missed many correct answers (FNs) as well as provided many false answers (FPs), sometimes even more than 20 FPs for a question.

Olelo did not return any results for many questions, and we believe that these might have been recognized as summary questions. As discussed above, our system still fails to return answers for concepts not properly covered by the MeSH ontology, but results are promising given the complexity of the task. More importantly, the manual evaluation shows that the user could receive at least one correct answers for 14% of the questions, while some answers could also have been found in the summary, for those questions for which only a summaries were returned.

5.3 Performance of semantic role labeling experiments

As of the date of the BioASQ submissions, our experiments on SRL were still in a preliminary phase. For the specific case of list questions, we could already show how a biomedical QA system could benefit from SRL. However, in general, we got the impression that SRL should not be used to design a QA system from scratch (as we tried in our experiments) but to improve our existing approaches. A major problem of our SRL approach

was its coverage: if no matching labels for a question were found, we need an alternative approach to it. Otherwise, the recall will be too low, as experienced in our experiments. For list questions, considering enumerations as a baseline approach was very helpful.

For yes/no questions, more sophisticated detection strategies based on negation might be applied to find out when the answer is "no" with higher precision. There might be further potential when analyzing occurrences of double negation or other sophisticated contextual information. A less "yes"-biased training dataset in the BioASQ challenge could also produce further insights. At least having training data with more "no"-samples might be desirable and allow more sophisticated approaches like machine learning. As stated before, the answer snippet selection strategy for the summarization task was not meant to be very promising. Nevertheless, the strategy could be combined with the current approach in the olelo system.

5.4 QA performance in a real-time scenario

Given the comparison of our systems to the other three available biomedical QA applications (cf. Section 4), we now present a discussion on the performance of the systems.

Olelo displayed high response times (19.85 seconds and 28.12 seconds) only for two questions, namely: "Where in the cell do we find the protein Cep135?" and "Intact macromolecular assemblies are analysed by advanced mass spectrometry. How large complexes (in molecular weight) have been studied?". The second question is indeed longer than usual questions in BioASQ. Even though AskHermes outperformed Olelo in both minimum and maximum time, our application has on average a lower response time, besides being able to return an answer to all questions (cf. below). Further, three of the questions with response time under 10 seconds in AskHermes were those which returned no results, which suggests that the processing might have been interrupted. Finally, processing in EAGLi takes far too much time.

We manually analyzed the answers provided for the questions by each system. For all questions, Olelo returned a summary as answer, and in four of these questions, the summaries contained at least one of the correct answers for the question, as provided in the BioASQ benchmark. For instance, the following sentence is the first one in the summary that the system returned: "Cep135 is a 135-kDa, coiled-coil centrosome protein important for microtubule organization in mammalian cells." (PubMed 14983524). It contains the answer (centrosome) for the question "Where in the cell do we find the protein Cep135?".

In contrast, AskHermes extracted the correct answer only for one question. Nevertheless, the answer was indeed given as the top ranked. EAGLi could not provide exact answers for none of the questions, instead, only relevant documents (titles) and their corresponding single selected passages were presented. Two of these top passages indeed contained the correct answer to the question. Some of the snippets that contained the answer, as returned by AskHermes and EAGLi, appeared at the far end of a very long results page. However, these were too far from the top ranked answers (or passages) to be read by the average user, in our opinion. Finally, we should notice that EAGLi restricts the size of the question up to 80 characters, which could result in some questions not being properly processed by the system.

Even though Olelo was not able to detect that the questions were of the factoid type, and thus generated summaries for all questions, these summaries contain a maximum of five sentences (default value). Thus, we believe that most users could indeed find those four correct answers by reading through the short paragraphs. Changes on our question processing component could allow our system to output more short answers, instead of summaries, for questions that are in fact of the the factoid type. Currently, it only returns exact answers when both the headword and semantic types are detected, in addition to the candidate answers being of this same semantic type.

6 Conclusions

In this work, we presented an assessment of our Olelo QA system for biomedicine. We considered both the current online state of the system as well as a future extension based on semantic role labeling. We presented an evaluation both in terms of response time and robustness, in comparison to other online QA systems, as well as automatic and manual evaluation of the exact answers based on the BioASQ dataset. Our results are promising, given the complexity of the QA task, and future work will focus on the improvement of our current

methods, integration of additional terminologies (e.g., for gene/proteins names) and support for additional question types (e.g., yes/no questions).

References

Sofia J. Athenikos and Hyoil Han. 2010. Biomedical question answering: A survey. *Computer Methods and Programs in Biomedicine* 99(1):1 – 24.

MichaelA Bauer and Daniel Berleant. 2012. Usability survey of biomedical question answering systems. *Human Genomics* 6(1):1–4.

Yonggang Cao, Feifan Liu, Pippa Simpson, Lamont D. Antieau, Andrew S. Bennett, James J. Cimino, John W. Ely, and Hong Yu. 2011. Askhermes: An online question answering system for complex clinical questions. *Journal of Biomedical Informatics* 44(2):277–288.

Sarah Cruchet, Arnaud Gaudinat, Thomas Rindflesch, and Celia Boyer. 2009. What about trust in the question answering world? In *Proceedings of the AMIA Annual Symposium*. San Francisco, USA, pages 1–5.

Daniel Dahlmeier and Hwee Tou Ng. 2010. Domain adaptation for semantic role labeling in the biomedical domain. *Bioinformatics* 26(8):1098.

Julien Gobeill, Arnaud Gaudinat, Emilie Pasche, Dina Vishnyakova, Pascale Gaudet, Amos Bairoch, and Patrick Ruch. 2015. Deep question answering for protein annotation. *Database* 2015:bav081.

Daniel Jurafsky and James H. Martin. 2013. *Speech and Language Processing*. Prentice Hall International, 2 revised edition.

Milena Kraus, Julian Niedermeier, Marcel Jankrift, Sören Tietböhl, Toni Stachewicz, Hendrik Folkerts, Matthias Uflacker, and Mariana Neves. 2017. Olelo: a web application for intuitive exploration of biomedical literature. *Nucleic Acids Research* gkx363 .

Mariana Neves. 2014. HPI in-memory-based database system in task 2b of bioasq. In *Working Notes for CLEF 2014 Conference, Sheffield, UK, September 15-18, 2014.*. pages 1337–1347.

Mariana Neves. 2015. HPI question answering system in the bioasq 2015 challenge. In *In Working Notes for CLEF 2015 Conference, Toulouse, France*.

Mariana Neves and Milena Kraus. 2016. BioMedLAT corpus: Annotation of the lexical answer type for biomedical questions. In *Open Knowledge Base and Question Answering Workshop at the 26th International Conference on Computational Linguistics (Coling)*.

Mariana Neves and Ulf Leser. 2015. Question answering for biology. *Methods* 74(0):36 – 46.

Frederik Schulze and Mariana Neves. 2016. Entity-supported summarization of biomedical abstracts. In *Fifth Workshop on Building and Evaluating Resources for Biomedical Text Mining at the 26th International Conference on Computational Linguistics (Coling)*.

Frederik Schulze, Ricarda Schler, Tim Draeger, Daniel Dummer, Alexander Ernst, Pedro Flemming, Cindy Perscheid, and Mariana Neves. 2016. Hpi question answering system in bioasq 2016. In *Proceedings of the Fourth BioASQ workshop at the Conference of the Association for Computational Linguistics*. pages 38–44.

George Tsatsaronis, Georgios Balikas, Prodromos Malakasiotis, Ioannis Partalas, Matthias Zschunke, Michael R Alvers, Dirk Weissenborn, Anastasia Krithara, Sergios Petridis, Dimitris Polychronopoulos, et al. 2015. An overview of the bioasq large-scale biomedical semantic indexing and question answering competition. *BMC bioinformatics* 16(1):138.

Georg Wiese, Dirk Weissenborn, and Mariana Neves. 2017. Neural question answering at bioasq 5b. In *In Biomedical Natural Language Processing (BioNLP) workshop at ACL*.

Clinical Event Detection with Hybrid Neural Architecture

Adyasha Maharana
Biomedical and Health Informatics
University of Washington, Seattle
adyasha@uw.edu

Meliha Yetisgen
Biomedical and Health Informatics
University of Washington, Seattle
melihay@uw.edu

Abstract

Event detection from clinical notes has been traditionally solved with rule based and statistical natural language processing (NLP) approaches that require extensive domain knowledge and feature engineering. In this paper, we have explored the feasibility of approaching this task with recurrent neural networks, clinical word embeddings and introduced a hybrid architecture to improve detection for entities with smaller representation in the dataset. A comparative analysis is also done which reveals the complementary behavior of neural networks and conditional random fields in clinical entity detection.

1 Introduction

Event detection from clinical notes is a well studied problem in biomedical informatics; yet, it is constantly evolving through research. Much of this research has been promoted by the i2b2 challenges (2010, 2012) and their publicly available datasets comprised of annotated discharge summaries. For the 2010 task, the notes were annotated for three types of events - *Problem*, *Test* and *Treatment*, which are predominantly noun phrases. (Uzuner et al., 2011) The task was made even more challenging in 2012 with the addition of three new entity classes - *Occurrence*, *Evidential* and *Clinical Department*. *Occurrence* and *Evidential* concepts are mostly verb phrases, with some examples being 'readmitted', 'diagnosed', 'seen in consultation', 'revealed' etc. Rule based and statistical NLP approaches such as Conditional Random Fields have been used at identifying these entities. These approaches require extensive domain knowledge and feature engineering. (Sun et al., 2013) In this paper, we explore discretized word embeddings as new features in structured inference and also implement a neural network architecture for clinical entity recognition. We defined a CRF baseline to compare the performance of our neural networks and performed a detailed error analysis.

2 Related Work

The best performing system on 2010 i2b2 corpus is a semi-supervised HMM (semi-Markov) model which scored 0.9244 (partial match F1-score) in the concept extraction track (Uzuner et al., 2011). Xu et al. (2012) divided the *Treatment* category into *Medication* and *Non-medication* concepts, and trained two separate conditional random field (CRF) classifiers for sentences with and without medication. With additional features, this system scored 0.9166 on event detection track in 2012 i2b2 challenge, taking the top spot. Tang et al. (2013) built a cascaded CRF system which scored 0.9013 on event detection and came a close second. Most of the other competing teams also employed CRF for this task along with Support Vector Machines or Maximum Entropy for classifying the event category, with the exception of Jindal and Roth (2013) who implemented a sentence-level inference strategy using Integer Quadratic Program. Sun et al. (2013) showed that these systems found it harder to identify *Clinical Department*, *Occurrence* and *Evidential* concepts.

With the surge in deep learning, there have been several new approaches to clinical event detection. Wu et al. (2015) used word embeddings as features in a CRF model and noted improvement in recall for the i2b2 2010 corpus. Chalapathy et al. (2016) implemented a bi-directional LSTM-CRF model with generic embeddings and reported no improvement over the top-performing system in 2010 i2b2 challenge. Jagannatha and Yu (2016a)

Proceedings of the BioNLP 2017 workshop, pages 351–355,
Vancouver, Canada, August 4, 2017. ©2017 Association for Computational Linguistics

tested a bi-directional LSTM framework initialized with pre-trained biomedical embeddings on an independent dataset and reported improvement over a CRF baseline. Recent results show that approximate skip-chain CRFs are more effective at capturing long-range dependencies in clinical text than recurrent neural networks (RNN) (Jagannatha and Yu, 2016b).

The 2012 i2b2 corpus has remained relatively unexplored in light of recent advances in NLP. We analyze the performance of recurrent neural networks for identification of clinical events from this dataset.

3 Methods

3.1 Dataset

The 2012 i2b2 corpus is made of 310 discharge summaries consisting of 178, 000 tokens annotated with clinical events, temporal expressions and temporal relations. The entire corpus is divided into training and test sets, containing 190 and 120 documents respectively. Each discharge summary has sections for clinical history and hospital course. Annotation of clinical events includes problems, tests, treatments, clinical departments, occurrences (admission, discharge) and evidences of information (patient *denies*, tests *revealed*). The inter-annotator agreement for event spans is 0.83 for exact match and 0.87 for partial match (Sun et al., 2013). *Clinical Department* and *Evidential* concepts are under-represented in training set with less than 1000 examples each.

3.2 Approach

3.2.1 Baseline

The best performing system in 2012 i2b2 challenge (Xu et al., 2013) requires additional annotation. So, we choose to replicate the second best performing system built by Tang et al. (2013) as our baseline. It is a cascaded CRF classifier, wherein the first CRF is trained on datasets released in 2010 & 2012 to classify for problem, test and treatment. The next CRF is trained on 2012 dataset to extract clinical department, occurrence and evidential concepts. This split in classes is performed to leverage the 2010 dataset which is annotated for the first three classes only. Precision, recall and F-measure (exact event span) for the original system is reported as 93.74%, 86.79% and 90.13% respectively. Our baseline system is built with the same cascaded configuration. The

following features are used: N-grams (± 2 context window), word-level orthographic information, syntactic features using MedPOST (Smith et al., 2004), discourse information using a statistical section chunker (Tepper et al., 2012) and semantic features from normalized UMLS concepts (CUIs and semantic types). Tang et al. (2013) employs several other lexical sources and NLP systems for additional features, such as MedLEE, KnowledgeMap and private dictionaries of clinical concepts. For lack of access, they have been left out of our baseline. We have implemented the baseline using CRFSuite package (Okazaki, 2007) and optimum parameters are selected through 5-fold cross-validation on the training set.

3.2.2 Word Embeddings

We use the publicly available source code of GloVe (Pennington et al., 2014) to extract word vectors of dimension 50 for 133,968 words from MIMIC-III. The MIMIC-III dataset (Johnson et al., 2016) contains 2,083,180 clinical notes including discharge summaries, ECG reports, radiology reports etc. Since we are dealing exclusively with discharge summaries in our task, GloVe is run only on the discharge summaries present in MIMIC. These vectors are unfit for direct use in structured prediction and are discretized using methods advocated by Guo et al. (2014).

3.2.3 Recurrent Neural Networks

The bi-directional LSTM-CRF neural architecture introduced by Lample et al. (2016) has been shown to excel on multi-lingual NER tasks. Among others, its components include a char-RNN that models word prefixes, suffixes and shape - features that are critical to NER. We initialize two instances of the complete network with the GloVe vectors extracted from MIMIC-III discharge summaries. First instance is trained to classify problem, test and treatment concepts only; second instance is trained for other three classes. 78.96% words in the training corpus are initialized with pre-trained embeddings. Results from both the networks are merged in a combination module for final evaluation of the end-to-end system. Overlaps are resolved by placing preference on predictions from the first instance.

3.2.4 Hybrid Architecture

The current of state-of-art for detecting problem, test and treatment concepts from clinical text is

System	TP	FP	FN	Precision	Recall	F1 Score
Baseline	13951	794	2517	**94.63**	84.71	89.40
Baseline + BinEmb	13982	818	2486	94.47	84.90	89.43
Baseline + ProtoEmb	14006	825	2460	94.43	85.06	89.50
Baseline + Brown Clusters	14129	843	2339	94.38	85.78	89.88
Baseline + Brown Clusters + ProtoEmb	14130	860	2338	94.26	85.78	89.82
RNN + random initialization	12370	3123	4098	79.84	75.12	77.38
RNN + MIMIC Embeddings	14315	1373	2153	91.25	**86.93**	89.31
CRF + RNN (Hybrid)	14236	936	2232	93.66	86.45	**89.91**

Table 1: 5-fold cross validation performance of various systems on 2012 i2b2 training set

Entity Class	System	TP	FP	FN	Precision	Recall	F1 Score
Problem	Baseline + Brown Clusters	4607	194	414	**95.96**	**91.72**	**93.79**
	RNN + Embeddings	4429	776	594	85.09	88.17	86.61
Test	Baseline + Brown Clusters	2355	100	242	**95.93**	**90.64**	**93.21**
	RNN + Embeddings	2182	342	415	86.45	83.98	85.20
Treatment	Baseline + Brown Clusters	3469	160	361	**95.62**	**90.57**	**93.03**
	RNN + Embeddings	3296	525	534	86.26	86.06	86.16
Occurrence	Baseline + Brown Clusters	2030	620	1256	76.60	61.78	68.40
	RNN + Embeddings	2042	510	1234	**79.51**	**62.14**	**70.82**
Evidential	Baseline + Brown Clusters	456	116	284	**79.72**	61.62	69.51
	RNN + Embeddings	497	134	243	78.76	**67.16**	**72.5**
Clinical Department	Baseline + Brown Clusters	741	122	256	**85.86**	74.32	79.68
	RNN + Embeddings	813	188	194	79.96	**82.05**	**80.99**

Table 2: Entity-level performance of best performing CRF system and RNN on 2012 i2b2 training set

based on CRF and it has been hard to improve on this baseline, even with neural networks. (Chalapathy et al., 2016) Cross-validation performance (presented in Table 2) reveals entity-level differences between CRF and RNN systems. So, we combine the merits of both approaches to create a hybrid end-to-end model. The exact configuration is discussed in the results section.

4 Evaluation Metrics and Results

We report the micro-averaged precision, recall and F1-score, for 'overlap' match of event spans as per the i2b2 evaluation script. TP, FP, FN counts of overall performance are calculated for entity spans, irrespective of entity tag. Systems are also evaluated for performance in individual entity classes and TP, FP, FN counts are compared between the CRF and RNN+Embedding systems. We perform five-fold cross validation for various configurations of the baseline and RNN systems on the training set. The results are presented in Table 1 and Table 2.

The best performing CRF system i.e. Baseline + Brown Clusters, achieves F1-score of 89.88. Except for brown clusters, additional features derived from distributional semantics, such as binarized word embeddings (BinEmb), prototype embeddings (ProtoEmb) contribute marginally to performance of the system. Pre-trained clinical embeddings improve F1 score by 11.93%, over random initialization of RNNs. In terms of recall, the RNN initialized with MIMIC embeddings is found to perform remarkably well without hand-engineered features. However, it fails to beat the CRF system at F1-score. Comparative analysis of individual entity classes reveals that the RNN improves recall for evidential and clinical department phrases by 5.44% and 8.32% respectively. It registers some drop in precision, but improves F1-score by up to 3%. Clearly, RNNs are better suited for detecting occurrence, evidential and clinical department phrases from clinical text.

Based on these results on the training set, we build the hybrid sequence tagger where the best performing CRF system is combined with RNN. The former is trained to tag problem, test and treatment and the latter is trained to tag rest of the three entity classes. The results are merged in a combination module and overlapping predictions are resolved by prioritizing the first three classes. We evaluate its performance on the i2b2 2012 test set. Results are listed in Table 3 and 4.

The hybrid model improves recall by 2.36% and F1-score by 0.56% over the best-performing CRF system. Dramatic improvement in recall (as high as 14%) is noted for some entities, but a similar drop in precision is observed.

System	TP	FP	FN	Precision	Recall	F1 Score
Tang et al. (2013)	-	-	-	93.74	86.79	90.13
Baseline + Brown Clusters	11664	647	1930	**94.74**	85.80	90.05
Hybrid CRF-RNN	11985	875	1609	93.20	**88.16**	**90.61**

Table 3: Performance of best performing CRF and Hybrid CRF-RNN on 2012 i2b2 test set

Entity Class	System	TP	FP	FN	Precision	Recall	F1 Score
Occurrence	Baseline + Brown Clusters	1509	489	991	**75.53**	60.36	67.10
	Hybrid	1565	563	935	73.54	**62.60**	**67.63**
Evidential	Baseline + Brown Clusters	370	76	226	**82.96**	62.08	71.02
	Hybrid CRF-RNN	446	177	150	71.59	**74.83**	**73.17**
Clinical Department	Baseline + Brown Clusters	557	109	176	**83.63**	75.99	79.63
	Hybrid CRF-RNN	657	234	76	73.74	**89.63**	**80.91**

Table 4: Entity-level performance of best performing CRF and Hybrid CRF-RNN on 2012 i2b2 test set

5 Discussion

The hybrid architecture serves as a concept extraction model with a predisposition for higher recall of clinical events, as compared to the CRF system which exhibits better precision in performance. On comparing errors, we found the %overlap between false negatives of CRF and RNN systems to be only about 52%. The CRF model is able to exploit semantic, syntactic and orthographic information among others, while RNNs are only initialized with limited semantic information. Automatic learning of syntactic structure and finer semantic correlations is inherent to recurrent neural architecture. However, this may be somewhat limited by our small corpus. This situation leads to subtle disparities in performance of both systems.

The RNN is able to detect clinical departments (which includes physician names, hospitals names and clinical departments) with good recall value in spite of being trained with only 997 data points. CRF has lowest recall for clinical department, among all classes that contain more noun phrases. The RNN confuses higher percentage of *Treatment* concepts as *Occurrence* than CRF, mostly those which are verb phrases like 'excised', 'intubated' etc. Instead of initializing all words with clinical embeddings, the performance of RNN may be improved by selectively initializing clinical terms only. This can be done by filtering for certain UMLS semantic groups/types and providing only those words with a pre-trained word vector. On the other hand, word embeddings help the RNN in handling unseen vocabulary effectively. For example, when RNN is trained to tag 'decreased' as occurrence, it tags the word 'weaned' correctly as occurrence in the test set. Under sim-

ilar conditions, CRF is unable to make the correct decision. Word vectors derived from a larger biomedical corpus may enable the RNN to make finer semantic distinctions.

Unlike RNN, CRF fails to recognize the occasional long phrases such as *'normal appearing portal veins and hepatic arteries'*, even under overlap matching criteria. We expect the LSTM cells in RNN to capture long-term dependencies from various ranges within a sentence, and our hypothesis is confirmed by the test results. The CRF operates within a pre-specified context window and is limited by its linear chain framework. With a skip chain CRF, this situation can be remedied.

6 Conclusion & Future Work

This paper evaluates various methods for using neural architecture in clinical entity recognition and minimizing feature engineering. Benefits are observed when the merits of structured prediction model and RNN are fused into a hybrid architecture after analysis of their cross-validation performance. The hybrid model's recall and F1 score surpass that of the state-of-art system we have used for replication. Through error analysis, we highlight some of the situations where RNNs fare better such as longer concept length, unseen clinical terms, semantically similar generic words, proper nouns etc.

In future work, we will attempt to integrate long-term dependencies within a sentence by implementing the skip chain CRF model and explore the efficient use of word embeddings for structured prediction. This clinical entity recognition model will also be extended to a temporal evaluation system.

References

Raghavendra Chalapathy, Ehsan Zare Borzeshi, and Massimo Piccardi. 2016. Bidirectional lstm-crf for clinical concept extraction. *arXiv preprint arXiv:1611.08373* .

Jiang Guo, Wanxiang Che, Haifeng Wang, and Ting Liu. 2014. Revisiting embedding features for simple semi-supervised learning. In *EMNLP*. pages 110–120.

Abhyuday N Jagannatha and Hong Yu. 2016a. Bidirectional rnn for medical event detection in electronic health records. In *Proceedings of the conference. Association for Computational Linguistics. North American Chapter. Meeting*. NIH Public Access, volume 2016, page 473.

Abhyuday N Jagannatha and Hong Yu. 2016b. Structured prediction models for rnn based sequence labeling in clinical text. In *Proceedings of the Conference on Empirical Methods in Natural Language Processing. Conference on Empirical Methods in Natural Language Processing*. NIH Public Access, volume 2016, page 856.

Prateek Jindal and Dan Roth. 2013. Extraction of events and temporal expressions from clinical narratives. *Journal of biomedical informatics* 46:S13–S19.

Alistair EW Johnson, Tom J Pollard, Lu Shen, Liwei H Lehman, Mengling Feng, Mohammad Ghassemi, Benjamin Moody, Peter Szolovits, Leo Anthony Celi, and Roger G Mark. 2016. Mimic-iii, a freely accessible critical care database. *Scientific data* 3.

Guillaume Lample, Miguel Ballesteros, Sandeep Subramanian, Kazuya Kawakami, and Chris Dyer. 2016. Neural architectures for named entity recognition. *arXiv preprint arXiv:1603.01360* .

Naoaki Okazaki. 2007. Crfsuite: a fast implementation of conditional random fields (crfs). http://www.chokkan.org/software/crfsuite/.

Jeffrey Pennington, Richard Socher, and Christopher D Manning. 2014. Glove: Global vectors for word representation. In *EMNLP*. volume 14, pages 1532–1543.

L Smith, Thomas Rindflesch, W John Wilbur, et al. 2004. Medpost: a part-of-speech tagger for biomedical text. *Bioinformatics* 20(14):2320–2321.

Weiyi Sun, Anna Rumshisky, and Ozlem Uzuner. 2013. Evaluating temporal relations in clinical text: 2012 i2b2 challenge. *Journal of the American Medical Informatics Association* 20(5):806–813.

Buzhou Tang, Yonghui Wu, Min Jiang, Yukun Chen, Joshua C Denny, and Hua Xu. 2013. A hybrid system for temporal information extraction from clinical text. *Journal of the American Medical Informatics Association* 20(5):828–835.

Michael Tepper, Daniel Capurro, Fei Xia, Lucy Vanderwende, and Meliha Yetisgen-Yildiz. 2012. Statistical section segmentation in free-text clinical records. In *LREC*. pages 2001–2008.

Özlem Uzuner, Brett R South, Shuying Shen, and Scott L DuVall. 2011. 2010 i2b2/va challenge on concepts, assertions, and relations in clinical text. *Journal of the American Medical Informatics Association* 18(5):552–556.

Yonghui Wu, Jun Xu, Min Jiang, Yaoyun Zhang, and Hua Xu. 2015. A study of neural word embeddings for named entity recognition in clinical text. In *AMIA Annual Symposium Proceedings*. American Medical Informatics Association, volume 2015, page 1326.

Yan Xu, Kai Hong, Junichi Tsujii, I Eric, and Chao Chang. 2012. Feature engineering combined with machine learning and rule-based methods for structured information extraction from narrative clinical discharge summaries. *Journal of the American Medical Informatics Association* 19(5):824–832.

Yan Xu, Yining Wang, Tianren Liu, Junichi Tsujii, I Eric, and Chao Chang. 2013. An end-to-end system to identify temporal relation in discharge summaries: 2012 i2b2 challenge. *Journal of the American Medical Informatics Association* 20(5):849–858.

Extracting Personal Medical Events for User Timeline Construction using Minimal Supervision

Aakanksha Naik
Language Technologies
Institute
Carnegie Mellon University
anaik@cs.cmu.edu

Chris Bogart
Institute for Software
Research
Carnegie Mellon University
cbogart@cs.cmu.edu

Carolyn Rose
Language Technologies
Institute
Carnegie Mellon University
cprose@cs.cmu.edu

Abstract

In this paper, we describe a system for automatic construction of user disease progression timelines from their posts in online support groups using minimal supervision. In recent years, several online support groups have been established which has led to a huge increase in the amount of patient-authored text available. Creating systems which can automatically extract important medical events and create disease progression timelines for users from such text can help in patient health monitoring as well as studying links between medical events and users' participation in support groups. Prior work in this domain has used manually constructed keyword sets to detect medical events. In this work, our aim is to perform medical event detection using minimal supervision in order to develop a more general timeline construction system. Our system achieves an accuracy of 55.17%, which is 92% of the performance achieved by a supervised baseline system.

1 Introduction

In recent years, the steady shift towards a consumer-centric paradigm in healthcare, in conjunction with the meteoric rise of social networking, has led to the establishment of several online support groups and an increasing amount of available patient-authored text. Analyzing this text can provide us an opportunity to study many important issues such as how important medical events affect people's lives and how important changes in their personal lives affect disease progression. We can also study how important medical events affect users' participation in these online communities.

To perform such analyses on large-scale data, there is a need to develop automated methods to extract important personal medical events and associate them with dates from user posts in online support groups. These extracted events and dates can then be used to construct medical event timelines for users and study links between user participation or posting behaviors in online support groups and important personal medical events (Wen and Rosé (2012)). Such automated methods can also be used for patient health monitoring. In this work, we propose a novel unsupervised approach to personal medical event extraction that achieves an accuracy of 55.17%, which is 92% of the performance of the most similar supervised approach on a cancer support forum corpus.

Prior work in personal medical event extraction (Wen and Rosé (2012)) from user posts uses manually constructed sets of keywords to detect medical events from text. This limits the generality of such systems, since using the system on a new corpus requires prior knowledge about types of medical events, and the vocabulary used by users to describe these events. To make them more general, we propose a data-driven personal medical event extraction pipeline which detects medical events with minimal supervision. This makes our system independent of the corpus on which it is used and reduces the manual effort required. We test the performance of our system on the task of constructing cancer event timelines from the dataset used by Wen and Rosé (2012). In spite of being almost completely unsupervised, our system reaches 92% of the performance achieved by a supervised baseline system.

The rest of paper is organized as follows. Section 2 describes prior work in event extraction and temporal resolution which we leverage, while sec-

Proceedings of the BioNLP 2017 workshop, pages 356–364,
Vancouver, Canada, August 4, 2017. ©2017 Association for Computational Linguistics

tion 3 describes our datasets. Section 4 introduces the architecture of our proposed system and section 5 talks about the system modules in more detail. Section 6 describes our experiments and evaluation, while section 7 presents a brief error analysis and describes possible future extensions. Section 8 concludes the paper.

2 Related Work

Event extraction is a well-studied topic in natural language processing. This has resulted in the development of several off-the-shelf tools for event extraction (Saurí et al. (2005), Chambers (2013), Derczynski et al. (2016)). All these tools have been developed for extraction of public events from news corpora. Some prior work has also studied extraction of public events from social media (Sakaki et al. (2010), Becker et al. (2010), Ritter et al. (2012)). However, in this work, we want to focus on extracting personal medical events for users from their posts on online support groups using minimal supervision.

There has been some prior work on personal event extraction from social media, especially twitter)Li and Cardie (2014); Li et al. (2014)). Li et al. (2014) developed a system for personal event extraction from twitter using minimal supervision. They used the presence of congratulations/ condolence speech acts to detect personal event mentions in tweets and clustering based on the Latent Dirichlet Algorithm (Blei et al. (2003)) to detect personal event types. However, they did not focus specifically on medical events. While we also want to build a system for personal event extraction from online support groups, our focus is on identifying medical events. Hence, the techniques used by Li et al. (2014) do not work very well for us. Online support groups are not as person-focused as twitter, so the presence of congratulations/ condolence speech acts is not a strong signal for personal medical event detection. Moreover, as we show in section 6, LDA is unable to perform well on personal medical event type detection. So, we use a different technique for event type detection, which is partly similar to the technique used by Huang et al. (2016). Our overall system pipeline for data-driven medical event detection with minimal supervision is partly inspired by Li et al. (2014).

On the other hand, there has not been extensive research on personal medical event extraction from online support groups. Wen and Rosé (2012) developed a system for medical event extraction from online support groups. Their system used manually constructed keyword sets for event extraction. We propose a minimally supervised medical event detection pipeline which can remove the need to create these manual keyword sets.

Since we want to create event timelines for users from their posts in online support groups, we also need to perform temporal expression detection and resolution as well as linking of temporal expressions to events. Temporal expression extraction and normalization is also a well-studied area and several off-the-shelf systems are available (Strötgen and Gertz (2010), Chang and Manning (2012), Derczynski et al. (2016)). Moreover, some systems perform both temporal resolution and linking of events with temporal expressions (Chambers (2013)). However, most of these systems are developed for news data and do not work very well with the informal writing style used on social media. But there have been some efforts to develop systems which work well for this space. Wen et al. (2013) developed a temporal tagger and resolver for informal temporal references on social media, but the system is not available for use. The HeidelTime system Strötgen and Gertz (2010) also has a "colloquial english" setting which works well for temporal resolution from social media data. To link events with temporal expressions, we use the heuristics proposed by Wen and Rosé (2012).

3 Dataset

We use two datasets in this paper. The first dataset comprises of all posts from two groups called "Knitters with Breast Cancer" and "Beginners Knit-Along" from Ravelry[1], a website for fiber arts enthusiasts. "Knitters with Breast Cancer" is one of the largest and most active breast cancer groups on Ravelry. This group was started in December 2008. As of December 2016, it had 426 members and 120,000 posts. "Beginners Knit-Along" is a group for knitting enthusiasts who have just started learning how to knit. This group was started in 2013. As of December 2016, it had 3274 members and 70,000 posts. The data from these groups is used to create a list of medical terms, based on vocabulary difference, which is used in the medical event extraction module. We

---

[1] https://www.ravelry.com/

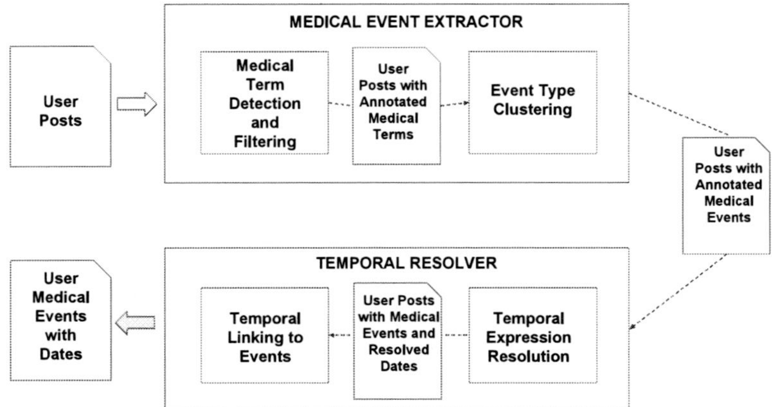

Figure 1: Architecture of proposed system pipeline

do not use this data for system evaluation.

In order to facilitate comparison with previous work, we use a different dataset to evaluate the performance of our system on the user timeline construction task. This dataset comprises of user posts from an online breast cancer community called breastcancer.org. This dataset is a subset [2] of the annotated dataset used by Wen and Rosé (2012). It comprises of major cancer events and associated dates for 50 users, along with all posts by these users. This dataset is much smaller in comparison to Knitters with Breast Cancer, comprising only of 3293 posts.

4 System Description

Our system pipeline is similar to the pipeline used by Wen and Rosé (2012), which is used as a baseline to compare our system performance. It consists of two main modules: medical event extractor and temporal resolver. However, there are a few differences from the baseline system. We do not use any filtering to remove sentences which do not contain mentions of self-reported events. Moreover, we use a different system for temporal expression extraction and normalization since the temporal resolved used by Wen and Rosé (2012) is not available for use. Because of these differences, we re-implement the baseline system in Wen and Rosé (2012) as described in section 6 to facilitate a fair comparison. After re-implementation, the only difference between the baseline and our system lies in the medical event extractor mod-

ule. Instead of using manually designed keyword sets for extracting sentences containing medical events, we use a data-driven medical event extraction pipeline with minimal supervision. Fig 1 shows the architecture of our proposed system. We explain the modules for our proposed system in more detail in subsequent sections.

5 Modules for Proposed System

Our timeline construction system consists of two main modules: medical event extractor and temporal resolver.

5.1 Medical Event Extractor

We propose a pipeline for data-driven medical event detection using minimal supervision. Our pipeline comprises of three stages:

- **Medical Term Detection**

- **Medical Term Filtering**

- **Event Type Clustering**

In the following sections, we describe the algorithms used in these stages in more detail. We present evaluation results for each stage in section 6.

5.1.1 Medical Term Detection

In this stage, our aim is to select sentences which may contain mentions of a user's personal medical events. we use a simple rule to perform this selection: if a sentence contains a medical term, it is selected as a candidate sentence for the second stage of the pipeline. We experiment with two different methods for medical term detection.

---

[2]We use this subset because we could not get access to the full dataset used in Wen and Rosé (2012)

The first method uses ADEPT MacLean and Heer (2013), a medical term recognizer developed specifically for patient-authored text, to detect the presence of medical terms in sentences. All sentences containing at least one medical term, as detected by ADEPT, are chosen as candidate sentences. The second method is based on vocabulary difference between an online support group and a non-illness related group (VOCAB). We create term vocabularies from two groups on Ravelry: Knitters with Breast Cancer, a breast cancer support group and Beginners' Knit-Along, a non illness-related group. We then create a list of terms which occur at least once in Knitters with Breast Cancer, but do not appear at all in Beginners' Knit-Along. All terms in this list are now considered to be medical terms. Choosing two groups focusing on different interests from the same online community to detect medical terms, mitigates the problem of Ravelry-specific terms (such as Raveler, Ravatar etc.) being mistakenly included in the list. Using this term list, we perform candidate sentence extraction by choosing all sentences which contain at least one of the terms from the list.

Candidate sentences chosen by both methods contain a lot of spurious sentences, since many spurious terms are marked as medical terms by these methods. Hence, the next stage in our pipeline filters these candidate sentence sets.

5.1.2 Medical Term Filtering

In this stage, we filter out spurious terms to improve the quality of the candidate sentence set. We first discuss major sources of errors for both medical term detection methods and then discuss some strategies we use to mitigate these errors. These strategies are used to filter medical terms detected by both systems, which in turn filters candidate sentences selected by both.

We face one major issue while running the ADEPT system on our data. The system manages to correctly identify most important medical terms from the text, but it also marks several words used in non-medical contexts as medical terms. For example, in the sentence "I must learn to speak more slowly than my brain thinks !", the word "brain" is marked as a medical term, even though it not being used in a medical context. Performing such filtering is difficult, but we observe that when use a combination of terms from both methods, some of these errors get mitigated.

k	Vocab Size
1	28136
5	6585
10	2833
20	1197

Table 1: Massive decrease medical term vocabulary size with increasing value of k (the frequency limit for filtering)

While the vocabulary difference-based method does not fall into such context-based errors, it has its own drawbacks. Several terms in the list created via vocabulary difference are URLs, user names, email addresses and telephone numbers. We observe that such spurious terms are very infrequent. Hence, we perform filtering by removing all terms which occur with a frequency lower than k in the Knitters with Breast Cancer group, from our medical term list. We experiment with different values of k. Table 1 shows the massive reduction in medical term vocabulary with increasing value of k.

For further comparison, we evaluate the performance of both methods on candidate sentence extraction. These experiments and results are discussed further in section 6. Based on these results, we use a combination of both methods to perform medical event detection and filtering in the final system.

5.1.3 Event Type Clustering

In this stage, we use clustering to perform medical event type detection. We consider all sentences from the filtered set provided by the previous stage to be sentences containing mentions of medical events. This is an oversimplification since a sentence may contain a medical term which may not correspond to a medical event. For example, in the sentence "My onc gave me the choice, saying she would rather ovrtreat than undertreat", there are several medical terms (onc, overtreat, undertreat) but none of them are associated with medical events. However, we still perform clustering on the entire set, since such medical terms which do not correspond to medical events form a separate set of clusters which are later discarded. After clustering, we manually label each cluster with the medical event it corresponds to, and use these clusters as keyword sets to only retain sentences corresponding to each medical event. These sets of sentences for each medical event correspond to what Wen and Rosé (2012) call "date sentences" and are used to extract the dates associated with these events.

We experiment with two methods for clustering. The first method uses Latent Dirichlet Allocation (Blei et al. (2003)) to cluster the candidate set of sentences. The use of this algorithm is motivated by the observation that people use similar expressions to describe the same medical events. However, as further discussed in section 6, we do not get good results using this algorithm.

The second method focuses on clustering medical terms instead of candidate sentences. We use a two-pass hierarchical clustering algorithm to cluster medical terms based on their Word2Vec (Mikolov et al. (2013)) embeddings. The word vectors used are pretrained on biomedical articles from Pubmed and PMC as well as English Wikipedia, in order to ensure enough domain specificity [3]. We also experiment with k-means clustering, but use agglomerative clustering for the final system due to better performance. In the first pass of agglomerative clustering, our main focus is on weeding out medical terms which are not linked to major cancer events. Hence, we run agglomerative clustering on all medical terms in this pass and manually inspect the produced clusters, discarding those which do not contain any terms corresponding to major cancer events. Thus, after the first pass, we are left with a list of terms, of which most are associated with major cancer events. This list of terms is then clustered during a second pass of agglomerative clustering. The final clusters produced by this pass are inspected and labeled with the cancer event that they correspond to. This method of clustering is able to identify better clusters, as discussed further in section 6. Hence, we use the keyword sets generated by this method for the final system.

5.2 Temporal Resolver

This module detects temporal expressions in every sentence, resolves those expressions to dates and then associates them with medical events. It has two phases: (1) temporal expression detection and resolution and (2) linking temporal expressions with events

---

³These pretrained word vectors are provided by http://bio.nlplab.org/

5.2.1 Temporal Expression Detection and Resolution

We use HeidelTime (Strötgen and Gertz (2010)), a state-of-the-art temporal expression extractor and resolver to perform temporal expression detection and resolution on all candidate sentences extracted by the medical event extraction module. We run this system with the colloquial English setting, since our data comes from online support groups. Post timestamps are provided as document creation times.

5.2.2 Linking Temporal Expressions with Events

We use the rules of thumb proposed by Wen and Rosé (2012) to resolve temporal ambiguities, such as multiple temporal expressions occurring in a single sentence, for the baseline system. When multiple temporal expressions occur in the same sentence, the expression nearest to the event word in the sentence is chosen as the correct one. When an event is associated with multiple dates for the same user, we choose the most frequent date as the correct one.

6 Experimental Results and Evaluation

In this section, we first present our evaluation of each module for the proposed event detection pipeline. We then describe the performance achieved by our end-to-end system on the task of constructing cancer event timelines for users.

6.1 Evaluation of the Medical Term Detection Module

In this section, we evaluate the performance of two techniques (ADEPT-based term detection and vocabulary-based term detection) used in the medical term detection module. Since our aim is to replace the supervised sentence extraction phase in Wen and Rosé (2012) with our unsupervised pipeline while incurring minimal performance loss, we perform a comparative evaluation of this module. We use the sentence set extracted using manually defined keyword sets used by Wen and Rosé (2012) as our gold data. We measure performance by computing precision and recall of candidate sentence sets extracted by both medical term detection methods (ADEPT and VOCAB).Table 2 presents the precision, recall and F1 scores for these methods. We also present the scores for candidate sentence extraction using our vocabulary-based method before frequency-based filtering to

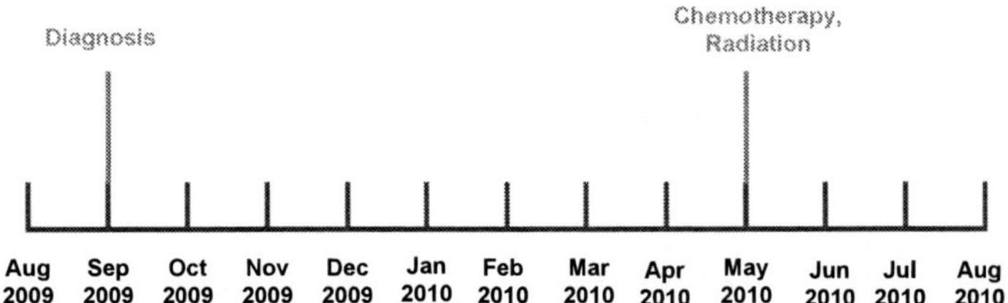

Figure 2: Sample cancer event timeline for a user constructed using events and dates extracted by our system

highlight the improvement achieved by filtering. As we can see from the table, each method has its own merits. ADEPT has extremely high precision but very poor recall, whereas our filtered term list improves significantly on precision while losing on recall. Hence, we combine both methods by only selecting sentences which contain terms marked as medical terms by both methods. As we can see from the table, this strategy works best, leading to an increase in precision without hurting recall [4]. This method is used in the final system.

6.2 Evaluation of the Clustering Module

We evaluate the performance of both methods (LDA and Word2Vec-based agglomerative clustering) used in the clustering module manually. We look at the words in each cluster generated by both clustering methods and label a cluster as corresponding to a certain cancer event, if most of the words in the cluster are associated with that event. For example, a cluster containing most diagnosis-related words ("diagnose", "diagnosis", "diagnosed" etc.) is labeled as the "Diagnosis" event. On manual inspection of the clusters detected by LDA, we observe that only four important cancer events (chemotherapy, radiation, mastectomy and diagnosis) out of eight major events are identified. The main reason behind the poor performance of this algorithm is that the candidate sentences being clustered are very short and do not provide enough contextual information for the algorithm. However, we cannot perform clustering on the entire posts, since a single post may describe multiple events. Moreover, most cancer events co-occur with similar words and this further

hampers LDA performance. On the other hand, a manual inspection of the clusters detected by Word2Vec-based agglomerative clustering detects six major cancer events (diagnosis, chemotherapy, radiation, reconstruction, metastasis, recurrence) very clearly. It also identifies a seventh event which is a mixture of words related to lumpectomy and mastectomy (it combines both these events into a single event). We present some keyword sets identified by this clustering algorithm for some cancer events below:

- **Chemotherapy:** chemotherapy, adjuvant, neoadjuvant, chemo, adriamycin, carboplatin, Taxol, herceptin, taxol, prednisone, Herceptin

- **Mastectomy/ Lumpectomy:** hysterectomy, lumpectomies, re-excision, mastectomy, Mastectomy, lumpectomy, mastectomies

As we can see from the above examples, these keyword sets are fairly coherent. [5]

6.3 End-to-End System Evaluation

To evaluate our end-to-end system, we test it on the user timeline construction task. As mentioned in section 3, we use a dataset consisting of all posts by a group of 50 users from breastcancer.org for this experiment. This dataset also contains date annotations for major cancer events for each user. However, this dataset is very small and only contains a total of 60 gold dates associated with cancer events, since dates pertaining to all cancer events for each user may not be available from their posting histories.

We compare the performance of our system with a re-implementation of the system described by Wen and Rosé (2012). We need to re-implement their system because we use different

---

[4]Increase in recall is observed because case-insensitive matching is used to find common terms selected by both ADEPT and VOCAB. This leads to the presence of some words selected only by one method in the final set. Such words however are case-variations of important words and must not be discarded

[5]It is difficult to peform a quantitative evaluation of the keyword sets since there is no gold standard

Method	Precision	Recall	F-Score
ADEPT	31.40	98.47	51.59
VOCAB (no filtering)	34.48	90.83	49.99
VOCAB (filtered)	47.46	84.17	60.45
ADEPT + VOCAB (filtered)	**50.32**	**99.03**	**66.73**

Table 2: Evaluation results for various methods used in the medical term detection module. For more details about these methods, refer to sections 5.1.1 and 5.1.2

System	Accuracy
(Wen and Rosé, 2012)	60
Our system	55.17

Table 3: Accuracy of date extraction for both systems on the cancer event timeline construction task

strategies for sentence filtering and temporal resolution, which can affect system performance. We use HeidelTime (Strötgen and Gertz, 2010) for temporal expression extraction and resolution. We also do not filter sentences which do not contain mentions of self-reported events. Hence, in order to facilitate a fair comparison between our system and (Wen and Rosé, 2012), we re-implement date sentence extraction (extraction of sentences containing medical events) as described in their paper, do not perform sentence filtering and use Heidel-Time for temporal resolution. This version of the system is used as our baseline. We do this in order to ensure that the only difference between both systems lies in medical event extraction, which is the main focus of our work. We measure system performance based on accuracy, which is computed as the number of dates correctly extracted by the system divided by the number of dates present in the gold data. Table 3 presents the performance of both systems. From this comparison, we can see that our medical event extraction pipeline, in spite of being almost completely unsupervised, is able to achieve almost 92% of the accuracy obtained by the baseline system which uses supervised medical event extraction. However, the accuracy of both systems is not high enough to be used in practice.

Fig 2 shows a sample cancer event timeline created for a user. These cancer event timelines for users can be used to visualize patient disease trajectories. They can also be used to visualize links between important cancer events and user posting trends in online support groups by plotting the number of posts made by the user in each month on the same timeline and observing whether users tend to post more/ less during these events.

7   Error Analysis and Future Work

Since our dataset contains only 60 gold dates, our system misses only 3 dates as compared to the baseline system. Though the results of our current system are encouraging, deeper analysis of the errors made by the end-to-end system as well as the event clusters detected by our pipeline presents many shortcomings which should be addressed in future work.

Our system manages to detect six out of eight cancer events, but it is unable to distinguish between lumpectomy and mastectomy. Because of this, our system extracts the same date for both events. Though this is a small source of errors for the current system because the dataset is very small, this may turn out to be a large source of errors for bigger datasets. This error also shows that some medical events may be extremely similar and our current system might not be able to tease them apart. It would be desirable to come up with better clustering techniques which can make such fine distinctions.

Another source of errors for our system arises from the use of word vectors trained on PubMed and PMC articles. Since the word vectors are trained on biomedical data they contain a lot of medical terminology, but they do not contain appropriate word vectors for a lot of colloquial medical terms used in online support groups (eg: "mets", "dx"). Hence such terms are not added to the correct cluster. For example, the words "metastatic" and "mets" appear in different clusters which is incorrect since they refer to the same event (metastasis). Transferring pre-trained word vectors from a biomedical corpus to data from an online support group can help mitigate this issue, which we plan to explore in the future.

An additional source of errors arises from the

rules used to link temporal expressions to events. While we have rules which take care of the situation in which multiple temporal expressions may occur in the same sentence as the event, we ignore scenarios in which multiple event words may occur in a sentence with a single temporal expression. The current temporal rules will assign that expression to all events, which may be wrong in certain cases. For example, in the sentence "Had lumpectomy in November 2000, but because margins were not clear, and another small tumor was found in the same breast, surgeon recommended modified radical mastectomy", two cancer events (lumpectomy, mastectomy) are mentioned with only one temporal expression (November 2000). The temporal expression will be linked to both events according to the current resolution rules, whereas it should only be linked to lumpectomy. Moreover, sometimes the sentences may contain exactly one cancer event and one temporal expression. However, the temporal expression still does not refer to the cancer event. For example, in the sentence "He died in Oct '04, right after my bc diagnosis.", the date October 2004 does not refer to the user's diagnosis event. These issues with temporal resolution impact the performances of both our system and the baseline system. Improving strategies for linking events with temporal expressions should help in tackling these issues.

8 Conclusion

In this paper, we propose a novel data-driven pipeline for personal medical event extraction from social media using minimal supervision, which is able to achieve 92% of the performance achieved by a supervised baseline. The extracted medical events can be used to study and identify links between user participation on online support groups and important medical events in their lives. While the results of our current system pipeline for personal medical event extraction are encouraging, there is a lot of scope for further improvement.

Acknowledgments

This research was supported in parts by NSF IIS 1546393 and NHLBI R01 HL122639/. We would like to acknowledge Miaomiao Wen and Hyeju Jang for providing access to the data from Wen and Rosé (2012) and Khyathi Chandu for her feedback on earlier drafts of this paper. We would also like to thank the anonymous reviewers for their constructive feedback.

References

Hila Becker, Mor Naaman, and Luis Gravano. 2010. Learning similarity metrics for event identification in social media. In *Proceedings of the third ACM international conference on Web search and data mining*. ACM, pages 291–300.

David M Blei, Andrew Y Ng, and Michael I Jordan. 2003. Latent dirichlet allocation. *Journal of machine Learning research* 3(Jan):993–1022.

Nathanael Chambers. 2013. Navytime: Event and time ordering from raw text. Technical report, DTIC Document.

Angel X Chang and Christopher D Manning. 2012. Sutime: A library for recognizing and normalizing time expressions. In *LREC*. volume 2012, pages 3735–3740.

Leon Derczynski, Jannik Strötgen, Diana Maynard, Mark A Greenwood, and Manuel Jung. 2016. Gatetime: Extraction of temporal expressions and event. In *10th Language Resources and Evaluation Conference*. European Language Resources Association (ELRA), pages 3702–3708.

Lifu Huang, T Cassidy, X Feng, H Ji, CR Voss, J Han, and A Sil. 2016. Liberal event extraction and event schema induction. In *Proceedings of the 54th Annual Meeting of the Association for Computational Linguistics: Human Language Technologies (ACL-16)*.

Jiwei Li and Claire Cardie. 2014. Timeline generation: Tracking individuals on twitter. In *Proceedings of the 23rd international conference on World wide web*. ACM, pages 643–652.

Jiwei Li, Alan Ritter, Claire Cardie, and Eduard H Hovy. 2014. Major life event extraction from twitter based on congratulations/condolences speech acts. In *EMNLP*. pages 1997–2007.

Diana Lynn MacLean and Jeffrey Heer. 2013. Identifying medical terms in patient-authored text: a crowdsourcing-based approach. *Journal of the American Medical Informatics Association* 20(6):1120–1127.

Tomas Mikolov, Kai Chen, Greg Corrado, and Jeffrey Dean. 2013. Efficient estimation of word representations in vector space. *arXiv preprint arXiv:1301.3781* .

Alan Ritter, Oren Etzioni, Sam Clark, et al. 2012. Open domain event extraction from twitter. In *Proceedings of the 18th ACM SIGKDD international conference on Knowledge discovery and data mining*. ACM, pages 1104–1112.

Takeshi Sakaki, Makoto Okazaki, and Yutaka Matsuo. 2010. Earthquake shakes twitter users: real-time event detection by social sensors. In *Proceedings of the 19th international conference on World wide web*. ACM, pages 851–860.

Roser Saurí, Robert Knippen, Marc Verhagen, and James Pustejovsky. 2005. Evita: a robust event recognizer for qa systems. In *Proceedings of the conference on Human Language Technology and Empirical Methods in Natural Language Processing*. Association for Computational Linguistics, pages 700–707.

Jannik Strötgen and Michael Gertz. 2010. Heideltime: High quality rule-based extraction and normalization of temporal expressions. In *Proceedings of the 5th International Workshop on Semantic Evaluation*. Association for Computational Linguistics, pages 321–324.

Miaomiao Wen and Carolyn Penstein Rosé. 2012. Understanding participant behavior trajectories in online health support groups using automatic extraction methods. In *Proceedings of the 17th ACM international conference on Supporting group work*. ACM, pages 179–188.

Miaomiao Wen, Zeyu Zheng, Hyeju Jang, Guang Xiang, and Carolyn Penstein Rosé. 2013. Extracting events with informal temporal references in personal histories in online communities. In *ACL (2)*. pages 836–842.

Detecting mentions of pain and acute confusion in Finnish clinical text

Hans Moen [1,2]*, **Kai Hakala**[1,3]*, **Farrokh Mehryary**[1,3], **Laura-Maria Peltonen**[2,4],
Tapio Salakoski[1,5], **Filip Ginter**[1], **Sanna Salanterä**[2,4]
1. Turku NLP Group, Department of Future Technologies, University of Turku, Finland.
2. Department of Nursing Science, University of Turku, Finland.
3. The University of Turku Graduate School (UTUGS), University of Turku, Finland.
4. Turku University Hospital, Finland.
5. Turku Centre for Computer Science (TUCS), Finland.
`firstname.lastname@utu.fi`

Abstract

We study and compare two different approaches to the task of automatic assignment of predefined classes to clinical free-text narratives. In the first approach this is treated as a traditional mention-level named-entity recognition task, while the second approach treats it as a sentence-level multi-label classification task. Performance comparison across these two approaches is conducted in the form of sentence-level evaluation and state-of-the-art methods for both approaches are evaluated. The experiments are done on two data sets consisting of Finnish clinical text, manually annotated with respect to the topics pain and acute confusion. Our results suggest that the mention-level named-entity recognition approach outperforms sentence-level classification overall, but the latter approach still manages to achieve the best prediction scores on several annotation classes.

1 Introduction

In relation to patient care in hospitals, clinicians document the administrated care on a regular basis. The documented information is stored as clinical notes in electronic health record (EHR) systems. In many countries and hospital districts, a substantial portion of the information that clinicians document concerning patient status, performed interventions, thoughts, uncertainties and plans are written in a narrative manner using (natural) free text. This means that much of the patient information is only found in free-text form, as opposed to structured or coded information (c.f.

standardized terminology, medications and diagnosis codes).

When it comes to information retrieval, management and secondary use, having the computer automatically identify and extract information from health records related to a given query or topic is desirable. This could, for example, be information about pain treatment given to a patient, or a patient group. Although free text is easy to produce by humans and allows for great flexibility and expressibility, it is challenging to have computers automatically classify and extract information from such text. The use of computers to automatically extract, label and structure information in free text is referred to as information extraction (Meystre et al., 2008), with named-entity recognition as a sub-task (Patawar Maithilee, 2015; Quimbaya et al., 2016). Due to the complexity of free text, this task is commonly approached using manually annotated text as training data for machine learning algorithms (see e.g. Velupillai and Kvist (2012)).

We present an ongoing work towards automated annotation of text, i.e. labelling with pre-defined classes/entity types, by first having the computer learn from a set of manually annotated clinical notes. The annotations concern two topics relevant to clinical care: *Pain* and *Acute Confusion*. To get a better insight into these topics and how this is being documented, two separate data sets have been manually annotated, one for each topic. For each of the two topics, a set of classes has been initially identified that reflect the information which the domain experts are interested in. An example sentence demonstrating the annotations is presented in Figure 1. The ultimate aim of this annotation work is to achieve improved documentation, assessment, handling and treatment of pain and acute confusion in hospitals (Heikkilä et al., 2016; Voyer et al., 2008). Now we want to inves-

---

*These authors contributed equally.

365

Proceedings of the BioNLP 2017 workshop, pages 365–372,
Vancouver, Canada, August 4, 2017. ©2017 Association for Computational Linguistics

tigate how to best train the computer to automatically detect and annotate mentions of these topics in new, unseen text by exploring various machine learning methods.

We address this by testing and comparing two different overall approaches:

- Named-entity recognition (NER), where we have the computer attempt to detect the mention-level annotation boundaries.

- Sentence classification (SC), where we have the computer attempt to label sentences based on the contained annotations.

The motivation for comparing these two approaches is that: (a) the experts are satisfied with having the computer identify and extract information on sentence level; and (b) we hypothesize that several classes, in particular those reflecting the more complex concepts, are easier for the computer to identify when approached as a sentence classification task. Further, we are not aware of any other work where a similar comparison has been reported. The methods and algorithms that we explore are based on state-of-the-art machine learning methods for NER and SC.

2   Data

Pain is something that most patients experience to various degrees during or related to a hospital stay. Pain experience is subjective and hence it can be challenging for clinicians to properly assess if, how and to what extent patients are experiencing pain. Acute confusion is a mental state that patients may enter as a result of serious illness, infections, intense pain, anesthesia, surgery and/or drug use. When clearly evident, this is commonly diagnosed as acute confusion or delirium (Fearing and Inouye, 2009), which is identified as a mental disorder that affects perception, cognitivity, memory, personality, mood, psychomotricity and the sleep-wake rhythm. However, it can be challenging to clearly identify acute confusion or delirium at the point of care, in particular the milder cases. Still, signs and symptoms can often be found in the free text that clinicians document (Voyer et al., 2008), and the same goes for pain (Gunningberg and Idvall, 2007).

Our annotated data consists of a random sample of 280 care episodes that were gathered from patients who had an open heart surgery and who were admitted to one university hospital in Finland during the years 2005-2009. This sample includes 1327 days of nursing narratives and 2156 notes written by physicians. The same sample was used as data sets for both topics (i.e. pain and acute confusion). An ethical approval and an organizational permission from the hospital district was obtained before the data collection.

Separate annotation schemes, reflecting the classes and guidelines for the annotation work, were iteratively developed based on the literature for both topics. For pain the annotation scheme has 15 classes while the acute confusion scheme has 37 classes (see supplementary materials for more details). The annotation schemes were initially tested and refined by having the annotators annotate a separate data set of another 100 care episodes (not included in this study). The annotation task was conducted by four persons working in pairs of two, so that all the text was annotated by (at least) two annotators. This team of annotators consisted of two domain experts and two non domain experts with an informatics background. At the end, the annotators analyzed the made annotations with respect to common consensus before producing the final annotated data sets used in this study. The annotations were conducted using the brat annotation tool (Stenetorp et al., 2012).

The two data sets were individually divided into *training* (60%), *development* (20%) and *test* (20%) sets. As preprocessing of the data we tokenize and enrich the text with linguistic information in the form of lemmas and part-of-speech (POS) tags for each token. For this we use the Finnish dependency parser (Haverinen et al., 2014).

For training of word embeddings (word-level semantic vectors), we used a large corpus consisting of both physician and nursing narratives, extracted from the same university hospital (in Finland). In total, this corpus consist of approximately 0.5M nursing narratives and 0.4M physician notes, which amounts to 136M tokens.

3   Experiment and Methods

Below (Section 3.1 and 3.2) we describe the methods, algorithm implementations and hyper parameters used in the two approaches, i.e., named-entity recognition (NER) and sentence classification (SC). In the Results section, Section 4, we compare the scores achieved by these two ap-

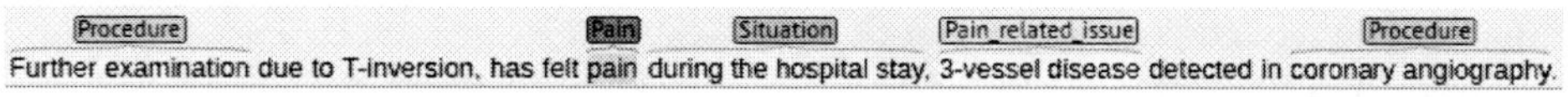

Figure 1: An artificial English example of the used pain annotations.

proaches for each of the two topics (i.e. pain and acute confusion).

3.1 Named-entity recognition (NER)

In this approach we focus on methods for predicting word-level annotation spans. More precisely we explore two such methods that have shown state-of-the-art performance in NER.

NERsuite Conditional random fields (CRFs) are a class of sequence modeling methods that have shown state-of-the-art performance in learning to identify biomedical named entities in text (Campos et al., 2013). We use a named-entity recognition toolkit called NERsuite (Cho et al., 2010), which is built on top of CRFsuite (Okazaki, 2007). For each of the two topics, one NERsuite model is trained using the corresponding training sets and the mentions are labeled using the common IOB tagging scheme. As training features, we use the original tokens, lemmas and POS tags. Although NERsuite allows the user to adjust regularization and label weight parameters, for this initial study we have used the default hyperparameters. It is worth noting that adjusting the regularization parameter is not as crucial for CRFs as it is for instance for support vector machines and strong results can be achieved even with the default values.

Several of the annotated entities have overlapping spans, e.g. the Finnish compound word *rintakipu* (chest pain) includes both *pain* and *location* mentions, but the standard CRF implementations are not able to do multi-label classification. Thus we form combination classes from the full spans of overlapping entities. This slightly distorts the annotated spans as the original mentions may have had only partial overlaps. Another option would have been to train separate models for each class, but as the number of classes is relatively high for both topics, this would have been very impractical.

CNN-BiLSTM-CRF The second method that we explore is an end-to-end neural model following the approach by Ma and Hovy (2016), which has produced state-of-the-art results for general domain English NER tasks. This model uses a CRF layer for the final predictions, but instead of relying on handcrafted features it utilizes a bidirectional recurrent neural network layer, with a long short-term memory (LSTM) (Hochreiter and Schmidhuber, 1997; Gers et al., 2000) chain, over input word embeddings. In addition to the input word embeddings, a convolutional layer is used over character embedding sequences to form another encoding for each token. Thus, this model is often called CNN-BiLSTM-CRF network. For training the model we use the example implementation provided by the authors [1].

Training the CNN-BiLSTM-CRF is computationally much more demanding then a standard CRF classifier and we have thus not ran an exhaustive hyperparameter search. Instead, we use the default values from the original paper except for setting the LSTM state dimensionality to 100 and learning rate to 0.05 as these produced slightly better results than the default values. The word embeddings are initialized with a word2vec (Mikolov et al., 2013) model trained on the large clinical Finnish text corpus.

3.2 Sentence classification (SC)

In this approach, we regard the task as a multi-label text classification task in which a sentence can be associated with multiple labels. For this task, we rely on artificial neural networks (ANN) since they have been shown to achieve state-of-the-art performance in text classification tasks (see e.g. Zhang et al. (2015); Tang et al. (2015)).

Neural network architecture We tried several neural network architectures, but report only the architecture that performed best. For both of the two topics, we apply a deep learning-based neural network architecture that use three separate LSTM chains: for the sequence of words, lemmas and POS tags.

The network has three separate channels for the words, lemmas and POS tags in the sentence. Each channel receives a sequence (words, lemmas or POS tags) as input. The items in the sequence are then mapped into their corresponding vector representations using a dedicated embedding look-up layer. The sequence of vectors is then input to an

---

[1]https://github.com/XuezheMax/LasagneNLP

LSTM chain and the last step-wise output of the chain is regarded as the representation of the sentence based on its words (or lemmas or POS tags).

Next, the outputs of the three channels are concatenated and the resulting vector is forwarded into the classification (decision) layer, which has a dimensionality equal to the number of annotation classes. The *sigmoid* activation function is applied on the output of the decision layer.

Training and optimization For implementation we use the Keras deep learning library (Chollet, 2015), with Theano tensor manipulation library (Bastien et al., 2012) as the back-end engine. We use *binary cross-entropy* as the objective function and the *Adam* optimization algorithm (Kingma and Ba, 2014) for training the network. We initialize the embeddings for words and lemmas with pre-trained vectors, trained using word2vec on the Finnish clinical corpus. For hyper-parameter optimization, we do a grid search and evaluate each model on the development set. To detect the best number of epochs needed for training, we use the *early stopping* method. Optimization is done against the *micro-averaged* F-score.

To avoid overfitting, we apply *dropout* (Srivastava et al., 2014) regularization with a rate of 20% on the input gates and with a rate of 1% on the recurrent connections of all LSTM units. In addition, we have set the dimensionality of the word, lemma and POS tag embeddings to 300 and the dimensionality of the LSTMs' output are also set to 300.

4   Results

We first evaluate the two NER methods on mention level using a strict offset matching criteria. The micro-averaged results are presented in Table 1. The NERsuite model achieves F-scores of 73.10 and 48.11 on the test sets of pain and acute confusion data set, respectively. Surprisingly the CNN-BiLSTM-CRF model is not able to reach the performance of the vanilla NERsuite on the pain dataset even though it is able to utilize pre-trained word embeddings. This might be due to the data sets being limited to open heart surgery patients and thus to a rather narrow vocabulary. Consequently we do not train CNN-BiLSTM-CRF on the confusion data. To analyse the performance of the NER approach in relation to the SC approach, we also convert the detected entity mentions to sentence-level predictions. For this the predictions

Approach	Precision	Recall	F-score
Pain			
NERsuite	87.29	62.88	73.10
CNN-BiLSTM-CRF	79.30	63.80	70.71
Acute confusion			
NERsuite	69.33	36.84	48.11

Table 1: Mention-level evaluation of the tested NER approaches on the test sets of the Pain and Acute confusion corpora. The reported numbers are micro-averaged over the various classes.

from the best performing method, i.e. NERsuite, is used.

Table 2 shows the sentence-level scores for both the NER and SC approach. The best performing neural network used in the SC approach achieves slightly inferior results compared to the NER approach (when evaluated on sentence level). This seems to somewhat falsify our hypothesis about sentence-level classification methods potentially performing better than mention-level NER methods when the task is approached as a sentence classification task. Still, in Table 3 we see that the SC approach achieves best overall prediction scores for several of the annotation classes (see also supplementary materials). Based on our analysis so far, it is difficult to say whether these classes (i.e. the concepts they represents) are more "complex" than the others, or if there are some other factors affecting the results. In an attempt to achieve better insight into this, we calculated the average annotation spans and vocabulary size associated with the different classes. However, these numbers did not show any clear trend.

Approach	Pain	Acute confusion
NER	78.61	59.41
SC	77.65	57.49

Table 2:   Micro-averaged F-scores for the different approaches on the test sets of the pain and acute confusion data sets. NERsuite was used to produce the NER scores.

The actual pain mentions which are divided into explicit, implicit and potential pain subcategories all achieve relatively high performance, implicit pain being the hardest to predict (see supplementary materials for more details). The other classes, which describe additional information about the pain mentions, are generally speaking harder to detect than the actual pain mentions. The acute confusion related entities seems to be much harder

Approach	Pain	Acute confusion
NER	8	11
SC	7	8
Equal performance	0	18

Table 3: Counts showing the number of classes that the various approaches performed best at predicting.

to predict due to the vague and sparse nature of these concepts.

5 Discussion and Future Work

In this study we have gathered the initial results for detecting mentions of pain and acute confusion in Finnish clinical text. We also use a relaxed evaluation based on sentence level predictions and experiment with approaches designed specifically for this definition. Surprisingly the NERsuite based mention-level approach outperforms all other tested methods, showing strong performance and being the best suited alternative for real-world applications. However, it might be that these two approaches are complementary.

As the used datasets are limited to open heart surgery patients, a critical future work direction will be assessing the generalizability of the trained models on larger sets of patient health records, and from other hospital units. This study also reveals that multiple classes in the annotation schemes, in particular for acute confusion, need more manual annotation data, i.e. more training examples, in order to be reliably detected in an automatic manner.

As many of the classes can be considered as descriptive attributes of the pain and acute confusion mentions, but the relations have not been annotated explicitly, another future work direction is to investigate how often these relations are ambiguous and whether the relation extraction could be solved in an unsupervised fashion.

Acknowledgments

Funding sources for this research were Academy of Finland and Tekes (Räätäli ("Tailor") project). We would like to thank the annotators, Pauliina Anttila, Timo Viljanen, Satu Poikajärvi and Kristiina Heikkilä. We would also like to thank Juho Heimonen for assisting us in preprocessing the clinical text.

References

Frédéric Bastien, Pascal Lamblin, Razvan Pascanu, James Bergstra, Ian Goodfellow, Arnaud Bergeron, Nicolas Bouchard, David Warde-Farley, and Yoshua Bengio. 2012. Theano: New features and speed improvements. *arXiv preprint arXiv:1211.5590 (2012)* .

David Campos, Sérgio Matos, and José Luís Oliveira. 2013. Gimli: open source and high-performance biomedical name recognition. *BMC Bioinformatics* 14(1):54.

HC Cho, N Okazaki, M Miwa, and J Tsujii. 2010. Nersuite: a named entity recognition toolkit. `http://nersuite.nlplab.org`. Last visited 20th April 2017.

Franois Chollet. 2015. Keras. `https://github.com/fchollet/keras`. Last visited 10th March 2017.

Michael A Fearing and Sharon K Inouye. 2009. Delirium. *Focus* 7(1):53–63. https://doi.org/10.1176/foc.7.1.foc53.

Felix A Gers, Jürgen Schmidhuber, and Fred Cummins. 2000. Learning to forget: Continual prediction with LSTM. *Neural Computation* 12(10):2451–2471.

Lena Gunningberg and Ewa Idvall. 2007. The quality of postoperative pain management from the perspectives of patients, nurses and patient records. *Journal of Nursing Management* 15(7):756–766.

Katri Haverinen, Jenna Nyblom, Timo Viljanen, Veronika Laippala, Samuel Kohonen, Anna Missil, Stina Ojala, Tapio Salakoski, and Filip Ginter. 2014. Building the essential resources for finnish: the turku dependency treebank. *Language Resources and Evaluation* 48:493–531. https://doi.org/10.1007/s10579-013-9244-1.

Kristiina Heikkilä, Laura-Maria Peltonen, and Sanna Salanterä. 2016. Postoperative pain documentation in a hospital setting: A topical review. *Scandinavian Journal of Pain* 11:77–89.

Sepp Hochreiter and Jürgen Schmidhuber. 1997. Long short-term memory. *Neural Computation* 9(8):1735–1780.

Diederik P. Kingma and Jimmy Ba. 2014. Adam: A method for stochastic optimization. *CoRR* abs/1412.6980. http://arxiv.org/abs/1412.6980.

Xuezhe Ma and Eduard Hovy. 2016. End-to-end sequence labeling via bi-directional LSTM-CNNs-CRF. *arXiv preprint arXiv:1603.01354* .

Stéphane M Meystre, Guergana K Savova, Karin C Kipper-Schuler, and John F Hurdle. 2008. Extracting information from textual documents in the electronic health record: A review of recent research. *Yearbook of Medical Informatics* 35:128–44.

Tomas Mikolov, Ilya Sutskever, Kai Chen, Greg S. Corrado, and Jeff Dean. 2013. Distributed representations of words and phrases and their compositionality. In *Advances in Neural Information Processing Systems 26*, pages 3111–3119.

Naoaki Okazaki. 2007. Crfsuite: A fast implementation of Conditional Random Fields (CRFs). http://www.chokkan.org/software/crfsuite/. Last visited 20th April 2017.

Potey M. A. Patawar Maithilee. 2015. Approaches to named entity recognition: A survey. *International Journal of Innovative Research in Computer and Communication Engineering* 3(12):12201–12208.

Alexandra Pomares Quimbaya, Alejandro Sierra Mnera, Rafael Andrs Gonzlez Rivera, Julin Camilo Daza Rodrguez, Oscar Mauricio Muoz Velandia, Angel Alberto Garcia Pea, and Cyril Labb. 2016. Named entity recognition over electronic health records through a combined dictionary-based approach. *Procedia Computer Science* 100:55 – 61. https://doi.org/http://dx.doi.org/10.1016/j.procs.2016.09.123.

Nitish Srivastava, Geoffrey Hinton, Alex Krizhevsky, Ilya Sutskever, and Ruslan Salakhutdinov. 2014. Dropout: A simple way to prevent neural networks from overfitting. *Journal of Machine Learning Research* 15:1929–1958. http://jmlr.org/papers/v15/srivastava14a.html.

Pontus Stenetorp, Sampo Pyysalo, Goran Topić, Tomoko Ohta, Sophia Ananiadou, and Jun'ichi Tsujii. 2012. brat: a Web-based Tool for NLP-Assisted Text Annotation. In *Proceedings of the Demonstrations at the 13th Conference of the European Chapter of the Association for Computational Linguistics*. Association for Computational Linguistics, pages 102–107.

Duyu Tang, Bing Qin, and Ting Liu. 2015. Document modeling with gated recurrent neural network for sentiment classification. In *Proceedings of the 2015 Conference on Empirical Methods in Natural Language Processing*. pages 1422–1432.

Sumithra Velupillai and Maria Kvist. 2012. Fine-grained certainty level annotations used for coarser-grained E-Health scenarios. In Alexander Gelbukh, editor, *Computational Linguistics and Intelligent Text Processing*, Springer Berlin Heidelberg, volume 7182 of *Lecture Notes in Computer Science*, pages 450–461. https://doi.org/10.1007/978-3-642-28601-8_38.

Philippe Voyer, Martin G Cole, Jane McCusker, Sylvie St-Jacques, and Johanne Laplante. 2008. Accuracy of nurse documentation of delirium symptoms in medical charts. *International Journal of Nursing Practice* 14(2):165–177. https://doi.org/10.1111/j.1440-172X.2008.00681.x.

Xiang Zhang, Junbo Zhao, and Yann LeCun. 2015. Character-level convolutional networks for text classification. In C. Cortes, N. D. Lawrence, D. D. Lee, M. Sugiyama, and R. Garnett, editors, *Advances in Neural Information Processing Systems 28*, Curran Associates, Inc., pages 649–657.

A   Supplemental Material

The supplementary material includes specific information about the annotation classes for the pain and acute confusion data sets, as well as the detailed evaluation of the studied methods.

English names	Finnish names	SC	NER
A recurrent situation	Toistuva_tilanne	83.60	86.10
Care plan	Suunnitelma	83.39	84.24
Implicit pain	Implisiittinen_kipu	74.64	73.68
Pain related issue	Kipuun_liittyva_asia	54.69	54.51
Location of pain	Sijainti	83.36	87.23
Pain	Kipu	92.04	93.23
Pain intensity	Voimakkuus	75.29	80.80
Pain management	Kivunhoito	85.36	86.97
Patient education	Ohjeistus	33.33	0.00
Potential pain	Potentiaalinen_kipu	93.69	95.58
Procedure	Toimenpide	64.73	63.90
Quality of pain	Laatu	59.33	71.07
Success of treatment	Hoidon_onnistuminen	73.74	64.56
Situation	Tilanne	37.21	28.57
Time	Aika	79.49	72.41
Micro-average		**77.65**	**78.61**

Table 4: Comparison of SC and NER for sentence classification, for pain corpus test set, evaluated on micro-averaged F1-scores.

English Names	Finnish Names	SC	NER
Abnormal level of consciousness	Muu_poikkeava_tajunnan_taso	28.57	0.00
Aggressiveness	Aggressiivisuus_vihaisuus	20.00	64.00
Appetite disturbance	Ruokahalun_hairio	57.35	59.02
Calming activity	Rauhoittelu	0.00	0.00
Confusion	Sekavuus	85.95	95.00
Delirium	Delirium	0.00	0.00
Delusion	Harhaisuus	34.29	27.59
Dementia	Dementia	0.00	0.00
Desorientation	Desorientaatio	66.67	89.86
Diagnosed	Diagnosoitu	0.00	0.00
Disturbance in ability to focus	Vaikea_kiinnittaa_huomiota	15.39	0.00
Disturbance in the quality of speech	Puheen_laadun_hairiot	40.00	29.41
Drowsy	Unelias	77.98	79.44
Falls - fall out of bed	Kaatuminen_Sangysta_tippuminen	0.00	0.00
Hyper-alert	Ylivalpas	0.00	0.00
Hyperactivity	Hyperaktiivisuus	68.71	70.23
Hypoactivity	Hypoaktiivisuus	26.67	22.22
Infusion line detachment	Letkun_irtoaminen	20.00	20.00
Memory disorder	Muistiongelma	73.24	80.00
Not awakable	Ei_herateltavissa	0.00	0.00
Orientation to time and place	Orientoiminen_aikaan_paikkaan	0.00	0.00
Other abnormal behavior	Muu_poikkeava_kayttaytyminen	0.00	0.00
Other affective disturbance	Muu_tunnehairio	64.52	42.25
Other care activity	Muu_hoitotoimenpide	0.00	0.00
Other cognitive disturbance	Muu_kognitiivinen_hairio	0.00	0.00
Other disturbance of attention	Muu_tarkkaavaisuuden_hairio	0.00	0.00
Other incident	Muu_haittatapahtuma	0.00	0.00
Other symptom	Muu_oire	0.00	0.00
Pain management	Kivunhoito	51.52	61.33
Problems with motor functions	Motoriikan_ongelmat	59.56	59.79
restraint - restraining	Lepositeet_sitominen	75.00	76.19
Sleep-wake disorder	Unirytmin_valverytmin_hairiot	54.32	48.65
Slow rate of speech - Speechlessness	Hidastunut_puhe_puhumattomuus	0.00	0.00
Substance induced delirium	Substance_induced_delirium	0.00	0.00
Unappropriate behaviour	Asiaankulumaton_kayttaytyminen	9.52	10.26
Uncertain	Epavarma	0.00	0.00
Unorganised thinking	Ajatuksenkulun_jarjestaytymattomyys	25.81	21.43
Micro-average		**57.49**	**59.41**

Table 5: Comparison of SC and NER for sentence classification, for acute confusion corpus test set, evaluated on micro-averaged F1-scores.

English Names	Finnish Names	Train	Devel	Test	Total
A recurrent situation	Toistuva_tilanne	589	215	210	1014
Care plan	Suunnitelma	517	176	170	863
Implicit pain	Implisiittinen_kipu	552	160	201	913
Pain related issue	Kipuun_liittyva_asia	1058	377	372	1807
Location of pain	Sijainti	1001	326	333	1660
Pain	Kipu	1655	536	549	2740
Pain intensity	Voimakkuus	1094	291	341	1726
Pain management	Kivunhoito	1158	368	419	1945
Patient education	Ohjeistus	11	3	4	18
Potential pain	Potentiaalinen_kipu	752	222	255	1229
Procedure	Toimenpide	1423	478	468	2369
Quality of pain	Laatu	323	100	125	548
Success of treatment	Hoidon_onnistuminen	226	75	102	403
Situation	Tilanne	286	85	82	453
Time	Aika	1257	386	426	2069
	Overall	11902	3798	4057	19757
	Tokens	437935	147444	153975	739354
	Sentences	71390	23470	25123	119983
	Documents	2084	697	702	3483

Table 6: Pain annotation counts per class.

English Names	Finnish Names	Train	Devel	Test	Total
Abnormal level of consciousness	Muu_poikkeava_tajunnan_taso	11	9	6	26
Aggressiveness	Aggressiivisuus_vihaisuus	24	5	16	45
Appetite disturbance	Ruokahalun_hairio	229	84	76	389
Calming activity	Rauhoittelu	6	4	6	16
Confusion	Sekavuus	131	45	60	236
Delirium	Delirium	4	1	1	6
Delusion	Harhaisuus	37	16	25	78
Dementia	Dementia	3	2	1	6
Desorientation	Desorientaatio	77	25	38	140
Diagnosed	Diagnosoitu	1	0	0	1
Disturbance in ability to focus	Vaikea_kiinnittaa_huomiota	29	8	12	49
Disturbance in the quality of speech	Puheen_laadun_hairiot	43	10	25	78
Drowsy	Unelias	275	88	115	478
Falls - fall out of bed	Kaatuminen_Sangysta_tippuminen	6	0	3	9
Hyper-alert	Ylivalpas	3	1	1	5
Hyperactivity	Hyperaktiivisuus	232	66	78	376
Hypoactivity	Hypoaktiivisuus	103	35	44	182
Infusion line detachment	Letkun_irtoaminen	15	4	9	28
Memory disorder	Muistiongelma	92	40	41	173
Not awakable	Ei_herateltavissa	15	7	6	28
Orientation to time and place	Orientoiminen_aikaan_paikkaan	6	0	0	6
Other abnormal behavior	Muu_poikkeava_kayttaytyminen	6	2	4	12
Other affective disturbance	Muu_tunnehairio	109	52	52	213
Other care activity	Muu_hoitotoimenpide	12	9	7	28
Other cognitive disturbance	Muu_kognitiivinen_hairio	23	4	5	32
Other disturbance of attention	Muu_tarkkaavaisuuden_hairio	10	1	3	14
Other incident	Muu_haittatapahtuma	10	3	5	18
Other symptom	Muu_oire	25	5	8	38
Pain management	Kivunhoito	118	39	40	197
Problems with motor functions	Motoriikan_ongelmat	329	93	117	539
restraint - restraining	Lepositeet_sitominen	25	8	13	46
Sleep-wake disorder	Unirytmin_valverytmin_hairiot	147	56	48	251
Slow rate of speech - Speechlessness	Hidastunut_puhe_puhumattomuus	25	11	11	47
Substance induced delirium	Substance_induced_delirium	1	0	0	1
Unappropriate behaviour	Asiaankulumaton_kayttaytyminen	81	22	33	136
Uncertain	Epavarma	1	0	0	1
Unorganised thinking	Ajatuksenkulun_jarjestaytymattomyys	62	17	24	103
	Overall	2326	772	933	4031
	Tokens	434542	149387	155425	739354
	Sentences	71146	23797	25040	119983
	Documents	2080	698	705	3483

Table 7: Acute confusion annotation counts per class.

A Multi-strategy Query Processing Approach for Biomedical Question Answering: USTB_PRIR at BioASQ 2017 Task 5B

Zan-Xia Jin, Bo-Wen Zhang*, Fan Fang, Le-Le Zhang and **Xu-Cheng Yin***

Pattern Recognition and Information Retrieval lab (PRIR)

Department of Computer Scicene, University of Science and Technology Beijing

bowenzhang@xs.ustb.edu.cn, xuchengyin@ustb.edu.cn

Abstract

This paper describes the participation of USTB_PRIR team in the 2017 BioASQ 5B on question answering, including document retrieval, snippet retrieval and concept retrieval task. We introduce different multimodal query processing strategies to enrich query terms and assign different weights to them. Specifically, sequential dependence model *(SDM)*, pseudo relevance feedback *(PRF)*, fielded sequential dependence model *(FSDM)* and Divergence from Randomness model *(D-FRM)* are respectively performed on different fields of PubMed articles, sentences extracted from relevant articles, the five terminologies or ontologies (MeSH, GO, Jochem, Uniprot and DO) to achieve better search performances. Preliminary results show that our systems outperform others in the document and snippet retrieval task in the first two batches.

1 Introduction

Due to the continuous growth of information produced in the biomedical domain, there is a particularly growing demand for biomedical QA from the general public, medical students, health care professionals and biomedical researchers (Zweigenbaum, 2003). They consult knowledge about the natures, the preventions or the treatments of diseases, or learn from research results of other researchers. To some extent, biomedical QA is one of the most significant applications of the existing real-world biomedical systems (Han and Athenikos, 2010).

Since 2013, BioASQ organizers has proposed a community-based shared task which aims to evaluate the current solutions of a variety of QA sub-tasks. Several benchmarks have been provided for researchers to evaluate their QA systems. BioASQ 2017 Task 5B challenge (Tsatsaronis et al., 2015a) is the fifth edition of the question answering task, of which the phase A requires the evaluated system to (i) semantically annotate the questions with concepts from a set of designated terminologies and ontologies (MeSH, GO, Jochem, Uniprot and DO); and (ii) retrieve relevant articles, text snippets, and RDF triples from designated article repositories and ontologies (PubMed/MEDLINE articles) with biomedical questions in natural language provided by biomedical professionals or researchers. The ground truth are manually annotated by these experts with some annotated tools. There are five batches of evaluation and in each batch participants are provided with 100 natural language questions and required to return at most 10 relevant documents, snippets, concepts to the questions within 24 hours.

Over the past decade, a variety of approaches have been proposed for biomedical question answering (Bauer and Berleant, 2012). Generally, a QA system typically consists of question processing, document processing, and answer processing phases, which are respectively in charge of 1) converting natural language questions into queries, 2) searching relevant documents, and 3) extracting, ranking candidate answers and formatting them into expected answer type. (Han and Athenikos, 2010; Holzinger et al., 2014). There are several studies concerning the improvements on query processing phase (Huang et al., 2006; Yu et al., 2005; Kobayashi and Shyu, 2006) and document processing phase (Cairns et al., 2011; Yu and Cao, 2008). However for answer processing phase, especially answer matching and ranking, only some simple approaches in previous BioASQ challenge have been proposed (Tsatsaronis et al., 2015a; Mao and Lu, 2015). According to the

Proceedings of the BioNLP 2017 workshop, pages 373–380,
Vancouver, Canada, August 4, 2017. ©2017 Association for Computational Linguistics

above researches, the most challenges of biomedical QA are three main issues, specifically 1) how to generate query terms appropriately from natural language questions, 2) how to match relevant documents or sentences when they use different expressions (maybe synonyms of keywords) and 3) how to measure and utilize the difference in importance of query terms.

In order to address these challenges, in this paper we propose a multi-strategy query processing approach which combines several mature query processing models according to the different characteristics of data sources, which is also actually the participation of our USTB_PRIR team in the BioASQ Task 5B phase A challenge[1]. Specifically, in order to extract proper keywords and generate queries, we perform stop-words removal, noun extraction with Pos-of-Tagger (POS) and stemming. For the missing issue caused by expressions, we utilize a thesaurus which is produced through computing the similarities between the vector representations of each pairs of words. Moreover for query keyword weighting, we take the word sequences, different fields of appearance, TF-IDF, etc into consideration for different BioASQ tasks. We evaluate our approach on the BioASQ 2016 and 2017 benchmarks for document, snippet, concept retrieval and experimental results demonstrate our method outperforms the baseline methods or other participants so far on document, snippet and concept retrieval tasks.

2 Related Work

The participants of previous BioASQ challenge have proposed several approaches for searching relevant documents, snippets and concepts for biomedical QA. One of the participants(Choi, 2015) proposed to utilize semantic concept enriched dependence model where the recognised UMLS concepts in the query are used as additional dependence features for ranking documents. Another team(Papanikolaou et al., 2014) developed a figure-inspired text retrieval method as a way of retrieving documents and text passages from biomedical publications. For matching relevant snippets, most participants works on similar methods of searching articles. An exception is the framework proposed by NCBI(Mao et al., 2014), which directly compute the cosine similarities between the questions and the sentences.

However, these methods focus on the matching function or the ranking process, which ignores the three challenges mentioned in the Introduction section. The natural language questions are too raw to be regarded as query keywords and the difference in importance of keywords should be considered. Some re-ranking or learning-to-rank based approaches works not well either for the same reason because they rely much on the initial ranking results.

3 Task 5B Phase A: Document Retrieval

3.1 The Framework Architecture

The framework of searching relevant documents is shown in Figure 1, which includes document preprocessing, query pre-processing, several ranking models based on query expansion and term weighting strategies.

3.2 Pre-Processing

3.2.1 Document Pre-processing

We download the entire database of MEDLINE updated in Feb 2017 through the FTP service of National Institutes of Health (NIH) which contains 26,759,010 citations. These documents are represented in JSON files which contains a variety of information, including journal information, contents of title, author, abstract and keywords, similar articles and comments. We analyze the resources and select the following fields to represent the documents: *ArticleTitle*, *AbstractText*, *Title*, *MedlineTA*, *NameOfSubstance*, *DescriptorName*, *QualifierName*, *Keyword* and *ISOAbbreviation*. These fields are extracted from the document resources and indexed with Galago, an open source search engine[2], which is developed as an improved JAVA version of Indri. We also perform stemming and stop-words removal work like other IR applications, however unfortunately, the performances seems worse during the training process. As a result, we decide not to utilize these strategies for document pre-processing.

3.2.2 Query Pre-Processing

As is mentioned above, one of the challenges is how to automatically generate the query terms from a natural language question. During query pre-processing, we carry out a series of work to extract the keywords of the user queries. There

---

[1]http://bioasq.org

[2]http://www.lemurproject.org/galago.php

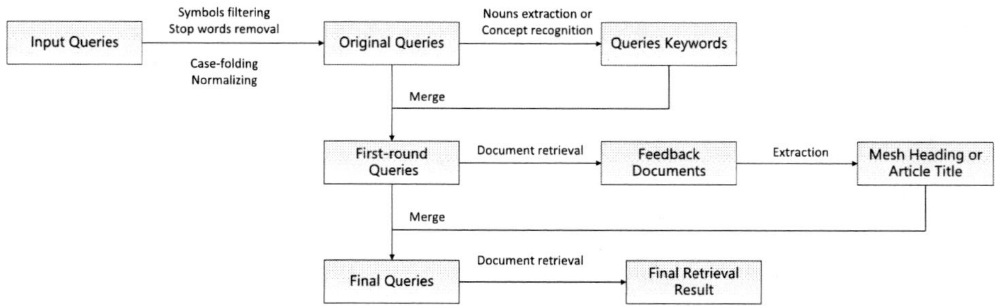

Figure 1: The whole framework architecture of query generation method based on multimodal document retrieval strategies

are several symbols which is unnecessary and unrelated to the requests so we filter out the symbols in the first step. Note that the symbols which may be a part of named entity cannot be removed. Afterwards, stop words like "what" or "are" are common in natural language questions and are not suitable to feed into search engine so they are removed according to a stop-words list. As usual, the query terms are case-folded and normalized. In addition, we used the Stanford-Postagger package to identify nouns from queries and the MetaMap to identify the biomedical concepts in query terms.

3.3 Ranking Models

3.3.1 Sequential Dependence Model

The traditional IR techniques in biomedical domain rely on a unigram Bag of Words (BoW) retrieval model. Each document in the collection of candidates, as well as each query, is represented by a set of words and the corresponding frequency based on the assumption that the appearance of each pair of words are independent. Different sequence of queries is regarded as the same. Consider an example of two documents that contain all query keywords. It is obvious that the document with the right sequence of terms appearing in query is more likely to meet the demand. Therefore, we introduce the Sequence Dependency Model (SDM) (Bonnefoy et al., 2012) to take the sequence information into account when computing the relevance between a document and a query.

SDM is a special case of the Markov Random Field (MRF) (Metzler and Croft, 2005). In order to capture the information of a sentence, this model extracts the phrases in different ways, and gives corresponding weights to different types of phrases to indicate their importance.

There are three features in the SDM to be considered: single-word features (a collection consists of single-word, Q_T), ordered bi-words phrase features (the two words in a phrase appearing in order, Q_O) and unordered window features (one or several words can be allowed appearing between the two words, Q_U). Generally, the potential function for unigrams (single-word feature) looks as follows:

$$f_T(q_i, D) = \log P(q_i|\theta_D) = \log \frac{tf_{q_i,D} + \mu \frac{cf_{q_i}}{|C|}}{|D| + \mu} \quad (1)$$

where q_i is a query term, D is a document, $tf_{q_i,D}$ is the frequency of q_i in D, $|D|$ is the document length, μ is a Dirichlet prior, that is usually set to the average document length in the collection, cf_{q_i} is the collection frequency of q_i and $|C|$ is the total number of terms in the collection. Similarly, for ordered and unordered bi-grams, the potential functions are respectively as follows:

$$\begin{aligned} f_O(q_i, q_{i+1}, D) &= \log P(\#1(q_i, q_{i+1})|\theta_D) \\ &= \log \frac{tf_{\#1(q_i,q_{i+1})} + \mu \frac{cf_{\#1(q_i,q_{i+1})}}{|C|}}{|D| + \mu} \end{aligned} \quad (2)$$

$$\begin{aligned} f_U(q_i, q_{i+1}, D) &= \log P(\#uwN(q_i, q_{i+1})|\theta_D) \\ &= \log \frac{tf_{\#uwN(q_i,q_{i+1})} + \mu \frac{cf_{\#1(q_i,q_{i+1})}}{|C|}}{|D| + \mu} \end{aligned} \quad (3)$$

where $\#1(q_i, q_{i+1})$ and $\#uwN(q_i, q_{i+1})$ are respectively the appearances of the exact phrase $q_i q_{i+1}$ and the term $q_i, q_{i+1}]$ within a window N terms. Hence, the scoring function of a document in SDM is the combination of the above three functions, shown as follows:

$$score_{\text{SDM}} = score_{\text{SDM}}(Q_T, Q_O, Q_U, D)$$
$$= \lambda_T \sum_{i=1}^{|Q|} f_T(q_i, D)$$
$$+ \lambda_O \sum_{i=1}^{|Q|-1} f_O(q_i, q_{i+1}, D) \qquad (4)$$
$$+ \lambda_U \sum_{i=1}^{|Q|-1} f_U(q_i, q_{i+1}, D)$$

$$score_{\text{FSDM}} = score_{\text{FSDM}}(Q_T, Q_O, Q_U, D)$$
$$= \lambda_T \sum_{i=1}^{|Q|} \tilde{f}_T(q_i, D)$$
$$+ \lambda_O \sum_{i=1}^{|Q|-1} \tilde{f}_O(q_i, q_{i+1}, D) \qquad (6)$$
$$+ \lambda_U \sum_{i=1}^{|Q|-1} \tilde{f}_U(q_i, q_{i+1}, D)$$

Where Q is a sequence of keywords extracted from a user query, D is a candidate document, q_i is the i-th query keyword of Q. f_T, f_O, f_U are the maximum likelihood estimations of the corresponding feature terms in document D. $\lambda_T, \lambda_O, \lambda_U$ are the features weights satisfy these conditions:

(1)$0 \leq \lambda_T, \lambda_O, \lambda_U \leq 1$ and $\lambda_T + \lambda_O + \lambda_U = 1$

(2)$\lambda_T \geq 0.6$

(3)$\lambda_O = 2\lambda_U$

Often, $\lambda_T = 0.85, \lambda_O = 0.1, \lambda_U = 0.05$.

3.3.2 Fielded Sequential Dependence Model

As is mentioned, the candidate documents are structured into several fields which contains different types of information. One of the limitations of standard SDM for structured document retrieval is that it considers term matches in different parts of a document as equally important (i.e. having the same contribution to the final relevance score of a document), thus disregarding the document structure.

To adapt the MRF framework to multi-fielded entity descriptions, we introduce (Zhiltsov et al., 2015)'s approach from their FSDM model to replace a single document language model $P(q_i|\theta_D)$ with a mixture of language models (MLM) for each document field. Consequently, the potential function for unigrams in case of FSDM is:

$$\tilde{f}_T(q_i, D) = \log \sum_j w_j P(q_i|\theta^j) \qquad (5)$$

where j represents the different fields, and the $P(q_i|\theta^j)$ is the language model in each individual field. Similarly, we can compute $\tilde{f}_O(q_i, q_{i+1}, D)$ and $\tilde{f}_O(q_i, q_{i+1}, D)$. Therefore, the scoring function of FSDM is as follows:

3.3.3 Pseudo Relevance Feedback

With the first-pass retrieval results, we assume that the initially retrieved top-K documents are relevant to questions, and their title and mesh fields contain relevant terms to the original query (Zhang et al., 2015a). Thus, for document retrieval, we use the Pseudo Relevance Feedback (PRF) to enrich query terms from the top-K document initially retrieved. The titles or mesh headings of the top-K documents are extracted and then added to the original query term set. However, the performance of PRF can be affected by the quality of the initial result, the number of pseudo-relevant documents (top K), the number of expansion terms, and the term re-weighting method applied. In our experiments, we use $K = 3$ and extract all the words in title or mesh headings as the expansion terms, which results in the best performance.

3.3.4 Multimodal Strategies Combination

Since there are several strategies to enrich query terms and optimize their weights, the final scoring function is expected to make full use of these strategies and combine these strategies effectively. We take the importance of nouns, sequence orders and crucial fields into consideration so our weight optimization of query terms is based on to noun extraction, sequential dependence model (SDM), Fielded sequential dependence model (FSDM), and Pseudo Relevance Feedback (PRF). According to massive experiments we find out that for some questions, the original queries, noun queries and enriched queries with PRF from relevant articles are all useful to some degree. Furthermore, we also find out that it is necessary to both search in the full text of the document, and to assign different weights to different fields at meanwhile. Hence, the final scoring function to search relevant documents is shown as follows:

Table 2: MAP performances compared with BioASQ Task 5B document retrieval participants.

System	Batch 1	Batch 2
sdm + NN + fsdm	0.1049	0.0850
sdm + NN + fsdm + PRF (mesh)	**0.1086**	0.0863
sdm + fsdm + PRF (mesh)	0.1032	0.0859
sdm + w2v	0.0928	**0.0874**
sdm	0.0952	0.0866
best of fdu	0.1072	0.0834
best of UNCC	0.1080	-
best of Olelo	0.0465	0.0318
best of KNU-SG	0.0413	0.0419
best of HPI	0.0307	0.0329
best of Others	0.0437	0.0265

$$score(Q, D) = \lambda_1 score_{\text{SDM}}(Q, D) + \lambda_2 score_{\text{FSDM}}(Q', D)$$
$$+ \lambda_3 score_{\text{SDM}}(Q'', D) \quad (7)$$

where Q is the original query term set after query pre-processing, Q' represents the noun query term set with noun extraction and the Q'' stands for the enriched query term set with PRF.

3.4 Experments

We evaluate our proposed method by using both the benchmark datasets from the previous BioASQ challenges and the current challenge. The optimization of all parameters, including the weighting paramters like w_j in FSDM function and hyper-parameters (e.g. λ_T, λ_O, λ_U, λ_1, λ_2, λ_3) are processed through tuning with the rules on training set (when evaluated on BioASQ Task 4B, the training set includes 800 questions from BioASQ 2B and 3B; for BioASQ Task 5B, the training set contains 500 more questions on BioASQ 4B). Table 1 provides the results of our experiments in BioASQ task 4B, and Table 2 provides the results of our experiments in BioASQ Task 5B. The $sdm + w2v$ approach refers to our previous approach in (Zhang et al., 2015b). Obviously, our proposed method shows greater performance compared with baseline, SDM and FSDM and outperform than other participants in current challenge.

4 Task 5B Phase A: Snippet Retrieval

4.1 The Framework Architecture

The framework of searching relevant snippets is shown in Figure 2, which includes pre-processing, some additional ranking models which is different from document retrieval.

4.2 Pre-Processing

The query pre-processing for snippets retrieval is the same to the strategies for document retrieval, which includes unnecessary symbol removal, stop-words removal, case-folding, noun extraction and concept extraction with Metamap.

For the snippet pre-processing, we choose the candidate snippets from the top-K documents of the best performed document retrieval approach on the basis of results of document retrieval. The sentences with the field ArticleTitle and the field abstract of these articles are separated through some specific rules, which can be regarded as "small documents". These sentences make up a pile of new files with unstructured text. They are then indexed by Galago for search in the next step.

4.3 Ranking Models

Different from document retrieval, the candidate snippets are represented in unstructured text, which makes some ranking models more difficult to utilize (e.g. FSDM). Moreover, since they are much shorter in length, they are more likely express similar meaning with different expressions (e.g. synonyms) which may emphasize the importance of the issue of recognizing these relevant results. Furthermore, the PRF method generally provides massive expansion query terms, which may affect the search performance of the short text so we give up applying PRF as query expansion method.

In addition, we introduce DFRM from (Clinchant and Gaussier, 2011) as an additional term weigting model to optimize the most appropriate weight for query terms.

4.3.1 Divergence from Randomness Model

The Divergence from Randomness models (DFRM) are based on this simple idea: "The more the divergence of the within-document term-frequency from its frequency within the collection, the more the information carried by the word t in the document d". In other words the term-weight is inversely related to the probability of term-frequency within the document d obtained by a model M of randomness:

$$\text{weight}(t|d) \propto -\log \text{Prob}_M(t \in d | \text{Collection}) \quad (8)$$

where the subscript M stands for the type of model of randomness employed to compute the probability. In order to choose the appropriate model M of randomness, we can use different urn models.

Table 1: MAP performances of system components on BioASQ Task 4B document retrieval.

	Batch 1	Batch 2	Batch 3	Batch 4	Batch 5
baseline	0.2056	0.2593	0.228	0.2324	0.2516
sdm	0.2214	0.2577	0.2436	0.2517	0.2935
fsdm	0.2156	0.2621	0.2228	0.2469	0.2728
sdm + fsdm	0.2269	0.2768	0.2447	0.2608	0.2968
sdm + NN + fsdm	0.2307	0.2741	0.2454	0.2632	0.2926
sdm + fsdm + PRF(title)	0.2337	0.2778	0.2455	0.265	0.2931
sdm + fsdm + PRF(mesh)	0.2372	0.2863	**0.2564**	0.2762	0.3019
sdm + NN + fsdm + PRF(mesh)	0.2436	0.2859	0.2465	0.2773	**0.3083**
sdm + NN + fsdm + PRF(title)	0.2377	0.2767	0.2429	0.2681	0.2985
sdm + NN + fsdm + mesh + PRF(mesh)	**0.2440**	**0.2876**	0.2505	**0.2821**	0.3059

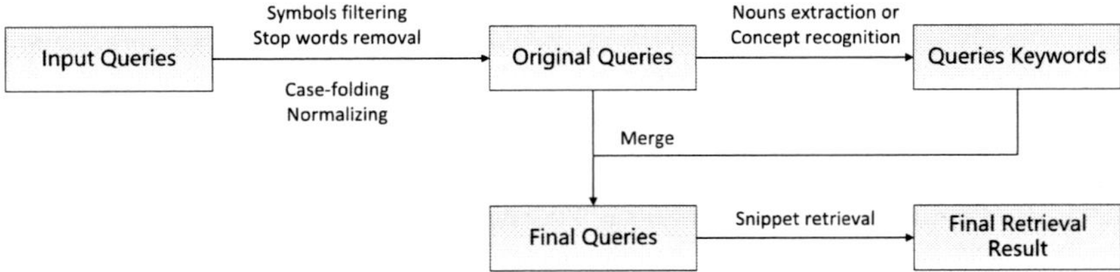

Figure 2: The whole framework architecture of query generation method based on multimodal snippet retrieval strategies

Table 3: Basic DFR Models.

D	Divergence approximation of the binomial
P	Approximation of the binomial
B_E	Bose-Einstein distribution
G	Geometric approximation of the Bose-Einstein
$I(N)$	Inverse Document Frequency model
$I(F)$	Inverse Term Frequency model
$I(n_e)$	Inverse Expected Document Frequency model

There are many ways to choose M, each of these provides a basic DFR model. The basic models are derived in Table 4.

If the model M is the binomial distribution, then the basic model is P and the value can be computed approximately as follows:

$$-\log \text{Prob}_P(t \in d|\text{Collection}) = -\log\binom{TF}{tf}p^{tf}q^{TF-tf} \tag{9}$$

where TF is the term-frequency of the term t in the collection, tf is the term-frequency of the term t in the document d, N is the number of documents in the collection, and p is $\frac{1}{N}$ and $q = 1 - p$.

4.3.2 Multimodal Strategies Combination

Similar to document retrieval, the final scoring function of snippet retrieval is expected to combine these strategies together effectively. Due to the reason of shorter text length the FSDM model cannot be used and the default IR language model performs not so satisfying for returning relevant snippets, we construct the merging scoring function to optimize the query term weights according to the Term Frequency−Inverse Document Frequency (TF-IDF), sequential dependence model (SDM) and Divergence from Randomness model (DFRM). As mentioned above, we control the length of queries to guarantee the performance, thus we no longer use PRF for snippets retrieval when merging the strategies. As the length of the queries decreases, the divergence of importance of each word becomes larger, so it is necessary to assign the weights of query terms according to the different importance. So we apply the DFRM method or the TF-IDF method along with SDM to achieve the results, which are respectively shown as follows:

$$\begin{aligned} score(Q, D) = &(1 - \lambda_1 - \lambda_2)score_{\text{SDM}}(Q, D) \\ &+ \lambda_1 score_{\text{TF-IDF}}(Q, D) \\ &+ \lambda_2 score_{\text{DFRM}}(Q, D) \end{aligned} \tag{10}$$

where the terms are weighted according to corresponding strategies through the following weighting function:

$$score_{\text{TF-IDF/DFRM}}(Q, D) = \sum_t score_{\text{TF-IDF/DFRM}}(t, D) \tag{11}$$

where t is the query term appearing in query Q. It is worth noting that when conducting the experiments we only consider $\lambda_1 = 0$ or $\lambda_2 = 0$ for tuning parameters.

Table 4: MAP performances of system components on BioASQ Task 4B snippet retrieval.

	Batch 1	Batch 2	Batch 3
baseline	0.1003	0.1361	0.1275
sdm	0.1047	0.1368	0.1338
sdm + PRF	0.1044	0.1370	0.1327
sdm + NN	0.1023	0.1402	0.1347
sdm + NN + PRF	0.1030	0.1357	0.1307
sdm + DFRM	**0.1193**	**0.1520**	**0.1469**
sdm + TF-IDF	0.1087	0.1424	0.1357

Table 5: MAP performances compared with BioASQ Task 5B snippet retrieval participants.

System	Batch 1	Batch 2
sdm + NN	0.0458	0.0811
sdm + NN + PRF(mesh)	0.0439	0.0716
sdm + DFRM	**0.0467**	**0.0898**
sdm + TF-IDF	0.0463	0.0874
sdm	0.0452	0.0736
best of fdu	-	0.0621
best of UNCC	-	-
best of Olelo	0.0260	0.0318
best of KNU-SG	0.0181	0.0362
best of HPI	0.0323	0.0335
best of Others	0.0249	0.0262

4.4 Experments

Similar to document retrieval, we evaluate the method on the first 3 batches from the previous BioASQ challenge and the current challenge. Similar to document retrieval, the optimization of all parameters are processed through tuning with the rules on training set. Table 4 provides the results of our experiments in BioASQ task 4B, and Table 5 provides the results of our experiments in BioASQ Task 5B. Obviously, our proposed merging strategy shows greater performance compared with various components and achieve better results than other particatants.

5 Task 5B Phase A: Concept Retrieval

Unlike the previous two tasks, the concept retrieval task is more like a named entity recognition task than an IR task. For each natural language question, participants are required to return relevant concepts from five ontologies or terminologies: MeSH, GO, Jochem, Uniprot and DO. In other words, the task aims at recognizing relevant biomedical concept within the question and matching them with the concepts in the data sources.

Since we have few experience in named entity recognition, we have to regard the task as an IR problem and design three query processing ap-

Table 6: MAP performances of system components on BioASQ Task 4B concept retrieval.

	Batch 1	Batch 2	Batch 3	Batch 4	Batch 5
Ours	0.1094	**0.1124**	**0.1386**	0.1174	0.1031
fdu	-	-	0.1566	0.1319	0.1004
HPI	0.0860	-	0.0863	0.0721	0.0439
oaqa	-	-	0.1067	0.1332	0.0915
auth	**0.1433**	0.0814	0.1361	**0.1376**	**0.1066**

proaches to generate appropriate query keywords for the web search services provided by BioASQ officials and implement the requested JSON file according to the examples in the guidelines (Neves, 2014). The five URLs of web services are utilized to post search requests for concepts and obtain search results. The request consists of two basic elements: keywords, the query to feed into search engine. Typically, this is a simple set of phrases separated by spaces acting as queries which may contain alphanumeric and punctuation characters; *page* and *concepts-per-page*, to control the number of results since the search engine may return thousands of concepts for one query. Thus, a pagination mechanism is used. Specifically, *Page* is a number representing the page (batch of concepts) to be retrieved, and *concepts-per-page* is a number representing the number of concepts per page (Tsatsaronis et al., 2015b).

For concept retrieval, noun extraction, synonym query expansion and pseudo relevance feedback are respectively used. On the purpose of obtaining the synonyms of query keywords, we download the vector representations of vocabularies produced through google word2vec tool (a word embedding tool to train word vectors on corpora), provided by BioASQ officials. We compute the cosine similarity between each query keyword and the word in the word list to find out the most semantic related words. These words are regarded as synonyms of the query keywords. We select the top 10 concepts as the submitted results ordered by descending predicted relevance score to the corresponding queries.

Since the results for this subtask will only be available after the manual assessment phase, we only evaluate the proposed method on the BioASQ 4B with other participants or any runs submitted off the evaluation. Table 6 provides the results of our experiments in BioASQ Task 4B and the statistics indicate our approach shows fairly good performance on all batches.

6 Conclusion

In this paper, we describe how to utilize multimodal query processing strategies for biomedical question answering applied to the participation of our USTB_PRIR team on phase A of BioASQ Task 5B. According to the official results, our system shows great robustness and effectiveness with competitive performance among the participating systems.

During the study of concept retrieval, we realize that named entity recognition of biomedical concepts may be helpful for the other tasks and so we may focus on utilizing this in the future.

References

Michael A Bauer and Daniel Berleant. 2012. Usability survey of biomedical question answering systems. *Human genomics*, 6(1):17.

Ludovic Bonnefoy, Romain Deveaud, Patrice Bellot, P Forner, J Karlgren, and C Womser-Hacker. 2012. Do social information help book search? In *INEX*, volume 12, pages 109–113.

Brian L Cairns, Rodney D Nielsen, James J Masanz, James H Martin, Martha S Palmer, Wayne H Ward, and Guergana K Savova. 2011. The mipacq clinical question answering system. In *AMIA Annual Symposium Proceedings*, volume 2011, page 171. American Medical Informatics Association.

Sungbin Choi. 2015. Snumedinfo at CLEF bioasq 2015. In *Working Notes of CLEF 2015 - Conference and Labs of the Evaluation forum, Toulouse, France, September 8-11, 2015*.

Stphane Clinchant and Eric Gaussier. 2011. Bridging language modeling and divergence from randomness models: A log-logistic model for ir.

Hyoil Han and Sofia J Athenikos. 2010. Biomedical question answering: a survey. *Computer Methods & Programs in Biomedicine*, 99(1):1–24.

Andreas Holzinger, Matthias Dehmer, and Igor Jurisica. 2014. Knowledge discovery and interactive data mining in bioinformatics-state-of-the-art, future challenges and research directions. *BMC bioinformatics*, 15(Suppl 6):I1.

X. Huang, J. Lin, and D Demnerfushman. 2006. Evaluation of pico as a knowledge representation for clinical questions. *AMIA ... Annual Symposium proceedings / AMIA Symposium. AMIA Symposium*, 2006:359–63.

T Kobayashi and C. R. Shyu. 2006. Representing clinical questions by semantic type for better classification. *Proceedings of the AMIA 2006 Symposium*, 2006:987.

Yuqing Mao and Zhiyong Lu. 2015. NCBI at the 2015 bioasq challenge task: Baseline results from mesh now. In *Working Notes of CLEF 2015 - Conference and Labs of the Evaluation forum, Toulouse, France, September 8-11, 2015*.

Yuqing Mao, Chih-Hsuan Wei, and Zhiyong Lu. 2014. NCBI at the 2014 bioasq challenge task: Large-scale biomedical semantic indexing and question answering. In *Working Notes for CLEF 2014 Conference, Sheffield, UK, September 15-18, 2014.*, pages 1319–1327.

Donald Metzler and W. Bruce Croft. 2005. A markov random field model for term dependencies. In *SIGIR*, pages 472–479.

Mariana L Neves. 2014. Hpi in-memory-based database system in task 2b of bioasq. In *CLEF (Working Notes)*, pages 1337–1347.

Yannis Papanikolaou, Dimitrios Dimitriadis, Grigorios Tsoumakas, Manos Laliotis, Nikos Markantonatos, and Ioannis P. Vlahavas. 2014. Ensemble approaches for large-scale multi-label classification and question answering in biomedicine. In *Working Notes for CLEF 2014 Conference, Sheffield, UK, September 15-18, 2014.*, pages 1348–1360.

George Tsatsaronis, Georgios Balikas, Prodromos Malakasiotis, Ioannis Partalas, Matthias Zschunke, Michael R. Alvers, Dirk Weissenborn, Anastasia Krithara, Sergios Petridis, Dimitris Polychronopoulos, Yannis Almirantis, John Pavlopoulos, Nicolas Baskiotis, Patrick Gallinari, Thierry Artières, Axel Ngonga, Norman Heino, Éric Gaussier, Liliana Barrio-Alvers, Michael Schroeder, Ion Androutsopoulos, and Georgios Paliouras. 2015a. An overview of the BIOASQ large-scale biomedical semantic indexing and question answering competition. *BMC Bioinformatics*, 16:138.

George Tsatsaronis, Georgios Balikas, Prodromos Malakasiotis, Ioannis Partalas, Matthias Zschunke, Michael R Alvers, Dirk Weissenborn, Anastasia Krithara, Sergios Petridis, Dimitris Polychronopoulos, et al. 2015b. An overview of the bioasq large-scale biomedical semantic indexing and question answering competition. *BMC bioinformatics*, 16(1):138.

H. Yu and Y. G. Cao. 2008. Automatically extracting information needs from ad hoc clinical questions. *Amia Annual Symposium Proceedings*, 2008:96–100.

Hong Yu, Carl Sable, and Hai Ran Zhu. 2005. Classifying medical questions based on an evidence taxonomy. In *National Conference on Artificial Intelligence*.

Yanchun Zhang, S Peng, R You, Z Xie, B Wang, and Shanfeng Zhu. 2015a. The fudan participation in the 2015 bioasq challenge: Large-scale biomedical semantic indexing and question answering. In *CEUR Workshop Proceedings*, volume 1391. CEUR Workshop Proceedings.

Zhijuan Zhang, Tiantian Liu, Bo-Wen Zhang, Yan Li, Chun Hua Zhao, Shao-Hui Feng, Xu-Cheng Yin, and Fang Zhou. 2015b. A generic retrieval system for biomedical literatures: Ustb at bioasq2015 question answering task. In *CLEF (Working Notes)*.

Nikita Zhiltsov, Alexander Kotov, and Fedor Nikolaev. 2015. Fielded sequential dependence model for ad-hoc entity retrieval in the web of data. In *International ACM SIGIR Conference on Research and Development in Information Retrieval*, pages 253–262.

Pierre Zweigenbaum. 2003. Question answering in biomedicine. In *Proceedings Workshop on Natural Language Processing for Question Answering, EACL*, volume 2005, pages 1–4. Citeseer.

Author Index